Canadian Mathematical Society
Société mathématique du Canada

Editors-in-Chief
Rédacteurs-en-chef
Jonathan Borwein
Peter Borwein

T0202468

Springer
New York
Berlin
Heidelberg
Hong Kong
London
Milan
Paris
Tokyo

CMS Books in Mathematics

Ouvrages de mathématiques de la SMC

1 HERMAN/KUČERA/ŠIMŠA Equations and Inequalities

2 ARNOLD Abelian Groups and Representations of Finite Partially Ordered Sets

3 BORWEIN/LEWIS Convex Analysis and Nonlinear Optimization

4 LEVIN/LUBINSKY Orthogonal Polynomials for Exponential Weights

5 KANE Reflection Groups and Invariant Theory

6 PHILLIPS Two Millennia of Mathematics

7 DEUTSCH Best Approximation in Inner Product Spaces

8 FABIAN ET AL. Functional Analysis and Infinite-Dimensional Geometry

9 KŘÍŽEK/LUCA/SOMER 17 Lectures on Fermat Numbers

10 BORWEIN Computational Excursions in Analysis and Number Theory

11 REED/SALES (Editors) Recent Advances in Algorithms and Combinatorics

12 HERMAN/KUČERA/ŠIMŠA Counting and Configurations

13 NAZARETH Differentiable Optimization and Equation Solving

14 PHILLIPS Interpolation and Approximation by Polynomials

15 BEN-ISRAEL/GREVILLE Generalized Inverses, Second Edition

16 ZHAO Dynamical Systems in Population Biology

Adi Ben-Israel   Thomas N.E. Greville

Generalized Inverses

Theory and Applications

Second Edition

Springer

Adi Ben-Israel
RUTCOR—Rutgers Center for
  Operations Research
Rutgers University
Piscataway, NJ 08854-8003
USA
bisrael@rutcor.rutgers.edu

Thomas N.E. Greville (deceased)

Editors-in-Chief
Rédacteurs-en-chef
Jonathan Borwein
Peter Borwein
Centre for Experimental and Constructive Mathematics
Department of Mathematics and Statistics
Simon Fraser University
Burnaby, British Columbia V5A 1S6
Canada
cbs-editors@cms.math.ca

With 1 figure.

Mathematics Subject Classification (2000): 15A09, 65Fxx, 47A05

Library of Congress Cataloging-in-Publication Data
Ben-Israel, Adi.
    Generalized inverses : theory and applications / Adi Ben-Israel, Thomas N.E. Greville.—
2nd ed.
        p. cm.—(CMS books in mathematics ; 15)
    Includes bibliographical references and index.

    1. Matrix inversion. I. Greville, T.N.E. (Thomas Nall Eden), 1910–1998   II. Title.
III. Series.
QA188.B46 2003
512.9′434—dc21   2002044506

ISBN 978-1-4419-1814-7   e-ISBN 978-0-387-21634-8

First edition published by Wiley-Interscience, 1974.

Printed in the United States of America.

9 8 7 6 5 4 3 2 1

www.springer-ny.com

Springer-Verlag   New York Berlin Heidelberg
A member of BertelsmannSpringer Science+Business Media GmbH

Preface to the Second Edition

The field of generalized inverses has grown much since the appearance of the first edition in 1974 and is still growing. I tried to account for these developments while maintaining the informal and leisurely style of the first edition. New material was added, including a preliminary chapter (Chapter 0), a chapter on applications (Chapter 8), an Appendix on the work of E.H. Moore, and new exercises and applications.

While preparing this volume I compiled a bibliography on generalized inverses, posted in the webpage of the *International Linear Algebra Society*

http://www.math.technion.ac.il/iic/research.html

This on-line bibliography, containing over 2000 items, will be updated from time to time. For reasons of space, many important works that appear in the on-line bibliography are not included in the bibliography of this book. I apologize to the authors of these works.

Many colleagues helped this effort. Special thanks go to R. Bapat, S. Campbell, J. Miao, S.K. Mitra, Y. Nievergelt, R. Puystjens, A. Sidi, G.-R. Wang, and Y. Wei.

Tom Greville, my friend and coauthor, passed away before this project started. His scholarship and style marked the first edition and are sadly missed.

I dedicate this book with love to my wife Yoki.

Piscataway, New Jersey   *Adi Ben-Israel*
January 2002

From the Preface to the First Edition

This book is intended to provide a survey of generalized inverses from a unified point of view, illustrating the theory with applications in many areas. It contains more than 450 exercises at different levels of difficulty, many of which are solved in detail. This feature makes it suitable either for reference and self–study or for use as a classroom text. It can be used profitably by graduate students or advanced undergraduates, only an elementary knowledge of linear algebra being assumed.

The book consists of an introduction and eight chapters, seven of which treat generalized inverses of finite matrices, while the eighth introduces generalized inverses of operators between Hilbert spaces. Numerical methods are considered in Chapter 7 and in Section 9.7.

While working in the area of generalized inverses, the authors have had the benefit of conversations and consultations with many colleagues. We would like to thank especially A. Charnes, R.E. Cline, P.J. Erdelsky, I. Erdélyi, J.B. Hawkins, A.S. Householder, A. Lent, C.C. MacDuffee, M.Z. Nashed, P.L. Odell, D.W. Showalter, and S. Zlobec. However, any errors that may have occurred are the sole responsibility of the authors.

This book is dedicated to Abraham Charnes and J. Barkley Rosser.

Haifa, Israel   *Adi Ben-Israel*
Madison, Wisconsin   *Thomas N.E. Greville*
September 1973

Contents

Preface to the Second Edition   v

From the Preface to the First Edition   vii

Glossary of Notation   xiii

Introduction   1
  1. The Inverse of a Nonsingular Matrix   1
  2. Generalized Inverses of Matrices   1
  3. Illustration: Solvability of Linear Systems   2
  4. Diversity of Generalized Inverses   3
  5. Preparation Expected of the Reader   4
  6. Historical Note   4
  7. Remarks on Notation   5
  Suggested Further Reading   5

Chapter 0. Preliminaries   6
  1. Scalars and Vectors   6
  2. Linear Transformations and Matrices   10
  3. Elementary Operations and Permutations   22
  4. The Hermite Normal Form and Related Items   23
  5. Determinants and Volume   28
  6. Some Multilinear Algebra   32
  7. The Jordan Normal Form   34
  8. The Smith Normal Form   38
  9. Nonnegative Matrices   39
  Suggested Further Reading   39

Chapter 1. Existence and Construction of Generalized Inverses   40
  1. The Penrose Equations   40
  2. Existence and Construction of $\{1\}$-Inverses   41
  3. Properties of $\{1\}$-Inverses   42
  4. Existence and Construction of $\{1,2\}$-Inverses   45
  5. Existence and Construction of $\{1,2,3\}$-, $\{1,2,4\}$-, and $\{1,2,3,4\}$-Inverses   46
  6. Explicit Formula for $A^\dagger$ 48
  7. Construction of $\{2\}$-Inverses of Prescribed Rank   49
  Notes on Terminology   51
  Suggested Further Reading   51

Chapter 2.   Linear Systems and Characterization of Generalized
                 Inverses   52
    1.   Solutions of Linear Systems   52
    2.   Characterization of $A\{1,3\}$ and $A\{1,4\}$ 55
    3.   Characterization of $A\{2\}$, $A\{1,2\}$, and Other Subsets of $A\{2\}$ 56
    4.   Idempotent Matrices and Projectors   58
    5.   Matrix Functions   65
    6.   Generalized Inverses with Prescribed Range and Null Space   71
    7.   Orthogonal Projections and Orthogonal Projectors   74
    8.   Efficient Characterization of Classes of Generalized Inverses   85
    9.   Restricted Generalized Inverses   88
    10. The Bott–Duffin Inverse   91
    11. An Application of $\{1\}$-Inverses in Interval Linear Programming   95
    12. A $\{1,2\}$-Inverse for the Integral Solution of Linear Equations   97
    13. An Application of the Bott–Duffin Inverse to Electrical
         Networks   99
    Suggested Further Reading   103

Chapter 3.   Minimal Properties of Generalized Inverses   104
    1.   Least-Squares Solutions of Inconsistent Linear Systems   104
    2.   Solutions of Minimum Norm   108
    3.   Tikhonov Regularization   114
    4.   Weighted Generalized Inverses   117
    5.   Least-Squares Solutions and Basic Solutions   122
    6.   Minors of the Moore–Penrose Inverse   127
    7.   Essentially Strictly Convex Norms and the Associated Projectors
         and Generalized Inverses   130
    8.   An Extremal Property of the Bott–Duffin Inverse with
         Application to Electrical Networks   149
    Suggested Further Reading   151

Chapter 4.   Spectral Generalized Inverses   152
    1.   Introduction   152
    2.   The Matrix Index   153
    3.   Spectral Inverse of a Diagonable Matrix   155
    4.   The Group Inverse   156
    5.   Spectral Properties of the Group Inverse   161
    6.   The Drazin Inverse   163
    7.   Spectral Properties of the Drazin Inverse   168
    8.   Index 1-Nilpotent Decomposition of a Square Matrix   169
    9.   Quasi-Commuting Inverses   171
    10. Other Spectral Generalized Inverses   172
    Suggested Further Reading   174

Chapter 5.   Generalized Inverses of Partitioned Matrices   175
    1.   Introduction   175
    2.   Partitioned Matrices and Linear Equations   175
    3.   Intersection of Manifolds   182

4. Common Solutions of Linear Equations and Generalized Inverses
of Partitioned Matrices   189
5. Generalized Inverses of Bordered Matrices   196
Suggested Further Reading   200

Chapter 6. A Spectral Theory for Rectangular Matrices   201
1. Introduction   201
2. The Singular Value Decomposition   205
3. The Schmidt Approximation Theorem   212
4. Partial Isometries and the Polar Decomposition Theorem   218
5. Principal Angles Between Subspaces   230
6. Perturbations   238
7. A Spectral Theory for Rectangular Matrices   242
8. Generalized Singular Value Decompositions   251
Suggested Further Reading   255

Chapter 7. Computational Aspects of Generalized Inverses   257
1. Introduction   257
2. Computation of Unrestricted $\{1\}$- and $\{1,2\}$-Inverses   258
3. Computation of Unrestricted $\{1,3\}$-Inverses   260
4. Computation of $\{2\}$-Inverses with Prescribed Range and Null
Space   261
5. Greville's Method and Related Results   263
6. Computation of Least-Squares Solutions   269
7. Iterative Methods for Computing $A^\dagger$ 270
Suggested Further Reading   281

Chapter 8. Miscellaneous Applications   282
1. Introduction   282
2. Parallel Sums   282
3. The Linear Statistical Model   284
4. Ridge Regression   292
5. An Application of $\{2\}$-Inverses in Iterative Methods for Solving
Nonlinear Equations   295
6. Linear Systems Theory   302
7. Application of the Group Inverse in Finite Markov Chains   303
8. An Application of the Drazin Inverse to Difference Equations   310
9. Matrix Volume and the Change-of-Variables Formula in
Integration   313
10. An Application of the Matrix Volume in Probability   323
Suggested Further Reading   328

Chapter 9. Generalized Inverses of Linear Operators between Hilbert
Spaces   330
1. Introduction   330
2. Hilbert Spaces and Operators: Preliminaries and Notation   330
3. Generalized Inverses of Linear Operators Between Hilbert
Spaces   336
4. Generalized Inverses of Linear Integral Operators   344
5. Generalized Inverses of Linear Differential Operators   348

6. Minimal Properties of Generalized Inverses   356
7. Series and Integral Representations and Iterative Computation
      of Generalized Inverses   363
8. Frames   367
Suggested Further Reading   368

Appendix A.   The Moore of the Moore–Penrose Inverse   370
1. Introduction   370
2. The 1920 Lecture to the American Mathematical Society   371
3. The General Reciprocal in *General Analysis*   372

Bibliography   375

Subject Index   409

Author Index   415

Glossary of Notation

$\Gamma(p)$ – Gamma function, 320

$\eta(\mathbf{u}, \mathbf{v}, \mathbf{w})$, 96

$\gamma(T)$, 334

$\lambda^\dagger$ – Moore–Penrose inverse of the scalar λ, 43

$\lambda(A)$ – spectrum of A, 13

$\langle \alpha \rangle$ – smallest integer $\geq \alpha$, 278

$\mu(A, B)$, 251

$\mu_{W,Q}(A)$, 254

$\nu(\lambda)$ – index of eigenvalue λ, 36

$\overline{1, n}$ – the index set $\{1, 2, \ldots, n\}$, 5

π^{-1} – permutation inverse to π, 22

$\pi_i^{(t)}$ – probability of $\mathbf{X}_t = i$, 305

$\rho(A)$ – spectral radius of A, 20

$\sigma(A)$ – singular values of A (see footnote, p. 13), 14

$\sigma_j(A)$ – the jth singular value of A, 14

$\tau(i)$ – period of state i, 304

A/A_{11} – Schur complement of A_{11} in A, 30

$A \succ O$, 80

$A\{1\}_{T,S}$ – $\{1\}$-inverses of A associated with T, S, 71

$A\{i, j, \ldots, k\}_s$ – matrices in $A\{i, j, \ldots, k\}$ of rank s, 56

A^* – adjoint of A, 12

$A_{(L)}^{(-1)}$ – Bott–Duffin inverse of A with respect to L, 92

$A^{1/2}$ – square root of A, 222

A^D – Drazin inverse of A, 163, 164

$A\{2\}_{T,S}$ – $\{2\}$-inverses with range T, null space S, 73

$A\{i, j, \ldots, k\}$ – $\{i, j, \ldots, k\}$-inverses of A, 40

$A_{\alpha,\beta}^{(-1)}$ – α-β generalized inverse of A, 134

$A_{T,S}^{(1)}$ – a $\{1\}$-inverse of A associated with T, S, 71

$A^{(i,j,\ldots,k)}$ – an $\{i, j, \ldots, k\}$-inverse of A, 40

$A^\#$ – group inverse of A, 156

$A^\dagger$ – Moore–Penrose inverse of A, 40

$\| A \|_\infty$ – ∞-norm of a matrix, 20

$\| A \|_{\alpha,\beta}$ – least upper bound of A with respect to $\{\alpha, \beta\}$, 143

$\widehat{A}$, 98

$\widetilde{A}$ – perturbation of A, 238

$\| A \|_1$ – 1-norm of a matrix, 20

$\| A \|_2$ – spectral norm of a matrix, 20

$A : B$ – Anderson–Duffin parallel sum of A, B, 283

$A \otimes B$ – Kronecker product of A, B, 53

$A \succcurlyeq B$ – Löwner ordering, 80, 286, 287

$A \overset{*}{<} B$ – $*$-order, 84

$A \mp B$ – Rao–Mitra parallel sum of A, B, 283

$A[\beta \leftarrow I_\alpha]$, 128

$\| A \|_F$ – Frobenius norm, 19

$A[I, *]$, 10

A_{I*}, 10

$A[I, J]$, 10

A_{IJ}, 10

$A[*, J]$, 10

A_{*J}, 10

$A[j \leftarrow \mathbf{b}]$ – A with j^{th}-column replaced by $\mathbf{b}$, 30

$A_{(k)}$ – best rank-k approximation of A, 213

$A^{\langle k \rangle}$ – generalized k^{th} power of A, 249

$A^{(N)}$ – nilpotent part of A, 170

$\| A \|_p$ – p-norm of a matrix, 20

$A_{[S]}$ – restriction of A to S, 89

$A^{(S)}$ – S-inverse of A, 173

$A_{\{U,V\}}$ – matrix representation of A with respect to $\{U, V\}$, 11

$A_{\{V\}}$ – matrix representation of A with respect to $\{V, V\}$, 11

$A_{(W,Q)}^{(1,2)}$ – $\{W, Q\}$ weighted $\{1, 2\}$-inverse of A, 119, 121, 255

$\mathcal{B}(\mathcal{H}_1, \mathcal{H}_2)$ – bounded operators in $\mathcal{L}(\mathcal{H}_1, \mathcal{H}_2)$, 332

$B(p, q)$ – Beta function, 321

$B(\mathbf{x}^0, r)$ – ball with center $\mathbf{x}_0$ and radius r, 296

$\mathbb{C}$ – complex field, 6

$C[a, b]$ – continuous functions on $[a, b]$, 348

$\mathcal{C}(\mathcal{H}_1, \mathcal{H}_2)$ – closed operators in $\mathcal{L}(\mathcal{H}_1, \mathcal{H}_2)$, 332

xiii

$C_k(A)$ – k compound matrix, 32
$\mathbb{C}^{m \times n}$ – $m \times n$ complex matrices, 10
$\mathbb{C}_r^{m \times n}$ – $m \times n$ complex matrices with rank r, 23
$\mathbb{C}^n$ – n-dimensional complex vector space, 6
cond(A) – condition number of A, 204
$\cos\{L, M\}$, 233
Cov $\mathbf{X}$ – covariance of $\mathbf{X}$, 284
$C(T)$, 331

$\mathcal{D}_+$ – positive diagonal matrices, 126
$d(A)$ – diagonal elements in UDV^*-decomposition, 209
det A – determinant of A, 28
diag $(a_{11}, \ldots, a_{pp})$ – diagonal matrix, 10
dist(L, M) – distance between L, M, 233
$D(T)$, 331

$\mathbf{e}$ – vector of ones, 303
$E^i(\alpha)$, $E^{ij}(\beta)$, E^{ij} – elementary operations of types 1,2,3 respectively, 22
$\mathcal{E}_n$ – standard basis of $\mathbb{C}^n$, 11
EP – matrices A with $R(A) = R(A^*)$, 157
EP_r, 157
E $\mathbf{X}$ – expected value of $\mathbf{X}$, 284
ext B – extension of B to $\mathbb{C}^n$, 89

$\mathbb{F}$ – field, 6
$\widehat{f}(x_1, \ldots, x_{n-1}, p)$, 316
$f_{ij}^{(n)}$ – probability of first transition $i \to j$ in n^{th} step, 304
$\mathcal{F}(A)$ – functions $f : \mathbb{C} \to \mathbb{C}$ analytic on $\lambda(A)$, 68
fl – floating point, 106
$\mathbb{F}^{m \times n}$ – $m \times n$ matrices over $\mathbb{F}$, 10
$\mathbb{F}^n$ – n-dimensional vector space over $\mathbb{F}$, 6

$G(\mathbf{x}_1, \ldots, \mathbf{x}_n)$ – Gram matrix, 29
$G(T)$, 331
$G^{-1}(T)$, 332

$\mathcal{H}, \mathcal{H}_1, \mathcal{H}_2$ – Hilbert spaces, 330
$\mathcal{H}_{\xi, p}$ – hyperplane, 315

$i \rightleftarrows j$ – states i, j communicate, 303
$\mathcal{I}(A)$, 29
Ind A – index of A, 153
IP$(\mathbf{a}, \mathbf{b}, \mathbf{c}, A)$, 95

$J_\mathbf{f}(\mathbf{x})$ – Jacobian matrix of f at $\mathbf{x}$, 295
$J_\mathbf{x}$ – Jacobian matrix at $\mathbf{x}$, 295
$\mathcal{J}(A)$, 29
$J_k(\lambda)$ – Jordan block, 34

$L^\perp$ – orthogonal complement of L, 12, 330

$\mathcal{L}(\mathbb{C}^n, \mathbb{C}^m)$, 11
$\mathcal{L}(\mathcal{H}_1, \mathcal{H}_2)$, 331
LHS$(i.j)$, 5
$L \oplus M$ – direct sum of L, M, 6, 331
$L \overset{\perp}{\oplus} M$ – orthogonal direct sum of L, M, 12, 331
$\text{lub}_{\alpha, \beta}(A)$, 143
$\mathcal{L}(U, V)$ – linear transformations from U to V, 10

$\mathcal{N}(A)$, 29
$N(A, B)$ – matrices X with $AXB = O$, 110
$N(T)$ – null space of T, 11, 331

$p_{ij}^{(n)}$ – n-step transition probability, 303
PD_n – $n \times n$ positive definite matrices, 13, 80
P_π – permutation matrix, 22
P_L – orthogonal projector on L, 74
$P_{L, \phi}$ – ϕ-metric projector on L, 132
$P_{L, \phi}^{-1}(\ell)$ – inverse image of ℓ under $P_{L, \phi}$, 133
$P_{L, M}$ – projector on L along M, 59
PSD_n – $n \times n$ positive semidefinite matrices, 13, 80

$Q(\alpha)$ – projective bound of α, 144
$Q_{k, n}$ – increasing k sequences in $\overline{1, n}$, 10

$R(\lambda, A)$ – resolvent of A, 246
$\mathbb{R}$ – real field, 6
$\widehat{R}(\lambda, A)$ – generalized resolvent of A, 246
$R(A, B)$ – matrices AXB for some X, 110
$\Re$ – real part, 8
RHS$(i.j)$, 5
R_k – residual, 270
$R(\lambda, A)$ – resolvent of A, 70
$R(L, M)$ – coefficient of inclination between L, M, 230
$r(L, M)$ – dimension of inclination between L, M, 230
$\mathbb{R}^{m \times n}$ – $m \times n$ real matrices, 10
$\mathbb{R}_r^{m \times n}$ – $m \times n$ real matrices with rank r, 23
$\mathbb{R}^n$ – n-dimensional real vector space, 6
$\mathbb{R}_j^n$ – basic subspace, 236
$R(T)$ – range of T, 11, 331
RV – random variable, 5, 323

$\mathcal{S}$ – function space, 348
sign π – sign of permutation π, 23
$\sin\{L, M\}$, 233
S_n – symmetric group (permutations of order n), 22

$(T_2)_{[D(T_1)]}$ – restriction of T_2 to $D(T_1)$, 332

$T \succcurlyeq O$, 334

T^*, 333

T_r – restriction of T, 342

$T_S^\dagger$ – the $N(S)$-restricted pseudoinverse of T, 362

$T_e^\dagger$ – extremal inverse, 358

T^q – Tseng inverse, 336

$U^{n \times n}$ – $n \times n$ unitary matrices, 201

$\text{vec}(X)$ – vector made of rows of X, 54

$\text{vol } A$ – volume of matrix A, 29

$W_\ell^{m \times n}$ – partial isometries in $\mathbb{C}_\ell^{m \times n}$, 227

$\| \mathbf{x} \|$ – norm of $\mathbf{x}$, 7

$\| \mathbf{x} \|_Q$ – ellipsoidal norm of $\mathbf{x}$, 8

$\langle X, Y \rangle$ – inner product on $\mathbb{C}^{m \times n}$, 110

$\angle \{ \mathbf{x}, \mathbf{y} \}$ – angle between $\mathbf{x}, \mathbf{y}$, 8

$\langle \mathbf{x}, \mathbf{y} \rangle$ – inner product of $\mathbf{x}, \mathbf{y}$, 7, 330

$\langle \mathbf{x}, \mathbf{y} \rangle_Q$ – the inner product $\mathbf{y}^* Q \mathbf{x}$, 8

$(\mathbf{y}, X\boldsymbol{\beta}, V^2)$ – linear model, 285

$\mathbb{Z}$ – ring of integers, 38

$\mathbb{Z}^m$, 38

$\mathbb{Z}^{m \times n}$, 38

$\mathbb{Z}_r^{m \times n}$, 38

Introduction

1. The Inverse of a Nonsingular Matrix

It is well known that every nonsingular matrix A has a unique inverse, denoted by A^{-1}, such that

$$A A^{-1} = A^{-1}A = I, \tag{1}$$

where I is the identity matrix. Of the numerous properties of the inverse matrix, we mention a few. Thus,

$$(A^{-1})^{-1} = A,$$
$$(A^T)^{-1} = (A^{-1})^T,$$
$$(A^*)^{-1} = (A^{-1})^*,$$
$$(AB)^{-1} = B^{-1}A^{-1},$$

where A^T and A^*, respectively, denote the transpose and conjugate transpose of A. It will be recalled that a real or complex number λ is called an eigenvalue of a square matrix A, and a nonzero vector $\mathbf{x}$ is called an eigenvector of A corresponding to λ, if

$$A\mathbf{x} = \lambda\mathbf{x}.$$

Another property of the inverse A^{-1} is that its eigenvalues are the reciprocals of those of A.

2. Generalized Inverses of Matrices

A matrix has an inverse only if it is square, and even then only if it is nonsingular or, in other words, if its columns (or rows) are linearly independent. In recent years needs have been felt in numerous areas of applied mathematics for some kind of partial inverse of a matrix that is singular or even rectangular. By a *generalized inverse* of a given matrix A we shall mean a matrix X associated in some way with A that:

   (i) exists for a class of matrices larger than the class of nonsingular matrices;

   (ii) has some of the properties of the usual inverse; and

   (iii) reduces to the usual inverse when A is nonsingular.

Some writers have used the term "pseudoinverse" rather than "generalized inverse."

As an illustration of part (iii) of our description of a generalized inverse, consider a definition used by a number of writers (e.g., Rohde [**704**])to the effect that a generalized inverse of A is any matrix satisfying

$$AXA = A. \tag{2}$$

If A were nonsingular, multiplication by A^{-1} both on the left and on the right would give, at once,

$$X = A^{-1}.$$

3. Illustration: Solvability of Linear Systems

Probably the most familiar application of matrices is to the solution of systems of simultaneous linear equations. Let

$$A\mathbf{x} = \mathbf{b} \tag{3}$$

be such a system, where $\mathbf{b}$ is a given vector and $\mathbf{x}$ is an unknown vector. If A is nonsingular, there is a unique solution for $\mathbf{x}$ given by

$$\mathbf{x} = A^{-1}\mathbf{b}.$$

In the general case, when A may be singular or rectangular, there may sometimes be no solutions or a multiplicity of solutions.

The existence of a vector $\mathbf{x}$ satisfying (3) is tantamount to the statement that $\mathbf{b}$ is some linear combination of the columns of A. If A is $m \times n$ and of rank less than m, this may not be the case. If it is, there is some vector $\mathbf{h}$ such that

$$\mathbf{b} = A\mathbf{h}.$$

Now, if X is some matrix satisfying (2), and if we take

$$\mathbf{x} = X\mathbf{b},$$

we have

$$A\mathbf{x} = AX\mathbf{b} = AXA\mathbf{h} = A\mathbf{h} = \mathbf{b},$$

and so this $\mathbf{x}$ satisfies (3).

In the general case, however, when (3) may have many solutions, we may desire not just one solution but a characterization of all solutions. It has been shown (Bjerhammar [103], Penrose [635]) that, if X is any matrix satisfying $AXA = A$, then $A\mathbf{x} = \mathbf{b}$ has a solution if and only if

$$AX\mathbf{b} = \mathbf{b},$$

in which case the general solution is

$$\mathbf{x} = X\mathbf{b} + (I - XA)\mathbf{y}, \tag{4}$$

where $\mathbf{y}$ is arbitrary.

We shall see later that for every matrix A there exist one or more matrices satisfying (2).

Exercises

EX. 1. If A is nonsingular and has an eigenvalue λ, and $\mathbf{x}$ is a corresponding eigenvector, show that λ^{-1} is an eigenvalue of A^{-1} with the same eigenvector $\mathbf{x}$.

EX. 2. For any square A, let a "generalized inverse" be defined as any matrix X satisfying $A^{k+1}X = A^k$ for some positive integer k. Show that $X = A^{-1}$ if A is nonsingular.

Ex. 3. If X satisfies $AXA = A$, show that $A\mathbf{x} = \mathbf{b}$ has a solution if and only if $AX\mathbf{b} = \mathbf{b}$.

Ex. 4. Show that (4) is the general solution of $A\mathbf{x} = \mathbf{b}$. [*Hint:* First show that it is a solution; then show that every solution can be expressed in this form. Let $\mathbf{x}$ be any solution; then write $\mathbf{x} = XA\mathbf{x} + (I - XA)\mathbf{x}$.]

Ex. 5. If A is an $m \times n$ matrix of zeros, what is the class of matrices X satisfying $AXA = A$?

Ex. 6. Let A be an $m \times n$ matrix whose elements are all zeros except the $(i, j)^{\text{th}}$ element, which is equal to 1. What is the class of matrices X satisfying (2)?

Ex. 7. Let A be given, and let X have the property that $\mathbf{x} = X\mathbf{b}$ is a solution of $A\mathbf{x} = \mathbf{b}$ for *all* $\mathbf{b}$ such that a solution exists. Show that X satisfies $AXA = A$.

4. Diversity of Generalized Inverses

From Exercises 3, 4, and 7 the reader will perceive that, for a given matrix A, the matrix equation $AXA = A$ alone characterizes those generalized inverses X that are of use in analyzing the solutions of the linear system $A\mathbf{x} = \mathbf{b}$. For other purposes, other relationships play an essential role. Thus, if we are concerned with least-squares properties, (2) is not enough and must be supplemented by further relations. There results a more restricted class of generalized inverses.

If we are interested in spectral properties (i.e., those relating to eigenvalues and eigenvectors), consideration is necessarily limited to square matrices, since only these have eigenvalues and eigenvectors. In this connection, we shall see that (2) plays a role only for a restricted class of matrices A and must be supplanted, in the general case, by other relations.

Thus, unlike the case of the nonsingular matrix, which has a single unique inverse for all purposes, there are different generalized inverses for different purposes. For some purposes, as in the examples of solutions of linear systems, there is not a unique inverse, but any matrix of a certain class will do.

This book does not pretend to be exhaustive, but seeks to develop and describe in a natural sequence the most interesting and useful kinds of generalized inverses and their properties. For the most part, the discussion is limited to generalized inverses of finite matrices, but extensions to infinite-dimensional spaces and to differential and integral operators are briefly introduced in Chapter 9. Generalized inverses on rings and semigroups are not discussed; the interested reader is referred to Bhaskara Rao [94], Drazin [233], Foulis [284], and Munn [587].

The literature on generalized inverses has become so extensive that it would be impossible to do justice to it in a book of moderate size. We have been forced to make a selection of topics to be covered, and it is inevitable that not everyone will agree with the choices we have made. We apologize to those authors whose work has been slighted. A virtually complete bibliography as of 1976 is found in Nashed and Rall [597]. An on-line bibliography is posted in the webpage of the International Linear Algebra Society

http://www.math.technion.ac.il/iic/research.html

5. Preparation Expected of the Reader

It is assumed that the reader has a knowledge of linear algebra that would normally result from completion of an introductory course in the subject. In particular, vector spaces will be extensively utilized. Except in Chapter 9, which deals with Hilbert spaces, the vector spaces and linear transformations used are finite-dimensional, real or complex. Familiarity with these topics is assumed, say at the level of Halmos [365] or Noble [615], see also Chapter 0 below.

6. Historical Note

The concept of a generalized inverse seems to have been first mentioned in print in 1903 by Fredholm [290], where a particular generalized inverse (called by him "pseudoinverse") of an integral operator was given. The class of all pseudoinverses was characterized in 1912 by Hurwitz [435], who used the finite dimensionality of the null spaces of the Fredholm operators to give a simple algebraic construction (see, e.g., Exercises 9.18–9.19). Generalized inverses of differential operators, already implicit in Hilbert's discussion in 1904 of generalized Green functions, [418], were consequently studied by numerous authors, in particular, Myller (1906), Westfall (1909), Bounitzky [124] in 1909, Elliott (1928), and Reid (1931). For a history of this subject see the excellent survey by Reid [685].

Generalized inverses of differential and integral operators thus antedated the generalized inverses of matrices, whose existence was first noted by E.H. Moore, who defined a unique inverse (called by him the "general reciprocal") for every finite matrix (square or rectangular). Although his first publication on the subject [575], an abstract of a talk given at a meeting of the American Mathematical Society, appeared in 1920, his results are thought to have been obtained much earlier. One writer, [496, p. 676], has assigned the date 1906. Details were published, [576], only in 1935 after Moore's death. A summary of Moore's work on the *general reciprocal* is given in Appendix A. Little notice was taken of Moore's discovery for 30 years after its first publication, during which time generalized inverses were given for matrices by Siegel [762] in 1937, and for operators by Tseng ([816]–1933, [819],[817],[818]–1949), Murray and von Neumann [589] in 1936, Atkinson ([27]–1952, [28]–1953) and others. Revival of interest in the subject in the 1950s centered around the least squares properties (not mentioned by Moore) of certain generalized inverses. These properties were recognized in 1951 by Bjerhammar, who rediscovered Moore's inverse and also noted the relationship of generalized inverses to solutions of linear systems (Bjerhammar [102], [101], [103]). In 1955 Penrose [635]sharpened and extended Bjerhammar's results on linear systems, and showed that Moore's inverse, for a given matrix A, is the unique matrix X satisfying the four equations (1)–(4) of Chapter 1. The latter discovery has been so important and fruitful that this unique inverse (called by some writers *the* generalized inverse) is now commonly called the *Moore–Penrose inverse*.

Since 1955 thousands of papers on various aspects of generalized inverses and their applications have appeared. In view of the vast scope of this literature, we shall not attempt to trace the history of the subject further, but the subsequent chapters will include selected references on particular items.

7. Remarks on Notation

Equation j of Chapter i is denoted by (j) in Chapter i, and by $(i.j)$ in other chapters. Theorem j of Chapter i is called Theorem j in Chapter i, and Theorem $i.j$ in other chapters. Similar conventions apply to Sections, Corollaries, Lemmas, Definitions, etc.

Many sections are followed by Exercises, some of them solved. Exercises are denoted by "Ex." (e.g., Ex. j, Ex. $i.j$), to distinguish from Examples (e.g., Example j, Example $i.j$) that appear inside sections.

Some of the abbreviations used in this book:

$\overline{k,\ell}$ – the index set $\{k, k+1, \ldots, \ell\}$; in particular,

$\overline{1,n}$ – the index set $\{1, 2, \ldots, n\}$;

BLUE – best linear unbiased estimator;

e.s.c. – essentially strictly convex;

LHS$(i.j)$ – the left-hand side of equation $(i.j)$;

LUE – linear unbiased estimator;

MSE – mean square error;

o.n. – orthonormal;

PD – positive definite;

PSD – positive semidefinite;

RHS$(i.j)$ – the right-hand side of equation $(i.j)$;

RRE – ridge regression estimator;

RV – random variable;

SVD – singular value decomposition; and

TLS – total least squares.

Suggested Further Reading

SECTION 2. A ring $\mathcal{R}$ is called *regular* if for every $A \in \mathcal{R}$ there exists an $X \in \mathcal{R}$ satisfying $AXA = A$. See von Neumann [838], [841, p. 90], Murray and von Neumann [589, p. 299], McCoy [538], Hartwig [379].

SECTION 4. For generalized inverses in abstract algebraic setting see also Davis and Robinson [215], Gabriel [291], [292], [293], Hansen and Robinson [373], Hartwig [379], Munn and Penrose [588], Pearl [634], Rabson [662], Rado [663].

Preliminaries

For ease of reference we collect here facts, definitions, and notations that are used in successive chapters. This chapter can be skipped in first reading.

1. Scalars and Vectors

1.1. *Scalars* are denoted by lowercase letters: $x, y, \lambda, \ldots$. We use mostly the *complex field* $\mathbb{C}$, and specialize to the *real field* $\mathbb{R}$ as necessary. A generic field is denoted by $\mathbb{F}$.

1.2. *Vectors* are denoted by bold letters: $\mathbf{x}, \mathbf{y}, \boldsymbol{\lambda}, \ldots$. Vector spaces are finite-dimensional, except in Chapter 9. The n-dimensional vector space over a field $\mathbb{F}$ is denoted by $\mathbb{F}^n$, in particular, $\mathbb{C}^n$ $[\mathbb{R}^n]$ denote the n-dimensional complex [real] vector space.

A vector $\mathbf{x} \in \mathbb{F}^n$ is written in a column form

$$\mathbf{x} = \begin{bmatrix} x_1 \\ \vdots \\ x_n \end{bmatrix}, \text{ or } \mathbf{x} = (x_i), \quad i \in \overline{1, n}, \quad x_i \in \mathbb{F}.$$

The n-dimensional vector $\mathbf{e}_i$ with components

$$\delta_{ij} = \begin{cases} 1, & \text{if } i = j, \\ 0, & \text{otherwise,} \end{cases}$$

is called the ith *unit vector* of $\mathbb{F}^n$. The set $\mathcal{E}_n$ of unit vectors $\{\mathbf{e}_1, \mathbf{e}_2, \ldots, \mathbf{e}_n\}$ is called the *standard basis* of $\mathbb{F}^n$.

1.3. The *sum* of two sets L, M in $\mathbb{C}^n$, denoted by $L + M$, is defined as

$$L + M = \{\mathbf{y} + \mathbf{z} : \mathbf{y} \in L, \, \mathbf{z} \in M\}.$$

If L and M are subspaces of $\mathbb{C}^n$, then $L + M$ is also a subspace of $\mathbb{C}^n$. If, in addition, $L \cap M = \{\mathbf{0}\}$, i.e., the only vector common to L and M is the zero vector, then $L + M$ is called the *direct sum* of L and M, denoted by $L \oplus M$. Two subspaces L and M of $\mathbb{C}^n$ are called *complementary* if

$$\mathbb{C}^n = L \oplus M. \tag{1}$$

When this is the case (see Ex. 1 below), every $\mathbf{x} \in \mathbb{C}^n$ can be expressed uniquely as a sum

$$\mathbf{x} = \mathbf{y} + \mathbf{z} \quad (\mathbf{y} \in L, \mathbf{z} \in M). \tag{2}$$

We shall then call $\mathbf{y}$ the *projection of $\mathbf{x}$ on L along M*.

1.4. *Inner product.* Let V be a complex vector space. An *inner product* is a function: $V \times V \to \mathbb{C}$, denoted by $\langle \mathbf{x}, \mathbf{y} \rangle$, that satisfies:

(I1) $\langle \alpha \mathbf{x} + \mathbf{y}, \mathbf{z} \rangle = \alpha \langle \mathbf{x}, \mathbf{z} \rangle + \langle \mathbf{y}, \mathbf{z} \rangle$ (*linearity*);

(I2) $\langle \mathbf{x}, \mathbf{y} \rangle = \overline{\langle \mathbf{y}, \mathbf{x} \rangle}$ (*Hermitian symmetry*); and

(I3) $\langle \mathbf{x}, \mathbf{x} \rangle \geq 0$, $\langle \mathbf{x}, \mathbf{x} \rangle = 0$ if and only if $\mathbf{x} = \mathbf{0}$ (*positivity*);

for all $\mathbf{x}, \mathbf{y}, \mathbf{z} \in V$ and $\alpha \in \mathbb{C}$.

Note:

(a) For all $\mathbf{x}, \mathbf{y} \in V$ and $\alpha \in \mathbb{C}$, $\langle \mathbf{x}, \alpha \mathbf{y} \rangle = \overline{\alpha} \langle \mathbf{x}, \mathbf{y} \rangle$ by (I1)–(I2).

(b) Condition (I2) states, in particular, that $\langle \mathbf{x}, \mathbf{x} \rangle$ is real for all $\mathbf{x} \in V$.

(c) The *if* part in (I3) follows from (I1) with $\alpha = 0$, $\mathbf{y} = \mathbf{0}$.

The *standard inner product* in $\mathbb{C}^n$ is

$$\mathbf{y}^* \mathbf{x} = \sum_{i=1}^{n} x_i \, \overline{y_i}, \tag{3}$$

for all $\mathbf{x} = (x_i)$ and $\mathbf{y} = (y_i)$ in $\mathbb{C}^n$. See Exs. 2–4.

1.5. Let V be a complex vector space. A (*vector*) *norm* is a function: $V \to \mathbb{R}$, denoted by $\|\mathbf{x}\|$, that satisfies:

(N1) $\|\mathbf{x}\| \geq 0$, $\|\mathbf{x}\| = 0$ if and only if $\mathbf{x} = \mathbf{0}$ (*positivity*);

(N2) $\|\alpha \mathbf{x}\| = |\alpha| \, \|\mathbf{x}\|$ (*positive homogeneity*); and

(N3) $\|\mathbf{x} + \mathbf{y}\| \leq \|\mathbf{x}\| + \|\mathbf{y}\|$ (*triangle inequality*);

for all $\mathbf{x}, \mathbf{y} \in V$ and $\alpha \in \mathbb{C}$.

Note:

(a) The *if* part of (N1) follows from (N2).

(b) $\|\mathbf{x}\|$ is interpreted as the *length* of the vector $\mathbf{x}$. Inequality (N3) then states, in $\mathbb{R}^2$, that the length of any side of a triangle is no greater than the sum of lengths of the other two sides.

See Exs. 3–11.

Exercises

EX. 1. *Direct sums.* Let L and M be subspaces of a vector space V. Then the following statements are equivalent:

(a) $V = L \oplus M$.

(b) Every vector $\mathbf{x} \in V$ is uniquely represented as

$$\mathbf{x} = \mathbf{y} + \mathbf{z} \quad (\mathbf{y} \in L, \ \mathbf{z} \in M).$$

(c) $\dim V = \dim L + \dim M$, $L \cap M = \{\mathbf{0}\}$.

(d) If $\{\mathbf{x}_1, \mathbf{x}_2, \ldots, \mathbf{x}_l\}$ and $\{\mathbf{y}_1, \mathbf{y}_2, \ldots, \mathbf{y}_m\}$ are bases for L and M, respectively, then $\{\mathbf{x}_1, \mathbf{x}_2, \ldots, \mathbf{x}_l, \mathbf{y}_1, \mathbf{y}_2, \ldots, \mathbf{y}_m\}$ is a basis for V.

EX. 2. *The Cauchy–Schwartz inequality.* For any $\mathbf{x}, \mathbf{y} \in \mathbb{C}^n$

$$|\langle \mathbf{x}, \mathbf{y} \rangle| \leq \sqrt{\langle \mathbf{x}, \mathbf{x} \rangle} \sqrt{\langle \mathbf{y}, \mathbf{y} \rangle} \tag{4}$$

with equality if and only if $\mathbf{x} = \lambda \mathbf{y}$ for some $\lambda \in \mathbb{C}$.

PROOF. For any complex z,

$$0 \leq \langle \mathbf{x} + z\mathbf{y}, \mathbf{x} + z\mathbf{y} \rangle, \quad \text{by (I3)},$$
$$= \langle \mathbf{y}, \mathbf{y} \rangle |z|^2 + z\langle \mathbf{y}, \mathbf{x} \rangle + \overline{z}\langle \mathbf{x}, \mathbf{y} \rangle + \langle \mathbf{x}, \mathbf{x} \rangle, \quad \text{by (I1)–(I2)},$$
$$= \langle \mathbf{y}, \mathbf{y} \rangle |z|^2 + 2\Re\{z\langle \mathbf{x}, \mathbf{y} \rangle\} + \langle \mathbf{x}, \mathbf{x} \rangle,$$
$$\leq \langle \mathbf{y}, \mathbf{y} \rangle |z|^2 + 2|z||\langle \mathbf{x}, \mathbf{y} \rangle| + \langle \mathbf{x}, \mathbf{x} \rangle. \tag{5}$$

Here $\Re$ denotes *real part*. The quadratic equation RHS(5) $= 0$ can have at most one solution $|z|$, proving that $|\langle \mathbf{x}, \mathbf{y} \rangle|^2 \leq \langle \mathbf{x}, \mathbf{x} \rangle \langle \mathbf{y}, \mathbf{y} \rangle$, with equality if and only if $\mathbf{x} + z\mathbf{y} = \mathbf{0}$ for some $z \in \mathbb{C}$. $\qquad\square$

Ex. 3. If $\langle \mathbf{x}, \mathbf{y} \rangle$ is an inner product on $\mathbb{C}^n$, then

$$\|\mathbf{x}\| := \sqrt{\langle \mathbf{x}, \mathbf{x} \rangle} \tag{6}$$

is a norm on $\mathbb{C}^n$. The *Euclidean norm* in $\mathbb{C}^n$

$$\|\mathbf{x}\| = \sqrt{\sum_{j=1}^{n} |x|^2}, \tag{7}$$

corresponds to the standard inner product. [*Hint*: Use (4) to verify the triangle inequality (N3) in §1.5 above.]

Ex. 4. Show that to every inner product $f : \mathbb{C}^n \times \mathbb{C}^n \to \mathbb{C}$ there corresponds a unique positive definite matrix $Q = [q_{ij}] \in \mathbb{C}^{n \times n}$ such that

$$f(\mathbf{x}, \mathbf{y}) = \mathbf{y}^* Q \mathbf{x} = \sum_{i=1}^{n} \sum_{j=1}^{n} \overline{y_i}\, q_{ij} x_j. \tag{8}$$

The inner product (8) is denoted by $\langle \mathbf{x}, \mathbf{y} \rangle_Q$. It induces a norm, by Ex. 3,

$$\|\mathbf{x}\|_Q = \sqrt{\mathbf{x}^* Q \mathbf{x}},$$

called *ellipsoidal*, or *weighted Euclidean* norm. The standard inner product (3), and the Euclidean norm, correspond to the special case $Q = I$.

SOLUTION. The inner product f and the positive definite matrix $Q = [q_{ij}]$ completely determine each other by

$$f(\mathbf{e}_i, \mathbf{e}_j) = q_{ij}, \quad (i, j \in \overline{1, n}),$$

where $\mathbf{e}_i$ is the ith unit vector. $\qquad\square$

Ex. 5. Given an inner product $\langle \mathbf{x}, \mathbf{y} \rangle$ and the corresponding norm $\|\mathbf{x}\| = \langle \mathbf{x}, \mathbf{x} \rangle^{1/2}$, the *angle* between two vectors $\mathbf{x}, \mathbf{y} \in \mathbb{R}^n$, denoted by $\angle\{\mathbf{x}, \mathbf{y}\}$, is defined by

$$\cos \angle\{\mathbf{x}, \mathbf{y}\} = \frac{\langle \mathbf{x}, \mathbf{y} \rangle}{\|\mathbf{x}\|\|\mathbf{y}\|}. \tag{9}$$

Two vectors $\mathbf{x}, \mathbf{y} \in \mathbb{R}^n$ are *orthogonal* if $\langle \mathbf{x}, \mathbf{y} \rangle = 0$. Although it is not obvious how to define angles between vectors in $\mathbb{C}^n$, see, e.g., Scharnhorst [**725**], we define orthogonality by the same condition, $\langle \mathbf{x}, \mathbf{y} \rangle = 0$, as in the real case.

Ex. 6. Let $\langle \cdot, \cdot \rangle$ be an inner product on $\mathbb{C}^n$. A set $\{\mathbf{v}_1, \ldots, \mathbf{v}_k\}$ of $\mathbb{C}^n$ is called *orthonormal* (abbreviated o.n.) if

$$\langle \mathbf{v}_i, \mathbf{v}_j \rangle = \delta_{ij}, \quad \text{for all } i, j \in \overline{1, k}. \tag{10}$$

(a) An o.n. set is linearly independent.

(b) If $\mathcal{B} = \{\mathbf{v}_1, \ldots, \mathbf{v}_n\}$ is an o.n. basis of $\mathbb{C}^n$, then for all $\mathbf{x} \in \mathbb{C}^n$,

$$\mathbf{x} = \sum_{j=1}^{n} \xi_j \mathbf{v}_j, \quad \text{with } \xi_j = \langle \mathbf{x}, \mathbf{v}_j \rangle, \tag{11}$$

and

$$\langle \mathbf{x}, \mathbf{x} \rangle = \sum_{j=1}^{n} |\xi_j|^2. \tag{12}$$

Ex. 7. *Gram–Schmidt orthonormalization.* Let $\mathcal{A} = \{\mathbf{a}_1, \mathbf{a}_2, \ldots, \mathbf{a}_n\} \subset \mathbb{C}^m$ be a set of vectors spanning a subspace L, $L = \{\sum_{i=1}^n \alpha_i \, \mathbf{a}_i : \alpha_i \in \mathbb{C}\}$. Then an o.n. basis $\mathcal{Q} = \{\mathbf{q}_1, \mathbf{q}_2, \ldots, \mathbf{q}_r\}$ of L is computed using the *Gram–Schmidt orthonormalization* process (abbreviated GSO) as follows.

$$\mathbf{q}_1 = \frac{\mathbf{a}_{c_1}}{\|\mathbf{a}_{c_1}\|}, \quad \text{if } \mathbf{a}_{c_1} \neq \mathbf{0} = \mathbf{a}_j \text{ for } 1 \leq j < c_1, \tag{13a}$$

$$\mathbf{x}_j = \mathbf{a}_j - \sum_{\ell=1}^{k-1} \langle \mathbf{a}_j, \mathbf{q}_\ell \rangle \, \mathbf{q}_\ell, \quad j = c_{k-1}+1, c_{k-1}+2, \ldots, c_k, \tag{13b}$$

and

$$\mathbf{q}_k = \frac{\mathbf{x}_{c_k}}{\|\mathbf{x}_{c_k}\|}, \quad \text{if } \mathbf{x}_{c_k} \neq \mathbf{0} = \mathbf{x}_j \text{ for } c_{k-1}+1 \leq j < c_k, \; k = 2, \ldots, r. \tag{13c}$$

The integer r found by the GSO process is the dimension of the subspace L. The integers $\{c_1, \ldots, c_r\}$ are the indices of a maximal linearly independent subset $\{\mathbf{a}_{c_1}, \ldots, \mathbf{a}_{c_r}\}$ of $\mathcal{A}$.

Ex. 8. Let $\| \; \|_{(1)}$, $\| \; \|_{(2)}$ be two norms on $\mathbb{C}^n$ and let α_1, α_2 be positive scalars. Show that the following functions:

(a) $\max\{\|\mathbf{x}\|_{(1)}, \|\mathbf{x}\|_{(2)}\}$;
(b) $\alpha_1 \|\mathbf{x}\|_{(1)} + \alpha_2 \|\mathbf{x}\|_{(2)}$;

are norms on $\mathbb{C}^n$.

Ex. 9. *The ℓ_p-norms.* For any $p \geq 1$ the function

$$\|\mathbf{x}\|_p = \left(\sum_{j=1}^n |x_j|^p \right)^{1/p} \tag{14}$$

is a norm on $\mathbb{C}^n$, called the *ℓ_p-norm.*
Hint: The statement that (14) satisfies (N3) for $p \geq 1$ is the classical Minkowski inequality; see, e.g., Beckenbach and Bellman [**55**].

Ex. 10. The most popular ℓ_p-norms are the choices $p = 1, 2$, and ∞,

$$\|\mathbf{x}\|_1 = \sum_{j=1}^n |x_j|, \quad \text{the } \ell_1\text{-norm,} \tag{14.1}$$

$$\|\mathbf{x}\|_2 = \left(\sum_{j=1}^n |x_j|^2 \right)^{1/2}, \quad \text{the } \ell_2\text{-norm or the } Euclidean \; norm, \tag{14.2}$$

$$\|\mathbf{x}\|_\infty = \max\{|x_j| : j \in \overline{1,n}\}, \quad \text{the } \ell_\infty\text{-norm or the } Tchebycheff \; norm. \tag{14.∞}$$

Is $\|\mathbf{x}\|_\infty = \lim_{p \to \infty} \|\mathbf{x}\|_p$?

Ex. 11. Let $\| \; \|_{(1)}$, $\| \; \|_{(2)}$ be any two norms on $\mathbb{C}^n$. Show that there exist positive scalars α, β such that

$$\alpha \|\mathbf{x}\|_{(1)} \leq \|\mathbf{x}\|_{(2)} \leq \beta \|\mathbf{x}\|_{(1)}, \tag{15}$$

for all $\mathbf{x} \in \mathbb{C}^n$.
Hint: $\alpha = \inf\{\|\mathbf{x}\|_{(2)} : \|\mathbf{x}\|_{(1)} = 1\}$, $\quad \beta = \sup\{\|\mathbf{x}\|_{(2)} : \|\mathbf{x}\|_{(1)} = 1\}$.

REMARK 1. Two norms $\| \; \|_{(1)}$ and $\| \; \|_{(2)}$ are called *equivalent* if there exist positive scalars α, β such that (15) holds for all $\mathbf{x} \in \mathbb{C}^n$. From Ex. 11, *any* two norms on $\mathbb{C}^n$ are equivalent. Therefore, if a sequence $\{\mathbf{x}_k\} \subset \mathbb{C}^n$ satisfies

$$\lim_{k \to \infty} \|\mathbf{x}_k\| = 0 \tag{16}$$

for some norm, then (16) holds for any norm. Topological concepts like convergence and continuity, defined by limiting expressions like (16), are therefore independent of the norm used in their definition. Thus we say that a sequence $\{\mathbf{x}_k\} \subset \mathbb{C}^n$ converges to a point $\mathbf{x}_\infty$ if

$$\lim_{k \to \infty} \|\mathbf{x}_k - \mathbf{x}_\infty\| = 0$$

for some norm.

2. Linear Transformations and Matrices

2.1.   The set of $m \times n$ matrices with elements in $\mathbb{F}$ is denoted $\mathbb{F}^{m \times n}$.

In particular, $\mathbb{C}^{m \times n}$ [$\mathbb{R}^{m \times n}$] denote the class of $m \times n$ complex [real] matrices.

A matrix $A \in \mathbb{F}^{m \times n}$ is *square* if $m = n$, *rectangular* otherwise.

The elements of a matrix $A \in \mathbb{F}^{m \times n}$ are denoted by a_{ij} or $A[i, j]$.

We denote by

$$Q_{k,n} = \{(i_1, i_2, \ldots, i_k) : 1 \leq i_1 < i_2 < \cdots < i_k \leq n\}.$$

the set of increasing sequences of k elements from $\overline{1, n}$, for given integers $0 < k \leq n$. For $A \in \mathbb{C}^{m \times n}$, $I \in Q_{p,m}$, $J \in Q_{q,n}$ we denote

A_{IJ} (or $A[I, J]$),   the $p \times q$ submatrix $(A[i, j])$, $i \in I$, $j \in J$,

A_{I*} (or $A[I, *]$),   the $p \times n$ submatrix $(A[i, j])$, $i \in I$, $j \in \overline{1, n}$,

A_{*J} (or $A[*, J]$),   the $m \times q$ submatrix $(A[i, j])$, $i \in \overline{1, m}$, $j \in J$.

The matrix A is:

diagonal if $A[i, j] = 0$ for $i \neq j$;

upper triangular if $A[i, j] = 0$ for $i > j$; and

lower triangular if $A[i, j] = 0$ for $i < j$.

An $m \times n$ diagonal matrix $A = [a_{ij}]$ is denoted $A = \text{diag}(a_{11}, \ldots, a_{pp})$ where $p = \min\{m, n\}$.

Given a matrix $A \in \mathbb{C}^{m \times n}$, its:

transpose is the matrix $A^T \in \mathbb{C}^{n \times m}$ with $A^T[i, j] = A[j, i]$ for all i, j; and its

conjugate transpose is the matrix $A^* \in \mathbb{C}^{n \times m}$ with $A^*[i, j] = \overline{A[j, i]}$ for all i, j.

A square matrix is:

Hermitian [*symmetric*] if $A = A^*$ [A is real, $A = A^T$];

normal if $AA^* = A^*A$; and

unitary [*orthogonal*] if $A^* = A^{-1}$ [A is real, $A^T = A^{-1}$].

2.2.   Given vector spaces U, V over a field $\mathbb{F}$, and a mapping $T: U \to V$, we say that T is *linear*, or a *linear transformation*, if $T(\alpha \mathbf{x} + \mathbf{y}) = \alpha T\mathbf{x} + T\mathbf{y}$, for all $\alpha \in \mathbb{F}$ and $\mathbf{x}, \mathbf{y} \in U$. The set of linear transformations from U to V is denoted $\mathcal{L}(U, V)$. It is a vector space with operations $T_1 + T_2$ and αT defined by

$$(T_1 + T_2)\mathbf{u} = T_1\mathbf{u} + T_2\mathbf{u}, \quad (\alpha T)\mathbf{u} = \alpha(T\mathbf{u}), \quad \forall \, \mathbf{u} \in U.$$

The zero element of $\mathcal{L}(U, V)$ is the transformation O mapping every $\mathbf{u} \in U$ into $\mathbf{0} \in V$. The identity mapping $I_U \in \mathcal{L}(U, U)$ is defined by $I_U\mathbf{u} = \mathbf{u}$, $\forall \, \mathbf{u} \in U$. We usually omit the subscript U, writing the identity as I.

2.3.   Let $T \in \mathcal{L}(U, V)$. For any $\mathbf{u} \in U$, the point $T\mathbf{u}$ in V is called the *image* of $\mathbf{u}$ (under T). The *range* of T, denoted $R(T)$ is the set of all its images

$$R(T) = \{\mathbf{v} \in V : \mathbf{v} = T\mathbf{u} \text{ for some } \mathbf{u} \in U\}.$$

For any $\mathbf{v} \in R(T)$, the *inverse image* $T^{-1}(\mathbf{v})$ is the set

$$T^{-1}(\mathbf{v}) = \{\mathbf{u} \in U : T\mathbf{u} = \mathbf{v}\}.$$

In particular, the *null space* of T, denoted by $N(T)$, is the inverse image of the zero vector $\mathbf{0} \in V$,

$$N(T) = \{\mathbf{u} \in U : T\mathbf{u} = \mathbf{0}\}.$$

2.4. $T \in \mathcal{L}(U, V)$ is *one-to-one* if for all $\mathbf{x}, \mathbf{y} \in U$, $\mathbf{x} \neq \mathbf{y} \implies T\mathbf{x} \neq T\mathbf{y}$ or, equivalently, if for every $\mathbf{v} \in R(T)$ the inverse image $T^{-1}\mathbf{v}$ is a singleton. T is *onto* if $R(T) = V$. If T is one-to-one and onto, it has an *inverse* $T^{-1} \in \mathcal{L}(V, U)$ such that

$$T^{-1}(T\mathbf{u}) = \mathbf{u} \quad \text{and} \quad T(T^{-1}\mathbf{v}) = \mathbf{v}, \ \forall \, \mathbf{u} \in U, \, \mathbf{v} \in V, \qquad (17a)$$

or, equivalently,

$$T^{-1}T = I_U, \ TT^{-1} = I_V, \qquad (17b)$$

in which case T is called *invertible* or *nonsingular*.

2.5.   Given:

- a linear transformation $A \in \mathcal{L}(\mathbb{C}^n, \mathbb{C}^m)$; and
- two bases $\mathcal{U} = \{\mathbf{u}_1, \dots, \mathbf{u}_m\}$ and $\mathcal{V} = \{\mathbf{v}_1, \dots, \mathbf{v}_n\}$ of $\mathbb{C}^m$ and $\mathbb{C}^n$, respectively;

the *matrix representation of A with respect to the bases* $\{U, V\}$ is the $m \times n$ matrix $A_{\{\mathcal{U}, \mathcal{V}\}} = [a_{ij}]$ determined (uniquely) by

$$A\mathbf{v}_j = \sum_{i=1}^{m} a_{ij}\,\mathbf{u}_i, \quad j \in \overline{1, n}. \qquad (18)$$

For any such pair of bases $\{\mathcal{U}, \mathcal{V}\}$, (18) is a one-to-one correspondence between the linear transformations $\mathcal{L}(\mathbb{C}^n, \mathbb{C}^m)$ and the matrices $\mathbb{C}^{m \times n}$, allowing the customary practice of using the same symbol A to denote both the linear transformation $A : \mathbb{C}^n \to \mathbb{C}^m$ and its matrix representation $A_{\{\mathcal{U}, \mathcal{V}\}}$.

If A is a linear transformation from $\mathbb{C}^n$ to itself, and $\mathcal{V} = \{\mathbf{v}_1, \dots, \mathbf{v}_n\}$ is a basis of $\mathbb{C}^n$, then the matrix representation $A_{\{\mathcal{V}, \mathcal{V}\}}$ is denoted simply by $A_{\{\mathcal{V}\}}$. It is the (unique) matrix $A_{\{\mathcal{V}\}} = [a_{ij}] \in \mathbb{C}^{n \times n}$ satisfying

$$A\mathbf{v}_j = \sum_{i=1}^{n} a_{ij}\,\mathbf{v}_i, \quad j \in \overline{1, n}. \qquad (19)$$

The *standard basis* of $\mathbb{C}^n$ is the basis $\mathcal{E}_n$ consisting of the n unit vectors

$$\mathcal{E}_n = \{\mathbf{e}_1, \dots, \mathbf{e}_n\}.$$

Unless otherwise noted, linear transformations $A \in \mathcal{L}(\mathbb{C}^n, \mathbb{C}^m)$ are represented in terms of the standard bases $\{\mathcal{E}_m, \mathcal{E}_n\}$.

For any $A \in \mathbb{C}^{m \times n}$ we denote, as in Section 2.3 above,

$$R(A) = \{\mathbf{y} \in \mathbb{C}^m : \mathbf{y} = A\mathbf{x} \text{ for some } \mathbf{x} \in \mathbb{C}^n\}, \quad \text{the } range \text{ of } A, \quad (20a)$$

$$N(A) = \{\mathbf{x} \in \mathbb{C}^n : A\mathbf{x} = \mathbf{0}\}, \quad \text{the } null\ space \text{ of } A. \quad (20b)$$

2.6. Let $\langle \cdot, \cdot \rangle$ denote the standard inner product. If $A \in \mathbb{C}^{m \times n}$ then

$$\langle A\mathbf{x}, \mathbf{y} \rangle = \langle \mathbf{x}, A^*\mathbf{y} \rangle, \quad \text{for all } \mathbf{x} \in \mathbb{C}^n, \mathbf{y} \in \mathbb{C}^m. \quad (21)$$

$H \in \mathbb{C}^{n \times n}$ is Hermitian if and only if

$$\langle H\mathbf{x}, \mathbf{y} \rangle = \langle \mathbf{x}, H\mathbf{y} \rangle, \quad \text{for all } \mathbf{x}, \mathbf{y} \in \mathbb{C}^n. \quad (22)$$

If $\langle A\mathbf{x}, \mathbf{x} \rangle = \langle \mathbf{x}, A\mathbf{x} \rangle$ for all $\mathbf{x}$, then A need not be Hermitian. Example: $A = \begin{pmatrix} 1 & 1 \\ 0 & 0 \end{pmatrix}$.

2.7. Let $\langle \cdot, \cdot \rangle_{\mathbb{C}^n}$ and $\langle \cdot, \cdot \rangle_{\mathbb{C}^m}$ be inner products on C^n and $\mathbb{C}^m$, respectively, and let $A \in \mathcal{L}(\mathbb{C}^n, \mathbb{C}^m)$. The *adjoint* of A, denoted by A^*, is the linear transformation $A^* \in \mathcal{L}(\mathbb{C}^m, \mathbb{C}^n)$ defined by

$$\langle A\mathbf{v}, \mathbf{u} \rangle_{\mathbb{C}^m} = \langle \mathbf{v}, A^*\mathbf{u} \rangle_{\mathbb{C}^n} \quad (23)$$

for all $\mathbf{v} \in \mathbb{C}^n$, $\mathbf{u} \in \mathbb{C}^m$. Unless otherwise stated, we use the standard inner product, in which case *adjoint = conjugate transpose*.

2.8. Given a subspace L of $\mathbb{C}^n$, define

$$L^\perp := \{\mathbf{x} \in \mathbb{C}^n : \mathbf{x} \text{ is orthogonal to every vector in } L\}. \quad (24)$$

Then $L^\perp$ is a subspace complementary to L. $L^\perp$ is called the *orthogonal complement* of L. If $M \subset L^\perp$ is a subspace, then $L \oplus M$ is called the *orthogonal direct sum* of L, M and denoted by $L \overset{\perp}{\oplus} M$. In particular, $\mathbb{C}^n$ is the orthogonal direct sum of $L, L^\perp$,

$$\mathbb{C}^n = L \overset{\perp}{\oplus} L^\perp. \quad (25)$$

With any matrix $A \in \mathbb{C}^{m \times n}$ there are associated four subspaces

$$N(A), R(A^*) \quad \text{in } \mathbb{C}^n,$$
$$N(A^*), R(A) \quad \text{in } \mathbb{C}^m.$$

An important result is that these pairs form orthogonal complements.

THEOREM 1. *For any* $A \in \mathbb{C}^{m \times n}$,

$$N(A) = R(A^*)^\perp, \quad (26)$$

$$N(A^*) = R(A)^\perp. \quad (27)$$

PROOF. Let $\mathbf{x} \in N(A)$. Then LHS(21) vanishes for all $\mathbf{y} \in \mathbb{C}^m$. It follows then that $\mathbf{x} \perp A^*\mathbf{y}$ for all $\mathbf{y} \in \mathbb{C}^m$ or, in other words, $\mathbf{x} \perp R(A^*)$. This proves that $N(A) \subset R(A^*)^\perp$.

Conversely, let $\mathbf{x} \in R(A^*)^\perp$, so that RHS(21) vanishes for all $\mathbf{y} \in \mathbb{C}^m$. This implies that $A\mathbf{x} \perp \mathbf{y}$ for all $\mathbf{y} \in \mathbb{C}^m$. Therefore $A\mathbf{x} = \mathbf{0}$. This proves that $R(A^*)^\perp \subset N(A)$, and completes the proof.

The dual relation (27) follows by reversing the roles of A, A^*. $\square$

2.9.   A *(matrix) norm* of $A \in \mathbb{C}^{m \times n}$, denoted by $\|A\|$, is defined as a function: $\mathbb{C}^{m \times n} \to \mathbb{R}$ that satisfies

$$\|A\| \geq 0, \quad \|A\| = 0 \quad \text{only if } A = O, \tag{M1}$$

$$\|\alpha A\| = |\alpha| \|A\|, \tag{M2}$$

$$\|A + B\| \leq \|A\| + \|B\|, \tag{M3}$$

for all $A, B \in \mathbb{C}^{m \times n}$, $\alpha \in \mathbb{C}$. If, in addition,

$$\|AB\| \leq \|A\| \|B\| \tag{M4}$$

whenever the matrix product AB is defined, then $\| \ \|$ is called a *multiplicative norm*. Some authors (see, e.g., Householder [**432**, Section 2.2]) define a matrix norm as having all four properties (M1)–(M4).

2.10.   If $A \in \mathbb{C}^{n \times n}$, $\mathbf{0} \neq \mathbf{x} \in \mathbb{C}^n$, and $\lambda \in \mathbb{C}$ are such that

$$A\mathbf{x} = \lambda\mathbf{x}, \tag{28}$$

then λ is an *eigenvalue* of A corresponding to the *eigenvector* $\mathbf{x}$. The set of eigenvalues of A is called its *spectrum*, and is denoted by[1] $\lambda(A)$. If λ is an eigenvalue of A, the subspace

$$\{\mathbf{x} \in \mathbb{C}^n : A\mathbf{x} = \lambda\mathbf{x}\} \tag{29}$$

is the corresponding *eigenspace* of A, its dimension is called the *geometric multiplicity* of the eigenvalue λ.

2.11.   If $H \in \mathbb{C}^{n \times n}$ is Hermitian, then:

(a) the eigenvalues of H are real;
(b) eigenvectors corresponding to different eigenvalues are orthogonal;
(c) there is an o.n. basis of $\mathbb{C}^n$ consisting of eigenvectors of H; and
(d) the eigenvalues of H, ordered by

$$\lambda_1 \geq \lambda_2 \geq \cdots \geq \lambda_n,$$

and corresponding eigenvectors,

$$H\mathbf{x}_j = \lambda_j \mathbf{x}_j, \quad j \in \overline{1,n},$$

can be computed recursively as

$$\lambda_1 = \max\left\{\langle H\mathbf{x}, \mathbf{x}\rangle : \|\mathbf{x}\| = 1\right\} = \langle H\mathbf{x}_1, \mathbf{x}_1\rangle,$$

$$\lambda_j = \max\left\{\langle H\mathbf{x}, \mathbf{x}\rangle : \|\mathbf{x}\| = 1, \quad \mathbf{x} \perp \{\mathbf{x}_1, \mathbf{x}_2, \ldots, \mathbf{x}_{j-1}\}\right\}$$

$$= \langle H\mathbf{x}_j, \mathbf{x}_j\rangle, \quad j \in \overline{2,n}.$$

2.12.   A Hermitian matrix $H \in \mathbb{C}^{n \times n}$ is *positive semidefinite* (PSD for short) if $\langle H\mathbf{x}, \mathbf{x}\rangle \geq 0$ for all $\mathbf{x}$ or, equivalently, if its eigenvalues are nonnegative. Similarly, H is called *positive definite* (PD for short), if $\langle H\mathbf{x}, \mathbf{x}\rangle > 0$ for all $\mathbf{x} \neq \mathbf{0}$ or, equivalently, if its eigenvalues are positive. The set of $n \times n$ PSD [PD] matrices is denoted by PSD_n [PD_n].

---

[1]The spectrum of A is often denoted elsewhere by $\sigma(A)$, a symbol reserved here for the *singular values* of A.

2.13. Let $A \in \mathbb{C}^{m \times n}$ and let the eigenvalues $\lambda_j(AA^*)$ of AA^* (which is PSD) be ordered by

$$\lambda_1(AA^*) \geq \cdots \geq \lambda_r(AA^*) > \lambda_{r+1}(AA^*) = \cdots = \lambda_n(AA^*) = 0. \qquad (30)$$

The *singular values* of A, denoted by $\sigma_j(A)$ or σ_j, $j \in \overline{1,r}$, are defined as

$$\sigma_j(A) = +\sqrt{\lambda_j(AA^*)}, \quad j \in \overline{1,r}, \qquad (31a)$$

or, equivalently,

$$\sigma_j(A) = +\sqrt{\lambda_j(A^*A)}, \quad j \in \overline{1,r}, \qquad (31b)$$

and are ordered, by (30),

$$\sigma_1 \geq \sigma_2 \geq \cdots \geq \sigma_r > 0. \qquad (32)$$

The *set of singular values* of A is denoted by $\sigma(A)$.

2.14. Let A and $\{\sigma_j\}$ be as above, let $\{\mathbf{u}_i : i \in \overline{1,m}\}$ be an o.n. basis of $\mathbb{C}^m$ made of eigenvectors of AA^*,

$$AA^*\mathbf{u}_i = \sigma_i^2\mathbf{u}_i, \quad i \in \overline{1,r}, \qquad (33a)$$

$$AA^*\mathbf{u}_i = \mathbf{0}, \qquad i \in \overline{r+1,m}, \qquad (33b)$$

let

$$\mathbf{v}_j = \frac{1}{\sigma_j} A^*\mathbf{u}_j, \ j \in \overline{1,r}, \qquad (34)$$

and let $\{\mathbf{v}_j : j \in \overline{r+1,n}\}$ be an o.n. set of vectors, orthogonal to $\{\mathbf{v}_j : j \in \overline{1,r}\}$. Then the set $\{\mathbf{v}_j : j \in \overline{1,n}\}$ is an o.n. basis of $\mathbb{C}^n$ consisting of eigenvectors of A^*A,

$$A^*A\mathbf{v}_j = \sigma_i^2\mathbf{v}_j, \quad j \in \overline{1,r}, \qquad (35a)$$

$$A^*A\mathbf{v}_j = \mathbf{0}, \qquad j \in \overline{r+1,n}. \qquad (35b)$$

Conversely, starting from an o.n. basis $\{\mathbf{v}_j : j \in \overline{1,n}\}$ of $\mathbb{C}^n$ satisfying (35a)–(35b), we construct an o.n. basis of $\mathbb{C}^m$ with (33a)–(33b) by

$$\mathbf{u}_i = \frac{1}{\sigma_i} A\mathbf{v}_j, \quad i \in \overline{1,r}, \qquad (36)$$

and completing to an o.n. set. See Theorem 6.1.

2.15. Let A, $\{\mathbf{u}_i : i \in \overline{1,m}\}$ and $\{\mathbf{v}_j : j \in \overline{1,n}\}$ be as above. Then (36) can be written as

$$A[\mathbf{v}_1 \ \cdots \ \mathbf{v}_r \vdots \mathbf{v}_{r+1} \ \cdots \ \mathbf{v}_n]$$

$$= [\mathbf{u}_1 \ \cdots \ \mathbf{u}_r \vdots \mathbf{u}_{r+1} \ \cdots \ \mathbf{u}_m]
\begin{bmatrix}
\sigma_1 & & & \vdots & \\
 & \ddots & & \vdots & O \\
 & & \sigma_r & \vdots & \\
\cdots & \cdots & \cdots & \cdots & \cdots \\
 & O & & \vdots & O
\end{bmatrix}$$

or

$$AV = U\Sigma, \quad \text{where } \Sigma = \begin{bmatrix} \sigma_1 & & & \vdots & \\ & \ddots & & \vdots & O \\ & & \sigma_r & \vdots & \\ \cdots & \cdots & \cdots & \cdots & \cdots \\ & O & & \vdots & O \end{bmatrix}, \tag{37}$$

and $U = [\mathbf{u}_1 \cdots \mathbf{u}_m]$, $V = [\mathbf{v}_1 \cdots \mathbf{v}_n]$. Since V is unitary,

$$A = U\Sigma V^*, \tag{38}$$

called a *singular value decomposition* (abbreviated SVD) of A. See Theorem 6.2.

Exercises

Ex. 12. Let L, M be subspaces of $\mathbb{C}^n$, with $\dim L \geq (k+1)$, $\dim M \leq k$. Then $L \cap M^\perp \neq \{\mathbf{0}\}$.
PROOF. Otherwise $L + M^\perp$ is a direct sum with dimension $= \dim L + \dim M^\perp \geq (k+1) + (n-k) > n$.   □

Ex. 13. *The QR factorization.* Let the o.n. set $\{\mathbf{q}_1, \ldots, \mathbf{q}_r\}$ be obtained from the set of vectors $\{\mathbf{a}_1, \ldots, \mathbf{a}_n\}$ by the GSO process described in Ex. 7, and let $\widetilde{Q} = [\mathbf{q}_1, \ldots, \mathbf{q}_r] \in \mathbb{C}^{m \times r}$ and $A = [\mathbf{a}_1, \ldots, \mathbf{a}_n] \in \mathbb{C}^{m \times n}$ be the corresponding matrices. Then

$$A = \widetilde{Q}\widetilde{R}, \tag{39}$$

where the columns of $\widetilde{Q}$ are an o.n. basis of $R(A)$ and $\widetilde{R} \in \mathbb{C}_r^{r \times n}$ is an upper triangular matrix. If $r < m$, it is possible to complete $\widetilde{Q}$ to a unitary matrix $Q = [\widetilde{Q}\ Z]$, where the columns of Z are an o.n. basis of $N(A^*)$. Then (39) can be written as

$$A = QR \tag{40}$$

where $R = \begin{bmatrix} \widetilde{R} \\ O \end{bmatrix}$ is upper triangular.

The expression (40) is called a *QR-factorization* of A. By analogy, we call (39) a *$\widetilde{Q}\widetilde{R}$-factorization* of A.

Ex. 14. Let U and V be finite-dimensional vector spaces over a field $\mathbb{F}$ and let $T \in \mathcal{L}(U, V)$. Then the null space $N(T)$ and range $R(T)$ are subspaces of U and V, respectively.
PROOF. L is a subspace of U if and only if

$$\mathbf{x}, \mathbf{y} \in L, \alpha \in \mathbb{F} \quad \Longrightarrow \quad \alpha \mathbf{x} + \mathbf{y} \in L.$$

If $\mathbf{x}, \mathbf{y} \in N(T)$, then $T(\mathbf{x} + \alpha \mathbf{y}) = T\mathbf{x} + \alpha T\mathbf{y} = \mathbf{0}$ for all $\alpha \in \mathbb{F}$, proving that $N(T)$ is a subspace of U. The proof that $R(T)$ is a subspace is similar.   □

Ex. 15. Let P_n be the set of polynomials with real coefficients, of degree $\leq n$,

$$P_n = \{\mathbf{p} : \mathbf{p}(x) = p_0 + p_1 x + \cdots + p_n x^n, \ p_i \in \mathbb{R}\}. \tag{41}$$

The name x of the variable in (41) is immaterial.

(a) Show that P_n is a vector space with the operations

$$\mathbf{p} + \mathbf{q} = \sum_{i=0}^{n} p_i x^i + \sum_{i=0}^{n} q_i x^i = \sum_{i=0}^{n} (p_i + q_i) x^i, \quad \alpha \mathbf{p} = \sum_{i=0}^{n} (\alpha p_i) x^i,$$

and the dimension of P_n is $n + 1$.

(b) The set of monomials $\mathcal{U}_n = \{1, x, x^2, \ldots, x^n\}$ is a basis of P_n. Let T be the differentiation operator, mapping a function $f(x)$ into its derivative $f'(x)$. Show that $T \in \mathcal{L}(P_n, P_{n-1})$. What are the range and null space of T? Find the representation of T with respect to the bases $\{\mathcal{U}_n, \mathcal{U}_{n-1}\}$.

(c) Let S be the integration operator, mapping a function $f(x)$ into its integral $\int f(x) \, dx$ with zero constant of integration. Show that $S \in \mathcal{L}(P_{n-1}, P_n)$. What are the range and null space of S? Find the representation of S with respect to $\{\mathcal{U}_{n-1}, \mathcal{U}_n\}$.

(d) Let $T_{\mathcal{U}_n, \mathcal{U}_{n-1}}$ and $S_{\mathcal{U}_{n-1}, \mathcal{U}_n}$ be the matrix representations in parts (b) and (c). What are the matrix products $T_{\{\mathcal{U}_n, \mathcal{U}_{n-1}\}} S_{\{\mathcal{U}_{n-1}, \mathcal{U}_n\}}$ and $S_{\{\mathcal{U}_{n-1}, \mathcal{U}_n\}} T_{\{\mathcal{U}_n, \mathcal{U}_{n-1}\}}$? Interpret these results in view of the fact that integration and differentiation are *inverse operations*.

Ex. 16. Let $\mathcal{U} = \{\mathbf{u}_1, \ldots, \mathbf{u}_m\}$ and $\mathcal{V} = \{\mathbf{v}_1, \ldots, \mathbf{v}_n\}$ be o.n. bases of $\mathbb{C}^m$ and $\mathbb{C}^n$, respectively. Then for any $A \in \mathcal{L}(\mathbb{C}^n, \mathbb{C}^m)$:

(a) The matrix representation $A_{\{\mathcal{U}, \mathcal{V}\}} = [a_{ij}]$ is given by

$$a_{ij} = \langle A\mathbf{v}_j, \mathbf{u}_i \rangle, \ \forall \, i, j.$$

where $\langle \cdot, \cdot \rangle$ is the inner product on $\mathbb{C}^m$ (in which $\mathcal{U}$ is an o.n. basis.)

(b) The adjoint A^* is represented by the matrix $A^*_{\{\mathcal{V}, \mathcal{U}\}} = [b_{k\ell}]$ where $b_{k\ell} = \overline{a_{\ell k}}$, i.e., the matrix $A^*_{\{\mathcal{V}, \mathcal{U}\}}$ is the conjugate transpose of $A_{\{\mathcal{U}, \mathcal{V}\}}$.

Ex. 17. *The simplest matrix representation.* Let $O \neq A \in \mathcal{L}(\mathbb{C}^n, \mathbb{C}^m)$. Then there exist o.n. bases $\mathcal{U} = \{\mathbf{u}_1, \ldots, \mathbf{u}_m\}$ and $\mathcal{V} = \{\mathbf{v}_1, \ldots, \mathbf{v}_n\}$ of $\mathbb{C}^m$ and $\mathbb{C}^n$, respectively, such that

$$A_{\{\mathcal{U}, \mathcal{V}\}} = \text{diag}\,(\sigma_1, \ldots, \sigma_r, 0, \ldots, 0) \in \mathbb{R}^{n \times m}, \tag{42}$$

a diagonal matrix, whose nonzero diagonal elements $\sigma_1 \geq \sigma_2 \geq \cdots \geq \sigma_r > 0$ are the singular values of A.

Ex. 18. Let $\mathcal{V} = \{\mathbf{v}_1, \ldots, \mathbf{v}_n\}$ and $\mathcal{W} = \{\mathbf{w}_1, \ldots, \mathbf{w}_n\}$ be two bases of $\mathbb{C}^n$. Show that there is a unique $n \times n$ matrix $S = [s_{ij}]$ such that

$$\mathbf{w}_j = \sum_{i=1}^{n} s_{ij} \, \mathbf{v}_i, \quad j \in \overline{1, n}, \tag{43}$$

and S is nonsingular. Using the rules of matrix multiplication we rewrite (43) as

$$[\mathbf{w}_1, \mathbf{w}_2, \ldots, \mathbf{w}_n] = [\mathbf{v}_1, \mathbf{v}_2, \ldots, \mathbf{v}_n] \begin{bmatrix} s_{11} & \cdots & s_{1n} \\ \vdots & & \vdots \\ s_{n1} & \cdots & s_{nn} \end{bmatrix} = [\mathbf{v}_1, \mathbf{v}_2, \ldots, \mathbf{v}_n] S, \tag{44}$$

i.e.,

$$[\mathbf{v}_1, \mathbf{v}_2, \ldots, \mathbf{v}_n] = [\mathbf{w}_1, \mathbf{w}_2, \ldots, \mathbf{w}_n] S^{-1}. \tag{45}$$

Ex. 19. *Similar matrices.* We recall that two square matrices A, B are called *similar* if

$$B = S^{-1} A S \tag{46}$$

for some nonsingular matrix S. If S in (46) is *unitary [orthogonal]*, then A, B are called called *unitarily similar [orthogonally similar]*.

Show that two $n \times n$ complex matrices are similar if and only if each is a matrix representation of the same linear transformation relative to a basis of $\mathbb{C}^n$.

PROOF. *If*: Let $\mathcal{V} = \{\mathbf{v}_1, \mathbf{v}_2, \ldots, \mathbf{v}_n\}$ and $\mathcal{W} = \{\mathbf{w}_1, \mathbf{w}_2, \ldots, \mathbf{w}_n\}$ be two bases of $\mathbb{C}^n$ and let $A_{\{\mathcal{V}\}}$ and $A_{\{\mathcal{W}\}}$ be the corresponding matrix representations of a given linear transformation $A : \mathbb{C}^n \to \mathbb{C}^n$. The bases $\mathcal{V}$ and $\mathcal{W}$ determine a (unique) nonsingular matrix $S = [s_{ij}]$ satisfying (43). Rewriting (19) as

$$A[\mathbf{v}_1, \mathbf{v}_2, \ldots, \mathbf{v}_n] = [\mathbf{v}_1, \mathbf{v}_2, \ldots, \mathbf{v}_n]A_{\{\mathcal{V}\}}. \tag{47}$$

we conclude, by substituting (45) in (47), that

$$A[\mathbf{w}_1, \mathbf{w}_2, \ldots, \mathbf{w}_n] = [\mathbf{w}_1, \mathbf{w}_2, \ldots, \mathbf{w}_n]S^{-1}A_{\{\mathcal{V}\}}S,$$

and by the uniqueness of the matrix representation,

$$A_{\{\mathcal{W}\}} = S^{-1}A_{\{\mathcal{V}\}}S.$$

Only if: Similarly proved.   □

EX. 20. *Schur triangularization.* Any $A \in \mathbb{C}^{n \times n}$ is unitarily similar to a triangular matrix. (For proof see, e.g., Marcus and Minc [**534**, p. 67]).

EX. 21. *Perron's approximate diagonalization.* Let $A \in \mathbb{C}^{n \times n}$. Then for any $\epsilon > 0$ there is a nonsingular matrix S such that $S^{-1}AS$ is a triangular matrix

$$S^{-1}AS = \begin{bmatrix} \lambda_1 & b_{12} & \cdots & \cdots & b_{1n} \\ 0 & \lambda_2 & & & \vdots \\ \vdots & \ddots & \ddots & & \vdots \\ \vdots & & \ddots & \ddots & \vdots \\ 0 & \cdots & \cdots & 0 & \lambda_n \end{bmatrix}$$

with the off-diagonal elements satisfying

$$\sum_{ij} |b_{ij}| \leq \epsilon \quad \text{(Bellman [\textbf{56}, p. 205]).}$$

EX. 22. A matrix in $\mathbb{C}^{n \times n}$ is:

(a) normal if and only if it is unitarily similar to a diagonal matrix; and

(b) Hermitian if and only if it is unitarily similar to a real diagonal matrix.

EX. 23. For any $n \geq 2$ there is an $n \times n$ real matrix which is not similar to a triangular matrix in $\mathbb{R}^{n \times n}$.

Hint. The diagonal elements of a triangular matrix are its eigenvalues.

EX. 24. Denote the transformation of bases (43) by $\mathcal{W} = \mathcal{V}S$. Let $\{\mathcal{U}, \mathcal{V}\}$ be bases of $\{\mathbb{C}^m, \mathbb{C}^n\}$, respectively, and let $\{\widetilde{\mathcal{U}}, \widetilde{\mathcal{V}}\}$ be another pair of bases, obtained by

$$\widetilde{\mathcal{U}} = \mathcal{U}S, \quad \widetilde{\mathcal{V}} = \mathcal{V}T,$$

where S and T are $m \times m$ and $n \times n$ matrices, respectively. Show that for any $A \in \mathcal{L}(\mathbb{C}^n, \mathbb{C}^m)$, the representations $A_{\{\mathcal{U}, \mathcal{V}\}}$ and $A_{\{\widetilde{\mathcal{U}}, \widetilde{\mathcal{V}}\}}$ are related by

$$A_{\{\widetilde{\mathcal{U}}, \widetilde{\mathcal{V}}\}} = S^{-1}A_{\{\mathcal{U}, \mathcal{V}\}}T. \tag{48}$$

PROOF. Similar to the proof of Ex. 19.   □

Ex. 25. *Equivalent matrices.* Two matrices A, B in $\mathbb{C}^{m \times n}$ are called *equivalent* if there are nonsingular matrices $S \in \mathbb{C}^{m \times m}$ and $T \in \mathbb{C}^{n \times n}$ such that

$$B = S^{-1}AT. \tag{49}$$

If S and T in (49) are unitary matrices, then A, B are called *unitarily equivalent.*

It follows from Ex. 24 that two matrices in $\mathbb{C}^{m \times n}$ are equivalent if, and only if, each is a matrix representation of the same linear transformation relative to a pair of bases of $\mathbb{C}^m$ and $\mathbb{C}^n$.

Ex. 26. Let $A \in \mathcal{L}(\mathbb{C}^n, \mathbb{C}^m)$ and $B \in \mathcal{L}(\mathbb{C}^p, \mathbb{C}^n)$, and let $\mathcal{U}, \mathcal{V}$, and $\mathcal{W}$ be bases of $\mathbb{C}^m, \mathbb{C}^n$, and $\mathbb{C}^p$, respectively. The *product* (or *composition*) of A and B, denoted by AB, is the transformation $\mathbb{C}^p \to \mathbb{C}^m$ defined by

$$(AB)\mathbf{w} = A(B\mathbf{w}), \quad \text{for all } \mathbf{w} \in \mathbb{C}^p.$$

(a) The transformation AB is linear, i.e., $(AB) \in \mathcal{L}(\mathbb{C}^p, \mathbb{C}^m)$.
(b) The matrix representation of AB relative to $\{\mathcal{U}, \mathcal{W}\}$ is

$$(AB)_{\{\mathcal{U}, \mathcal{W}\}} = A_{\{\mathcal{U}, \mathcal{V}\}} B_{\{\mathcal{V}, \mathcal{W}\}},$$

the (matrix) product of the corresponding matrix representations of A and B.

Ex. 27. The matrix representation of the identity transformation I in $\mathbb{C}^n$, relative to any basis, is the $n \times n$ identity matrix I.

Ex. 28. For any invertible $A \in \mathcal{L}(\mathbb{C}^n, \mathbb{C}^n)$ and any two bases $\{\mathcal{U}, \mathcal{V}\}$ of $\mathbb{C}^n$, the matrix representation of A^{-1} relative to $\{\mathcal{V}, \mathcal{U}\}$ is the inverse of the matrix $A_{\{\mathcal{U}, \mathcal{V}\}}$,

$$(A^{-1})_{\{\mathcal{V}, \mathcal{U}\}} = (A_{\{\mathcal{U}, \mathcal{V}\}})^{-1}.$$

PROOF. Follows from Exs. 26–27. □

Ex. 29. The real matrix $A = \left(\begin{smallmatrix} 0 & 1 \\ -1 & 0 \end{smallmatrix} \right)$ has the complex eigenvalue $\lambda = i$, with geometric multiplicity $= 2$, i.e., every nonzero $\mathbf{x} \in \mathbb{R}^2$ is an eigenvector.
PROOF.

$$\begin{bmatrix} 0 & 1 \\ -1 & 0 \end{bmatrix} \begin{bmatrix} x_1 \\ x_2 \end{bmatrix} = \lambda \begin{bmatrix} x_1 \\ x_2 \end{bmatrix} \quad \Longrightarrow \quad \lambda^2 = -1,$$

unless $x_1 = x_2 = 0$. □

Ex. 30. Let $A \in \mathcal{L}(\mathbb{C}^m, \mathbb{C}^n)$. A property shared by all matrix representations $A_{\{\mathcal{U}, \mathcal{V}\}}$ of A, as $\mathcal{U}$ and $\mathcal{V}$ range over all bases of $\mathbb{C}^m$ and $\mathbb{C}^n$, respectively, is an *intrinsic property* of the linear transformation A. Example: If A, B are similar matrices, they have the same determinant. The determinant is thus intrinsic to the linear transformation represented by A and B.

Given a matrix $A = (a_{ij}) \in \mathbb{C}^{m \times n}$, which of the following items are intrinsic properties of a linear transformation represented by A?

(a) if $m = n$:
    (a1) the eigenvalues of A;
    (a2) their geometric multiplicities;
    (a3) the eigenvectors of A;
(b) if m, n are not necessarily equal:
    (b1) the rank of A;
    (b2) the null space of A;
    (b3) $\sum_{i=1}^{m} \sum_{j=1}^{n} |a_{ij}|^2$.

Ex. 31. Let $\widetilde{\mathcal{U}}_n = \{\widetilde{p}_1, \ldots, \widetilde{p}_n\}$ be the set of partial sums of monomials

$$\widetilde{p}_k(x) = \sum_{i=0}^{k} x^i, \quad k \in \overline{1,n}.$$

(a) Show that $\widetilde{\mathcal{U}}_n$ is a basis of P_n, and determine the matrix A, such that $\widetilde{\mathcal{U}}_n = A\mathcal{U}_n$, where $\mathcal{U}_n$ is the basis of monomials, see Ex. 15.
(b) Calculate the representations of the differentiation operator (Ex. 15(b)) with respect to to the bases $\{\widetilde{\mathcal{U}}_n, \widetilde{\mathcal{U}}_{n-1}\}$, and verify (48).
(c) Same for the integration operator of Ex. 15(c).

Ex. 32. Let L and M be complementary subspaces of $\mathbb{C}^n$. Show that the projector $P_{L,M}$, which carries $\mathbf{x} \in \mathbb{C}^n$ into its projection on L along M, is a linear transformation (from $\mathbb{C}^n$ to L).

Ex. 33. Let L and M be complementary subspaces of $\mathbb{C}^n$, let $\mathbf{x} \in \mathbb{C}^n$, and let $\mathbf{y}$ be the projection of $\mathbf{x}$ on L along M. What is the unique expression for $\mathbf{x}$ as the sum of a vector in L and a vector in M? What, therefore, is $P_{L,M}\,\mathbf{y} = P_{L,M}^2\,\mathbf{x}$, the projection of $\mathbf{y}$ on L along M? Show, therefore, that the transformation $P_{L,M}$ is idempotent.

Ex. 34. *Matrix norms.* Show that the functions

$$\left(\sum_{i=1}^{m} \sum_{j=1}^{n} |a_{ij}|^2\right)^{1/2} = (\text{trace } A^*A)^{1/2} \tag{50}$$

and

$$\max\{|a_{ij}| : i \in \overline{1,m}, \, j \in \overline{1,n}\} \tag{51}$$

are matrix norms. The norm (50) is called the *Frobenius norm*, and denoted $\|A\|_F$. Which of these norms is multiplicative?

Ex. 35. *Consistent norms.* A vector norm $\|\ \|$ and a matrix norm $\|\ \|$ are called *consistent* if for any vector $\mathbf{x}$ and matrix A such that $A\mathbf{x}$ is defined,

$$\|A\mathbf{x}\| \leq \|A\|\|\mathbf{x}\|. \tag{52}$$

Given a vector norm $\|\ \|_*$ show that

$$\|A\|_* = \sup_{\mathbf{x} \neq 0} \frac{\|A\mathbf{x}\|_*}{\|\mathbf{x}\|_*} \tag{53}$$

is a multiplicative matrix norm consistent with $\|\mathbf{x}\|_*$ and that any other matrix norm $\|\ \|$ consistent with $\|\mathbf{x}\|_*$ satisfies

$$\|A\| \geq \|A\|_*, \quad \text{for all } A. \tag{54}$$

The norm $\|A\|_*$ defined by (53), is called the *matrix norm corresponding to the vector norm* $\|\mathbf{x}\|_*$, or the *bound* of A with respect to $K = \{\mathbf{x} : \|\mathbf{x}\|_* \leq 1\}$; see, e.g., Householder [**432**, Section 2.2] and Ex. 3.66 below.

Ex. 36. Show that (53) is the same as

$$\|A\|_* = \max_{\|\mathbf{x}\|_* \leq 1} \frac{\|A\mathbf{x}\|_*}{\|\mathbf{x}\|_*} = \max_{\|\mathbf{x}\|_* = 1} \|A\mathbf{x}\|_*. \tag{55}$$

Ex. 37. Given a multiplicative matrix norm, find a vector norm consistent with it.

Ex. 38. *Corresponding norms.*

Vector norm on $\mathbb{C}^n$ Corresponding matrix norm on $\mathbb{C}^{m \times n}$

$$(14) \quad \|\mathbf{x}\|_p = (\sum_{j=1}^{n} |x_j|^p)^{1/p}, \quad \|A\|_p = \max_{\|\mathbf{x}\|_p = 1} \|A\mathbf{x}\|_p ; \qquad (56)$$

$$(14.1) \quad \|\mathbf{x}\|_1 = \sum_{j=1}^{n} |x_j|, \quad \|A\|_1 = \max_{1 \le j \le n} \sum_{i=1}^{m} |a_{ij}|; \qquad (56.1)$$

$$(14.\infty) \quad \|\mathbf{x}\|_\infty = \max_{1 \le j \le n} |x_j|, \quad \|A\|_\infty = \max_{1 \le i \le m} \sum_{j=1}^{n} |a_{ij}|; \qquad (56.\infty)$$

$$(14.2) \quad \|\mathbf{x}\|_2 = (\sum_{j=1}^{n} |x_j|^2)^{1/2}, \quad \|A\|_2 = \max\{\sqrt{\lambda} : \lambda \text{ an eigenvalue of } A^*A\}.$$
$$\qquad (56.2)$$

Note that (56.2) is different from the Frobenius norm (50), which is the Euclidean norm of the mn-dimensional vector obtained by listing all components of A. The norm $\| \ \|_2$ given by (56.2) is called the *spectral norm*.

PROOF. Equation (56.1) follows from (55) since, for any $\mathbf{x} \in \mathbb{C}^n$,

$$\|A\mathbf{x}\|_1 = \sum_{i=1}^{m} |\sum_{j=1}^{n} a_{ij} x_j| \le \sum_{i=1}^{m} \sum_{j=1}^{n} |a_{ij}||x_j|$$

$$\le \sum_{j=1}^{n} |x_j| \sum_{i=1}^{m} |a_{ij}|$$

$$\le (\max_{1 \le j \le n} \sum_{i=1}^{m} |a_{ij}|)(\|\mathbf{x}\|_1)$$

with equality if $\mathbf{x}$ is the kth unit vector, where k is any j for which the maximum in (56) is attained

$$\sum_{i=1}^{m} |a_{ik}| = \max_{1 \le j \le n} \sum_{i=1}^{m} |a_{ij}|.$$

Equation (56.∞) is similarly proved and (56.2) is left as exercise. $\qquad \square$

Ex. 39. For any matrix norm $\| \ \|$ on $\mathbb{C}^{m \times n}$, consistent with some vector norm, the norm of the unit matrix satisfies

$$\|I_n\| \ge 1.$$

In particular, if $\| \ \|_*$ is a matrix norm, computed by (53) from a corresponding vector norm, then

$$\|I_n\|_* = 1. \qquad (57)$$

Ex. 40. A matrix norm $\| \ \|$ on $\mathbb{C}^{m \times n}$ is called *unitarily invariant* if, for any two unitary matrices $U \in \mathbb{C}^{m \times m}$ and $V \in \mathbb{C}^{n \times n}$,

$$\|UAV\| = \|A\|, \quad \text{for all } A \in \mathbb{C}^{m \times n}.$$

Show that the matrix norms (50) and (56.2) are unitarily invariant.

Ex. 41. *Spectral radius.* The *spectral radius* $\rho(A)$ of a square matrix $A \in \mathbb{C}^{n \times n}$ is the maximal value among the n moduli of the eigenvalues of A,

$$\rho(A) = \max\{|\lambda| : \lambda \in \lambda(A)\}. \qquad (58)$$

Let $\| \ \|$ be any multiplicative norm on $\mathbb{C}^{n\times n}$. Then, for any $A \in \mathbb{C}^{n\times n}$,

$$\rho(A) \leq \|A\|. \tag{59}$$

PROOF. Let $\| \ \|$ denote both a given multiplicative matrix norm and a vector norm consistent with it. Then $Ax = \lambda x \implies |\lambda|\|x\| = \|Ax\| \leq \|A\|\|x\|.$ $\square$

Ex. 42. For any $A \in \mathbb{C}^{n\times n}$ and any $\epsilon > 0$, there exists a multiplicative matrix norm $\| \ \|$ such that

$$\|A\| \leq \rho(A) + \epsilon \quad \text{(Householder [432, p. 46])}.$$

Ex. 43. If A is a square matrix,

$$\rho(A^k) = \rho^k(A), \quad k = 0, 1, \ldots. \tag{60}$$

Ex. 44. For any $A \in \mathbb{C}^{m\times n}$, the spectral norm $\| \ \|_2$ of (56.2) equals

$$\|A\|_2 = \rho^{1/2}(A^*A) = \rho^{1/2}(AA^*). \tag{61}$$

In particular, if A is Hermitian, then

$$\|A\|_2 = \rho(A). \tag{62}$$

In general, the spectral norm $\|A\|_2$ and the spectral radius $\rho(A)$ may be quite apart; see, e.g., Noble [615, p. 430].

Ex. 45. *Convergent matrices.* A square matrix A is called *convergent* if

$$A^k \to O, \quad \text{as } k \to \infty. \tag{63}$$

Show that $A \in \mathbb{C}^{n\times n}$ is convergent if and only if

$$\rho(A) < 1. \tag{64}$$

PROOF. *If:* From (64) and Ex. 42 it follows that there exists a multiplicative matrix norm $\| \ \|$ such that $\|A\| < 1$. Then

$$\|A^k\| \leq \|A\|^k \to 0, \quad \text{as } k \to \infty,$$

proving (63).
Only if: If $\rho(A) \geq 1$, then by (60), so is $\rho(A^k)$ for $k = 0, 1, \ldots$, contradicting (63). $\square$

Ex. 46. A square matrix A is convergent if and only if the sequence of partial sums

$$S_k = I + A + A^2 + \cdots + A^k = \sum_{j=0}^{k} A^j$$

converges, in which case it converges to $(I - A)^{-1}$, i.e.,

$$(I - A)^{-1} = I + A + A^2 + \cdots = \sum_{j=0}^{\infty} A^j \quad \text{(Householder [432, p. 54])}. \tag{65}$$

Ex. 47. Let A be convergent. Then

$$(I + A)^{-1} = I - A + A^2 - \cdots = \sum_{j=0}^{\infty}(-1)^j A^j. \tag{66}$$

Ex. 48. *Stein's Theorem.* A square matrix is convergent if and only if there exists a PD matrix H such that $H - A^*HA$ is also PD (Stein [776], Taussky [799]).

3. Elementary Operations and Permutations

3.1. *Elementary operations.* The following operations on a matrix:

(1) multiplying row i by a nonzero scalar α, denoted by $E^i(\alpha)$;
(2) adding β times row j to row i, denoted by $E^{ij}(\beta)$ (here β is any scalar); and
(3) interchanging rows i and j, denoted by E^{ij}, (here $i \neq j$);

are called *elementary row operations* of types 1, 2, and 3, respectively.[2]

Applying an elementary row operation to the identity matrix I_m results in an *elementary matrix* of the same type. We denote these elementary matrices also by $E^i(\alpha)$, $E^{ij}(\beta)$, and E^{ij}. Elementary matrices of types 1, 2 have only one row that is different from the corresponding row of the identity matrix of the same order. Examples for $m = 4$,

$$E^2(\alpha) = \begin{bmatrix} 1 & 0 & 0 & 0 \\ 0 & \alpha & 0 & 0 \\ 0 & 0 & 1 & 0 \\ 0 & 0 & 0 & 1 \end{bmatrix}, \quad E^{42}(\beta) = \begin{bmatrix} 1 & 0 & 0 & 0 \\ 0 & 1 & 0 & 0 \\ 0 & 0 & 1 & 0 \\ 0 & \beta & 0 & 1 \end{bmatrix}, \quad E^{13} = \begin{bmatrix} 0 & 0 & 1 & 0 \\ 0 & 1 & 0 & 0 \\ 1 & 0 & 0 & 0 \\ 0 & 0 & 0 & 1 \end{bmatrix}.$$

Elementary column operations and the corresponding elementary matrices are defined analogously.

Performing an elementary row [column] operation is the same as multiplying on the left [right] by the corresponding elementary matrix. For example, $E^{25}(-3)A$ is the matrix obtained from A by subtracting $3 \times$ row 5 from row 2.

3.2. *Permutations.* Given a positive integer n, a *permutation* of order n is a rearrangement of $\{1, 2, \dots, n\}$, i.e., a mapping: $\overline{1, n} \longrightarrow \overline{1, n}$. The set of such permutations is denoted by S_n. It contains:

(a) the identity permutation $\pi_0\{1, 2, \dots, n\} = \{1, 2, \dots, n\}$;
(b) with any two permutations π_1, π_2, their *product* $\pi_1 \pi_2$, defined as π_1 applied to $\{\pi_2(1), \pi_2(2), \dots, \pi_2(n)\}$; and
(c) with any permutation π, its *inverse*, denoted by π^{-1}, mapping $\{\pi(1), \pi(2), \dots, \pi(n)\}$ back to $\{1, 2, \dots, n\}$.

Thus S_n is a *group*, called the *symmetric group*.

Given a permutation $\pi \in S_n$, the corresponding *permutation matrix* P_π is defined as $P_\pi = [\delta_{\pi(i),j}]$, and the correspondence $\pi \longleftrightarrow P_\pi$ is one-to-one. For example,

$$\pi\{1, 2, 3\} = \{2, 3, 1\} \quad \longleftrightarrow \quad P_\pi = \begin{bmatrix} 0 & 1 & 0 \\ 0 & 0 & 1 \\ 1 & 0 & 0 \end{bmatrix}.$$

Products of permutations correspond to matrix products:

$$P_{\pi_1 \pi_2} = P_{\pi_1} P_{\pi_2}, \quad \forall\, \pi_1, \pi_2 \in S_n.$$

A *transposition* is a permutation that switches only a pair of elements, for example, $\pi\{1, 2, 3, 4\} = \{1, 4, 3, 2\}$. Every permutation $\pi \in S_n$ is a product of transpositions, generally in more than one way. However, the number of

---

[2]Only operations of types 1, 2 are necessary, see Ex. 49(b). Type 3 operations are introduced for convenience, because of their frequent use.

transpositions in such a product is always even or odd, depending only on π. Accordingly, a permutation π is called *even* or *odd*, if it is the product of an even or odd number of transpositions, respectively. The *sign* of the permutation π, denoted $\operatorname{sign} \pi$, is defined as

$$\operatorname{sign} \pi = \begin{cases} +1, & \text{if } \pi \text{ is even}, \\ -1, & \text{if } \pi \text{ is odd}. \end{cases}$$

The following table summarizes the situation for permutations of order 3:

Permutation π		Inverse π^{-1}	Product of transpositions	$\operatorname{sign} \pi$
π_0	$\{1,2,3\}$	π_0	$\pi_1\pi_1,\ \pi_2\pi_2$, etc.	$+1$
π_1	$\{1,3,2\}$	π_1	π_1	-1
π_2	$\{2,1,3\}$	π_2	π_2	-1
π_3	$\{2,3,1\}$	π_4	$\pi_1\pi_2$	$+1$
π_4	$\{3,1,2\}$	π_3	$\pi_2\pi_1$	$+1$
π_5	$\{3,2,1\}$	π_5	π_5	-1

Multiplying a matrix A by a permutation matrix P_π on the left [right] results in a permutation π $[\pi^{-1}]$ of the rows [columns] of A. For example,

$$\begin{bmatrix} 0 & 1 & 0 \\ 0 & 0 & 1 \\ 1 & 0 & 0 \end{bmatrix} \begin{bmatrix} a_{11} & a_{12} \\ a_{21} & a_{22} \\ a_{31} & a_{32} \end{bmatrix} = \begin{bmatrix} a_{21} & a_{22} \\ a_{31} & a_{32} \\ a_{11} & a_{12} \end{bmatrix},$$

$$\begin{bmatrix} b_{11} & b_{12} & b_{13} \\ b_{21} & b_{22} & b_{23} \end{bmatrix} \begin{bmatrix} 0 & 1 & 0 \\ 0 & 0 & 1 \\ 1 & 0 & 0 \end{bmatrix} = \begin{bmatrix} b_{13} & b_{11} & b_{12} \\ b_{23} & b_{21} & b_{22} \end{bmatrix}.$$

Exercises

Ex. 49. *Elementary operations.*

(a) The elementary matrices are nonsingular, and their inverses are

$$E^i(\alpha)^{-1} = E^i(1/\alpha), \quad E^{ij}(\beta)^{-1} = E^{ij}(-\beta), \quad (E^{ij})^{-1} = E^{ij}. \tag{67}$$

(b) Type 3 elementary operations are expressible in terms of the other two types:

$$E^{ij} = E^i(-1)E^{ji}(1)E^{ij}(-1)E^{ji}(1). \tag{68}$$

(c) Conclude from (b) that any permutation matrix is a product of elementary matrices of types 1, 2.

Ex. 50. Describe a recursive method for listing all $n!$ permutations in S_n.
Hint: If π is a permutation in S_{n-1}, mapping $\{1, 2, \dots, n-1\}$ to

$$\{\pi(1), \pi(2), \dots, \pi(n-1)\}, \tag{69}$$

then π gives rise to n permutations in S_n obtained by placing n in the "gaps" $\{\sqcup\pi(1) \sqcup \pi(2) \sqcup \dots \sqcup \pi(n-1)\sqcup\}$ of (69).

4. The Hermite Normal Form and Related Items

4.1. *Hermite normal form.* Let $\mathbb{C}_r^{m\times n}$ $[\mathbb{R}_r^{m\times n}]$ denote the class of $m \times n$ complex [real] matrices of rank r.

DEFINITION 1. (Marcus and Minc [**534**, § 3.6]). A matrix in $\mathbb{C}_r^{m \times n}$ is said to be in *Hermite normal form* (also called *reduced row-echelon form*) if:

(a) the first r rows contain at least one nonzero element; the remaining rows contain only zeros;

(b) there are r integers

$$1 \leq c_1 < c_2 < \cdots < c_r \leq n, \tag{70}$$

such that the first nonzero element in row $i \in \overline{1,r}$, appears in column c_i; and

(c) all other elements in column c_i are zero, $i \in \overline{1,r}$.   □

By a suitable permutation of its columns, a matrix $H \in \mathbb{C}_r^{m \times n}$ in Hermite normal form can be brought into the partitioned form

$$R = \begin{bmatrix} I_r & K \\ O & O \end{bmatrix}, \tag{71}$$

where O denotes a null matrix. Such a permutation of the columns of H can be interpreted as multiplication of H on the right by a suitable permutation matrix P. If P_j denotes the jth column of P, and $\mathbf{e}_j$ the jth column of I_n, we have

$$P_j = \mathbf{e}_k, \quad \text{where } k = c_j, \quad j \in \overline{1,r},$$

the remaining columns of P are the remaining unit vectors $\{\mathbf{e}_k : k \neq c_j, \ j \in \overline{1,r}\}$ in any order. In general, there are $(n - r)!$ different pairs $\{P, K\}$, corresponding to all arrangements of the last $n - r$ columns of P.

In particular cases, the partitioned form (71) may be suitably interpreted. If $R \in \mathbb{C}_r^{m \times n}$, then the two right-hand submatrices are absent in the case $r = n$, and the two lower submatrices are absent if $r = m$.

4.2.   *Gaussian elimination.* A *Gaussian elimination* is a sequence of elementary row operations that transform a given matrix to a desired form.

The Hermite normal form of a given matrix can be computed by Gaussian elimination. Transpositions of rows (i.e., elementary operations of type 3) are used, if necessary, to bring the nonzero rows to the top. The *pivots* of the elimination are the leading nonzeros in these rows. This is illustrated in Ex. 51 below.

Let $A \in \mathbb{C}^{m \times n}$ and let $E_k, E_{k-1}, \ldots, E_2, E_1$ be elementary row operations, and let P be a permutation matrix such that

$$E\,A\,P = \begin{bmatrix} I_r & K \\ O & O \end{bmatrix}, \tag{72}$$

where

$$E = E_k E_{k-1} \cdots E_2 E_1, \tag{73}$$

in which case A is determined to have rank r. Equation (72) can be rewritten as

$$A = E^{-1} \begin{bmatrix} I_r & K \\ O & O \end{bmatrix} P^{-1}. \tag{74}$$

4.3. *Bases for the range and null space of a matrix.* Let $A \in \mathbb{C}^{m \times n}$ and let $R(A)$ and $N(A)$ be as in (20).

A basis for $R(A)$ is useful in a number of applications, such as, for example, in the numerical computation of the Moore–Penrose inverse, and the group inverse to be discussed in Chapter 4.

The need for a basis of $N(A)$ is illustrated by the fact that the general solution of the linear inhomogeneous equation

$$A\mathbf{x} = \mathbf{b}$$

is the sum of any particular solution $\mathbf{x}_0$ and the general solution of the homogeneous equation

$$A\mathbf{x} = \mathbf{0}.$$

The latter general solution consists of all linear combinations of the elements of any basis for $N(A)$.

A further advantage of the Hermite normal form EA of A (and its column-permuted form EAP) is that from them bases for $R(A), N(A)$, and $R(A^*)$ can be read off directly.

A basis for $R(A)$ consists of the $c_1^{\text{th}}, c_2^{\text{th}}, \dots, c_r^{\text{th}}$ columns of A, where the $\{c_j : j \in \overline{1,r}\}$ are as in Definition 1. To see this, let P_1 denote the submatrix consisting of the first r columns of the permutation matrix P of (72). Then, because of the way in which these r columns of P were chosen,

$$EAP_1 = \begin{bmatrix} I_r \\ O \end{bmatrix}. \tag{75}$$

Now, AP_1 is an $m \times r$ matrix, and is of rank r, since RHS(75) is of rank r. But AP_1 is merely the submatrix of A consisting of the $c_1^{\text{th}}, c_2^{\text{th}}, \dots, c_r^{\text{th}}$ columns.

It follows from (74) that the columns of the $n \times (n - r)$ matrix

$$P \begin{bmatrix} -K \\ I_{n-r} \end{bmatrix} \tag{76}$$

are a basis for $N(A)$. (The reader should verify this.)

Moreover, it is evident that the first r rows of the Hermite normal form EA are linearly independent, and each is some linear combination of the rows of A. Thus, they are a basis for the space spanned by the rows of A. Consequently, if

$$EA = \begin{bmatrix} G \\ O \end{bmatrix}, \tag{77}$$

then the columns of the $n \times r$ matrix

$$G^* = P \begin{bmatrix} I_r \\ K^* \end{bmatrix}$$

are a basis for $R(A^*)$.

4.4. *Full-rank factorization.* A nonzero matrix can be expressed as the product of a matrix of full column rank and a matrix of full row rank. Such factorizations turn out to be a powerful tool in the study of generalized inverses.

LEMMA 1. *Let* $A \in \mathbb{C}_r^{m \times n}$, $r > 0$. *Then there exist matrices* $F \in \mathbb{C}_r^{m \times r}$ *and* $G \in \mathbb{C}_r^{r \times n}$, *such that*

$$A = FG. \tag{78}$$

PROOF. The $\widetilde{Q}\widetilde{R}$ factorization, Ex. 13, is a case in point. Let F be any matrix whose columns are a basis for $R(A)$. Then $F \in \mathbb{C}_r^{m \times r}$. The matrix $G \in \mathbb{C}^{r \times n}$ is then uniquely determined by (78), since every column of A is uniquely representable as a linear combination of the columns of F. Finally, $\operatorname{rank} G = r$, since

$$\operatorname{rank} G \geq \operatorname{rank} FG = r. \qquad \square$$

The columns of F can, in particular, be chosen as any maximal linearly independent set of columns of A. Also, G could be chosen first as any matrix whose rows are a basis for the space spanned by the rows of A, and then F is uniquely determined by (78).

A factorization (78) with the properties stated in Lemma 1 is called a *(full-)rank factorization* of A. When A is of full (column or row) rank, the most obvious factorization is a trivial one, one factor being a unit matrix.

A rank factorization of any matrix is easily read off from its Hermite normal form. Indeed, it was pointed out in §4.3 above that the first r rows of the Hermite normal form EA (i.e., the rows of the matrix G of (77)) form a basis for the space spanned by the rows of A. Thus, this G can also serve as the matrix G of (78). Consequently, (78) holds for some F. As in §4.3, let P_1 denote the submatrix of P consisting of the first r columns. Because of the way in which these r columns were constructed,

$$GP_1 = I_r.$$

Thus, multiplying (78) on the right by P_1 gives

$$F = AP_1,$$

and so (78) becomes

$$A = (AP_1)G, \tag{79}$$

where P_1 and G are as in §4.3. (Indeed it was already noted there that the columns of AP_1 are a basis for $R(A)$.)

Exercises

Ex. 51. *Transforming a matrix into Hermite normal form.* Let $A \in \mathbb{C}^{m \times n}$ and let $T_0 = [A \quad I_m]$. A matrix E transforming A into a Hermite normal form EA can be found by Gaussian elimination on T_0 where, after the elimination is completed,

$$ET_0 = [EA \quad E],$$

E being recorded as the right-hand $m \times m$ submatrix of ET_0. We illustrate this procedure for the matrix

$$A = \begin{bmatrix} 0 & 2i & i & 0 & 4+2i & 1 \\ 0 & 0 & 0 & -3 & -6 & -3-3i \\ 0 & 2 & 1 & 1 & 4-4i & 1 \end{bmatrix}, \tag{80}$$

marking the pivots by boxes,

$$T_0 = \begin{bmatrix} 0 & \boxed{2i} & i & 0 & 4+2i & 1 & \vdots & 1 & 0 & 0 \\ 0 & 0 & 0 & -3 & -6 & -3-3i & \vdots & 0 & 1 & 0 \\ 0 & 2 & 1 & 1 & 4-4i & 1 & \vdots & 0 & 0 & 1 \end{bmatrix},$$

$$T_1 = E^{31}(-2)E^1(\tfrac{1}{2i})T_0$$

$$= \begin{bmatrix} 0 & 1 & \tfrac{1}{2} & 0 & 1-2i & -\tfrac{1}{2}i & \vdots & -\tfrac{1}{2}i & 0 & 0 \\ 0 & 0 & 0 & \boxed{-3} & -6 & -3-3i & \vdots & 0 & 1 & 0 \\ 0 & 0 & 0 & 1 & 2 & 1+i & \vdots & i & 0 & 1 \end{bmatrix},$$

$$T_2 = E^{32}(-1)E^2(-\tfrac{1}{3})T_1$$

$$= \begin{bmatrix} 0 & 1 & \tfrac{1}{2} & 0 & 1-2i & -\tfrac{1}{2}i & \vdots & -\tfrac{1}{2}i & 0 & 0 \\ 0 & 0 & 0 & 1 & 2 & 1+i & \vdots & 0 & -\tfrac{1}{3} & 0 \\ 0 & 0 & 0 & 0 & 0 & 0 & \vdots & i & \tfrac{1}{3} & 1 \end{bmatrix}.$$

From $T_2 = [EA \quad E]$ we read the Hermite normal form

$$EA = \begin{bmatrix} 0 & 1 & \tfrac{1}{2} & 0 & 1-2i & -\tfrac{1}{2}i \\ 0 & 0 & 0 & 1 & 2 & 1+i \\ \cdots & \cdots & \cdots & \cdots & \cdots & \cdots \\ 0 & 0 & 0 & 0 & 0 & 0 \end{bmatrix}, \tag{81}$$

where

$$E = E^{32}(-1)E^2(-\tfrac{1}{3})E^{31}(-2)E^1(\tfrac{1}{2i}) = \begin{bmatrix} -\tfrac{1}{2}i & 0 & 0 \\ 0 & -\tfrac{1}{3} & 0 \\ i & \tfrac{1}{3} & 1 \end{bmatrix}, \tag{82}$$

and $r = \operatorname{rank} A = 2$.

EX. 52. (Ex. 51 continued). To bring the Hermite normal form (81) to the standard form (72), use

$$P = \begin{bmatrix} 0 & 0 & \vdots & 1 & 0 & 0 & 0 \\ 1 & 0 & \vdots & 0 & 0 & 0 & 0 \\ 0 & 0 & \vdots & 0 & 1 & 0 & 0 \\ 0 & 1 & \vdots & 0 & 0 & 0 & 0 \\ 0 & 0 & \vdots & 0 & 0 & 1 & 0 \\ 0 & 0 & \vdots & 0 & 0 & 0 & 1 \end{bmatrix},$$

to get $EAP = \begin{bmatrix} I_2 & K \\ O & O \end{bmatrix}$

with $K = \begin{bmatrix} 0 & \tfrac{1}{2} & 1-2i & -\tfrac{1}{2}i \\ 0 & 0 & 2 & 1+i \end{bmatrix}.$ $\tag{83}$

In this example there are 4! different pairs $\{P, K\}$, corresponding to all arrangements of the last four columns of P.

EX. 53. (Ex. 51 continued). Consider the matrix A of (80), and its Hermite normal form (81) where the two unit vectors of $\mathbb{C}^2$ appear in the *second* and *fourth* columns. Therefore, the second and fourth columns of A form a basis for $R(A)$.

Using (76) with P and K selected by (83), we find that the columns of the following matrix form a basis for $N(A)$:

$$P \begin{bmatrix} -K \\ I_{n-r} \end{bmatrix} = \begin{bmatrix} 0 & 0 & \vdots & 1 & 0 & 0 & 0 \\ 1 & 0 & \vdots & 0 & 0 & 0 & 0 \\ 0 & 0 & \vdots & 0 & 1 & 0 & 0 \\ 0 & 1 & \vdots & 0 & 0 & 0 & 0 \\ 0 & 0 & \vdots & 0 & 0 & 1 & 0 \\ 0 & 0 & \vdots & 0 & 0 & 0 & 1 \end{bmatrix} \begin{bmatrix} 0 & -\frac{1}{2} & -1+2i & \frac{1}{2}i \\ 0 & 0 & -2 & -1-i \\ \cdots & \cdots & \cdots\cdots & \cdots \\ 1 & 0 & 0 & 0 \\ 0 & 1 & 0 & 0 \\ 0 & 0 & 1 & 0 \\ 0 & 0 & 0 & 1 \end{bmatrix}$$

$$= \begin{bmatrix} 1 & 0 & 0 & 0 \\ 0 & -\frac{1}{2} & -1+2i & \frac{1}{2}i \\ 0 & 1 & 0 & 0 \\ 0 & 0 & -2 & -1-i \\ 0 & 0 & 1 & 0 \\ 0 & 0 & 0 & 1 \end{bmatrix}$$

Ex. 54. If $A \in \mathbb{C}^{m \times m}$ is nonsingular, then the permutation matrix P in (72) can be taken as the identity (i.e., permutation is unnecessary). Therefore $E = A^{-1}$ and

$$A = E_1^{-1} E_2^{-1} \cdots E_{k-1}^{-1} E_k^{-1}. \tag{84}$$

(a) Conclude that A is nonsingular if and only if it is a product of elementary row matrices.

(b) Compute the Hermite normal forms of

$$A = \begin{bmatrix} 1 & 2 & 3 \\ 4 & 5 & 6 \\ 7 & 8 & 8 \end{bmatrix} \quad \text{and} \quad B = \begin{bmatrix} 1 & 2 & 3 \\ 4 & 5 & 6 \\ 7 & 8 & 9 \end{bmatrix},$$

and illustrate (84).

Ex. 55. Consider the matrix A of (80). Using the results of Exs. 51–52, a rank factorization is computed by (79),

$$A = (AP_1)G = \begin{bmatrix} 2i & 0 \\ 0 & -3 \\ 2 & 1 \end{bmatrix} \begin{bmatrix} 0 & 1 & \frac{1}{2} & 0 & 1-2i & -\frac{1}{2}i \\ 0 & 0 & 0 & 1 & 2 & 1+i \end{bmatrix}. \tag{85}$$

Ex. 56. Given a nonzero $m \times n$ matrix $A = [\mathbf{a}_1 \ \ldots \ \mathbf{a}_n]$, a full-column rank submatrix $F = [\mathbf{a}_{c_1} \ \ldots \ \mathbf{a}_{c_r}]$ can be found by Gram–Schmidt orthonormalization, see Ex. 7. Indeed, applying the GSO process to the columns of A gives the integers $\{c_1, \ldots, c_r\}$.

Ex. 57. Use the GSO process to compute a $\widetilde{Q}\widetilde{R}$ factorization of the matrix A of (80).

5. Determinants and Volume

5.1. *Determinants.* The *determinant* of an $n \times n$ matrix $A = [a_{ij}]$, denoted $\det A$, is customarily defined as

$$\det A = \sum_{\pi \in S_n} \text{sign}\,\pi \prod_{i=1}^{n} a_{\pi(i),i}, \tag{86}$$

see, e.g., Marcus and Minc [**534**, §2.4]. We use here an alternative definition.

DEFINITION 2. The *determinant* is a function det: $\mathbb{C}^{n \times n} \to \mathbb{C}$ such that:

(a) $\det(E^i(\alpha)) = \alpha$, for all $\alpha \in \mathbb{C}$, $i \in \overline{1,n}$; and
(b) $\det(AB) = \det(A)\det(B)$, for all $A, B \in \mathbb{C}^{n \times n}$.

The reader is referred to Cullen and Gale [**212**] for proof that Definition 2 is equivalent to (86). See also Exs. 58–59 below.

The Binet–Cauchy formula. If $A \in \mathbb{C}^{k \times n}$, $B \in \mathbb{C}^{n \times k}$, then

$$\det(AB) = \sum_{I \in Q_{k,n}} \det A_{I*} \det B_{*I}. \tag{87}$$

For proof see, e.g., Gantmacher [**296**, Vol. I, p. 9].

5.2. *Gram matrices.* The *Gram matrix* of $\{\mathbf{x}_1, \mathbf{x}_2, \dots, \mathbf{x}_k\} \subset \mathbb{C}^n$ is the $k \times k$ matrix of inner products (unless otherwise noted, $\langle \cdot, \cdot \rangle$ is the standard inner product)

$$G(\mathbf{x}_1, \mathbf{x}_2, \dots, \mathbf{x}_k) = [\langle \mathbf{x}_i, \mathbf{x}_j \rangle]. \tag{88}$$

The determinant of $G(\mathbf{x}_1, \mathbf{x}_2, \dots, \mathbf{x}_k)$ is called the *Gramian* of $\{\mathbf{x}_1, \mathbf{x}_2, \dots, \mathbf{x}_k\}$.

If $X \in \mathbb{C}^{n \times k}$ is the matrix with columns $\{\mathbf{x}_1, \mathbf{x}_2, \dots, \mathbf{x}_k\}$, then by the Binet–Cauchy formula,

$$\det G(\mathbf{x}_1, \dots, \mathbf{x}_n) = \det X^* X = \sum_{I \in Q_{k,n}} |\det X_{I*}|^2.$$

5.3. *Volume.* The matrices in this section are real. Given $A \in \mathbb{R}_r^{m \times n}$, we denote

$$\mathcal{I}(A) = \{I \in Q_{r,m} : \text{ rank } A(I, *) = r\},$$
$$\mathcal{J}(A) = \{J \in Q_{r,n} : \text{ rank } A(*, J) = r\},$$
$$\mathcal{N}(A) = \{(I, J) : I \in Q_{r,m}, J \in Q_{r,n}, A(I, J) \text{ is nonsingular}\},$$

or $\mathcal{I}, \mathcal{J}, \mathcal{N}$ if A is understood. $\mathcal{I}$ and $\mathcal{J}$ are the index sets of maximal submatrices of full row rank and full column rank, respectively, and $\mathcal{N}$ is the index set of maximal nonsingular submatrices.

The (*r-dimensional*) *volume* of a matrix $A \in \mathbb{R}_r^{m \times n}$, denoted vol A or $\text{vol}_r A$, is defined as 0 if $r = 0$, and otherwise

$$\text{vol } A := \sqrt{\sum_{(I,J) \in \mathcal{N}(A)} \det^2 A_{IJ}}. \tag{89}$$

see also Ex. 69 below.

Exercises

Ex. 58. *Properties of determinants.*

(a) $\det E^{ij}(\beta) = 1$, for all $\beta \in \mathbb{C}$, $i, j \in \overline{1,n}$; and
(b) $\det E^{ij} = -1$, for all $i, j \in \overline{1,n}$. (*Hint:* Use (68) and Definition 2.)
(c) If A is nonsingular, and given as product (84) of elementary matrices, then

$$\det A = \det(E_1^{-1})\det(E_2^{-1}) \cdots \det(E_{k-1}^{-1})\det(E_k^{-1}). \tag{90}$$

(d) Use (90) to compute the determinant of A in Ex. 54(b).

EX. 59. *The Cramer rule.* Given a matrix A and a compatible vector $\mathbf{b}$, we denote by $A[j \leftarrow \mathbf{b}]$ the matrix obtained from A by replacing the jth column by $\mathbf{b}$.

Let $A \in \mathbb{C}^{n \times n}$ be nonsingular. Then for any $\mathbf{b} \in \mathbb{C}^n$, the solution $\mathbf{x} = [x_j]$ of

$$A\mathbf{x} = \mathbf{b} \tag{91}$$

is given by

$$x_j = \frac{\det A[j \leftarrow \mathbf{b}]}{\det A}, \quad j \in \overline{1,n}. \tag{92}$$

PROOF (Robinson [700]). Write $A\mathbf{x} = \mathbf{b}$ as

$$AI_n[j \leftarrow \mathbf{x}] = A[j \leftarrow \mathbf{b}], \quad j \in \overline{1,n},$$

and take determinants

$$\det A \, \det I_n[j \leftarrow \mathbf{x}] = \det A[j \leftarrow \mathbf{b}]. \tag{93}$$

Then (92) follows from (93) since

$$\det I_n[j \leftarrow \mathbf{x}] = x_j. \qquad \square$$

See an extension of Cramer's rule in Corollary 5.6.

EX. 60. *The Hadamard inequality.* Let $A = [\mathbf{a_1} \ \mathbf{a_2} \ \cdots \ \mathbf{a_n}] \in \mathbb{C}^{n \times n}$. Then

$$|\det A| \le \prod_{i=1}^{n} \|\mathbf{a}_i\|_2, \tag{94}$$

with equality if and only if the set of columns $\{\mathbf{a}_i\}$ is orthogonal or if one of the columns is zero.

PROOF. LHS(94) = the volume of the parallelepiped defined by the columns of $A \le$ the volume of the cube with sides of lengths $\|\mathbf{a}_i\|_2 = $ RHS(94). $\qquad \square$

EX. 61. *The Schur complement* (Schur [732]). Let the matrix A be partitioned as

$$A = \begin{bmatrix} A_{11} & A_{12} \\ A_{21} & A_{22} \end{bmatrix}, \quad \text{with } A_{11} \text{ nonsingular.} \tag{95}$$

The *Schur complement* of A_{11} in A, denoted A/A_{11}, is defined by

$$A/A_{11} := A_{22} - A_{21} A_{11}^{-1} A_{12}. \tag{96}$$

(a) If A is square, its determinant is

$$\det A = \det A_{11} \det(A/A_{11}). \tag{97}$$

(b) *The quotient property.* If A_{11} is further partitioned as

$$A_{11} = \begin{bmatrix} E & F \\ G & H \end{bmatrix}, \quad \text{with } E \text{ nonsingular,} \tag{98}$$

then

$$A/A_{11} = (A/E)/(A_{11}/E) \quad \text{(Crabtree and Haynsworth [210]).} \tag{99}$$

(c) If A, A_{11}, and A_{22} are nonsingular, then

$$A^{-1} = \begin{bmatrix} (A/A_{22})^{-1} & -A_{11}^{-1} A_{12}(A/A_{11})^{-1} \\ -A_{22}^{-1} A_{21}(A/A_{22})^{-1} & (A/A_{11})^{-1} \end{bmatrix}. \tag{100}$$

PROOF. (a) follows from

$$\begin{bmatrix} A_{11} & A_{12} \\ A_{21} & A_{22} \end{bmatrix} \begin{bmatrix} I & -A_{11}^{-1}A_{12} \\ O & I \end{bmatrix} = \begin{bmatrix} A_{11} & O \\ A_{21} & A/A_{11} \end{bmatrix}.$$

(b) is left for the reader, after noting that A_{11}/E is nonsingular (since A_{11} and E are nonsingular, and $\det A_{11} = \det E \det(A_{11}/E)$ by (97)).

(c) is verified by multiplying A and RHS(100).   □

EX. 62. The set $S = \{\mathbf{x}_1, \dots, \mathbf{x}_k\}$ are linearly independent if, and only if, the Gram matrix $G(S)$ is nonsingular.

PROOF. If S is linearly dependent, then

$$\sum_{i=1}^{k} \alpha_i \mathbf{x}_i = \mathbf{0} \tag{101}$$

has a nonzero solution $(\alpha_1, \dots, \alpha_k)$. Therefore

$$\sum_{i=1}^{k} \alpha_i \langle \mathbf{x}_i, \mathbf{x}_j \rangle = 0, \quad j \in \overline{1,k}, \tag{102}$$

showing $G(S)$ is singular. Conversely, writing (102) as

$$\langle \sum_{i=1}^{k} \alpha_i \mathbf{x}_i, \mathbf{x}_j \rangle = 0, \quad j \in \overline{1,k},$$

multiplying by α_j, and summing we get

$$\langle \sum_{i=1}^{k} \alpha_i \mathbf{x}_i, \sum_{i=1}^{k} \alpha_i \mathbf{x}_i \rangle = 0$$

proving (101).   □

EX. 63. For any set of vectors $S = \{\mathbf{x}_1, \dots, \mathbf{x}_k\} \subset \mathbb{C}^n$, $\det G(S)$ is independent of the order of the vectors. Moreover, $\det G(S) \geq 0$ and $\det G(S) = 0$ if and only if the set S is linearly dependent.

EX. 64. Let $A \in \mathbb{C}_r^{m \times n}$ have a rank-factorization $A = CR$, $C \in \mathbb{C}_r^{m \times r}$, $R \in \mathbb{C}_r^{r \times n}$. Then:

(a) $\mathcal{I}(A) = \mathcal{I}(C)$.
(b) $\mathcal{J}(A) = \mathcal{J}(R)$.
(c) $\mathcal{N}(A) = \mathcal{I}(A) \times \mathcal{J}(A)$.

PROOF. (a) and (b) are obvious, as is $\mathcal{N}(A) \subset \mathcal{I}(A) \times \mathcal{J}(A)$. The converse

$$\mathcal{N}(A) \supset \mathcal{I}(A) \times \mathcal{J}(A),$$

follows since every A_{IJ} is the product

$$A_{IJ} = C_{I*}R_{*J}. \tag{103}$$
    □

EX. 65. If the matrix C is of full-column rank r then, by the Binet–Cauchy theorem,

$$\mathrm{vol}_r^2(C) = \det C^T C, \tag{104}$$

the Gramian of the columns of C. Similarly, if R is of full-row rank r,

$$\mathrm{vol}_r^2(R) = \det RR^T. \tag{105}$$

Ex. 66. Let $A \in \mathbb{R}_r^{m \times n}$, $r > 0$, and let $A = CR$ be any rank factorization. Then

$$\text{vol}_r^2(A) = \sum_{I \in \mathcal{I}} \text{vol}_r^2(A_{I*}), \tag{106a}$$

$$= \sum_{J \in \mathcal{J}} \text{vol}_r^2(A_{*J}), \tag{106b}$$

$$= \text{vol}_r^2(C) \, \text{vol}_r^2(R). \tag{106c}$$

Proof. Follows from Definition (89), Ex. 64(c), and (103). $\qquad\square$

Ex. 67. *A generalized Hadamard inequality.* Let $A \in \mathbb{R}_r^{m \times n}$ be partitioned into two matrices $A = (A_1, A_2)$, $A_i \in \mathbb{R}_{r_i}^{m \times n_i}$, $i = 1, 2$, with $r_1 + r_2 = r$. Then

$$\text{vol}_r A \leq \text{vol}_{r_1} A_1 \, \text{vol}_{r_2} A_2, \tag{107}$$

with equality if and only if the columns of A_1 are orthogonal to those of A_2.
Proof. The full-rank case $n_i = r_i$, $i = 1, 2$, was proved in [**296**, Vol. I, p. 254]. The general case follows since every $m \times r$ submatrix of rank r has r_1 columns from A_1 and r_2 columns from A_2. $\qquad\square$

A statement of (107) in terms of the principal angles [**3**] between $R(A_1)$ and $R(A_2)$ is given in [**549**].

6. Some Multilinear Algebra

The setting of multilinear algebra is natural for the matrix volume, allowing simplification of statements and proofs.

Let $V = \mathbb{R}^n$, $U = \mathbb{R}^m$. We use the same letter to denote both a linear transformation in $\mathcal{L}(V, U)$ and its matrix representation with respect to fixed bases in V and U.

Let $\bigwedge^k V$ be the kth *exterior space* over V, spanned by *exterior products* $\mathbf{x}_1 \wedge \cdots \wedge \mathbf{x}_k$ of elements $\mathbf{x}_i \in V$, see, e.g., [**531**], [**532**], and [**585**]. For $A \in \mathbb{R}_r^{m \times n}$, $r > 0$ and $k = 1, \ldots, r$, the k *compound matrix* $C_k(A)$ is an element of $\mathcal{L}(\bigwedge^k V, \bigwedge^k U)$, given by

$$A\mathbf{x}_1 \wedge \cdots \wedge A\mathbf{x}_k = C_k(A)(\mathbf{x}_1 \wedge \cdots \wedge \mathbf{x}_k), \quad \forall \, \mathbf{x}_i \in V, \tag{108}$$

see, e.g., [**532**, §4.2, p. 94]. Then $C_k(A)$ is an $\binom{m}{k} \times \binom{n}{k}$ matrix of rank $\binom{r}{k}$, see Ex. 6.22.

To any r-dimensional subspace $W \subset V$ there corresponds a unique one-dimensional subspace $\bigwedge^r W \subset \bigwedge^r V$, spanned by the exterior product

$$\mathbf{w}^\wedge = \mathbf{w}_1 \wedge \cdots \wedge \mathbf{w}_r, \tag{109}$$

where $\{\mathbf{w}_1, \ldots, \mathbf{w}_r\}$ is any basis of W, e.g., [**585**]. The $\binom{n}{r}$ components of $\mathbf{w}^\wedge$ (determined up to a multiplicative constant) are the *Plücker coordinates* of W.

Results relating volumes, compound matrices, and full-rank factorizations are collected in the following lemma. The proofs are omitted.

Lemma 2 (Volume and Compounds). *Let* $r > 0$, $A \in \mathbb{R}_r^{m \times n}$, $C \in \mathbb{R}_r^{m \times r}$ *have columns* $\mathbf{c}^{(j)}$ *and let* $R \in \mathbb{R}_r^{r \times n}$ *have rows* $\mathbf{r}_{(i)}$. *Then,*

$$C_r(R)(\mathbf{r}_{(1)} \wedge \cdots \wedge \mathbf{r}_{(r)}) = \text{vol}^2 R, \tag{110a}$$

$$C_r(C^T)(\mathbf{c}^{(1)} \wedge \cdots \wedge \mathbf{c}^{(r)}) = \text{vol}^2 C. \tag{110b}$$

If $A = CR$ is a rank factorization of A, then

$$C_r(A) = (\mathbf{c}^{(1)} \wedge \cdots \wedge \mathbf{c}^{(r)})(\mathbf{r}_{(1)} \wedge \cdots \wedge \mathbf{r}_{(r)}) \tag{110c}$$

is a full-rank factorization[3] of $C_r(A)$. Moreover, the volume of A is given by the inner product,

$$\langle \mathbf{c}^{(1)} \wedge \cdots \wedge \mathbf{c}^{(r)}, A\mathbf{r}_{(1)} \wedge \cdots \wedge A\mathbf{r}_{(r)} \rangle = \mathrm{vol}^2 A, \tag{111}$$

and

$$\mathrm{vol}_r^2 A = \mathrm{vol}_1^2 C_r(A) = \mathrm{vol}_1^2 C_r(C) \ \mathrm{vol}_1^2 C_r(R). \tag{112}$$

Exercises

Ex. 68. Consider the 3×3 matrix of rank 2, with a full-rank factorization

$$A = \begin{bmatrix} 1 & 2 & 3 \\ 4 & 5 & 6 \\ 7 & 8 & 9 \end{bmatrix} = CR = \begin{bmatrix} 1 & 2 \\ 4 & 5 \\ 7 & 8 \end{bmatrix} \begin{bmatrix} 1 & 0 & -1 \\ 0 & 1 & 2 \end{bmatrix}. \tag{113}$$

Then the 2-compound matrix is

$$\begin{aligned} C_2(A) &= \begin{bmatrix} -3 & -6 & -3 \\ -6 & -12 & -6 \\ -3 & -6 & -3 \end{bmatrix} = C_2(C)\,C_2(R) \\ &= \begin{bmatrix} -3 \\ -6 \\ -3 \end{bmatrix} \begin{bmatrix} 1 & 2 & 1 \end{bmatrix}, \end{aligned}$$

a full-rank factorization. The volume of A is calculated by (112),

$$\mathrm{vol}_2^2 A = \mathrm{vol}_1^2(C) \ \mathrm{vol}_1^2(R) = (9 + 36 + 9)(1 + 4 + 1) = 324.$$

Ex. 69. For $k = 1, \ldots, r$, the *k-volume* of A is defined as the Frobenius norm of the kth compound matrix $C_k(A)$,

$$\mathrm{vol}_k A := \sqrt{\sum_{I \in Q_{k,m}, J \in Q_{k,n}} |\det A_{IJ}|^2} \tag{114a}$$

or, equivalently,

$$\mathrm{vol}_k A = \sqrt{\sum_{I \in Q_{k,r}} \left(\prod_{i \in I} \sigma_i^2(A) \right)}, \tag{114b}$$

the square root of the kth symmetric function of $\{\sigma_1^2(A), \cdots, \sigma_r^2(A)\}$. We use the convention

$$\mathrm{vol}_k A := 0, \quad \text{for } k = 0 \text{ or } k > \mathrm{rank}\, A. \tag{114c}$$

Ex. 70. Let $S \in \mathbb{R}^{m \times m}$, $A \in \mathbb{R}_m^{m \times n}$. Then

$$\mathrm{vol}_m(SA) = |\det S|\, \mathrm{vol}\, A. \tag{115}$$

---

[3]This is a restatement of (103).

PROOF. If S is singular, then both sides of (115) are zero. Let S be nonsingular. Then $\mathrm{rank}(SA) = m$, and

$$\mathrm{vol}_m (SA) = \sqrt{\sum_{J \in Q_{m,n}} \det^2 (SA)_{*J}}$$

$$= \sqrt{\sum_{J \in Q_{m,n}} \det^2 S \det^2 A_{*J}}$$

$$= |\det S| \, \mathrm{vol} \, A. \qquad \square$$

7. The Jordan Normal Form

Let the matrices $A \in \mathbb{C}^{n \times n}$, $X \in \mathbb{C}^{n \times k}_k$, and the scalar $\lambda \in \mathbb{C}$ satisfy

$$AX = XJ_k(\lambda), \quad \text{where} \quad J_k(\lambda) = \begin{bmatrix} \lambda & 1 & 0 & \cdots & 0 \\ 0 & \lambda & 1 & \ddots & \vdots \\ \vdots & \ddots & \ddots & \ddots & 0 \\ \vdots & & \ddots & \lambda & 1 \\ 0 & \cdots & \cdots & 0 & \lambda \end{bmatrix} \in \mathbb{C}^{k \times k}, \quad (116)$$

or, writing X by its columns, $X = [\mathbf{x}_1 \ \mathbf{x}_2 \ \cdots \ \mathbf{x}_k]$,

$$A\mathbf{x}_1 = \lambda \mathbf{x}_1, \tag{117}$$

$$A\mathbf{x}_j = \lambda x_j + \mathbf{x}_{j-1}, \ j = 2, \dots, k. \tag{118}$$

It follows, for $j \in \overline{1,k}$, that

$$(A - \lambda I)^j \mathbf{x}_j = \mathbf{0}, \quad (A - \lambda I)^{j-1} \mathbf{x}_j = \mathbf{x}_1 \neq \mathbf{0}, \tag{119}$$

where we interpret $(A - \lambda I)^0$ as I. The vector $\mathbf{x}_1$ is therefore an *eigenvector* of A corresponding to the *eigenvalue* λ. We call $\mathbf{x}_j$ a λ-*vector* of A of grade j or, following Wilkinson [**872**, p. 43], a *principal vector*[4] of A of *grade j* associated with the eigenvalue λ. Evidently, principal vectors are a generalization of eigenvectors. In fact, an eigenvector is a principal vector of grade 1.

The matrix $J_k(\lambda)$ in (116) is called a $(k \times k)$ *Jordan block* corresponding to the eigenvalue λ. The following theorem, stated without proof (that can be found in linear algebra texts, see, e.g., [**495**, Chapter 6]), is of central importance:

THEOREM 2 (The Jordan Normal Form). *Any matrix $A \in \mathbb{C}^{n \times n}$ is similar to a block diagonal matrix J with Jordan blocks on its diagonal, i.e., there exists a nonsingular matrix X such that*

$$X^{-1}AX = J = \begin{bmatrix} J_{k_1}(\lambda_1) & O & \cdots & O \\ O & J_{k_2}(\lambda_2) & \cdots & O \\ \vdots & \vdots & \ddots & \vdots \\ O & O & \cdots & J_{k_p}(\lambda_p) \end{bmatrix}, \tag{120}$$

and the matrix J is unique up to a rearrangement of its blocks. $\qquad \square$

---

[4]The vectors $\mathbf{x}_j$, $j \in \overline{2,k}$, are sometimes called *generalized eigenvectors* associated with λ, see, e.g., [**678**, p. 74].

The matrix J is the *Jordan normal form* of A. The scalars $\{\lambda_1, \ldots, \lambda_p\}$ in (120) are the eigenvalues of A. Each Jordan block $J_{k_j}(\lambda_j)$ corresponds to k_j principal vectors, of which one is an eigenvector.

Writing (120) as

$$A = XJX^{-1}, \tag{121}$$

we verify

$$A^s = XJ^s X^{-1}, \quad \text{for all integers } s \geq 0, \tag{122}$$

and, for any polynomial p,

$$p(A) = Xp(J)X^{-1}. \tag{123}$$

Using (120) and (127) below we verify

$$(A - \lambda_1 I)^{k_1} = X \begin{bmatrix} O & O & \cdots & O \\ O & ? & \cdots & O \\ \vdots & \vdots & \ddots & \vdots \\ O & O & \cdots & ? \end{bmatrix} X^{-1},$$

where exact knowledge of the ? blocks is not needed. Continuing in this fashion we prove

THEOREM 3 (The Cayley–Hamilton Theorem). *For A as above,*

$$(A - \lambda_1 I)^{k_1}(A - \lambda_2 I)^{k_2} \cdots (A - \lambda_p I)^{k_p} = O. \tag{124}$$

PROOF.

LHS(124) =

$$X \begin{bmatrix} O & O & \cdots & O \\ O & ? & \cdots & O \\ \vdots & \vdots & \ddots & \vdots \\ O & O & \cdots & ? \end{bmatrix} \begin{bmatrix} ? & O & \cdots & O \\ O & O & \cdots & O \\ \vdots & \vdots & \ddots & \vdots \\ O & O & \cdots & ? \end{bmatrix} \cdots \begin{bmatrix} ? & O & \cdots & O \\ O & ? & \cdots & O \\ \vdots & \vdots & \ddots & \vdots \\ O & O & \cdots & O \end{bmatrix} X^{-1}.$$

□

This result can be stated in terms of the polynomial

$$c(\lambda) = (\lambda - \lambda_1)^{k_1}(\lambda - \lambda_2)^{k_2} \cdots (\lambda - \lambda_p)^{k_p}, \tag{125}$$

called the *characteristic polynomial* of A, see Ex. 72. Indeed, LHS(124) is obtained by substituting A for the variable λ, and replacing λ_i by $\lambda_i I$. The n roots of the characteristic polynomial of A are the eigenvalues of A, counting multiplicities. The Cayley–Hamilton theorem states that a matrix A satisfies the polynomial equation $c(A) = O$, where $c(\lambda)$ is its characteristic polynomial.

If an eigenvalue λ_i is repeated in q Jordan blocks,

$$J_{j_1}(\lambda_i), J_{j_2}(\lambda_i), \ldots, J_{j_q}(\lambda_i), \quad \text{with} \quad j_1 \geq j_2 \geq \cdots \geq j_q,$$

then the characteristic polynomial is the product of factors

$$c(\lambda) = c_1(\lambda)\, c_2(\lambda) \cdots c_p(\lambda), \quad \text{with} \quad \begin{cases} c_i(\lambda) = (\lambda - \lambda_i)^{a_i}, \\ a_i = j_1 + j_2 + \cdots + j_q. \end{cases}$$

The exponent a_i is called the *algebraic multiplicity* of the eigenvalue λ_i. It is the sum of dimensions of the Jordan blocks corresponding to λ_i. The dimension of the largest block, j_1, is called the *index* of the eigenvalue λ_i, and denoted $\nu(\lambda_i)$. By (127) below it is the smallest integer k such that

$$(J_{j_1}(\lambda_i) - \lambda_i\, I_{j_1})^k = O, \ (J_{j_2}(\lambda_i) - \lambda_i\, I_{j_2})^k = O, \ \cdots , \ (J_{j_q}(\lambda_i) - \lambda_i\, I_{j_q})^k = O,$$

see also Ex. 82.

Let $m_i(\lambda) = (\lambda - \lambda_i)^{\nu(\lambda_i)}$. Then the polynomial

$$m(\lambda) = m_1(\lambda)\, m_2(\lambda) \cdots m_p(\lambda)$$

satisfies $m(A) = O$, and has the smallest degree among such polynomials. It is called the *minimal polynomial* of A.

Exercises

Ex. 71. Let $A \in \mathbb{C}^{n \times n}$ have the Jordan form (120). Then the following statements are equivalent:

    (a) A is nonsingular;

    (b) J is nonsingular; and

    (c) 0 is not an eigenvalue of A.

If these hold, then

$$A^{-1} = X J^{-1} X^{-1}, \tag{126}$$

and (122) holds for all integers s, if we interpret A^{-s} as $\left(A^{-1}\right)^s$.

Ex. 72. Let $A \in \mathbb{C}^{n \times n}$. Then the characteristic polynomial of A is $c(\lambda) = (-1)^n \det(A - \lambda I)$.

Ex. 73. Let $A \in \mathbb{C}^{n \times n}$. Then λ is an eigenvalue of A if and only if $\overline{\lambda}$ is an eigenvalue of A^*.

PROOF. $\det(A^* - \overline{\lambda}\, I) = \overline{\det(A - \lambda\, I)}$. $\square$

Ex. 74. Let A be given in Jordan form

$$A = X \begin{bmatrix} J_3(\lambda_1) & O & O & O & O & O \\ O & J_2(\lambda_1) & O & O & O & O \\ O & O & J_2(\lambda_1) & O & O & O \\ O & O & O & J_1(\lambda_1) & O & O \\ O & O & O & O & J_2(\lambda_2) & O \\ O & O & O & O & O & J_2(\lambda_2) \end{bmatrix} X^{-1}.$$

Then the characteristic polynomial of A is $c(\lambda) = (\lambda - \lambda_1)^8(\lambda - \lambda_2)^4$ and the minimal polynomial is $m(\lambda) = (\lambda - \lambda_1)^3(\lambda - \lambda_2)^2$. The algebraic multiplicity of λ_1 is 8, its geometric multiplicity is 4 (every Jordan block contributes an eigenvector), and its index is 3.

Ex. 75. A matrix N is *nilpotent* if $N^k = O$ for some integer $k \geq 0$. The smallest such k is called the *index of nilpotency* of N. Let $J_k(\lambda)$ be a Jordan block and let $j \in \overline{1, k}$. Then

$$(J_k(\lambda) - \lambda\, I_k)^j = J_k(0)^j =$$

$$= \begin{bmatrix} 0 & 1 & \cdots & \cdots & 0 \\ \vdots & \ddots & 1 & & \vdots \\ \vdots & & \ddots & \ddots & \vdots \\ \vdots & & & \ddots & 1 \\ 0 & \cdots & \cdots & & 0 \end{bmatrix}^j = \begin{bmatrix} 0 & \cdots & 1 & \cdots & 0 \\ \vdots & & & \ddots & \vdots \\ \vdots & & & & 1 \\ \vdots & & & & \vdots \\ 0 & \cdots & \cdots & \cdots & 0 \end{bmatrix}, \tag{127}$$

with ones in positions $\{(i, i+j) : i \in \overline{1, k-j}\}$, zeros elsewhere. In particular, $(J_k(\lambda) - \lambda I_k)^{k-1} \neq O = (J_k(\lambda) - \lambda I_k)^k$, showing that $(J_k(\lambda) - \lambda I_k)$ is nilpotent with index k.

Ex. 76. Let $J_k(\lambda)$ be a Jordan block and let m be a nonnegative integer. Show that the power $(J_k(\lambda))^m$ is

$$(J_k(\lambda))^m = \begin{bmatrix} \lambda^m & m\lambda^{m-1} & \binom{m}{2}\lambda^{m-2} & \cdots & \binom{m}{k-1}\lambda^{m-k+1} \\ 0 & \lambda^m & m\lambda^{m-1} & \ddots & \vdots \\ \vdots & & \ddots & \ddots & \binom{m}{2}\lambda^{m-2} \\ \vdots & & & \lambda^m & m\lambda^{m-1} \\ 0 & \cdots & & 0 & \lambda^m \end{bmatrix} \tag{128a}$$

$$= \sum_{j=0}^{k-1} \binom{m}{j} \lambda^{m-j} (J_k(\lambda) - \lambda I_k)^j \tag{128b}$$

$$= \sum_{j=0}^{k-1} \frac{p^{(j)}(\lambda)}{j!} (J_k(\lambda) - \lambda I_k)^j, \tag{128c}$$

where $p(\lambda) = \lambda^m$, $p^{(j)}$ is the jth derivative, and in (128), $\binom{m}{\ell}$ is interpreted as zero if $m < \ell$.

Ex. 77. Let $J_k(\lambda)$ be a Jordan block and let $p(\lambda)$ be a polynomial. Then $p(J_k(\lambda))$ is defined by using (128)

$$p(J_k(\lambda)) = \begin{bmatrix} p(\lambda) & \frac{1}{1!}p'(\lambda) & \frac{1}{2!}p''(\lambda) & \cdots & \frac{1}{(k-1)!}p^{(k-1)}(\lambda) \\ 0 & p(\lambda) & \frac{1}{1!}p'(\lambda) & \ddots & \vdots \\ \vdots & & \ddots & \ddots & \frac{1}{2!}p''(\lambda) \\ \vdots & & & p(\lambda) & \frac{1}{1!}p'(\lambda) \\ 0 & \cdots & & 0 & p(\lambda) \end{bmatrix} \tag{129}$$

$$= \sum_{j=0}^{k-1} \frac{p^{(j)}(\lambda)}{j!} (J_k(\lambda) - \lambda I_k)^j, \quad \text{as in (128c)}.$$

Ex. 78. ([**365**, p. 104]). Let P_n be the set of polynomials with real coefficients, of degree $\leq n$, and let T be the differentiation operator $Tf(x) = f'(x)$, see Ex. 15(b). The solution of $f'(x) = \lambda f(x)$ is $f(x) = e^{\lambda x}$, which is a polynomial only if $\lambda = 0$, the only eigenvalue of T. The geometric multiplicity of this eigenvalue is 1, its algebraic multiplicity is $n + 1$.

Ex. 79. Let the $n \times n$ matrix A have the characteristic polynomial

$$c(\lambda) = \lambda^n + c_{n-1}\lambda^{n-1} + \cdots + c_2\lambda^2 + c_1\lambda + c_0.$$

Then A is nonsingular if and only if $c_0 \neq 0$, in which case

$$A^{-1} = -\frac{1}{c_0}\left(A^{n-1} + c_{n-1}A^{n-2} + \cdots + c_2 A + c_1 I\right).$$

Ex. 80. Let A be a nonsingular matrix. Show that its minimal polynomial can be written in the form

$$m(\lambda) = c(1 - \lambda\, q(\lambda)) \tag{130}$$

where $c \neq 0$ and q is a polynomial, in which case

$$A^{-1} = q(A). \tag{131}$$

See also Ex. 6.87 below.

Ex. 81. Let the 2×2 matrix A have eigenvalues $\pm i$. Find A^{-1}.
Hint: Here $c(\lambda) = m(\lambda) = \lambda^2 + 1$. Use Ex. 79 or Ex. 80.

Ex. 82. For a given eigenvalue λ, the maximal grade of the λ-vectors of A is the index of λ.

8. The Smith Normal Form

Let $\mathbb{Z}$ denote the *ring of integers* $0, \pm 1, \pm 2, \ldots$ and let:
  $\mathbb{Z}^m$ be the m-dimensional vector space over Z;
  $\mathbb{Z}^{m \times n}$ be the $m \times n$ matrices over $\mathbb{Z}$; and
  $\mathbb{Z}_r^{m \times n}$ be the same with rank r.
Any vector in $\mathbb{Z}^m$ will be called an *integral vector*. Similarly, any element of $\mathbb{Z}^{m \times n}$ will be called an *integral matrix*.

A nonsingular matrix $A \in \mathbb{Z}^{n \times n}$ whose inverse A^{-1} is also in $\mathbb{Z}^{n \times n}$ is called a *unit matrix*; e.g., Marcus and Minc [**534**, p. 42].

Two matrices $A, S \in \mathbb{Z}^{m \times n}$ are said to be *equivalent over* $\mathbb{Z}$ if there exist two unit matrices $P \in \mathbb{Z}^{m \times m}$ and $Q \in \mathbb{Z}^{n \times n}$ such that

$$PAQ = S. \tag{132}$$

THEOREM 4. *Let* $A \in \mathbb{Z}_r^{m \times n}$. *Then* A *is equivalent over* $\mathbb{Z}$ *to a matrix* $S = [s_{ij}] \in \mathbb{Z}_r^{m \times n}$ *such that:*
  (a) $s_{ii} \neq 0, \quad i \in \overline{1, r}$;
  (b) $s_{ij} = 0$ *otherwise; and*
  (c) s_{ii} *divides* $s_{i+1,i+1}$ *for* $i \in \overline{1, r-1}$.

REMARK. S is called the *Smith normal form* of A, and its nonzero elements s_{ii} $(i \in \overline{1, r})$ are the *invariant factors* of A; see, e.g., Marcus and Minc [**534**, pp. 42–44].
PROOF. The proof given in Marcus and Minc [**534**, p. 44] is constructive and describes an algorithm to:
  (i) find the greatest common divisor of the elements of A;
  (ii) bring it to position $(1, 1)$; and
  (iii) make zeros of all other elements in the first row and column.
This is done, in an obvious way, by using a sequence of elementary row and column operations consisting of

$$\text{interchanging two rows [columns],} \tag{133}$$

$$\begin{array}{l} \text{subtracting an integer multiple of one row [column]} \\ \text{from another row [column].} \end{array} \tag{134}$$

The matrix $B = [b_{ij}]$ so obtained is equivalent over $\mathbb{Z}$ to A, and
  b_{11} divides b_{ij} $(i > 1, j > 1)$;
  $b_{i1} = b_{1j} = 0$ $(i > 1, j > 1)$.

Setting $s_{11} = b_{11}$, one repeats the algorithm for $(m-1) \times (n-1)$ matrix $[b_{ij}]$ $(i > 1, j > 1)$, etc.

The algorithm is repeated r times and stops when the bottom right $(m - r) \times (n - r)$ submatrix is zero, giving the Smith normal form.

The unit matrix $P[Q]$ in (132) is the product of all the elementary row [column] operators, in the right order. $\square$

Exercises

EX. 83. Two matrices $A, B \in \mathbb{Z}^{m \times n}$ are equivalent over $\mathbb{Z}$ if and only if B can be obtained from A by a sequence of elementary row and column operations (133)–(134).

EX. 84. Describe in detail the algorithm mentioned in the proof of Theorem 4.

9. Nonnegative Matrices

A matrix $A = [a_{ij}] \in \mathbb{R}^{n \times n}$ is:

(a) *nonnegative* if all $a_{ij} \geq 0$; and

(b) *reducible* if there is a permutation matrix Q such that

$$Q^T A Q = \begin{bmatrix} A_{11} & A_{12} \\ O & A_{22} \end{bmatrix}, \tag{135}$$

where the submatrices A_{11}, A_{22} are square, and is otherwise *irreducible*.

THEOREM 5 (The Perron–Frobenius Theorem). *If $A \in \mathbb{R}^{n \times n}$ is nonnegative and irreducible then:*

(a) *A has a positive eigenvalue, ρ, equal to the spectral radius of A.*

(b) *ρ has algebraic multiplicity 1.*

(c) *There is a positive eigenvector corresponding to ρ.*

Suggested Further Reading

SECTION 4.4. Bhimasankaram [97], Hartwig [380].

SECTION 5. Schur complements: Carlson [166], Cottle [207], Horn and Johnson [428], Ouellette [624].

SECTION 6. Finzel [273], Marcus [531],[532], Mostow and Sampson [584], Mostow, Sampson, and Meyer [585], Niu [613].

SECTION 9. Berman and Plemmons [91], Lancaster and Tismenetsky [495, Chapter 15].

CHAPTER 1

Existence and Construction of Generalized Inverses

1. The Penrose Equations

In 1955 Penrose [635] showed that, for every finite matrix A (square or rectangular) of real or complex elements, there is a unique matrix X satisfying the four equations (that we call the *Penrose equations*)

$$AXA = A, \tag{1}$$

$$XAX = X, \tag{2}$$

$$(AX)^* = AX, \tag{3}$$

$$(XA)^* = XA, \tag{4}$$

where A^* denotes the conjugate transpose of A. Because this unique generalized inverse had previously been studied (though defined in a different way) by E.H. Moore [575], [576], it is commonly known as the *Moore–Penrose inverse*, and is often denoted by $A^\dagger$.

If A is nonsingular, then $X = A^{-1}$ trivially satisfies the four equations. It follows that the Moore–Penrose inverse of a nonsingular matrix is the same as the ordinary inverse.

Throughout this book we shall be much concerned with generalized inverses that satisfy some, but not all, of the four Penrose equations. As we shall wish to deal with a number of different subsets of the set of four equations, we need a convenient notation for a generalized inverse satisfying certain specified equations (see also *Notes on Terminology* in p. 51.)

DEFINITION 1. For any $A \in \mathbb{C}^{m \times n}$, let $A\{i, j, \ldots, k\}$ denote the set of matrices $X \in \mathbb{C}^{n \times m}$ which satisfy equations $(i), (j), \ldots, (k)$ from among equations (1)–(4). A matrix $X \in A\{i, j, \ldots, k\}$ is called an $\{i, j, \ldots, k\}$-*inverse* of A, and also denoted by $A^{(i,j,\ldots,k)}$.

In Chapter 4 we shall extend the scope of this notation by enlarging the set of four matrix equations to include several further equations, applicable only to square matrices, that will play an essential role in the study of generalized inverses having spectral properties.

Exercises

Ex. 1. If $A\{1, 2, 3, 4\}$ is nonempty, then it consists of a single element (Penrose [635]).

PROOF. Let $X, Y \in A\{1, 2, 3, 4\}$. Then

$$X = X(AX)^* = XX^*A^* = X(AX)^*(AY)^*$$
$$= XAY = (XA)^*(YA)^*Y = A^*Y^*Y$$
$$= (YA)^*Y = Y.$$

□

EX. 2. By means of a (trivial) example, show that $A\{2, 3, 4\}$ is nonempty.

2. Existence and Construction of {1}-Inverses

It is easy to construct a {1}-inverse of the matrix $R \in \mathbb{C}_r^{m \times n}$ given by

$$R = \begin{bmatrix} I_r & K \\ O & O \end{bmatrix}. \tag{0.71}$$

For any $L \in \mathbb{C}^{(n-r) \times (m-r)}$, the $n \times m$ matrix

$$S = \begin{bmatrix} I_r & O \\ O & L \end{bmatrix}$$

is a {1}-inverse of (0.71). If R is of full column [row] rank, the two lower [right-hand] submatrices are interpreted as absent.

The construction of {1}-inverses for an arbitrary $A \in \mathbb{C}^{m \times n}$ is simplified by transforming A into a Hermite normal form, as shown in the following theorem:

THEOREM 1. *Let $A \in \mathbb{C}_r^{m \times n}$ and let $E \in \mathbb{C}_m^{m \times m}$ and $P \in \mathbb{C}_n^{n \times n}$ be such that*

$$EAP = \begin{bmatrix} I_r & K \\ O & O \end{bmatrix}. \tag{0.72}$$

Then, for any $L \in \mathbb{C}^{(n-r) \times (m-r)}$, the $n \times m$ matrix

$$X = P \begin{bmatrix} I_r & O \\ O & L \end{bmatrix} E \tag{5}$$

is a {1}-inverse of A. The partitioned matrices in (0.72) and (5) must be suitably interpreted in case $r = m$ or $r = n$.

PROOF. Rewriting (0.72) as

$$A = E^{-1} \begin{bmatrix} I_r & K \\ O & O \end{bmatrix} P^{-1}, \tag{0.74}$$

it is easily verified that any X given by (5) satisfies $AXA = A$.   □

In the trivial case of $r = 0$, when A is therefore the $m \times n$ null matrix, *any* $n \times m$ matrix is a {1}-inverse.

We note that since P and E are both nonsingular, the rank of X is the rank of the partitioned matrix in RHS(5). In view of the form of the latter matrix,

$$\text{rank } X = r + \text{rank } L. \tag{6}$$

Since L is arbitrary, it follows that a {1}-inverse of A exists having any rank between r and $\min\{m, n\}$, inclusive (see also Fisher [274]).

Theorem 1 shows that every finite matrix with elements in the complex field has a {1}-inverse, and suggests how such an inverse can be constructed.

Exercises

Ex. 3. What is the Hermite normal form of a nonsingular matrix A? In this case, what is the matrix E and what is its relationship to A? What is the permutation matrix P? What is the matrix X given by (5)?

Ex. 4. An $m \times n$ matrix A has all its elements equal to 0 except for the $(i,j)^{\text{th}}$ element, which is 1. What is the Hermite normal form? Show that E can be taken as a permutation matrix. What are the simplest choices of E and P? (By "simplest" we mean having the smallest number of elements different from the corresponding elements of the unit matrix of the same order.) Using these choices of E and P, but regarding L as entirely arbitrary, what is the form of the resulting matrix X given by (5)? Is this X the most general $\{1\}$-inverse of A? (See Exercise 6, Introduction, and Exercise 11 below.)

Ex. 5. Show that every square matrix has a nonsingular $\{1\}$-inverse.

Ex. 6. *Computing a $\{1\}$-inverse.* This is demonstrated for the matrix A of (0.80), using (5) with E as computed in (0.82), and an arbitrary $L \in \mathbb{C}^{(n-r) \times (m-r)}$. Using the permutation matrix P selected in (0.83), and the corresponding submatrix K, we write

$$
EAP = \begin{bmatrix} 1 & 0 & \vdots & 0 & \frac{1}{2} & 1-2i & -\frac{1}{2}i \\ 0 & 1 & \vdots & 0 & 0 & 2 & 1+i \\ \cdots & \cdots & \vdots & \cdots & \cdots & \cdots & \cdots \\ 0 & 0 & \vdots & 0 & 0 & 0 & 0 \end{bmatrix}, \quad \text{and take} \quad L = \begin{bmatrix} \alpha \\ \beta \\ \gamma \\ \delta \end{bmatrix} \in \mathbb{C}^{4 \times 1},
$$

since $m = 3$, $n = 6$, $r = 2$. A $\{1\}$-inverse of A is, by (5),

$$
X = P \begin{bmatrix} I_r & O \\ O & L \end{bmatrix} E
$$

$$
= \begin{bmatrix} 0 & 0 & \vdots & 1 & 0 & 0 & 0 \\ 1 & 0 & \vdots & 0 & 0 & 0 & 0 \\ 0 & 0 & \vdots & 0 & 1 & 0 & 0 \\ 0 & 1 & \vdots & 0 & 0 & 0 & 0 \\ 0 & 0 & \vdots & 0 & 0 & 1 & 0 \\ 0 & 0 & \vdots & 0 & 0 & 0 & 1 \end{bmatrix} \begin{bmatrix} 1 & 0 & \vdots & 0 \\ 0 & 1 & \vdots & 0 \\ 0 & 0 & \vdots & \alpha \\ 0 & 0 & \vdots & \beta \\ 0 & 0 & \vdots & \gamma \\ 0 & 0 & \vdots & \delta \end{bmatrix} \begin{bmatrix} -\frac{1}{2}i & 0 & 0 \\ 0 & -\frac{1}{3} & 0 \\ i & \frac{1}{3} & 1 \end{bmatrix}
$$

$$
= \begin{bmatrix} i\alpha & \frac{1}{3}\alpha & \alpha \\ -\frac{1}{2}i & 0 & 0 \\ i\beta & \frac{1}{3}\beta & \beta \\ 0 & -\frac{1}{3} & 0 \\ i\gamma & \frac{1}{3}\gamma & \gamma \\ i\delta & \frac{1}{3}\delta & \delta \end{bmatrix}. \tag{7}
$$

Note that, in general, the scalars $i\alpha, i\beta, i\gamma, i\delta$ are not pure imaginaries since $\alpha, \beta, \gamma, \delta$ are complex.

3. Properties of $\{1\}$-Inverses

Certain properties of $\{1\}$-inverses are given in Lemma 1. For a given matrix A, we denote any $\{1\}$-inverse by $A^{(1)}$. Note that, in general, $A^{(1)}$ is not a uniquely defined matrix (see Ex. 8 below). For any scalar λ we define $\lambda^\dagger$ by

$$\lambda^\dagger = \begin{cases} \lambda^{-1}, & \text{if } \lambda \neq 0, \\ 0, & \text{if } \lambda = 0. \end{cases} \tag{8}$$

It will be recalled that a square matrix E is called *idempotent* if $E^2 = E$. Idempotent matrices are intimately related to generalized inverses, and their properties are considered in some detail in Chapter 2.

LEMMA 1. *Let* $A \in \mathbb{C}_r^{m \times n}$, $\lambda \in \mathbb{C}$. *Then:*

(a) $(A^{(1)})^* \in A^*\{1\}$.
(b) *If* A *is nonsingular,* $A^{(1)} = A^{-1}$ *uniquely (see also Ex. 7 below).*
(c) $\lambda^\dagger A^{(1)} \in (\lambda A)\{1\}$.
(d) $\operatorname{rank} A^{(1)} \geq \operatorname{rank} A$.
(e) *If* S *and* T *are nonsingular,* $T^{-1} A^{(1)} S^{-1} \in SAT\{1\}$.
(f) $AA^{(1)}$ *and* $A^{(1)}A$ *are idempotent and have the same rank as* A.

PROOF. These are immediate consequences of the defining relation (1); (d) and the latter part of (f) depend on the fact that the rank of a product of matrices does not exceed the rank of any factor. $\square$

If an $m \times n$ matrix A is of full-column rank, its {1}-inverses are its left inverses. If it is of full row rank, its {1}-inverses are its right inverses.

LEMMA 2. *Let* $A \in \mathbb{C}_r^{m \times n}$. *Then:*

(a) $A^{(1)}A = I_n$ *if and only if* $r = n$.
(b) $AA^{(1)} = I_m$ *if and only if* $r = m$.

PROOF. (a) *If:* Let $A \in \mathbb{C}_r^{m \times n}$. Then the $n \times n$ matrix $A^{(1)}A$ is, by Lemma 1(f), idempotent and nonsingular. Multiplying $(A^{(1)}A)^2 = A^{(1)}A$ by $(A^{(1)}A)^{-1}$ gives $A^{(1)}A = I_n$.
Only if: $A^{(1)}A = I_n \implies \operatorname{rank} A^{(1)}A = n \implies \operatorname{rank} A = n$, by Lemma 1(f).
(b) Similarly proved. $\square$

Exercises

EX. 7. Let $A = FHG$ where F is of full-column rank and G is of full-row rank. Then $\operatorname{rank} A = \operatorname{rank} H$. (*Hint:* Use Lemma 2.)

EX. 8. Show that A is nonsingular if and only if it has a unique {1}-inverse, which then coincides with A^{-1}.
PROOF. For any $\mathbf{x} \in N(A)$ $[\mathbf{y} \in N(A^*)]$, adding $\mathbf{x}$ $[\mathbf{y}^*]$ to any column [row] of an $X \in A\{1\}$ gives another {1}-inverse of A. The uniqueness of the {1}-inverse is therefore equivalent to

$$N(A) = \{\mathbf{0}\}, \quad N(A^*) = \{\mathbf{0}\},$$

i.e., to the nonsingularity of A. $\square$

EX. 9. Show that if $A^{(1)} \in A\{1\}$, then $R(AA^{(1)}) = R(A)$, $N(AA^{(1)}) = N(A)$, and $R((A^{(1)}A)^*) = R(A^*)$.
PROOF. We have

$$R(A) \supset R(AA^{(1)}) \supset R(AA^{(1)}A) = R(A),$$

from which the first result follows.
    Similarly,

$$N(A) \subset N(A^{(1)}A) \subset N(AA^{(1)}A) = N(A)$$

yields the second equation.

Finally, by Lemma 1(a),

$$R(A^*) \supset R(A^*(A^{(1)})^*) = R((A^{(1)}A)^*) \supset R(A^*(A^{(1)})^*A^*) = R(A^*). \qquad \Box$$

Ex. 10. More generally, show that $R(AB) = R(A)$ if and only if rank $AB =$ rank A and $N(AB) = N(B)$ if and only if rank $AB =$ rank B.

PROOF. Evidently, $R(A) \supset R(AB)$, and these two subspaces are identical if and only if they have the same dimension. But, the rank of any matrix is the dimension of its range.

Similarly, $N(B) \subset N(AB)$. Now, the nullity of any matrix is the dimension of its null space, and also the number of columns minus the rank. Thus, $N(B) = N(AB)$ if and only if B and AB have the same nullity, which is equivalent, in this case, to having the same rank, since the two matrices have the same number of columns. $\qquad \Box$

Ex. 11. The answer to the last question in Ex. 4 indicates that, for particular choices of E and P, one does not get all the $\{1\}$-inverses of A merely by varying L in (5). Note, however, that Theorem 1 does not require P to be a permutation matrix. Could one get all the $\{1\}$-inverses by considering all nonsingular P and Q such that

$$QAP = \begin{bmatrix} I_r & O \\ O & O \end{bmatrix} ? \tag{9}$$

Given $A \in \mathbb{C}_r^{m \times n}$, show that $X \in A\{1\}$ if and only if

$$X = P \begin{bmatrix} I_r & O \\ O & L \end{bmatrix} Q \tag{10}$$

for some L and for some nonsingular P and Q satisfying (9).

SOLUTION. If (9) and (10) hold, X is a $\{1\}$-inverse of A by Theorem 1.

On the other hand, let $AXA = A$. Then, both AX and XA are idempotent and of rank r, by Lemma 1(f). Since any idempotent matrix E satisfies $E(E-I) = O$, its only eigenvalues are 0 and 1. Thus, the Jordan canonical forms of both AX and XA are of the form

$$\begin{bmatrix} I_r & O \\ O & O \end{bmatrix},$$

being of orders m and n, respectively. Therefore, there exist nonsingular P and R such that

$$R^{-1}AXR = \begin{bmatrix} I_r & O \\ O & O \end{bmatrix}, \quad P^{-1}XAP = \begin{bmatrix} I_r & O \\ O & O \end{bmatrix}.$$

Thus,

$$R^{-1}AP = R^{-1}AXAXAP = (R^{-1}AXR)R^{-1}AP(P^{-1}XAP)$$

$$= \begin{bmatrix} I_r & O \\ O & O \end{bmatrix} R^{-1}AP \begin{bmatrix} I_r & O \\ O & O \end{bmatrix}.$$

It follows that $R^{-1}AP$ is of the form

$$R^{-1}AP = \begin{bmatrix} H & O \\ O & O \end{bmatrix},$$

where $H \in \mathbb{C}_r^{r \times r}$, i.e., nonsingular. Let

$$Q = \begin{bmatrix} H^{-1} & O \\ O & I_{m-r} \end{bmatrix} R^{-1}.$$

Then (9) is satisfied. Consider the matrix $P^{-1}XQ^{-1}$. We have

$$\begin{bmatrix} I_r & O \\ O & O \end{bmatrix} (P^{-1}XQ^{-1}) = (QAP)(P^{-1}XQ^{-1}) = QAXQ^{-1}$$

$$= \begin{bmatrix} H^{-1} & O \\ O & I_{m-r} \end{bmatrix} \begin{bmatrix} I_r & O \\ O & O \end{bmatrix} \begin{bmatrix} H & O \\ O & I_{m-r} \end{bmatrix} = \begin{bmatrix} I_r & O \\ O & O \end{bmatrix}$$

and

$$(P^{-1}XQ^{-1}) \begin{bmatrix} I_r & O \\ O & O \end{bmatrix} = (P^{-1}XQ^{-1})(QAP) = P^{-1}XAP$$

$$= \begin{bmatrix} I_r & O \\ O & O \end{bmatrix}.$$

From the latter two equations it follows that

$$P^{-1}XQ^{-1} = \begin{bmatrix} I_r & O \\ O & L \end{bmatrix}$$

for some L. But this is equivalent to (10).   □

4. Existence and Construction of {1,2}-Inverses

It was first noted by Bjerhammar [**103**] that the existence of a {1}-inverse of a matrix A implies the existence of a {1,2}-inverse. This easily verified observation is stated as a lemma for convenience of reference.

LEMMA 3. *Let* $Y, Z \in A\{1\}$, *and let*

$$X = YAZ.$$

Then $X \in A\{1,2\}$.

Since the matrices A and X occur symmetrically in (1) and (2), $X \in A\{1,2\}$ and $A \in X\{1,2\}$ are equivalent statements, and in either case we can say that A and X are {1,2}-inverses *of each other.*

From (1) and (2) and the fact that the rank of a product of matrices does not exceed the rank of any factor, it follows at once that if A and X are {1,2}-inverses of each other, they have the same rank. Less obvious is the fact, first noted by Bjerhammar [**103**], that if X is a {1}-inverse of A and of the same rank as A, it is a {1,2}-inverse of A.

THEOREM 2 (Bjerhammar). *Given* A *and* $X \in A\{1\}$, $X \in A\{1,2\}$ *if and only if* rank X = rank A.

PROOF. *If:* Clearly $R(XA) \subset R(X)$. But rank XA = rank A by Lemma 1(f) and so, if rank X = rank A, $R(XA) = R(X)$ by Ex. 10. Thus,

$$XAY = X$$

for some Y. Premultiplication by A gives

$$AX = AXAY = AY,$$

and therefore

$$XAX = X.$$

Only if: This follows at once from (1) and (2).   □

An equivalent statement is the following:

COROLLARY 1. *Any two of the following three statements imply the third:*

$$X \in A\{1\},$$
$$X \in A\{2\},$$
$$\operatorname{rank} X = \operatorname{rank} A. \qquad \square$$

In view of Theorem 2, (6) shows that the $\{1\}$-inverse obtained from the Hermite normal form is a $\{1,2\}$-inverse if we take $L = O$. In other words,

$$X = P \begin{bmatrix} I_r & O \\ O & O \end{bmatrix} E \qquad (11)$$

is a $\{1,2\}$-inverse of A where P and E are nonsingular and satisfy (0.72).

Exercises

EX. 12. Show that (5) gives a $\{1,2\}$-inverse of A if and only if $L = O$.

EX. 13. Let $A = [a_{ij}] \in \mathbb{C}^{m \times n}$ be nonzero and upper triangular, i.e., $a_{ij} = 0$ if $i > j$. Find a $\{1,2\}$-inverse of A.
SOLUTION. Let P, Q be permutation matrices such that

$$QAP = \begin{bmatrix} T & K \\ O & O \end{bmatrix}$$

where T is upper triangular and nonsingular (the K block, or the zero blocks, are absent if A is of full-rank.) Then

$$X = P \begin{bmatrix} T^{-1} & O \\ O & O \end{bmatrix} Q$$

is a $\{1,2\}$-inverse of A (again, some zero blocks are absent if A is of full-rank.) Note that the inverse T^{-1} is obtained from T by back substitution. $\qquad \square$

5. Existence and Construction of $\{1,2,3\}$-, $\{1,2,4\}$-, and $\{1,2,3,4\}$-Inverses

Just as Bjerhammar [103] showed that the existence of a $\{1\}$-inverse implies the existence of a $\{1,2\}$-inverse, Urquhart [824] has shown that the existence of a $\{1\}$-inverse of every finite matrix with elements in $\mathbb{C}$ implies the existence of a $\{1,2,3\}$-inverse and a $\{1,2,4\}$-inverse of every such matrix. However, in order to show the nonemptiness of $A\{1,2,3\}$ and $A\{1,2,4\}$ for any given A, we shall utilize the $\{1\}$-inverse not of A itself but of a related matrix. For that purpose we shall need the following lemma:

LEMMA 4. *For any finite matrix A,*

$$\operatorname{rank} AA^* = \operatorname{rank} A = \operatorname{rank} A^*A.$$

PROOF. If $A \in \mathbb{C}^{m \times n}$, both A and AA^* have m rows. Now, the rank of any m-rowed matrix is equal to m minus the number of independent linear relations among its rows. To show that $\operatorname{rank} AA^* = \operatorname{rank} A$, it is sufficient, therefore, to show that every linear relation among the rows of A holds for the corresponding rows of AA^*, and vice versa. Any nontrivial linear relation among the rows of a matrix H is equivalent to the existence of a nonzero row vector $\mathbf{x}^*$ such that $\mathbf{x}^* H = \mathbf{0}$. Now, evidently,

$$\mathbf{x}^* A = \mathbf{0} \implies \mathbf{x}^* AA^* = \mathbf{0},$$

and, conversely,

$$x^* A A^* = 0 \quad \Longrightarrow \quad 0 = x^* A A^* x = (A^* x)^* A^* x$$
$$\Longrightarrow \quad A^* x = 0 \quad \Longrightarrow \quad x^* A = 0.$$

Here we have used the fact that, for any column vector y of complex elements $y^* y$ is the sum of squares of the absolute values of the elements, and this sum vanishes only if every element is zero.

Finally, applying this result to the matrix A^* gives rank $A^* A =$ rank A^* and, of course, rank $A^* =$ rank A. $\qquad \square$

COROLLARY 2. *For any finite matrix A,*

$$R(AA^*) = R(A) \ and \ N(AA^*) = N(A).$$

PROOF. This follows from Lemma 4 and Ex. 10. $\qquad \square$

Using the preceding lemma, we can now prove the following theorem:

THEOREM 3 (Urquhart [824]). *For every finite matrix A with complex elements,*

$$Y = (A^* A)^{(1)} A^* \in A\{1,2,3\} \tag{12a}$$

and

$$Z = A^* (A A^*)^{(1)} \in A\{1,2,4\}. \tag{12b}$$

PROOF. Applying Corollary 2 to A^* gives

$$R(A^* A) = R(A^*),$$

and so,

$$A^* = A^* A U \tag{13}$$

for some U. Taking conjugate transpose gives

$$A = U^* A^* A. \tag{14}$$

Consequently,

$$AYA = U^* A^* A (A^* A)^{(1)} A^* A = U^* A^* A = A.$$

Thus, $Y \in A\{1\}$. But rank $Y \geq$ rank A by Lemma 1(d) and rank $Y \leq$ rank $A^* =$ rank A by the definition of Y. Therefore

$$\text{rank } Y = \text{rank } A,$$

and, by Theorem 2, $Y \in A\{1,2\}$. Finally, (13) and (14) give

$$AY = U^* A^* A (A^* A)^{(1)} A^* A U = U^* A^* A U,$$

which is clearly Hermitian. Thus, (12a) is established.

Relation (12b) is similarly proved. $\qquad \square$

A $\{1,2\}$-inverse of a matrix A is, of course, a $\{2\}$-inverse and, similarly, a $\{1,2,3\}$-inverse is also a $\{1,3\}$-inverse and a $\{2,3\}$-inverse. Thus, if we can establish the existence of a $\{1,2,3,4\}$-inverse, we will have demonstrated the existence of an $\{i,j,\ldots,k\}$-inverse for all possible choices of one, two, or three integers $i,j,\ldots,k$ from the set $\{1,2,3,4\}$. It was shown in Ex. 1 that if a $\{1,2,3,4\}$-inverse exists, it is unique. We know, as a matter of fact, that it does exist, because it is the well-known Moore–Penrose inverse, $A^\dagger$. However, we have not yet proved this. This is done in the next theorem.

THEOREM 4 (Urquhart [**824**]). *For any finite matrix A of complex elements,*

$$A^{(1,4)}AA^{(1,3)} = A^{\dagger}. \qquad (15)$$

PROOF. Let X denote LHS(15). It follows at once from Lemma 3 that $X \in A\{1,2\}$. Moreover, (15) gives

$$AX = AA^{(1,3)}, \quad XA = A^{(1,4)}A.$$

But, both $AA^{(1,3)}$ and $A^{(1,4)}A$ are Hermitian, by the definition of $A^{(1,3)}$ and $A^{(1,4)}$. Thus

$$X \in A\{1,2,3,4\}.$$

However, by Ex. 1, $A\{1,2,3,4\}$ contains at most a single element. Therefore, it contains exactly one element, namely $A^{\dagger}$ and $X = A^{\dagger}$. $\qquad \square$

6. Explicit Formula for $A^{\dagger}$

C.C. MacDuffee apparently was the first to point out, in private communications about 1959, that a full-rank factorization of a matrix A leads to an explicit formula for its Moore–Penrose inverse, $A^{\dagger}$.

THEOREM 5 (MacDuffee). *If $A \in \mathbb{C}_r^{m \times n}$, $r > 0$, has a full-rank factorization*

$$A = FG, \qquad (16)$$

then

$$A^{\dagger} = G^*(F^*AG^*)^{-1}F^*. \qquad (17)$$

PROOF. First, we must show that F^*AG^* is nonsingular. By (16),

$$F^*AG^* = (F^*F)(GG^*), \qquad (18)$$

and both factors of the right member are $r \times r$ matrices. Also, by Lemma 4, both are of rank r. Thus, F^*AG^* is the product of two nonsingular matrices and, therefore, nonsingular. Moreover, (18) gives

$$(F^*AG^*)^{-1} = (G^*G)^{-1}(F^*F)^{-1}.$$

Denoting by X the right member of (17), we now have

$$X = G^*(GG^*)^{-1}(F^*F)^{-1}F^*, \qquad (19)$$

and it is easily verified that this expression for X satisfies the Penrose equations (1)–(4). As $A^{\dagger}$ is the sole element of $A\{1,2,3,4\}$, (17) is therefore established. $\qquad \square$

Exercises

Ex. 14. Theorem 5 gives an alternative proof of the existence of the $\{1,2,3,4\}$-inverse (previously established by Theorem 4). However, Theorem 5 excludes the case $r = 0$. Complete the alternative existence proof by showing that if $r = 0$, (2) has a unique solution for X, and this X satisfies (1), (3), and (4).

Ex. 15. Compute $A^{\dagger}$ for the matrix A of (0.80).

Ex. 16. What is the most general $\{1,2\}$-inverse of the special matrix A of Ex. 4? What is its Moore–Penrose inverse?

Ex. 17. Show that if $A = FG$ is a rank factorization, then

$$A^{\dagger} = G^{\dagger}F^{\dagger}. \qquad (20)$$

Ex. 18. Show that for every matrix A:

(a) $(A^\dagger)^\dagger = A$;

(b) $(A^*)^\dagger = (A^\dagger)^*$;

(c) $(A^T)^\dagger = (A^\dagger)^T$;

(d) $A^\dagger = (A^*A)^\dagger A^* = A^*(AA^*)^\dagger$.

Ex. 19. If $\mathbf{a}$ and $\mathbf{b}$ are column vectors, then

(a) $\mathbf{a}^\dagger = (\mathbf{a}^*\mathbf{a})^\dagger \mathbf{a}^*$;

(b) $(\mathbf{ab}^*)^\dagger = (\mathbf{a}^*\mathbf{a})^\dagger (\mathbf{b}^*\mathbf{b})^\dagger \mathbf{ba}^*$.

Ex. 20. Show that if H is Hermitian and idempotent, $H^\dagger = H$.

Ex. 21. Show that $H^\dagger = H$ if and only if H^2 is Hermitian and idempotent and rank H^2 = rank H.

Ex. 22. If $D = \operatorname{diag}(d_1, d_2, \dots, d_n)$, show that $D^\dagger = \operatorname{diag}(d_1^\dagger, d_2^\dagger, \dots, d_n^\dagger)$.

Ex. 23. Let $J_k(0)$ be a Jordan block corresponding to the eigenvalue zero. Then $(J_k(0))^\dagger = (J_k(0))^T$, showing that, for a square matrix A, $A^\dagger$ is in general not a polynomial in A (if it were, then A and $A^\dagger$ would commute).

Ex. 24. Let $A, B \in \mathbb{C}^{n \times n}$ be similar, i.e., $B = S^{-1}AS$ for some nonsingular S. Then, in general, $B^\dagger \neq S^{-1}A^\dagger S$.

Ex. 25. If U and V are unitary matrices, show that

$$(UAV)^\dagger = V^*A^\dagger U^*$$

for any matrix A for which the product UAV is defined.

In particular, if $A, B \in \mathbb{C}^{n \times n}$ are unitarily similar, i.e., $B = U^{-1}AU$ for some unitary matrix U, then $B^\dagger = U^{-1}A^\dagger U$.

7. Construction of {2}-Inverses of Prescribed Rank

Following the proof of Theorem 1, we described A.G. Fisher's construction of a {1}-inverse of a given $A \in \mathbb{C}_r^{m \times n}$ having any prescribed rank between r and $\min(m, n)$, inclusive. From (2) it is easily deduced that

$$\operatorname{rank} A^{(2)} \leq r.$$

We note also that the $n \times m$ null matrix is a {2}-inverse of rank 0, and any $A^{(1,2)}$ is a {2}-inverse of rank r, by Theorem 2. For $r > 1$, is there a construction analogous to Fisher's for a {2}-inverse of rank s for arbitrary s between 0 and r? Using full-rank factorization, we can readily answer the question in the affirmative.

Let $X_0 \in A\{1, 2\}$ have a rank factorization

$$X_0 = YZ.$$

Then, $Y \in \mathbb{C}_r^{m \times r}$ and $Z \in \mathbb{C}_r^{r \times n}$, and (2) becomes

$$YZAYZ = YZ.$$

In view of Lemma 2, multiplication on the left by $Y^{(1)}$ and on the right by $Z^{(1)}$ gives (see Stewart [**780**])

$$ZAY = I_r. \tag{21}$$

Let Y_s denote the submatrix of Y consisting of the first s columns and let Z_s denote the submatrix of Z consisting of the first s rows. Then, both Y_s and Z_s are of full rank s, and it follows from (21) that

$$Z_s A Y_s = I_s. \tag{22}$$

Now, let

$$X_s = Y_s Z_s.$$

Then, rank $X_s = s$, by Ex. 7 and (22) gives

$$X_s A X_s = X_s.$$

Exercises

Ex. 26. For

$$A = \begin{bmatrix} 1 & 0 & 0 & 1 \\ 1 & 1 & 0 & 0 \\ 0 & 1 & 1 & 0 \\ 0 & 0 & 1 & 1 \end{bmatrix}$$

find elements of $A\{2\}$ of ranks 1, 2, and 3, respectively.

Ex. 27. With A as in Ex. 26, find a $\{2\}$-inverse of rank 2 having zero elements in the last two rows and the last two columns.

Ex. 28. Show that there is at most one matrix X satisfying the three equations $AX = B$, $XA = D$, $XAX = X$ (Cline; see Cline and Greville [**202**]).

Ex. 29. Let $A = FG$ be a rank factorization of $A \in \mathbb{C}_r^{m \times n}$, i.e., $F \in \mathbb{C}_r^{m \times r}$, $G \in \mathbb{C}_r^{r \times n}$. Then:

(a) $G^{(i)} F^{(1)} \in A\{i\}$ $(i = 1, 2, 4)$;   (b) $G^{(1)} F^{(j)} \in A\{j\}$ $(j = 1, 2, 3)$.

PROOF.

(a), $i = 1$:

$$FGG^{(1)} F^{(1)} FG = FG,$$

since

$$F^{(1)} F = GG^{(1)} = I_r, \quad \text{by Lemma 2.}$$

(a), $i = 2$:

$$G^{(2)} F^{(1)} FGG^{(2)} F^{(1)} = G^{(2)} F^{(1)},$$

since

$$F^{(1)} F = I_r, \quad G^{(2)} GG^{(2)} = G^{(2)}.$$

(a), $i = 4$:

$$G^{(4)} F^{(1)} FG = G^{(4)} G = (G^{(4)} G)^*.$$

(b)   Similarly proved, with the roles of F and G interchanged.   □

Ex. 30. Let A, F, G be as in Ex. 29. Then

$$A^\dagger = G^\dagger F^{(1,3)} = G^{(1,4)} F^\dagger.$$

Notes on Terminology

Some writers have adopted descriptive names to designate various classes of generalized inverses. However there is a notable lack of uniformity and consistency in the use of these terms by different writers. Thus, $X \in A\{1\}$ is called a *generalized inverse* (Rao [**671**]), or *pseudoinverse* (Sheffield [**749**]), or *inverse* (Bjerhammar [**103**]). $X \in A\{1, 2\}$ is called a *semi-inverse* (Frame [**285**]), or *reciprocal inverse* (Bjerhammar), or *reflexive generalized inverse* (Rohde [**705**]). $X \in A\{1, 2, 3\}$ is called a *weak generalized inverse* (Goldman and Zelen [**301**]). $X \in A\{1, 2, 3, 4\}$ is called the *general reciprocal* (Moore [**575**], [**576**]), or *generalized inverse* (Penrose [**635**]), or *pseudoinverse* (Greville [**324**]), or *natural inverse* (Lanczos [**497**, p. 124]), or *Moore–Penrose inverse* (Ben-Israel and Charnes [**77**]). In view of this diversity of terminology, the unambiguous notation adopted here is considered preferable. This notation also emphasizes the lack of uniqueness of many of the generalized inverses considered.

Suggested Further Reading

SECTION 1. Urquhart [**825**].

SECTION 2. Rao [**671**], Sheffield [**749**].

SECTION 3. Rao [**670**], [**673**].

SECTION 4. Deutsch [**228**], Frame [**285**], Greville [**330**], Hartwig [**378**], Przeworska–Rolewicz and Rolewicz [**653**].

SECTION 5. Hearon and Evans [**410**], Rao [**673**], Sibuya [**757**].

SECTION 6. Sakallioğlu and Akdeniz [**721**].

CHAPTER 2

Linear Systems and Characterization of Generalized Inverses

1. Solutions of Linear Systems

As already indicated in Section 3 of the Introduction, the principal application of $\{1\}$-inverses is to the solution of linear systems, where they are used in much the same way as ordinary inverses in the nonsingular case. The main result of this section is the following theorem of Penrose [635], to whom the proof is also due.

THEOREM 1. *Let $A \in \mathbb{C}^{m \times n}$, $B \in \mathbb{C}^{p \times q}$, $D \in \mathbb{C}^{m \times q}$. Then the matrix equation*

$$AXB = D \tag{1}$$

is consistent if and only if, for some $A^{(1)}, B^{(1)}$,

$$AA^{(1)}DB^{(1)}B = D, \tag{2}$$

in which case the general solution is

$$X = A^{(1)}DB^{(1)} + Y - A^{(1)}AYBB^{(1)} \tag{3}$$

for arbitrary $Y \in \mathbb{C}^{n \times p}$.

PROOF. If (2) holds, then $X = A^{(1)}DB^{(1)}$ is a solution of (1). Conversely, if X is any solution of (1), then

$$D = AXB = AA^{(1)}AXBB^{(1)}B = AA^{(1)}DB^{(1)}B.$$

Moreover, it follows from (2) and the definition of $A^{(1)}$ and $B^{(1)}$ that every matrix X of the form (3) satisfies (1). On the other hand, let X be any solution of (1). Then, clearly

$$X = A^{(1)}DB^{(1)} + X - A^{(1)}AXBB^{(1)},$$

which is of the form (3). □

The following characterization of the set $A\{1\}$, in terms of an arbitrary element $A^{(1)}$ of the set, is due essentially to Bjerhammar [103].

COROLLARY 1. *Let $A \in \mathbb{C}^{m \times n}$, $A^{(1)} \in A\{1\}$. Then*

$$A\{1\} = \{A^{(1)} + Z - A^{(1)}AZAA^{(1)} : Z \in \mathbb{C}^{n \times m}\}. \tag{4}$$

PROOF. The set described in RHS(4) is obtained by writing $Y = A^{(1)} + Z$ in the set of solutions of $AXA = A$ as given by Theorem 1. □

Specializing Theorem 1 to ordinary systems of linear equations gives:

COROLLARY 2. *Let $A \in \mathbb{C}^{m \times n}$, $\mathbf{b} \in \mathbb{C}^m$. Then the equation*

$$A\mathbf{x} = \mathbf{b} \tag{5}$$

is consistent if and only if, for some $A^{(1)}$,

$$AA^{(1)}\mathbf{b} = \mathbf{b}, \tag{6}$$

in which case the general solution of (5) is[1]

$$\mathbf{x} = A^{(1)}\mathbf{b} + (I - A^{(1)}A)\mathbf{y} \tag{7}$$

for arbitrary $\mathbf{y} \in \mathbb{C}^n$. □

The following theorem appears in the doctoral dissertation of C.A. Rohde [**704**], who attributes it to R.C. Bose. It is an alternative characterization of $A\{1\}$.

THEOREM 2. *Let $A \in \mathbb{C}^{m \times n}$, $X \in \mathbb{C}^{n \times m}$. Then $X \in A\{1\}$ if and only if, for all $\mathbf{b}$ such that $A\mathbf{x} = \mathbf{b}$ is consistent, $\mathbf{x} = X\mathbf{b}$ is a solution.*

PROOF. *If*: Let $\mathbf{a}_j$ denote the jth column of A. Then

$$A\mathbf{x} = \mathbf{a}_j$$

is consistent and $X\mathbf{a}_j$ is a solution, i.e.,

$$AX\mathbf{a}_j = \mathbf{a}_j \quad (j \in \overline{1,n}).$$

Therefore

$$AXA = A.$$

Only if: This follows from (6). □

Exercises

EX. 1. Consider the matrix A of (0.80) and the vector

$$\mathbf{b} = \begin{bmatrix} 14 + 5i \\ -15 + 3i \\ 10 - 15i \end{bmatrix}.$$

Use (1.7) to show that the general solution of $A\mathbf{x} = \mathbf{b}$ can be written in the form

$$\mathbf{x} = \begin{bmatrix} 0 \\ \frac{5}{2} - 7i \\ 0 \\ 5 - i \\ 0 \\ 0 \end{bmatrix} + \begin{bmatrix} 1 & 0 & 0 & 0 & 0 & 0 \\ 0 & 0 & -\frac{1}{2} & 0 & -1+2i & \frac{1}{2}i \\ 0 & 0 & 1 & 0 & 0 & 0 \\ 0 & 0 & 0 & 0 & -2 & -1-i \\ 0 & 0 & 0 & 0 & 1 & 0 \\ 0 & 0 & 0 & 0 & 0 & 1 \end{bmatrix} \begin{bmatrix} y_1 \\ y_2 \\ y_3 \\ y_4 \\ y_5 \\ y_6 \end{bmatrix}$$

where $y_1, y_2, \ldots, y_6$ are arbitrary.

Note: y_2 and y_4 do not matter, since they multiply zero columns, showing the general solution to have four degrees of freedom, in agreement with (1.7).

EX. 2. *Kronecker products.* The *Kronecker product* $A \otimes B$ of the two matrices $A = (a_{ij}) \in \mathbb{C}^{m \times n}$, $B \in \mathbb{C}^{p \times q}$ is the $mp \times nq$ matrix expressible in partitioned form as

$$A \otimes B = \begin{bmatrix} a_{11}B & a_{12}B & \cdots & a_{1n}B \\ a_{21}B & a_{22}B & \cdots & a_{2n}B \\ \cdots & \cdots & \cdots & \cdots \\ a_{m1}B & a_{m2}B & \cdots & a_{mn}B \end{bmatrix}.$$

---

[1]See also Theorem 7.1, p. 258.

The properties of this product (e.g., Marcus and Minc [**534**]) include

$$(A \otimes B)^* = A^* \otimes B^*, \quad (A \otimes B)^T = A^T \otimes B^T, \tag{8}$$

and

$$(A \otimes B)(P \otimes Q) = AP \otimes BQ \tag{9}$$

for every A, B, P, Q for which the above products are refined.

An important application of the Kronecker product is rewriting a matrix equation

$$AXB = D \tag{1}$$

as a vector equation. For any $X = (x_{ij}) \in \mathbb{C}^{m \times n}$, let the vector $\text{vec}(X) = (v_k) \in \mathbb{C}^{mn}$ be the vector obtained by listing the elements of X by rows. In other words,

$$v_{n(i-1)+j} = x_{ij} \quad (i \in \overline{1, m}; j \in \overline{1, n}).$$

For example,

$$\text{vec} \begin{bmatrix} 1 & 2 \\ 3 & 4 \end{bmatrix} = \begin{bmatrix} 1 \\ 2 \\ 3 \\ 4 \end{bmatrix}.$$

The energetic reader should now verify that

$$\text{vec}(AXB) = (A \otimes B^T) \text{vec}(X). \tag{10}$$

By using (10), the matrix equation (1) can be rewritten as the vector equation

$$(A \otimes B^T) \text{vec}(X) = \text{vec}(D). \tag{11}$$

Theorem 1 must therefore be equivalent to Corollary 2 applied to the vector equation (11). To demonstrate this we need the following two results:

$$A^{(1)} \otimes B^{(1)} \in (A \otimes B)\{1\} \quad \text{(follows from (9))}, \tag{12}$$

$$(A^{(1)})^T \in A^T\{1\}. \tag{13}$$

Now (1) is consistent if and only if (11) is consistent, and the latter statement

$$\iff (A \otimes B^T)(A \otimes B^T)^{(1)} \text{vec}(D) = \text{vec}(D) \quad \text{(by Corollary 2)},$$

$$\iff (A \otimes B^T)(A^{(1)} \otimes (B^{(1)})^T) \text{vec}(D) = \text{vec}(D) \quad \text{(by (12), (13))},$$

$$\iff (AA^{(1)} \otimes (B^{(1)}B)^T) \text{vec}(D) = \text{vec}(D) \quad \text{(by (9))},$$

$$\iff AA^{(1)}DB^{(1)}B = D \quad \text{(by (10))}.$$

The other statements of Theorem 1 can be similarly shown to follow from their counterparts in Corollary 2. The two results are thus equivalent.

Ex. 3. $(A \otimes B)^\dagger = A^\dagger \otimes B^\dagger$ (Greville [**326**]).

PROOF. Upon replacing A by $A \otimes B$ and X by $A^\dagger \otimes B^\dagger$ in (1.1)–(1.4) and making use of (8) and (9), it is easily verified that (1.1)–(1.4) are satisfied.   □

Ex. 4. The matrix equations

$$AX = B, \quad XD = E, \tag{14}$$

have a common solution if and only if each equation separately has a solution and

$$AE = BD.$$

PROOF. (Penrose [**635**]). *If*: For any $A^{(1)}, D^{(1)}$,

$$X = A^{(1)}B + ED^{(1)} - A^{(1)}AED^{(1)}$$

is a common solution of both equations (14) provided $AE = BD$ and

$$AA^{(1)}B = B, \quad ED^{(1)}D = E.$$

By Theorem 1, the latter two equations are equivalent to the consistency of equations (14) considered separately.

Only if: Obvious. $\square$

EX. 5. Let equations (14) have a common solution $X_0 \in \mathbb{C}^{m \times n}$. Then, show that the general solution is

$$X = X_0 + (I - A^{(1)}A)Y(I - DD^{(1)}) \tag{15}$$

for arbitrary $A^{(1)} \in A\{1\}, D^{(1)} \in D\{1\}, Y \in \mathbb{C}^{m \times n}$.

Hint: First, show that RHS(15) is a common solution. Then, if X is any common solution, evaluate RHS(15) for $Y = X - X_0$.

2. Characterization of $A\{1,3\}$ and $A\{1,4\}$

The set $A\{1\}$ is completely characterized in Corollary 1. Let us now turn our attention to $A\{1,3\}$. The key to its characterization is the following theorem:

THEOREM 3. *The set $A\{1,3\}$ consists of all solutions for X of*

$$AX = AA^{(1,3)}, \tag{16}$$

where $A^{(1,3)}$ is an arbitrary element of $A\{1,3\}$.

PROOF. If X satisfies (16), then clearly

$$AXA = AA^{(1,3)}A = A,$$

and, moreover, AX is Hermitian since $AA^{(1,3)}$ is Hermitian by definition. Thus, $X \in A\{1,3\}$.

On the other hand, if $X \in A\{1,3\}$, then

$$AA^{(1,3)} = AXAA^{(1,3)} = (AX)^* AA^{(1,3)} = X^* A^* (A^{(1,3)})^* A^*$$
$$= X^* A^* = AX,$$

where we have used Lemma 1.1(a). $\square$

COROLLARY 3. *Let $A \in \mathbb{C}^{m \times n}, A^{(1,3)} \in A\{1,3\}$. Then*

$$A\{1,3\} = \{A^{(1,3)} + (I - A^{(1,3)}A)Z : Z \in \mathbb{C}^{n \times m}\}. \tag{17}$$

PROOF. Applying Theorem 1 to (16) and substituting $Z + A^{(1,3)}$ for Y gives (17). $\square$

The following theorem and its corollary are obtained in a manner analogous to the proofs of Theorem 3 and Corollary 3.

THEOREM 4. *The set $A\{1,4\}$ consists of all solutions for X of*

$$XA = A^{(1,4)}A.$$

COROLLARY 4. *Let $A \in \mathbb{C}^{m \times n}, A^{(1,4)} \in A\{1,4\}$. Then*

$$A\{1,4\} = \{A^{(1,4)} + Y(I - AA^{(1,4)}) : Y \in \mathbb{C}^{n \times m}\}.$$

Other characterizations of $A\{1,3\}$ and $A\{1,4\}$ based on their least-squares properties will be given in Chapter 3.

Exercises

EX. 6. Prove Theorem 4 and Corollary 4.

Ex. 7. For the matrix A of (0.80), show that $A\{1,3\}$ is the set of matrices X of the form

$$
X = \tfrac{1}{38}
\begin{bmatrix}
0 & 0 & 0 \\
-10i & 3 & 9 \\
0 & 0 & 0 \\
2i & -12 & 2 \\
0 & 0 & 0 \\
0 & 0 & 0
\end{bmatrix}
+
\begin{bmatrix}
1 & 0 & 0 & 0 & 0 & 0 \\
0 & 0 & -\tfrac{1}{2} & 0 & -1+2i & \tfrac{1}{2}i \\
0 & 0 & 1 & 0 & 0 & 0 \\
0 & 0 & 0 & 0 & -2 & -1-i \\
0 & 0 & 0 & 0 & 1 & 0 \\
0 & 0 & 0 & 0 & 0 & 1
\end{bmatrix}
Z,
$$

where Z is an arbitrary element of $\mathbb{C}^{6\times 3}$.

Ex. 8. For the matrix A of (0.80), show that $A\{1,4\}$ is the set of matrices Y of the form

$$
Y = \tfrac{1}{276}
\begin{bmatrix}
0 & 0 & 0 \\
0 & 20-18i & 42 \\
0 & 10-9i & 21 \\
0 & -29-9i & -9-27i \\
0 & -2+4i & 24+30i \\
0 & -29+30i & -36+3i
\end{bmatrix}
+ Z
\begin{bmatrix}
1 & -\tfrac{1}{3}i & -i \\
0 & 0 & 0 \\
0 & 0 & 0
\end{bmatrix},
$$

where Z is an arbitrary element of $\mathbb{C}^{6\times 3}$.

Ex. 9. Using Theorem 1.4 and the results of Exs. 7 and 8, calculate $A^{\dagger}$. (Since any $A^{(1,4)}$ and $A^{(1,3)}$ will do, choose the simplest.)

Ex. 10. Give an alternative proof of Theorem 1.4, using Theorem 3 and 4. (*Hint*: Take $X = A^{\dagger}$.)

Ex. 11. By applying Ex. 5 show that if $A \in \mathbb{C}^{m\times n}$ and $A^{(1,3,4)} \in A\{1,3,4\}$, then

$$
A\{1,3,4\} = \{A^{(1,3,4)} + (I - A^{(1,3,4)}A)Y(I - AA^{(1,3,4)}) : Y \in \mathbb{C}^{n\times m}\}.
$$

Ex. 12. Show that if $A \in \mathbb{C}^{m\times n}$ and $A^{(1,2,3)} \in A\{1,2,3\}$, then

$$
A\{1,2,3\} = \{A^{(1,2,3)} + (I - A^{(1,2,3)}A)ZA^{(1,2,3)} : Z \in \mathbb{C}^{n\times m}\}.
$$

Ex. 13. Similarly, show that if $A \in \mathbb{C}^{m\times n}$ and $A^{(1,2,4)} \in A\{1,2,4\}$, then

$$
A\{1,2,4\} = \{A^{(1,2,4)} + A^{(1,2,4)}Z(I - AA^{(1,2,4)}) : Z \in \mathbb{C}^{m\times m}\}.
$$

3. Characterization of $A\{2\}$, $A\{1,2\}$, and Other Subsets of $A\{2\}$

Since

$$
XAX = X \tag{1.2}
$$

involves X nonlinearly, a characterization of $A\{2\}$ is not obtained by merely applying Theorem 1. However, such a characterization can be reached by using a full-rank factorization of X. The rank of X will play an important role, and it will be convenient to let $A\{i,j,\dots,k\}_s$ denote the subset of $A\{i,j,\dots,k\}$ consisting of matrices of rank s.

We remark that the sets $A\{2\}_0$, $A\{2,3\}_0$, $A\{2,4\}_0$ and $A\{2,3,4\}_0$ are identical and contain a single element. For $A \in \mathbb{C}^{m\times n}$ this sole element is the $n \times m$ matrix of zeros. Having thus disposed of the case of $s = 0$, we shall consider only positive s in the remainder of this section.

The following theorem has been stated by G.W. Stewart [780], who attributes it to R.E. Funderlic.

THEOREM 5. *Let $A \in \mathbb{C}_r^{m \times n}$ and $0 < s \le r$. Then*

$$A\{2\}_s = \{YZ : Y \in \mathbb{C}^{n \times s}, Z \in \mathbb{C}^{s \times m}, ZAY = I_s\}. \tag{18}$$

PROOF. Let

$$X = YZ, \tag{19}$$

where the conditions on Y and Z in RHS(18) are satisfied. Then Y and Z are of rank s and X is of rank s by Ex. 1.7. Moreover,

$$XAX = YZAYZ = YZ = X.$$

On the other hand, let $X \in A\{2\}_s$ and let (19) be a full-rank factorization. Then $Y \in \mathbb{C}_s^{n \times s}$, $Z \in \mathbb{C}_s^{s \times m}$ and

$$YZAYZ = YZ. \tag{20}$$

Moreover, if $Y^{(1)}$ and $Z^{(1)}$ are any $\{1\}$-inverses, then by Lemma 1.2

$$Y^{(1)}Y = ZZ^{(1)} = I_s.$$

Thus, multiplying (20) on the left by $Y^{(1)}$ and on the right by $Z^{(1)}$ gives

$$ZAY = I_s. \qquad \square$$

COROLLARY 5. *Let $A \in \mathbb{C}_r^{m \times n}$. Then*

$$A\{1,2\} = \{YZ : Y \in \mathbb{C}^{n \times r}, Z \in \mathbb{C}^{r \times m}, ZAY = I_r\}.$$

PROOF. By Theorem 1.2,

$$A\{1,2\} = A\{2\}_r. \qquad \square$$

The relation $ZAY = I_s$ of (18) implies that $Z \in (AY)\{1,2,4\}$. This remark suggests the approach to the characterization of $A\{2,3\}$ on which the following theorem is based.

THEOREM 6. *Let $A \in \mathbb{C}_r^{m \times n}$ and $0 < s \le r$. Then*

$$A\{2,3\}_s = \{Y(AY)^\dagger : AY \in \mathbb{C}_s^{m \times s}\}.$$

PROOF. Let $X = Y(AY)^\dagger$, where $AY \in \mathbb{C}_s^{m \times s}$. Then we have

$$AX = AY(AY)^\dagger. \tag{21}$$

The right member is Hermitian by (1.3), and

$$XAX = Y(AY)^\dagger AY(AY)^\dagger = Y(AY)^\dagger = X.$$

Thus, $X \in A\{2,3\}$. Finally, since $X \in A\{2\}$, $A \in X\{1\}$, (21) and Lemma 1.1(f) give

$$s = \operatorname{rank} AY = \operatorname{rank} AX = \operatorname{rank} X.$$

On the other hand, let $X \in A\{2,3\}_s$. Then AX is Hermitian and idempotent and is of rank s by Lemma 1.1(f), since $A \in X\{1\}$. By Ex. 1.20

$$(AX)^\dagger = AX,$$

and so

$$X(AX)^\dagger = XAX = X.$$

Thus X is of the form described in the theorem. $\qquad \square$

The following theorem is proved in an analogous fashion:

THEOREM 7. *Let $A \in \mathbb{C}_r^{m \times n}$ and $0 < s \le r$. Then*

$$A\{2,4\}_s = \{(YA)^\dagger Y : YA \in \mathbb{C}_s^{s \times m}\}.$$

Exercises

EX. 14. Could Theorem 6 be sharpened by replacing $(AY)^\dagger$ by $(AY)^{(i,j,k)}$ for some i, j, k? (Which properties are actually used in the proof?) Note that AY is of full column rank; what bearing, if any, does this have on the answer to the question?

EX. 15. Show that, if $A \in \mathbb{C}_r^{m \times n}$,

$$A\{1, 2, 3\} = \{Y(AY)^\dagger : AY \in \mathbb{C}_r^{m \times r}\},$$

$$A\{1, 2, 4\} = \{(YA)^\dagger Y : YA \in \mathbb{C}_r^{r \times m}\}.$$

(Compare these results with Exs. 12 and 13.)

EX. 16. The characterization of $A\{2, 3, 4\}$ is more difficult, and will be postponed until later in this chapter. Show, however, that if rank $A = 1$, $A\{2, 3, 4\}$ contains exactly two elements, $A^\dagger$ and O.

4. Idempotent Matrices and Projectors

A comparison of (1) in the Introduction with Lemma 1.1(f) suggests that the role played by the unit matrix in connection with the ordinary inverse of a nonsingular matrix is, in a sense, assumed by idempotent matrices in relation to generalized inverses. As the properties of idempotent matrices are likely to be treated in a cursory fashion in an introductory course in linear algebra, some of them are listed in the following lemma:

LEMMA 1. *Let $E \in \mathbb{C}^{n \times n}$ be idempotent. Then:*

(a) *E^* and $I - E$ are idempotent.*
(b) *The eigenvalues of E are 0 and 1. The multiplicity of the eigenvalue 1 is* rank E.
(c) rank $E = $ trace E.
(d) *$E(I - E) = (I - E)E = O$.*
(e) *$E\mathbf{x} = \mathbf{x}$ if and only if $\mathbf{x} \in R(E)$.*
(f) *$E \in E\{1, 2\}$.*
(g) *$N(E) = R(I - E)$.*

PROOF. Parts (a) to (f) are immediate consequences of the definition of idempotency: (c) follows from (b) and the fact that the trace of any square matrix is the sum of its eigenvalues counting multiplicities; (g) is obtained by applying Corollary 2 to the equation $E\mathbf{x} = \mathbf{0}$. □

LEMMA 2 (Langenhop [**499**]). *Let a square matrix have the full-rank factorization*

$$E = FG.$$

Then E is idempotent if and only if $GF = I$.

PROOF. If $GF = I$, then clearly

$$(FG)^2 = FGFG = FG. \tag{22}$$

On the other hand, since F is of full column rank and G is of full row rank,

$$F^{(1)}F = GG^{(1)} = I$$

by Lemma 1.2. Thus if (22) holds, multiplication on the left by $F^{(1)}$ and on the right by $G^{(1)}$ gives $GF = I$. □

Let $P_{L,M}$ denote the transformation that carries any $\mathbf{x} \in \mathbb{C}^n$ into its projection on L along M; see §0.1.3. It is easily verified that this transformation is linear (see Ex. 0.33). We shall call the transformation $P_{L,M}$ the *projector on L along M*, or, *oblique projector*.

We recall, see §0.2.5, that every linear transformation from one finite-dimensional vector space to another can be represented by a matrix, which is uniquely determined by the linear transformation and by the choice of bases for the spaces involved. Except where otherwise specified, the basis for any finite-dimensional vector space, used in this book, is the standard basis of unit vectors. Having thus fixed the bases, there is a one-to-one correspondence between $\mathbb{C}^{m \times n}$, the $m \times n$ *complex matrices*, and $\mathcal{L}(\mathbb{C}^n, \mathbb{C}^m)$, the *space of linear transformations mapping* $\mathbb{C}^n$ *into* $\mathbb{C}^m$. This correspondence permits using the same symbol, say A, to denote both the linear transformation $A \in \mathcal{L}(\mathbb{C}^n, \mathbb{C}^m)$ and its matrix representation $A \in \mathbb{C}^{m \times n}$. Thus the matrix–vector equation

$$A\mathbf{x} = \mathbf{y} \quad (A \in \mathbb{C}^{m \times n}, \mathbf{x} \in \mathbb{C}^n, \mathbf{y} \in \mathbb{C}^m)$$

can equally be regarded as a statement that the linear transformation A maps $\mathbf{x}$ into $\mathbf{y}$.

In particular, linear transformations mapping $\mathbb{C}^n$ into itself are represented by the square matrices of order n. Specializing further, the next theorem establishes a one-to-one correspondence between the idempotent matrices of order n and the projectors $P_{L,M}$ where $L \oplus M = \mathbb{C}^n$. Moreover, for any two complementary subspaces L and M, a method for computing $P_{L,M}$ is given by (27) below.

THEOREM 8. *For every idempotent matrix $E \in \mathbb{C}^{n \times n}$, $R(E)$ and $N(E)$ are complementary subspaces with*

$$E = P_{R(E),N(E)}. \tag{23}$$

Conversely, if L and M are complementary subspaces, there is a unique idempotent $P_{L,M}$ such that $R(P_{L,M}) = L$, $N(P_{L,M}) = M$.

PROOF. Let E be idempotent of order n. Then it follows from Lemma 1(e) and (g), and from the equation

$$\mathbf{x} = E\mathbf{x} + (I - E)\mathbf{x}, \tag{24}$$

that $\mathbb{C}^n$ is the sum of $R(E)$ and $N(E)$. Moreover, $R(E) \cap N(E) = \{\mathbf{0}\}$, since

$$E\mathbf{x} = (I - E)\mathbf{y} \quad \Longrightarrow \quad E\mathbf{x} = E^2\mathbf{x} = E(I - E)\mathbf{y} = \mathbf{0},$$

by Lemma 1(d). Thus, $R(E)$ and $N(E)$ are complementary and (24) shows that, for every $\mathbf{x}$, $E\mathbf{x}$ is the projection of $\mathbf{x}$ on $R(E)$ along $N(E)$. This establishes (23).

On the other hand, let $\{\mathbf{x}_1, \mathbf{x}_2, \dots, \mathbf{x}_\ell\}$ and $\{\mathbf{y}_1, \mathbf{y}_2, \dots, \mathbf{y}_m\}$ be any two bases for L and M, respectively. Then $P_{L,M}$, if it exists, is uniquely determined by

$$\begin{cases} P_{L,M}\, \mathbf{x}_i = \mathbf{x}_i, & (i \in \overline{1,\ell}), \\ P_{L,M}\, \mathbf{y}_i = \mathbf{0}, & (i \in \overline{1,m}). \end{cases} \tag{25}$$

Let $X = [\mathbf{x}_1 \quad \mathbf{x}_2 \quad \cdots \quad \mathbf{x}_\ell]$ denote the matrix whose columns are the vectors $\mathbf{x}_i$. Similarly, let $Y = [\mathbf{y}_1 \quad \mathbf{y}_2 \quad \cdots \quad \mathbf{y}_m]$. Then (25) is equivalent to

$$P_{L,M}\, [X \quad Y] = [X \quad O]. \tag{26}$$

Since $[X \quad Y]$ is nonsingular, the unique solution of (26), and therefore of (25), is

$$P_{L,M} = [X \quad O][X \quad Y]^{-1}. \tag{27}$$

Since (25) implies

$$P_{L,M}[X \quad O] = [X \quad O],$$

$P_{L,M}$ as given by (27) is clearly idempotent.   □

The relation between the direct sum (0.1) and the projector[2] $P_{L,M}$ is given in the following:

COROLLARY 6. *Let L and M be complementary subspaces of $\mathbb{C}^n$. Then, for every $\mathbf{x} \in \mathbb{C}^n$, the unique decomposition (0.2) is given by*

$$P_{L,M}\mathbf{x} = \mathbf{y}, \quad (I - P_{L,M})\mathbf{x} = \mathbf{z}.$$

If $A^{(1)} \in A\{1\}$, we know from Lemma 1.1(f) that both $AA^{(1)}$ and $A^{(1)}A$ are idempotent and, therefore, are projectors. It is of interest to find out what we can say about the subspaces associated with these projectors. In fact, we already know from Ex. 1.9 that

$$R(AA^{(1)}) = R(A), \quad N(A^{(1)}A) = N(A), \quad R((A^{(1)}A)^*) = R(A^*). \tag{28}$$

The following is an immediate consequence of these results:

COROLLARY 7. *If A and X are $\{1,2\}$-inverses of each other, AX is the projector on $R(A)$ along $N(X)$, and XA is the projector on $R(X)$ along $N(A)$.*

An important application of projectors is to the class of diagonable matrices. (The reader will recall that a square matrix is called *diagonable* if it is similar to a diagonal matrix.) It is easily verified that a matrix $A \in \mathbb{C}^{n \times n}$ is diagonable if and only if it has n linearly independent eigenvectors. The latter fact will be used in the proof of the following theorem, which expresses an arbitrary diagonable matrix as a linear combination of projectors.

THEOREM 9 (Spectral Theorem for Diagonable Matrices). *Let $A \in \mathbb{C}^{n \times n}$ with s distinct eigenvalues $\lambda_1, \lambda_2, \ldots, \lambda_s$. Then A is diagonable if*

---

[2]Our use of the term "projector" to denote either the linear transformation $P_{L,M}$ or its idempotent matrix representation is not standard in the literature. Many writers have used "projection" in the same sense. The latter usage, however, seems to us to lead to undesirable ambiguity, since "projection" also describes the image $P_{L,M}\mathbf{x}$ of the vector $\mathbf{x}$ under the transformation $P_{L,M}$. The use of "projection" in the sense of "image" is clearly much older (e.g., in elementary geometry) than its use in the sense of "transformation." "Projector" describes more accurately than "projection" what is meant here, and has been used in this sense by Afriat [3], de Boor [117], Bourbaki [125, Ch. I, Def. 6, p. 16], [126, Ch. VIII, Section 1], Greville [323], Przeworska–Rolewicz and Rolewicz [653], Schwerdtfeger [735], and Ward, Boullion, and Lewis [851]. Still other writers use "projector" to designate the orthogonal projector to be discussed in Section 7. This is true of Householder [432], Yosida [882], Kantorovich and Akilov [467], and numerous other Russian writers. We are indebted to de Boor for several of the preceding references.

and only if there exist projectors $E_1, E_2, \ldots, E_s$ such that

$$E_i E_j = \delta_{ij} E_i, \tag{29a}$$

$$I_n = \sum_{i=1}^{s} E_i, \tag{29b}$$

$$A = \sum_{i=1}^{s} \lambda_i E_i. \tag{29c}$$

PROOF. *If*: For $i \in \overline{1, s}$, let $r_i = \operatorname{rank} E_i$ and let $X_i \in \mathbb{C}^{n \times r_i}$ be a matrix whose columns are a basis for $R(E_i)$. Let

$$X = [X_1 \quad X_2 \quad \cdots \quad X_s].$$

Then, by Lemma 1(c), the number of columns of X is

$$\sum_{i=1}^{s} r_i = \sum_{i=1}^{s} \operatorname{trace} E_i = \operatorname{trace} \sum_{i=1}^{s} E_i = \operatorname{trace} I_n = n,$$

by (29b). Thus X is square of order n. By the definition of X_i, there exists for each i a Y_i such that

$$E_i = X_i Y_i.$$

Let

$$Y = \begin{bmatrix} Y_1 \\ Y_2 \\ \vdots \\ Y_s \end{bmatrix}.$$

Then

$$XY = \sum_{i=1}^{s} X_i Y_i = \sum_{i=1}^{s} E_i = I_n,$$

by (29b). Therefore X is nonsingular. By Lemma 1(e),

$$E_i X_i = X_i,$$

and therefore, by (29a) and (29c),

$$AX = \sum_{i=1}^{s} \lambda_i E_i X_i = [\lambda_1 X_1 \quad \lambda_2 X_2 \quad \cdots \quad \lambda_s X_s]$$

$$= XD, \tag{30}$$

where

$$D = \operatorname{diag} (\lambda_1 I_{r_1}, \lambda_2 I_{r_2}, \ldots, \lambda_s I_{r_s}). \tag{31}$$

Since X is nonsingular, it follows from (30) that A and D are similar.

Only if: If A is diagonable,

$$AX = XD, \tag{32}$$

where X is nonsingular and D can be represented in the form (31). Let X be partitioned by columns into $X_1, X_2, \ldots, X_s$ in conformity with the diagonal blocks of D and, for $i = 1, 2, \ldots, s$, let

$$E_i = [O \quad \cdots \quad O \quad X_i \quad O \quad \cdots \quad O] X^{-1}.$$

In other words, $E_i = \widetilde{X_i} X^{-1}$, where $\widetilde{X_i}$ denotes the matrix obtained from X by replacing all its columns except the columns of X_i by columns of zeros. It is then easily verified that E_i is idempotent and that (29a) and (29b) hold. Finally,

$$\sum_{i=1}^{s} \lambda_i E_i = [\lambda_1 X_1 \quad \lambda_2 X_2 \quad \cdots \quad \lambda_s X_s] X^{-1} = X D X^{-1} = A,$$

by (32).   □

The idempotent matrices $\{E_i : i \in \overline{1,s}\}$ (shown in Ex. 24 below to be uniquely determined by the diagonable matrix A) are called its *principal idempotents* or *Frobenius covariants*. Relation (29c) is called the *spectral decomposition* of A. Further properties of this decomposition are studied in Exs. 24–26.

Note that $R(E_i)$ is the *eigenspace* of A (space spanned by the eigenvectors) associated with the eigenvalue λ_i while, because of (29a), $N(E_i)$ is the direct sum of the eigenspaces associated with all eigenvalues of A other than λ_i.

Exercises

Ex. 17. Show that $I_n\{2\}$ consists of all idempotent matrices of order n.

Ex. 18. If E is idempotent, $X \in E\{2\}$ and $R(X) \subset R(E)$ show that X is idempotent.

Ex. 19. Let $E \in \mathbb{C}_r^{n \times n}$. Then E is idempotent if and only if its Jordan canonical form can be written as

$$\begin{bmatrix} I_r & O \\ O & O \end{bmatrix}.$$

Ex. 20. Show that $P_{L,M} A = A$ if and only if $R(A) \subset L$ and $A P_{L,M} = A$ if and only if $N(A) \supset M$.

Ex. 21. $AB(AB)^{(1)}A = A$ if and only if $\operatorname{rank} AB = \operatorname{rank} A$ and $B(AB)^{(1)}AB = B$ if and only if $\operatorname{rank} AB = \operatorname{rank} B$. (*Hint*: Use Exs. 20, 1.9, and 1.10.)

Ex. 22. A matrix $A \in \mathbb{C}^{n \times n}$ is diagonable if and only if it has n linearly independent eigenvectors.

PROOF. Diagonability of A is equivalent to the existence of a nonsingular matrix X such that $X^{-1}AX = D$, which in turn is equivalent to $AX = XD$. But the latter equation expresses the fact that each column of X is an eigenvector of A, and X is nonsingular if and only if its columns are linearly independent.   □

Ex. 23. Show that $I - P_{L,M} = P_{M,L}$.

Ex. 24. *Principal idempotents*. Let $A \in \mathbb{C}^{n \times n}$ be a diagonable matrix with s distinct eigenvalues $\lambda_1, \lambda_2, \dots, \lambda_s$. Then the idempotents $E_1, E_2, \dots, E_s$ satisfying (29a)–(29c) are uniquely determined by A.
PROOF. Let $\{F_i : i \in \overline{1,s}\}$ be any idempotent matrices satisfying

$$F_i F_j = O, \quad \text{if } i \neq j, \tag{29a*}$$

$$I_n = \sum_{i=1}^{s} F_i, \tag{29b*}$$

$$A = \sum_{i=1}^{s} \lambda_i F_i. \tag{29c*}$$

From (29a) and (29c) it follows that

$$E_i A = A E_i = \lambda_i E_i \quad (i \in \overline{1, s}). \tag{33}$$

Similarly, from (29a∗) and (29c∗),

$$F_i A = A F_i = \lambda_i F_i \quad (i \in \overline{1, s}), \tag{33∗}$$

so that

$$E_i(AF_j) = \lambda_j E_i F_j$$

and

$$(E_i A) F_j = \lambda_i E_i F_j,$$

proving that

$$E_i F_j = O, \quad \text{if } i \neq j. \tag{34}$$

The uniqueness of $\{E_i : i \in \overline{1, s}\}$ now follows:

$$\begin{aligned}
E_i &= E_i \sum_{j=1}^{s} F_j, \quad \text{by (29b∗)}, \\
&= E_i F_i, \quad \text{by (34)}, \\
&= \Big(\sum_{j=1}^{s} E_j\Big) F_i, \quad \text{by (34)}, \\
&= F_i, \quad \text{by (29b)}. \qquad \square
\end{aligned}$$

Ex. 25. Let $A \in \mathbb{C}^{n \times n}$ be a diagonable matrix with s distinct eigenvalues $\lambda_1, \lambda_2, \ldots, \lambda_s$. Then the principal idempotents of A are given by

$$E_i = \frac{p_i(A)}{p_i(\lambda_i)} \quad (i \in \overline{1, s}), \tag{35}$$

where

$$p_i(\lambda) = \prod_{\substack{j=1 \\ j \neq i}}^{s} (\lambda - \lambda_j). \tag{36}$$

PROOF. Let G_i ($i \in \overline{1, s}$) denote RHS(35) and let $E_1, E_2, \ldots, E_s$ be the principal idempotents of A. For any $i, j \in \overline{1, s}$,

$$\begin{aligned}
G_i E_j &= \frac{1}{p_i(\lambda_i)} \prod_{\substack{h=1 \\ h \neq i}}^{s} (A - \lambda_h I) E_j \\
&= \frac{1}{p_i(\lambda_i)} \prod_{\substack{h=1 \\ h \neq i}}^{s} (\lambda_j - \lambda_h I) E_j, \quad \text{by (33)}, \\
&= \begin{cases} O, & \text{if } i \neq j, \\ E_i, & \text{if } i = j. \end{cases}
\end{aligned}$$

Therefore, $G_i = G_i \sum_{j=1}^{s} E_j = E_i$ ($i \in \overline{1, s}$). $\qquad \square$

Ex. 26. Let A be a diagonable matrix with p distinct eigenvalues λ_i and principal idempotents E_i, $i \in \overline{1, s}$. Then:

(a) If $p(\lambda)$ is any polynomial,

$$p(A) = \sum_{i=1}^{s} f(\lambda_i) E_i.$$

(b) Any matrix commutes with A if and only if it commutes with every E_i ($i \in \overline{1, s}$).

PROOF. (a) Follows from (29a), (29b), and (29c).

(b) Follows from (29c) and (35) which express A as a linear combination of the $\{E_i : i \in \overline{1, s}\}$ and each E_i as a polynomial in A. $\square$

See Corollary 8 for polynomials in general square matrices.

EX. 27. Prove the following analog of Theorem 5 for $\{1\}$-inverses: Let $A \in \mathbb{C}_r^{m \times n}$ with $r < s \le \min(m, n)$. Then

$$A\{1\}_s = \left\{ YZ : Y \in \mathbb{C}_s^{n \times s}, Z \in \mathbb{C}_s^{s \times m}, ZAY = \begin{bmatrix} I_r & O \\ O & O \end{bmatrix} \right\}. \tag{37}$$

PROOF. Let $X = YZ$, where the conditions on Y and Z in RHS(37) are satisfied. Then rank $X = s$ by Ex. 1.7. Let

$$Y = [Y_1 \ Y_2], \quad Z = \begin{bmatrix} Z_1 \\ Z_2 \end{bmatrix},$$

where Y_1 denotes the first r columns of Y and Z_1 the first r rows of Z. Then (37) gives

$$Z_1 A Y_1 = I_r, \quad Z_1 A Y_2 = O. \tag{38}$$

Let $X_1 = Y_1 Z_1$. Then it follows from the first equation (38) that $X_1 \in A\{2\}$. Since by Ex. 1.7, rank $X_1 = r = \text{rank } A$, $X_1 \in A\{1\}$ by Theorem 1.2. Thus

$$AXA = AX_1AXA = AY_1(Z_1AY)ZA = AY_1[I_r \ O]\begin{bmatrix} Z_1 \\ Z_2 \end{bmatrix}A$$
$$= AY_1 Z_1 A = AX_1 A = A.$$

On the other hand, let $X \in A\{1\}_s$ and let $X = UV$ be a full-rank factorization. Then $U \in \mathbb{C}_s^{n \times s}$, $V \in \mathbb{C}_s^{s \times m}$, and

$$VAUVAU = VAU$$

and so VAU is idempotent and is of rank r by Ex. 1.7. Thus, by Ex. 19, there is a nonsingular T such that

$$TVVAUT^{-1} = \begin{bmatrix} I_r & O \\ O & O \end{bmatrix}.$$

If we now take

$$Y = UT^{-1}, \quad Z = TV,$$

then

$$Y \in \mathbb{C}_s^{n \times s}, \quad Z \in \mathbb{C}_s^{s \times m},$$
$$ZAY = \begin{bmatrix} I_r & O \\ O & O \end{bmatrix}, \quad \text{and} \quad YZ = UV = X. \qquad \square$$

5. Matrix Functions

Let $f(z)$ be a complex scalar function or a mapping $f : \mathbb{C} \to \mathbb{C}$. If $A \in \mathbb{C}^{n \times n}$ and if f is analytic in some open set containing $\lambda(A)$, then a corresponding matrix function $f(A)$ can be defined. We study here matrix functions, and the correspondence

$$f(z) \longleftrightarrow f(A). \tag{39}$$

As an example, consider the (scalar) inhomogeneous linear differential equation

$$\dot{x} + ax = b(t),$$

and its general solution

$$x(t) = e^{-at}y + e^{-at} \int^t e^{as}b(s)ds,$$

where y is arbitrary. The corresponding (vector) differential equation, with given $A \in \mathbb{C}^{n \times n}$ and $\mathbf{b}(t) \in \mathbb{C}^n$,

$$\dot{\mathbf{x}}(t) + A\mathbf{x}(t) = \mathbf{b}(t) \tag{40}$$

has the analogous general solution

$$\mathbf{x}(t) = e^{-At}\mathbf{y} + e^{-At} \int^t e^{As}\mathbf{b}(s)\,ds \tag{41}$$

with arbitrary $\mathbf{y}$. The matrix function e^{-At} plays here the same role as the scalar function e^{-at}, see Ex. 31.

If a matrix $A \in \mathbb{C}^{n \times n}$ is diagonable and if a function f is defined in $\lambda(A)$, then a reasonable definition of the matrix function $f(A)$ is, by Theorem 9,

$$f(A) = \sum_{\lambda \in \lambda(A)} f(\lambda)E_\lambda, \tag{42}$$

where E_λ is the principal idempotent associated with λ. We study here matrix functions for general matrices and obtain (42) as a special case.

Let $A \in \mathbb{C}^{n \times n}$ have s distinct eigenvalues $\{\lambda_1, \dots, \lambda_s\}$, and a Jordan form

$$A = XJX^{-1} = XJY$$

$$= [X_1 \quad X_2 \quad \cdots \quad X_s] \begin{bmatrix} J(\lambda_1) & O & \cdots & O \\ O & J(\lambda_2) & \cdots & O \\ \vdots & \vdots & \ddots & \vdots \\ O & O & \cdots & J(\lambda_s) \end{bmatrix} \begin{bmatrix} Y_1 \\ Y_2 \\ \vdots \\ Y_s \end{bmatrix}$$

$$= \sum_{i=1}^{s} X_i J(\lambda_i) Y_i, \tag{43}$$

where X and $Y = X^{-1}$ are partitioned in agreement with the partition of J and

$$
J(\lambda_i) = \begin{bmatrix} J_{j_1}(\lambda_i) & O & \cdots & O \\ O & J_{j_2}(\lambda_i) & \cdots & O \\ \vdots & \vdots & \ddots & \vdots \\ O & O & \cdots & J_{j_q}(\lambda_i) \end{bmatrix}, \quad j_1 \geq j_2 \geq \cdots \geq j_q, \quad (44)
$$

are the Jordan blocks corresponding to λ_i, $i \in \overline{1, s}$. Let

$$
E_{\lambda_i} := X_i Y_i. \tag{45}
$$

Since $AX_i = X_i J(\lambda_i)$, it follows that $AE_i = X_i J(\lambda_i) Y_i$ and, therefore, by (43),

$$
A = \sum_{i=1}^{s} \lambda_i E_i + \sum_{i=1}^{s} (A - \lambda_i I) E_i. \tag{46}
$$

Note that the second terms $(A - \lambda_i I) E_i$ in (46) are zero for eigenvalues λ_i of index 1.

An analog of Theorem 9 for general square matrices is:

THEOREM 10 (Spectral Theorem for Square Matrices). *Let the matrix* $A \in \mathbb{C}^{n \times n}$ *have* s *distinct eigenvalues. Then there exist* s *unique projectors* $\{E_\lambda : \lambda \in \lambda(A)\}$ *such that*

$$
E_\lambda E_\mu = \delta_{\lambda\mu} E_\lambda, \tag{29a}
$$

$$
I_n = \sum_{\lambda \in \lambda(A)} E_\lambda, \tag{29b}
$$

$$
A = \sum_{\lambda \in \lambda(A)} \lambda E_\lambda + \sum_{\lambda \in \lambda(A)} (A - \lambda I) E_\lambda, \tag{46}
$$

$$
AE_\lambda = E_\lambda A, \quad \text{for all } \lambda \in \lambda(A), \tag{47}
$$

$$
E_\lambda (A - \lambda I)^k = O, \quad \text{for all } \lambda \in \lambda(A), \ k \geq \nu(\lambda). \tag{48}
$$

PROOF. By definition and properties of the inverse matrix, the matrices $E_{\lambda_i} = X_i Y_i$ are projectors and satisfy (29a), (29b), and (46). The commutativity (47) is a consequence of (29b) and (46), see also Ex. 29. Equation (48) follows from Ex. 0.75. Uniqueness is proved as in Ex. 24.   □

As in the previous section, the projectors $\{E_\lambda : \lambda \in \lambda(A)\}$ are called the *principal idempotents* or *Frobenius covariants* of A, and (46) is called the *spectral decomposition* of A. The principal idempotents of A are polynomials in A, see Ex. 29.

EXAMPLE 1. Let

$$
A = \begin{bmatrix} 0 & -2 & 0 & -5 & 2 \\ 0 & -1 & 0 & -2 & 0 \\ 0 & 0 & 0 & 2 & 0 \\ 0 & 1 & 0 & 2 & 0 \\ -2 & -2 & 1 & -8 & 4 \end{bmatrix}.
$$

Then the Jordan form of A is $A = XJX^{-1}$, with:

$$
J = \begin{bmatrix} J_1(1) & & \\ & J_2(2) & \\ & & J_2(0) \end{bmatrix} = \begin{bmatrix} 1 & & & & \\ & 2 & 1 & & \\ & 0 & 2 & & \\ & & & 0 & 1 \\ & & & 0 & 0 \end{bmatrix} \quad \text{(zero blocks omitted)},
$$

$$
X = \begin{bmatrix} 1 & -2 & 1 & 1 & 0 \\ -1 & 0 & 0 & 0 & -2 \\ 2 & 0 & 0 & 2 & 0 \\ 1 & 0 & 0 & 0 & 1 \\ 2 & -2 & 0 & 0 & 1 \end{bmatrix} = \begin{bmatrix} X_1 & X_2 & X_0 \end{bmatrix}.
$$

Then

$$
Y = X^{-1} = \begin{bmatrix} 0 & 1 & 0 & 2 & 0 \\ \cdots & \cdots & \cdots & \cdots & \cdots \\ 0 & \frac{1}{2} & 0 & \frac{3}{2} & -\frac{1}{2} \\ 1 & 1 & -\frac{1}{2} & 3 & -1 \\ \cdots & \cdots & \cdots & \cdots & \cdots \\ 0 & -1 & \frac{1}{2} & -2 & 0 \\ 0 & -1 & 0 & -1 & 0 \end{bmatrix} = \begin{bmatrix} Y_1 \\ \cdots \\ Y_2 \\ \cdots \\ Y_0 \end{bmatrix}.
$$

and the projectors E_0, E_1, E_2 are, by (45),

$$
E_0 = X_0 Y_0 = \begin{bmatrix} 0 & -1 & \frac{1}{2} & -2 & 0 \\ 0 & 2 & 0 & 2 & 0 \\ 0 & -2 & 1 & -4 & 0 \\ 0 & -1 & 0 & -1 & 0 \\ 0 & -1 & 0 & -1 & 0 \end{bmatrix}, \quad E_1 = X_1 Y_1 = \begin{bmatrix} 0 & 1 & 0 & 2 & 0 \\ 0 & -1 & 0 & -2 & 0 \\ 0 & 2 & 0 & 4 & 0 \\ 0 & 1 & 0 & 2 & 0 \\ 0 & 2 & 0 & 4 & 0 \end{bmatrix},
$$

$$
E_2 = X_2 Y_2 = \begin{bmatrix} 1 & 0 & -\frac{1}{2} & 0 & 0 \\ 0 & 0 & 0 & 0 & 0 \\ 0 & 0 & 0 & 0 & 0 \\ 0 & 0 & 0 & 0 & 0 \\ 0 & -1 & 0 & -3 & 1 \end{bmatrix}.
$$

One can verify all statements of Theorem 10. In particular, (46) is verified by

$$
A = AE_0 + E_1 + 2E_2 + (A - 2I)E_2.
$$

See also Ex. 30 below.

For any polynomial p, the matrix $p(A)$, can now be computed.

COROLLARY 8. *Let $A \in \mathbb{C}^{n \times n}$ and let p be a polynomial with complex coefficients. Then*

$$
p(A) = \sum_{\lambda \in \lambda(A)} E_\lambda \sum_{k=0}^{\nu(\lambda)-1} \frac{p^{(k)}(\lambda)}{k!} (A - \lambda I_n)^k, \tag{49}
$$

where $\nu(\lambda)$ is the index of λ.

PROOF. Use (0.123) and Ex. 0.77. □

Matrix functions are defined analogously.

DEFINITION 1. For any $A \in \mathbb{C}^{n \times n}$ with spectrum $\lambda(A)$, let $\mathcal{F}(A)$ denote the class of all functions $f : \mathbb{C} \to \mathbb{C}$ which are analytic in some open set containing $\lambda(A)$. For any scalar function $f \in \mathcal{F}(A)$, the *corresponding matrix function* $f(A)$ is defined by

$$f(A) = \sum_{\lambda \in \lambda(A)} E_\lambda \sum_{k=0}^{\nu(\lambda)-1} \frac{f^{(k)}(\lambda)}{k!} (A - \lambda I_n)^k. \tag{50}$$

Note that (46) is a special case, with $f(\lambda) = \lambda$.

Definition (50) is equivalent to

$$f(A) = p(A), \tag{51a}$$

where $p(\lambda)$ is a polynomial satisfying

$$p^{(k)}(\lambda) = f^{(k)}(\lambda), \quad k = 0, 1, \dots, \nu(\lambda) - 1, \quad \text{for each } \lambda \in \lambda(A), \tag{51b}$$

i.e., $p(\lambda)$ is the polynomial interpolating the *spectral data* $\{f^{(k)}(\lambda) : \lambda \in \lambda(A), \ k = 0, \dots, \nu(\lambda) - 1\}$. Given the spectral data, the interpolating polynomial of degree $\leq \sum_{\lambda \in \lambda(a)} \nu(\lambda) - 1$ is unique; see Ex. 35.

If $f : \mathbb{C} \to \mathbb{C}$ is analytic in an open set U and the boundary Γ of U is a closed rectifiable Jordan curve, oriented in the customary way, then, for any $\lambda \in U$, *Cauchy's formula* gives

$$f(\lambda) = \frac{1}{2\pi i} \int_\Gamma \frac{f(z)}{z - \lambda} dz. \tag{52}$$

An analogous expression for matrix functions is given in the following:

THEOREM 11 (Cartan, see [**691**, p. 399]). *Let $A \in \mathbb{C}^{n \times n}$, let $f \in \mathcal{F}(A)$ be analytic in a domain containing the closure of an open set U containing $\lambda(A)$, and let the boundary Γ of U consist of finitely many closed rectifiable Jordan curves, oriented in the customary way. Then $f(A)$ is a Riemann contour integral over Γ,*

$$f(A) = \frac{1}{2\pi i} \int_\Gamma f(z)(zI - A)^{-1} dz. \tag{53}$$

PROOF. Let A have eigenvalues $\{\lambda_1, \dots, \lambda_s\}$ with corresponding indices $\{\nu_1, \dots, \nu_s\}$ and let $\Gamma = \bigcup_{j=1}^s \Gamma_j$ where Γ_j surrounds λ_j and no other eigenvalue. Substituting (57) in (53),

$$\frac{1}{2\pi i} \int_\Gamma f(z)(zI - A)^{-1} dz = \sum_{j=1}^s E_{\lambda_j} \sum_{k=0}^{\nu_j - 1} (A - \lambda_j I)^k \int_{\Gamma_j} \frac{dz}{(z - \lambda_j)^{k+1}}$$

$$= \sum_{j=1}^s E_{\lambda_j} \sum_{k=0}^{\nu_j - 1} (A - \lambda_j I)^k \frac{f^{(k)}(\lambda_j)}{k!},$$

by Cauchy's formula. $\square$

Exercises

EX. 28. (Dunford and Schwartz [**246**, Theorem 3, p. 556]). Let p and q be polynomials and let $A \in \mathbb{C}^{n \times n}$. Then $p(A) = q(A)$ if and only if $p - q$ has a zero of order $\nu(\lambda)$ at each point $\lambda \in \lambda(A)$.

Ex. 29. (Schwerdtfeger [**734**]). Let $A \in \mathbb{C}^{n \times n}$ have distinct eigenvalues $\lambda_1, \ldots, \lambda_s$ with respective indices $\nu_1, \ldots, \nu_s$, and let $m(z)$ be the minimal polynomial of A. Then the principal idempotents E_{λ_i} are polynomials of A,

$$E_{\lambda_i} = g_i(A),$$

where the g_i's are defined from the partial fractions expansion

$$\frac{1}{m(z)} = \sum_{j=1}^{s} \frac{h_j(z)}{(z - \lambda_j)^{\nu_j}}, \quad \text{as} \quad g_i(z) = \frac{h_i(z)m(z)}{(z - \lambda_i)^{\nu_i}}.$$

Ex. 30. For the matrix A of Example 1, the minimal polynomial is $m(z) = z^2(z - 1)(z - 2)^2$ and

$$\frac{1}{m(z)} = -\left(\frac{1 + 2z}{4}\right)\frac{1}{z^2} + \frac{1}{z - 1} + \left(\frac{5 - 2z}{4}\right)\frac{1}{(z - 2)^2}.$$

Therefore,

$$g_0(z) = -\left(\frac{1 + 2z}{4}\right)(z - 1)(z - 2)^2, \quad E_0 = g_0(A)$$
$$= -\tfrac{1}{4}(2A + I)(A - I)(A - 2I)^2,$$
$$g_1(z) = z^2(z - 2)^2, \quad E_1 = g_1(A) = A^2(A - 2I)^2,$$
$$g_2(z) = \left(\frac{5 - 2z}{4}\right)z^2(z - 1), \quad E_2 = g_2(A)$$
$$= -\tfrac{1}{4}(2A - 5I)A^2(A - I).$$

Ex. 31. The exponential function e^{At} is, by (50),

$$e^{At} = \sum_{\lambda \in \lambda(A)} E_\lambda \sum_{k=0}^{\nu(\lambda)-1} \frac{e^{\lambda t}t^k}{k!}(A - \lambda I)^k. \tag{54}$$

For any $A \in \mathbb{C}^{n \times n}$, and $s, t \in \mathbb{C}$:
  (a) $e^{As}e^{At} = e^{A(s+t)}$; and
  (b) e^{At} is nonsingular and $\left(e^{At}\right)^{-1} = e^{-At}$.

Ex. 32. We compute e^{At}, for A of Example 1, in two ways:
Method 1: Definition (50). Given the projectors E_0, E_1, E_2 (computed in Example 1 or Ex. 30),

$$f(A) = f(0)E_0 + f'(0)AE_0 + f(1)E_1 + f(2)E_2 + f'(2)(A - 2I)E_2,$$
$$\text{for any } f \in \mathcal{F}(A).$$
$$\therefore \quad e^{At} = E_0 + tAE_0 + e^t E_1 + e^{2t} E_2 + te^{2t}(A - 2I)E_2.$$

Method 2: Interpolating polynomial. The eigenvalues of At are $\lambda_0 = 0, \lambda_1 = t, \lambda_2 = 2t$, with indices $\nu_0 = 2, \nu_1 = 1, \nu_2 = 2$. The polynomial $p(z) = \alpha + \beta z + \gamma z^2 + \delta z^3 + \epsilon z^4$ interpolating $f(t) = e^{\lambda t}$ is given by

$$f(0) = 1 = \alpha,$$
$$f'(0) = t = \beta,$$
$$f(t) = e^t = \alpha + \beta + \gamma + \delta + \epsilon,$$
$$f(2t) = e^{2t} = \alpha + 2\beta + 4\gamma + 8\delta + 16\epsilon,$$
$$f'(2t) = te^{2t} = \beta + 4\gamma + 12\delta + 32\epsilon,$$

a system that can be solved for $\alpha, \beta, \gamma, \delta, \epsilon$. Then

$$e^{At} = \alpha I + \beta A + \gamma A^2 + \delta A^3 + \epsilon A^4.$$

See also Moler and Van Loan [**573**].

EX. 33. (Dunford and Schwartz [**246**, Lemma 6, p. 568]). Let $A \in \mathbb{C}^{n \times n}$. The *resolvent* of A, denoted $R(\lambda, A)$, is defined by

$$R(\lambda, A) = (\lambda I - A)^{-1}, \quad \forall \, \lambda \notin \lambda(A). \tag{55}$$

It satisfies the following identity, known as the *resolvent equation*:

$$R(\lambda, A) - R(\mu, A) = (\mu - \lambda) R(\lambda, A) R(\mu, A). \tag{56}$$

PROOF. Multiply both sides of

$$(\mu I - A)(\lambda I - A) \left[R(\lambda, A) - R(\mu, A) \right] = (\mu - \lambda) I$$

by $R(\lambda, A) R(\mu, A)$. □

EX. 34. Let $A \in \mathbb{C}^{n \times n}$ have spectrum $\lambda(A) = \{\lambda_1, \lambda_2, \ldots, \lambda_s\}$ and let $z \notin \lambda(A)$. Then

$$(zI - A)^{-1} = \sum_{j=1}^{s} \sum_{k=0}^{\nu(\lambda_j)-1} \frac{(A - \lambda_j I)^k}{(z - \lambda_j)^{k+1}} \, E_{\lambda_j}. \tag{57}$$

EX. 35. (Lagrange interpolation). Let $A \in \mathbb{C}^{n \times n}$ have s distinct eigenvalues $\lambda_1, \ldots, \lambda_s$ with respective indices $\nu_1, \ldots, \nu_s$, $\sum_{i=1}^{s} \nu_i = \nu \leq n$. The interpolating polynomial of degree $\leq \nu - 1$ is obtained as follows:

(a) If $s = 1$, then

$$p(z) = f(\lambda_1) + \sum_{i=1}^{\nu_1 - 1} \frac{f^{(i)}(\lambda_1)}{i!} (z - \lambda_1)^i.$$

(b) If $\nu_1 = \nu_2 = \cdots = \nu_s = 1$, then

$$p(z) = \sum_{i=1}^{s} f(\lambda_i) \frac{\prod\limits_{j \neq i} (z - \lambda_j)}{\prod\limits_{j \neq i} (\lambda_i - \lambda_j)}.$$

(c) In the general case, the interpolating polynomial $p(z)$ is obtained from

$$p(z) = \sum_{i=1}^{s} \prod_{j \neq i} (z - \lambda_i)^{\nu_i} \sum_{k=0}^{\nu_i - 1} \frac{1}{k!} q_j^{(k)}(\lambda_j)(z - \lambda_j)^k,$$

where

$$q_j(z) = \frac{p(z)}{\prod\limits_{i \neq j} (z - \lambda_i)^{\nu_i}}.$$

EX. 36. (Fantappiè [**268**]). The correspondence

$$f(z) \longleftrightarrow f(A) \tag{39}$$

to be useful, must satisfy certain formal conditions. The following four conditions are due to Fantappiè:

I   If $f(z) = k$, then $f(A) = kI$.
II   If $f(z) = z$, then $f(A) = A$.
III   If $f(z) = g(z) + h(z)$, then $f(A) = g(A) + h(A)$.
IV   If $f(z) = g(z)h(z)$, then $f(A) = g(A)h(A)$.

A fifth condition serves to assure consistency of compositions of matrix functions:

V   If $f(z) = h(g(z))$, then $f(A) = h(g(A))$.

Matrix functions given by Definition 1 satisfy the above conditions, see Rinehart [**691**] and Robinson [**696**].

6. Generalized Inverses with Prescribed Range and Null Space

Let $A \in \mathbb{C}^{m \times n}$ and let $A^{(1)}$ be an arbitrary element of $A\{1\}$. Let $R(A) = L$ and let $N(A) = M$. By Lemma 1.1(f), $AA^{(1)}$ and $A^{(1)}A$ are idempotent. By (28) and Theorem 8,

$$AA^{(1)} = P_{L,S}, \quad A^{(1)}A = P_{T,M},$$

where S is some subspace of $\mathbb{C}^m$ complementary to L, and T is some subspace of $\mathbb{C}^n$ complementary to M.

If we choose arbitrary subspaces S and T complementary to L and M, respectively, does there exist a $\{1\}$-inverse $A^{(1)}$ such that $N(AA^{(1)}) = S$ and $R(A^{(1)}A) = T$? The following theorem (parts of which have appeared previously in the works of Robinson [697], Langenhop [499], and Milne [555]) answers the question in the affirmative.

THEOREM 12. *Let* $A \in \mathbb{C}_r^{m \times n}$, $R(A) = L$, $N(A) = M$, $L \oplus S = \mathbb{C}^m$, *and* $M \oplus T = \mathbb{C}^n$. *Then:*

(a) X *is a* $\{1\}$*-inverse of* A *such that* $N(AX) = S$ *and* $R(XA) = T$ *if and only if*

$$AX = P_{L,S}, \quad XA = P_{T,M}. \tag{58}$$

(b) *The general solution of* (58) *is*

$$X = P_{T,M}A^{(1)}P_{L,S} + (I_n - A^{(1)}A)Y(I_m - AA^{(1)}), \tag{59}$$

where $A^{(1)}$ *is a fixed (but arbitrary) element of* $A\{1\}$ *and* Y *is an arbitrary element of* $\mathbb{C}^{n \times m}$.

(c) $A_{T,S}^{(1,2)} = P_{T,M}A^{(1)}P_{L,S}$ *is the unique* $\{1,2\}$*-inverse of* A *having range* T *and null space* S.

PROOF. (a) The "if" part of the statement follows at once from Theorem 8 and Lemma 1(e), the "only if" part from Lemma 1.1(f), (28) and Theorem 8.

(b) By repeated use of Ex. 20, along with (28), we can easily verify that (58) is satisfied by $X = P_{T,M}A^{(1)}P_{L,S}$. The result then follows from Ex. 5.

(c) Since $P_{T,M}A^{(1)}P_{L,S}$ is a $\{1\}$-inverse of A, its rank is at least r by Lemma 1.1(d), while its rank does not exceed r, since rank $P_{L,S} = r$ by (58) and Lemma 1.1(f). Thus it has the same rank as A, and is therefore a $\{1,2\}$-inverse, by Theorem 1.2. It follows from parts (a) and (b) that it has the required range and null space.

On the other hand, a $\{1,2\}$-inverse of A having range T and null space S satisfies (58) and also

$$XAX = X. \tag{1.2}$$

By Ex. 1.28, these three equations have at most one common solution. $\square$

COROLLARY 9. *Under the hypotheses of Theorem 12, let* $A_{T,S}^{(1)}$ *be some* $\{1\}$*-inverse of* A *such that* $R(A_{T,S}^{(1)}A) = T$, $N(AA_{T,S}^{(1)}) = S$, *and let* $A\{1\}_{T,S}$ *denote the class of such* $\{1\}$*-inverses of* A. *Then*

$$A\{1\}_{T,S} = \{A_{T,S}^{(1)} + (I_n - A_{T,S}^{(1)}A)Y(I_m - AA_{T,S}^{(1)}) : Y \in \mathbb{C}^{n \times m}\}. \tag{60}$$

For a subspace L of $\mathbb{C}^m$, a complementary subspace of particular interest is the orthogonal complement, denoted by $L^\perp$, which consists of all vectors in $\mathbb{C}^m$ orthogonal to L. If, in Theorem 12, we take $S = L^\perp$ and

$T = M^\perp$, the class of $\{1\}$-inverses given by (60) is the class of $\{1, 3, 4\}$-inverses and $A_{T,S}^{(1,2)} = A^\dagger$.

The formulas in Theorem 12 generally are not convenient for computational purposes. When this is the case, the following theorem (which extends results due to: Urquhart [824]) may be resorted to:

THEOREM 13. *Let $A \in \mathbb{C}_r^{m \times n}$, $U \in \mathbb{C}^{n \times p}$, $V \in \mathbb{C}^{q \times m}$, and*

$$X = U(VAU)^{(1)}V,$$

where $(VAU)^{(1)}$ is a fixed, but arbitrary element of $(VAU)\{1\}$. Then:

(a) $X \in A\{1\}$ *if and only if* $\operatorname{rank} VAU = r$.
(b) $X \in A\{2\}$ *and $R(X) = R(U)$ if and only if* $\operatorname{rank} VAU = \operatorname{rank} U$.
(c) $X \in A\{2\}$ *and $N(X) = N(V)$ if and only if* $\operatorname{rank} VAU = \operatorname{rank} V$.
(d) $X = A_{R(U),N(V)}^{(1,2)}$ *if and only if* $\operatorname{rank} U = \operatorname{rank} V = \operatorname{rank} VAU = r$.

PROOF. (a) *If*: We have $\operatorname{rank} AU = r$, since

$$r = \operatorname{rank} VAU \le \operatorname{rank} AU \le \operatorname{rank} A = r.$$

Therefore, by Ex. 1.10, $R(AU) = R(A)$ and so $A = AUY$ for some Y. Thus by Ex. 21,

$$AXA = AU(VAU)^{(1)}VAUY = AUY = A.$$

Only if: Since $X \in A\{1\}$,

$$A = AXAXA = AU(VAU)^{(1)}VAU(VAU)^{(1)}VA,$$

and therefore $\operatorname{rank} VAU = \operatorname{rank} A = r$.

(b) *If*: By Ex. 21,

$$XAU = U(VAU)^{(1)}VAU = U,$$

from which it follows that $XAX = X$ and also $\operatorname{rank} X = \operatorname{rank} U$. By Ex. 1.10, $R(X) = R(U)$.

Only if: Since $X \in A\{2\}$,

$$X = XAX = U(VAU)^{(1)}VAU(VAU)^{(1)}V.$$

Therefore

$$\operatorname{rank} X \le \operatorname{rank} VAU \le \operatorname{rank} U = \operatorname{rank} X.$$

(c) Similar to (b).
(d) Follows from (a), (b), and (c).   □

Note that if we require only a $\{1\}$-inverse X such that $R(X) \subset R(U)$ and $N(X) \supset N(V)$, part (a) of the theorem is sufficient.

Theorem 13 can be used to prove the following modified analog of Theorem 12(c) for all $\{2\}$-inverses, and not merely $\{1, 2\}$-inverses.

THEOREM 14. *Let $A \in \mathbb{C}_r^{m \times n}$, let T be a subspace of $\mathbb{C}^n$ of dimension $s \le r$, and let S be a subspace of $\mathbb{C}^m$ of dimension $m - s$. Then, A has a $\{2\}$-inverse X such that $R(X) = T$ and $N(X) = S$ if and only if*

$$AT \oplus S = \mathbb{C}^m, \tag{61}$$

in which case X is unique.

PROOF. *If*: Let the columns of $U \in \mathbb{C}_s^{n \times s}$ be a basis for T and let the columns of $V^* \in \mathbb{C}_s^{m \times s}$ be a basis for $S^\perp$. Then the columns of AU span AT. Since it follows from (61) that dim $AT = s$,

$$\text{rank } AU = s. \tag{62}$$

A further consequence of (61) is

$$AT \cap S = \{\mathbf{0}\}. \tag{63}$$

Moreover, the $s \times s$ matrix VAU is nonsingular (i.e., of rank s) because

$$VAU\mathbf{y} = \mathbf{0} \quad \Longrightarrow \quad AU\mathbf{y} \perp S^\perp \quad \Longrightarrow \quad AU\mathbf{y} \in S$$
$$\Longrightarrow \quad AU\mathbf{y} = \mathbf{0}, \quad \text{(by (63))},$$
$$\Longrightarrow \quad \mathbf{y} = \mathbf{0}, \quad \text{(by (62))}.$$

Therefore, by Theorem 13,

$$X = U(VAU)^{-1}V$$

is a {2}-inverse of A having range T and null space S (see also Stewart [**780**]). *Only if*: Since $A \in X\{1\}$, AX is idempotent by Lemma 1.1(f). Moreover, $AT = R(AX)$ and $S = N(X) = N(AX)$ by (28). Thus (61) follows from Theorem 8. *Proof of uniqueness*: Let X_1, X_2 be {2}-inverses of A having range T and null space S. By Lemma 1.1(f) and (28), X_1A is a projector with range T and AX_2 is a projector with null space S. Thus, by Ex. 20,

$$X_2 = (X_1A)X_2 = X_1(AX_2) = X_1. \qquad \square$$

COROLLARY 10. *Let* $A \in \mathbb{C}_r^{m \times n}$, *let* T *be a subspace of* $\mathbb{C}^n$ *of dimension* r, *and let* S *be a subspace of* $\mathbb{C}^m$ *of dimension* $m - r$. *Then, the following three statements are equivalent:*

(a) $AT \oplus S = \mathbb{C}^m$.

(b) $R(A) \oplus S = \mathbb{C}^m$ *and* $N(A) \oplus T = \mathbb{C}^n$.

(c) *There exists an* $X \in A\{1, 2\}$ *such that* $R(X) = T$ *and* $N(X) = S$.

The set of {2}-inverses of A with range T and null space S is denoted $A\{2\}_{T,S}$.

Exercises

Ex. 37. Show that $A_{T,S}^{(1,2)}$ is the unique matrix X satisfying the three equations

$$AX = P_{L,S}, \quad XA = P_{T,M}, \quad XP_{L,S} = X.$$

(For the Moore–Penrose inverse this was shown by Petryshyn [**641**]. Compare Ex. 1.28.)

Ex. 38. For any given matrix A, $A^\dagger$ is the unique matrix $X \in A\{1,2\}$ such that $R(X) = R(A^*)$ and $N(X) = N(A^*)$.

Ex. 39. Derive the formula of Mitra [**559**] and Zlobec [**891**],

$$A^\dagger = A^*YA^*,$$

where Y is an arbitrary element of $(A^*AA^*)\{1\}$.

Ex. 40. Derive the formula of Decell [**224**],

$$A^\dagger = A^*XA^*YA^*,$$

where X and Y are any {1}-inverses of AA^* and A^*A, respectively.

EX. 41. Penrose [635] showed that the Moore–Penrose inverse of a product of two Hermitian idempotent matrices is idempotent. Prove this, using Zlobec's formula (Ex. 39).

EX. 42. Let A be the matrix of (0.80) and let

$$S = R\left(\begin{bmatrix} 0 \\ 0 \\ 1 \end{bmatrix}\right), \quad T = R\left(\begin{bmatrix} 0 & 0 \\ 0 & 0 \\ 1 & 0 \\ 0 & 1 \\ 0 & 0 \\ 0 & 0 \end{bmatrix}\right).$$

Calculate $A_{T,S}^{(1,2)}$.

EX. 43. If E is idempotent and the columns of F and G^* are bases for $R(E)$ and $R(E^*)$, respectively, show that $E = F(GF)^{-1}G$.

EX. 44. If A is square and $A = FG$ is a full-rank factorization, show that A has a $\{1,2\}$-inverse X with $R(X) = R(A)$ and $N(X) = N(A)$ if and only if GF is nonsingular, in which case $X = F(GF)^{-2}G$ (Cline [201]).

7. Orthogonal Projections and Orthogonal Projectors

Given a vector $\mathbf{x} \in \mathbb{C}^n$ and a subspace L of $\mathbb{C}^n$, there is in L a unique vector $\mathbf{u_x}$ that is "closest" to $\mathbf{x}$ in the sense that the "distance" $\|\mathbf{x} - \mathbf{u}\|$ is smaller for $\mathbf{u} = \mathbf{u_x}$ than for any other $\mathbf{u} \in L$. Here, $\|\mathbf{v}\|$ denotes the *Euclidean norm* of the vector $\mathbf{v}$,

$$\|\mathbf{v}\| = +\sqrt{\langle \mathbf{v}, \mathbf{v} \rangle} = +\sqrt{\mathbf{v}^*\mathbf{v}} = +\sqrt{\sum_{j=1}^{n} |v_j|^2},$$

where $\langle \mathbf{v}, \mathbf{w} \rangle$ denotes the *standard inner product*, defined for $\mathbf{v}, \mathbf{w} \in \mathbb{C}^n$ by

$$\langle \mathbf{v}, \mathbf{w} \rangle = \mathbf{w}^*\mathbf{v} = \sum_{j=1}^{n} \overline{w}_j v_j.$$

Not surprisingly, the vector $\mathbf{u_x}$ that is "closest" to $\mathbf{x}$ of all vectors in L is uniquely characterized (see Ex. 47) by the fact that $\mathbf{x} - \mathbf{u_x}$ is orthogonal to $\mathbf{u_x}$, which we shall denote by

$$\mathbf{x} - \mathbf{u_x} \perp \mathbf{u_x}.$$

We shall therefore call the "closest" vector $\mathbf{u_x}$ the *orthogonal projection* of $\mathbf{x}$ on L. The transformation that carries each $\mathbf{x} \in \mathbb{C}^n$ into its orthogonal projection on L we shall denote by P_L and shall call the *orthogonal projector* on L. Comparison with the earlier definition of the projector on L along M (see Section 4) shows that the orthogonal projector on L is the same as the projector on L along $L^\perp$. (As previously noted, some writers call the orthogonal projector on L simply the projector on L.)

Being a particular case of the more general projector, the orthogonal projector is representable by a square matrix which, in this case, is not only idempotent but also Hermitian.

In order to prove this, we shall need the relation

$$N(A) = R(A^*)^\perp, \tag{0.26}$$

which, in fact, arises frequently in the study of generalized inverses.

Let L and M be complementary orthogonal subspaces of $\mathbb{C}^n$ and consider the matrix $P_{L,M}^*$. By Lemma 1(a), it is idempotent and therefore a projector, by Theorem 8. By the use of (0.26) and its dual

$$N(A^*) = R(A)^{\perp} \tag{0.27}$$

it is readily found that

$$R(P_{L,M}^*) = M^{\perp}, \quad N(P_{L,M}^*) = L^{\perp}.$$

Thus, by Theorem 8,

$$P_{L,M}^* = P_{M^{\perp},L^{\perp}} \tag{64}$$

from which the next lemma follows easily.

LEMMA 3. *Let* $\mathbb{C}^n = L \oplus M$. *Then* $M = L^{\perp}$ *if and only if* $P_{L,M}$ *is Hermitian.*

Just as there is a one-to-one correspondence between projectors and idempotent matrices, Lemma 3 shows that there is a one-to-one correspondence between orthogonal projectors and Hermitian idempotents. Matrices of the latter class have many striking properties, some of which are noted in the remainder of this section (including the Exercises).

For any subspace L for which a basis is available, it is easy to construct the matrix P_L. The basis must first be orthonormalized (e.g., by Gram–Schmidt orthogonalization). Let $\{\mathbf{x}_1, \mathbf{x}_2, \ldots, \mathbf{x}_l\}$ be an o.n. basis for L. Then

$$P_L = \sum_{j=1}^{l} \mathbf{x}_j \mathbf{x}_j^*. \tag{65}$$

The reader should verify that RHS(65) is the orthogonal projector on L and that (27) reduces to (65) if $M = L^{\perp}$ and the basis is o.n..

In the preceding section diagonable matrices were studied in relation to projectors. The same relations will now be shown to hold between normal matrices (a subclass of diagonable matrices) and orthogonal projectors. This constitutes the *spectral theory for normal matrices*. We recall that a square matrix A is called *normal* if it commutes with its conjugate transpose

$$AA^* = A^*A.$$

It is well known that every normal matrix is diagonable. A normal matrix A also has the property (see Ex. 54) that the eigenvalues of A^* are the conjugates of those of A, and every eigenvector of A associated with the eigenvalue λ is also an eigenvector of A^* associated with the eigenvalue $\bar{\lambda}$.

The following spectral theorem relates normal matrices to orthogonal projectors, in the same way that diagonable matrices and projectors are related in Theorem 9.

THEOREM 15 (Spectral Theorem for Normal Matrices). *Let* $A \in \mathbb{C}^{n \times n}$ *with k distinct eigenvalues* $\lambda_1, \lambda_2, \ldots, \lambda_k$. *Then A is normal if and only if* *there exist orthogonal projectors $E_1, E_2, \ldots, E_k$ such that*

$$E_i E_j = O, \quad if \ i \neq j, \tag{66}$$

$$I_n = \sum_{i=1}^{k} E_i, \tag{67}$$

$$A = \sum_{i=1}^{k} \lambda_i E_i. \tag{68}$$

PROOF. *If*: Let A be given by (68) where the principal idempotents are Hermitian. Then

$$AA^* = \left(\sum_{i=1}^{k} \lambda_i E_i \right) \left(\sum_{j=1}^{k} \bar{\lambda}_j E_j \right)$$

$$= \sum_{i=1}^{k} |\lambda_i|^2 E_i = A^* A.$$

Only if: Since A is normal, it is diagonable; let $E_1, E_2, \ldots, E_k$ be its principal idempotents. We must show that they are Hermitian. By Ex. 54, $R(E_i)$, the eigenspace of A associated with the eigenvalue λ_i is the same as the eigenspace of A^* associated with $\bar{\lambda}_i$. Because of (66), the null spaces of corresponding principal idempotents of A and A^* are also the same (for a given $i = h$, $N(E_h)$ is the direct sum of the eigenspaces $R(E_i)$ for all $i \neq h$, i.e.,

$$N(E_h) = \sum_{\substack{i=1 \\ i \neq h}}^{k} \oplus R(E_i) \quad (h \in \overline{1,k}).$$

Therefore, A and A^* have the same principal idempotents, by Theorem 8. Consequently,

$$A^* = \sum_{i=1}^{k} \bar{\lambda}_i E_i,$$

by Theorem 9. But taking conjugate transposes in (68) gives

$$A^* = \sum_{i=1}^{k} \bar{\lambda}_i E_i^*,$$

and it is easily seen that the idempotents E_i^* satisfy (66) and (67). Since the spectral decomposition is unique by Ex. 24, we must have

$$E_i = E_i^*, \quad i \in \overline{1,k}. \qquad \square$$

Exercises

EX. 45. Let A, B be matrices and let $\mathbf{a}, \mathbf{b}$ be vectors of appropriate dimensions. Then

$$A\mathbf{x} = \mathbf{a} \quad \Longrightarrow \quad B\mathbf{x} = \mathbf{b}$$

if and only if there is a matrix Y such that

$$B = YA, \quad \mathbf{b} = Y\mathbf{a}.$$

PROOF. *If*: Obvious.
   Only if: The general solution of $A\mathbf{x} = \mathbf{a}$ is

$$\mathbf{x} = A^\dagger \mathbf{a} + P_{N(A)} \mathbf{y}, \quad \mathbf{y} \text{ arbitrary.}$$

Substituting in $B\mathbf{x} = \mathbf{b}$ we get

$$BA^\dagger \mathbf{a} + BP_{N(A)}\mathbf{y} = \mathbf{b}, \quad \mathbf{y} \text{ arbitrary.}$$

Therefore $BP_{N(A)} = O$, i.e., $B = YA$ for some Y and $\mathbf{b} = Y\mathbf{a}$ follows. $\qquad\square$

EX. 46. *Orthogonal subspaces* or the *Pythagorean theorem.* Let Y and Z be subspaces of $\mathbb{C}^n$. Then $Y \perp Z$ if and only if

$$\|\mathbf{y} + \mathbf{z}\|^2 = \|\mathbf{y}\|^2 + \|\mathbf{z}\|^2, \quad \text{for all } \mathbf{y} \in Y, \mathbf{z} \in Z. \tag{69}$$

PROOF. *If*: Let $\mathbf{y} \in Y, \mathbf{z} \in Z$. Then (69) implies that

$$\langle \mathbf{y}, \mathbf{y} \rangle + \langle \mathbf{z}, \mathbf{z} \rangle = \|\mathbf{y}\|^2 + \|\mathbf{z}\|^2 = \|\mathbf{y} + \mathbf{z}\|^2$$
$$= \langle \mathbf{y} + \mathbf{z}, \mathbf{y} + \mathbf{z} \rangle = \langle \mathbf{y}, \mathbf{y} \rangle + \langle \mathbf{z}, \mathbf{z} \rangle + \langle \mathbf{y}, \mathbf{z} \rangle + \langle \mathbf{z}, \mathbf{y} \rangle$$

and, therefore,

$$\langle \mathbf{y}, \mathbf{z} \rangle + \langle \mathbf{z}, \mathbf{y} \rangle = 0. \tag{70}$$

Now, since Z is a subspace, $i\mathbf{z} \in Z$, and replacing $\mathbf{z}$ by $i\mathbf{z}$ in (70) gives

$$0 = \langle \mathbf{y}, i\mathbf{z} \rangle + \langle i\mathbf{z}, \mathbf{y} \rangle = i\langle \mathbf{y}, \mathbf{z} \rangle - i\langle \mathbf{z}, \mathbf{y} \rangle. \tag{71}$$

(Here we have used the fact that $\langle \alpha\mathbf{v}, \mathbf{w} \rangle = \alpha\langle \mathbf{v}, \mathbf{w} \rangle$ and $\langle \mathbf{v}, \beta\mathbf{w} \rangle = \bar{\beta}\langle \mathbf{v}, \mathbf{w} \rangle$.) It follows from (71) that

$$\langle \mathbf{y}, \mathbf{y} \rangle - \langle \mathbf{z}, \mathbf{z} \rangle = 0,$$

which, in conjunction with (70), gives

$$\langle \mathbf{y}, \mathbf{y} \rangle = \langle \mathbf{z}, \mathbf{z} \rangle = 0,$$

i.e., $\mathbf{y} \perp \mathbf{z}$.

Only if: Let $Y \perp Z$. Then, for arbitrary $\mathbf{y} \in Y, \mathbf{z} \in Z$,

$$\|\mathbf{y} + \mathbf{z}\|^2 = \langle \mathbf{y} + \mathbf{z}, \mathbf{y} + \mathbf{z} \rangle$$
$$= \langle \mathbf{y}, \mathbf{y} \rangle + \langle \mathbf{z}, \mathbf{z} \rangle, \quad \text{since } \langle \mathbf{y}, \mathbf{z} \rangle = \langle \mathbf{z}, \mathbf{y} \rangle = 0,$$
$$= \|\mathbf{y}\|^2 + \|\mathbf{z}\|^2. \qquad\square$$

EX. 47. *Orthogonal projections.* Let L be a subspace of $\mathbb{C}^n$. Then, for every $\mathbf{x} \in \mathbb{C}^n$, there is a unique vector $\mathbf{u_x}$ in L such that, for all $\mathbf{u} \in L$ different from $\mathbf{u_x}$,

$$\|\mathbf{x} - \mathbf{u_x}\| < \|\mathbf{x} - \mathbf{u}\|.$$

Among the vectors $\mathbf{u} \in L$, $\mathbf{u_x}$ is uniquely characterized by the fact that

$$\mathbf{x} - \mathbf{u_x} \perp \mathbf{u_x}.$$

PROOF. Let $\mathbf{x} \in \mathbb{C}^n$. Since L and $L^\perp$ are complementary subspaces, there exist uniquely determined vectors $\mathbf{x}_1 \in L$, $\mathbf{x}_2 \in L^\perp$ such that

$$\mathbf{x} = \mathbf{x}_1 + \mathbf{x}_2. \tag{72}$$

Therefore, for arbitrary $\mathbf{u} \in L$,

$$\|\mathbf{x} - \mathbf{u}\|^2 = \|\mathbf{x}_1 + \mathbf{x}_2 - \mathbf{u}\|^2$$
$$= \|\mathbf{x}_1 - \mathbf{u}\|^2 + \|\mathbf{x}_2\|^2, \tag{73}$$

by Ex. 46, since $\mathbf{x}_1 - \mathbf{u} \in L$, $\mathbf{x}_2 \in L^\perp$. Consequently, there is a unique $\mathbf{u} \in L$, namely $\mathbf{u_x} = \mathbf{x}_1$, for which (73) is smallest.

By the uniqueness of the decomposition (72), $\mathbf{u_x} = \mathbf{x}_1$ is the only vector $\mathbf{u} \in L$ satisfying

$$\mathbf{x} - \mathbf{u} \perp \mathbf{u}. \qquad\square$$

Ex. 48. Let L be a subspace of $\mathbb{C}^n$, let $V \in \mathbb{C}^{n \times r}$ be a matrix whose columns $\{v_1, \ldots, v_r\}$ are a basis of L, and let $G(v_1, \ldots, v_r)$ be the corresponding Gram matrix, see § 0.5.2. Then, for any $x \in \mathbb{C}^n$,

$$P_L x = \sum_{j=1}^{r} \xi_j \, v_j \qquad (74)$$

where

$$\xi_j = \frac{\det V^* V[j \leftarrow x]}{\det G(v_1, \ldots, v_r)}, \quad j \in \overline{1,r}. \qquad (75)$$

PROOF. Since $(x - P_L x) \perp L$ we get from (74) the system

$$\langle x - \sum_{j=1}^{r} \xi_j \, v_j, v_k \rangle = 0, \quad k \in \overline{1,r},$$

or

$$\sum_{j=1}^{r} \xi_j \, \langle v_j, v_k \rangle = \langle x, v_k \rangle, \quad k \in \overline{1,r},$$

whose solution, by Cramer's rule, is (75).   □

Ex. 49. Let L and $\{v_1, \ldots, v_r\}$ be as in Ex. 48. Then the Euclidean norm of $P_{L^\perp} x$ is given by

$$\|P_{L^\perp} x\|^2 = \frac{\det G(v_1, \ldots, v_r, x)}{\det G(v_1, \ldots, v_r)}. \qquad (76)$$

For proof see Ex. 51 below, or Gantmacher [**296**, Vol. I, p. 250].

Ex. 50. Simplify the results of Exs. 48–49 if the basis $\{v_1, \ldots, v_r\}$ is o.n.

Ex. 51. *Orthogonal projections.* Let L be a subspace of $\mathbb{R}^n$ of dimension r and let L be spanned by vectors $\{v_1, \ldots, v_k\}$. Let $x \in \mathbb{R}^n$ be written as $x = x_L + x_{L^\perp}$ where $x_L \in L$ and $x_{L^\perp}$ is orthogonal to L. Then

$$\|x_{L^\perp}\| = \frac{\mathrm{vol}_{r+1}(v_1, \ldots, v_k, x)}{\mathrm{vol}_r(v_1, \ldots, v_k)}, \qquad (77)$$

where $(v_1, \ldots, v_k)$ is the matrix with v_j as columns.
PROOF. If $x \in L$, then both sides of (77) are zero. If $x \notin L$, then

$$\mathrm{vol}_{r+1}^2(v_1, \ldots, v_k, x) = \mathrm{vol}_{r+1}^2(v_1, \ldots, v_k, x_{L^\perp}),$$

$$\text{by properties of determinants,}$$

$$= \mathrm{vol}_r^2(v_1, \ldots, v_k) \, \mathrm{vol}_1^2(v_{L^\perp}), \quad \text{by Ex. 0.67,}$$

which completes the proof since $\mathrm{vol}_1^2(x_{L^\perp}) = \|x_{L^\perp}\|^2$.   □
Note that (76) is a special case of (77).

Ex. 52. Let $x \in \mathbb{C}^n$ and let L be an arbitrary subspace of $\mathbb{C}^n$. Then

$$\|P_L x\| \leq \|x\|, \qquad (78)$$

with equality if and only if $x \in L$. See also Ex. 67.
PROOF. We have

$$x = P_L x + (I - P_L)x = P_L x + P_{L^\perp} x,$$

by Ex. 23. Then, by Ex. 46,

$$\|x\|^2 = \|P_L x\|^2 + \|P_{L^\perp} x\|^2,$$

from which (78) follows.
Equality holds in (78) if and only if $P_{L^\perp} x = 0$, which is equivalent to $x \in L$.   □

EX. 53. Let A be a square singular matrix, let $\{u_1, u_2, \ldots, u_n\}$ and $\{x_1, x_2, \ldots, x_n\}$ be o.n. bases of $N(A^*)$ and $N(A)$, respectively, and let $\{\alpha_1, \alpha_2, \ldots, \alpha_n\}$ be nonzero scalars. Then the matrix

$$A_0 = A + \sum_{i=1}^{n} \alpha_i\, u_i x_i^*$$

is nonsingular and its inverse is

$$A_0^{-1} = A^\dagger + \sum_{i=1}^{n} \frac{1}{\alpha_i}\, x_i u_i^*.$$

PROOF. Let X denote the expression given for A_0^{-1}. Then, from $x_i^* x_j = \delta_{ij}$ $(i, j \in \overline{1, n})$, it follows that

$$
\begin{aligned}
A_0 X &= AA^\dagger + \sum_{i=1}^{n} x_i\, x_i^* \\
&= AA^\dagger + P_{N(A^*)}, \quad \text{by (65)}, \\
&= AA^\dagger + (I_n - AA^\dagger), \quad \text{by Lemma 1(g)}, \\
&= I_n.
\end{aligned}
$$

Therefore, A_0 is nonsingular and $X = A_0^{-1}$.   □

EX. 54. If A is normal, $Ax = \lambda x$ if and only if $A^* x = \bar{\lambda} x$.

EX. 55. If L is a subspace of $\mathbb{C}^n$ and the columns of F are a basis for L, show that

$$P_L = FF^\dagger = F(F^* F)^{-1} F^*.$$

(This may be simpler computationally than orthonormalizing the basis and using (65).)

EX. 56. Let L be a subspace of $\mathbb{C}^n$. Then

$$P_{L^\perp} = I_n - P_L.$$

(See Ex. 23.)

EX. 57. Let $A \in \mathbb{C}^{m \times n}$, $X \in \mathbb{C}^{n \times m}$. Then $X \in A\{2\}$ if and only if it is of the form

$$X = (EAF)^\dagger,$$

where E and F are suitable Hermitian idempotents (Greville [**330**]).

PROOF. *If*: By Ex. 38,

$$R((EAF)^\dagger) \subset R(F), \quad N((EAF)^\dagger) \supset N(E).$$

Therefore, by Ex. 20,

$$X = (EAF)^\dagger = F(EAF)^\dagger = (EAF)^\dagger E.$$

Consequently,

$$XAX = (EAF)^\dagger EAF(EAF)^\dagger = (EAF)^\dagger = X.$$

Only if: By Theorem 12(c) and Ex. 38,

$$X^\dagger = P_{R(X^*)} A P_{R(X)},$$

and, therefore, by Ex. 1.18,

$$X = \left(P_{R(X^*)} A P_{R(X)} \right)^\dagger. \tag{79}$$

□

REMARK. Equation (79) states that if $X \in A\{2\}$, then X is the Moore–Penrose inverse of a modification of A obtained by projecting its columns on $R(X^*)$ and its rows on $R(X)$.

EX. 58. It follows from Exs. 37 and 1.28 that, for arbitrary A, $A^\dagger$ is the unique matrix X satisfying

$$AX = P_{R(A)}, \quad XA = P_{R(A^*)}, \quad XAX = X.$$

EX. 59. By means of Exs. 58 and 20, derive (79) directly from $XAX = X$ without using Theorem 12(c).

EX. 60. Prove the following amplification of Penrose's result stated in Ex. 41: A square matrix E is idempotent if and only if it can be expressed in the form

$$E = (FG)^\dagger$$

where F and G are Hermitian idempotents. (*Hint*: Use Ex. 17.)

In particular, derive the formula (Greville [**330**])

$$P_{L,M} = (P_{M^\perp} P_L)^\dagger = ((I - P_M)P_L)^\dagger. \tag{80}$$

EX. 61. Let S and T be subspaces of $\mathbb{C}^m$ and $\mathbb{C}^n$, respectively, such that

$$AT \oplus S = \mathbb{C}^m,$$

and let $A^{(2)}_{T,S}$ denote the unique $\{2\}$-inverse of A having range T and null space S (see Theorem 14). Then

$$A^{(2)}_{T,S} = (P_{S^\perp} A P_T)^\dagger.$$

EX. 62. Show that $P_L + P_M$ is an orthogonal projector if and only if $L \perp M$, in which case

$$P_L + P_M = P_{L+M}.$$

EX. 63. Show that $P_L P_M$ is an orthogonal projector if and only if P_L and P_M commute, in which case

$$P_L P_M = P_{L \cap M}.$$

EX. 64. Show that $L = L \cap M \oplus L \cap M^\perp$ if and only if P_L and P_M commute.

EX. 65. For any matrix $A \in \mathbb{C}^{n \times n}$ we denote by $A \succcurlyeq O$ the fact

$$\langle \mathbf{x}, A\mathbf{x} \rangle \geq 0, \quad \text{for all } \mathbf{x} \in \mathbb{C}^n. \tag{81}$$

For any two matrices $A, B \in \mathbb{C}^{n \times n}$, $A \succcurlyeq B$ denotes that $(A - B) \succcurlyeq O$. This is called the *Löwner ordering* on $\mathbb{C}^{n \times n}$. It is transitive ($A \succcurlyeq B, B \succcurlyeq C \implies A \succcurlyeq C$) and is antisymmetric ($A \succcurlyeq B, B \succcurlyeq A \implies A = B$) if A, B are Hermitian, see Chipman [**186**].

Similarly, we denote by $A \succ O$ the fact that

$$\langle \mathbf{x}, A\mathbf{x} \rangle > 0, \quad \text{for all } \mathbf{0} \neq \mathbf{x} \in \mathbb{C}^n. \tag{82}$$

A Hermitian matrix $H \in \mathbb{C}^{n \times n}$ is:

  positive semidefinite (PSD) if $H \succcurlyeq O$; and
  positive definite (PD) if $H \succ O$.

The set of $n \times n$ PSD [PD] matrices is denoted by PSD_n [PD_n].

If $A, B \in \mathbb{C}^{n \times n}$ are Hermitian, then $A \succcurlyeq B$ is equivalent to $A - B = L^* L$ for some $L \in \mathbb{C}^{n \times n}$ [**379**, p. 223]. See also Ex. 79 below.

EX. 66. Let P_L and P_M be orthogonal projectors on the subspaces L and M of $\mathbb{C}^n$, respectively. Then the following statements are equivalent:

(a) $P_L - P_M$ is an orthogonal projector;   (b) $P_L \succcurlyeq P_M$;
(c) $\|P_L \mathbf{x}\| \geq \|P_M \mathbf{x}\|$ for all $\mathbf{x} \in \mathbb{C}^n$;   (d) $M \subset L$;
(e) $P_L P_M = P_M$;   (f) $P_M P_L = P_M$.

Ex. 67. Let $P \in \mathbb{C}^{n \times n}$ be a projector. Then P is an orthogonal projector if and only if

$$\|P\mathbf{x}\| \leq \|\mathbf{x}\|, \quad \text{for all } \mathbf{x} \in \mathbb{C}^n. \tag{83}$$

PROOF. P is an orthogonal projector if and only if $I - P$ is an orthogonal projector. By the equivalence of statements (a) and (c) in Ex. 66, $I - P$ is an orthogonal projector if and only if (83) holds.   □

Note that for any non-Hermitian idempotent P (i.e., for any projector P which is not an orthogonal projector) there is by this exercise a vector $\mathbf{x}$ whose length is increased when multiplied by P, i.e., $\|P\mathbf{x}\| > \|\mathbf{x}\|$. For $P = \left(\begin{smallmatrix} 1 & 1 \\ 0 & 0 \end{smallmatrix}\right)$ such a vector is $\mathbf{x} = \left(\begin{smallmatrix} 1 \\ 1 \end{smallmatrix}\right)$.

Ex. 68. Let $P \in \mathbb{C}^{n \times n}$. Then P is an orthogonal projector if and only if

$$P = P^* P.$$

Ex. 69. It may be asked to what extent the results of Exs. 62–64 carry over to general projectors. This question is explored in this and the two following exercises. Let

$$\mathbb{C}^n = L \oplus M = Q \oplus S.$$

Then show that $P_{L,M} + P_{Q,S}$ is a projector if and only if $M \supset Q$ and $S \supset L$, in which case

$$P_{L,M} + P_{Q,S} = P_{L+Q, M \cap S}.$$

SOLUTION. Let $P_1 = P_{L,M}$, $P_2 = P_{Q,S}$. Then

$$(P_1 + P_2)^2 = P_1 + P_2 + P_1 P_2 + P_2 P_1.$$

Therefore, $P_1 + P_2$ is a projector if and only if

$$P_1 P_2 + P_2 P_1 = O. \tag{84}$$

Now, if $M \supset Q$ and $S \supset L$, each term of LHS(84) is O.

On the other hand, if (84) holds, multiplication by P_1 on the left and on the right, respectively, gives

$$P_1 P_2 + P_1 P_2 P_1 = O = P_1 P_2 P_1 + P_2 P_1.$$

Subtraction then yields

$$P_1 P_2 - P_2 P_1 = O, \tag{85}$$

and (84) and (85) together imply

$$P_1 P_2 = P_2 P_1 = O,$$

from which it follows by Lemma 1(e) that $M \supset Q$ and $S \supset L$. It is then fairly easy to show that

$$P_1 + P_2 = P_{L+Q, M \cap S}.$$

Ex. 70. With L, M, Q, S as in Ex. 69 show that if $P_{L,M}$ and $P_{Q,S}$ commute, then

$$P_{L,M} P_{Q,S} = P_{Q,S} P_{L,M} = P_{L \cap Q, M+S}. \tag{86}$$

EX. 71. If only one of the products in (86) is equal to the projector on the right, it does not necessarily follow that the other product is the same. Instead we have the following result: With L, M, Q, S as in Ex. 69, $P_{L,M} P_{Q,S} = P_{L \cap Q, M+S}$ if and only if $Q = L \cap Q \oplus M \cap Q$. Similarly, $P_{Q,S} P_{L,M} = P_{L \cap Q, M+S}$ if and only if $L = L \cap Q \oplus L \cap S$.

PROOF. Since $L \cap M = \{0\}$, $(L \cap Q) \cap (M \cap Q) = \{0\}$. Therefore $L \cap Q + M \cap Q = L \cap Q \oplus M \cap Q$. Since $M + S \supset M + Q$ and $L + S \supset L \cap Q$, Ex. 69 gives

$$P_{L \cap Q, M+S} + P_{M \cap Q, L+S} = P_{T,U},$$

where $T = L \cap Q \oplus M \cap Q$, $U = (L + S) \cap (M + S)$. Clearly $Q \supset T$ and $U \supset S$. Multiplying on the left by $P_{L,M}$ gives

$$P_{L,M} P_{T,U} = P_{L \cap Q, M+S}. \tag{87}$$

Thus, if $T = Q$, we have $U = S$, and

$$P_{L,M} P_{Q,S} = P_{L \cap Q, M+S}. \tag{88}$$

On the other hand, if (88) holds, (87) and (88) give

$$P_{Q,S} = P_{T,U} + H, \tag{89}$$

where $P_{L,M} H = O$. This implies $R(H) \subset M$. Also, since $T \subset Q$, (89) implies $R(H) \subset Q$ and, therefore, $R(H) \subset M \cap Q$. Consequently, $R(H) \subset T$ and therefore (89) gives $P_{T,U} P_{Q,S} = P_{Q,S}$. This implies rank $P_{Q,S} \leq$ rank $P_{T,U}$. Since $Q \supset T$ it follows that $T = Q$. This proves the first statement and the proof of the second statement is similar. □

EX. 72. The characterization of $A\{2,3,4\}$ was postponed until orthogonal projectors had been studied. This will now be dealt with in three stages in this exercise and in Exs. 73 and 75. If E is Hermitian idempotent show that $X \in E\{2,3,4\}$ if and only if X is Hermitian idempotent and $R(X) \subset R(E)$.

PROOF. *If*: Since $R(X) \subset R(E)$, $EX = X$ by Lemma 1(e), and taking conjugate transposes gives $XE = X$. Since X is Hermitian, EX and XE are Hermitian. Finally, $XEX = X^2 = X$, since X is idempotent. Thus, $X \in E\{2,3,4\}$.

Only if: Let $X \in E\{2,3,4\}$. Then $X = XEX = EX^*X$. Therefore $R(X) \subset R(E)$. Then $EX = X$ by Lemma 1(e). But EX is Hermitian idempotent, since $X \in E\{2,3\}$. Therefore X is Hermitian idempotent. □

EX. 73. Let H be Hermitian and PSD, and let

$$H = \sum_{i=1}^{k} \lambda_i E_i. \tag{90}$$

be its spectral decomposition as in (68), with orthogonal projectors as the principal idempotents. Then $X \in H\{2,3,4\}$ if and only if

$$X = \sum_{i=1}^{k} \lambda_i^\dagger F_i, \tag{91}$$

where, for each i, $F_i \in E_i\{2,3,4\}$.

PROOF. *If*: Since E_i is Hermitian idempotent, $R(F_i) \subset R(E_i)$ by Ex. 72. Therefore (29a) gives

$$E_i F_j = F_j E_i = O \quad (i \neq j), \tag{92}$$

and, by Lemma 1(e),

$$E_i F_i = F_i E_i = F_i \quad (i \in \overline{1,k}).$$

Consequently,

$$HX = \sum_{\substack{i=1 \\ \lambda_i \neq 0}}^{k} F_i = XH.$$

Since each F_i is Hermitian by Ex. 72, $HX = XH$ is Hermitian. Now,

$$F_i F_j = F_i E_j F_j = O \quad (i \neq j),$$

by (92) and, therefore,

$$XHX = \sum_{i=1}^{k} \lambda_i^\dagger F_i^2 = X$$

by (91), since each F_i is idempotent.

Only if: Let $X \in H\{2,3,4\}$. Then, by (29b),

$$X = IXI = \sum_{i=1}^{k} \sum_{j=1}^{k} E_i X E_j. \tag{93}$$

Now, (90) gives

$$HX = \sum_{i=1}^{k} \lambda_i E_i X = \sum_{i=1}^{k} \lambda_i X^* E_i, \tag{94}$$

since $HX = X^* H$. Similarly,

$$XH = \sum_{i=1}^{k} \lambda_i X E_i = \sum_{i=1}^{k} \lambda_i E_i X^*. \tag{95}$$

Multiplying by E_s on the left and by E_t on the right in both (94) and (95) and making use of (29a) and the idempotency of E_s and E_t gives

$$\lambda_s E_s X E_t = \lambda_t E_s X^* E_t, \tag{96}$$

$$\lambda_t E_s X E_t = \lambda_s E_s X^* E_t \quad (s, t \in \overline{1,k}). \tag{97}$$

Adding and subtracting (96) and (97) gives

$$(\lambda_s + \lambda_t) E_s X E_t = (\lambda_s + \lambda_t) E_s X^* E_t, \tag{98}$$

$$(\lambda_s - \lambda_t) E_s X E_t = -(\lambda_s - \lambda_t) E_s X^* E_t. \tag{99}$$

The λ_i are distinct, and are also nonnegative because H is Hermitian and PSD. Thus, if $s \neq t$, neither of the quantities $\lambda_s + \lambda_t$ and $\lambda_s - \lambda_t$ vanishes. Therefore, (98) and (99) give

$$E_s X E_t = E_s X^* E_t = -E_s X E_t = O \quad (s \neq t). \tag{100}$$

Consequently, (93) reduces to

$$X = \sum_{i=1}^{k} E_i X E_i. \tag{101}$$

Now, (90) gives

$$X = XHX = \sum_{i=1}^{k} \lambda_i X E_i X,$$

and, therefore, by (100),

$$E_s X E_s = \lambda_s E_s X E_s X E_s = \lambda_s (E_s X E_s)^2, \tag{102}$$

from which it follows that $E_s X E_s = O$ if $\lambda_s = 0$. Now, take

$$F_i = \lambda_i E_i X E_i \quad (i \in \overline{1, k}). \tag{103}$$

Then (101) becomes (91), and we have only to show that $F_i \in E_i\{2, 3, 4\}$. This is trivially true for that i, if any, such that $\lambda_i = 0$. For other i, we deduce from (96) that it is idempotent. Finally, (103) gives $R(F_i) \subset R(E_i)$, and the desired conclusion follows from Ex. 72. □

Ex. 74. Prove the following corollary of Ex. 73. If H is Hermitian PSD and $X \in H\{2, 3, 4\}$, then X is Hermitian PSD, and every nonzero eigenvalue of X is the reciprocal of an eigenvalue of H.

Ex. 75. For every $A \in \mathbb{C}^{m \times n}$,

$$A\{2, 3, 4\} = \{Y A^* : Y \in (A^* A)\{2, 3, 4\}\}.$$

Ex. 76. $A\{2, 3, 4\}$ is a finite set if and only if the nonzero eigenvalues of $A^* A$ are distinct (i.e., each eigenspace associated with a nonzero eigenvalue of $A^* A$ is of dimension 1). If this is the case and if there are k such eigenvalues, $A\{2, 3, 4\}$ contains exactly 2^k elements.

Ex. 77. Show that the matrix

$$A = \tfrac{1}{10} \begin{bmatrix} 9 - 3i & 12 - 4i & 10 - 10i \\ 3 - 3i & 4 - 4i & 0 \\ 6 + 6i & 8 + 8i & 0 \\ 6 & 8 & 0 \end{bmatrix}$$

has exactly four $\{2, 3, 4\}$-inverses, namely,

$$X_1 = A^\dagger = \tfrac{1}{70} \begin{bmatrix} 0 & 6 + 6i & 12 - 12i & 12 \\ 0 & 8 + 8i & 16 - 16i & 16 \\ 35 + 35i & -5 - 15i & -30 + 10i & -20 - 10i \end{bmatrix},$$

$$X_2 = \tfrac{1}{60} \begin{bmatrix} -9 - 3i & 3 + 3i & 6 - 6i & 6 \\ -12 - 4i & 4 + 4i & 8 - 8i & 8 \\ 25 + 25i & -5 - 15i & -30 + 10i & -20 - 10i \end{bmatrix},$$

$$X_3 = \tfrac{1}{420} \begin{bmatrix} 63 + 21i & 15 + 15i & 30 - 30i & 30 \\ 84 + 28i & 20 + 20i & 40 - 40i & 40 \\ 35 + 35i & 5 + 15i & 30 - 10i & 20 + 10i \end{bmatrix},$$

$$X_4 = O.$$

Ex. 78. *The ∗-order.* A partial order on $\mathbb{C}^{m \times n}$, called the *∗-order* and denoted $\overset{*}{<}$, is defined as follows:

$$A \overset{*}{<} B \iff \begin{cases} AA^* = AB^*, \\ A^* A = A^* B. \end{cases} \tag{104}$$

The ∗-order was introduced by Drazin [234] for semigroups with involution. For $A, B \in \mathbb{C}^{m \times n}$, the following statements are equivalent:

(a) $A \overset{*}{<} B$.

(b) $AA^\dagger B = A = B A^\dagger A$.

(c) $A^\dagger \overset{*}{<} B^\dagger$.

Ex. 79. (Hartwig [**382**]). Exercise 78(c) shows that the Moore–Penrose inverse preserves the ∗-ordering. This is not the case for the Löwner-ordering, as shown by the example

$$B = \begin{pmatrix} 1 & 1 \\ 0 & 0 \end{pmatrix}, \quad A = \begin{pmatrix} 0 & 1 \\ 0 & 0 \end{pmatrix} \quad \text{with} \quad B \succcurlyeq A, \quad \text{but } B^\dagger - A^\dagger = \tfrac{1}{2}\begin{pmatrix} 1 & 0 \\ -1 & 0 \end{pmatrix}.$$

In fact, the Moore–Penrose inverse may reverse the Löwner-ordering even for PSD matrices. A precise statement is:

Let $A, B \in \mathrm{PSD}_n$ with $B \succcurlyeq A$. Then:

(a) $A^\dagger \succcurlyeq B^\dagger \iff R(A) \subset R(B)$.
(b) $B^\dagger \succcurlyeq A^\dagger \iff A^2 = AB$.

8. Efficient Characterization of Classes of Generalized Inverses

In the preceding sections, characterizations of certain classes of generalized inverses of a given matrix have been given. Most of these characterizations involve one or more matrices with arbitrary elements. In general, the number of such arbitrary elements far exceeds the actual number of degrees of freedom available.

For example, in Section 1 we obtained the characterization

$$A\{1\} = \{A^{(1)} + Z - A^{(1)}AZAA^{(1)} : Z \in \mathbb{C}^{n\times m}\}. \tag{4}$$

Now, as Z ranges over the entire class $\mathbb{C}^{n\times m}$, every $\{1\}$-inverse of A will be obtained repeatedly an infinite number of times unless A is a matrix of zeros. In fact, the expression in RHS(4) is unchanged if Z is replaced by $Z + A^{(1)}AWAA^{(1)}$, where W is an arbitrary element of $\mathbb{C}^{n\times m}$. We shall now see how in some cases this redundancy in the number of arbitrary parameters can be eliminated. The cases of particular interest are $A\{1\}$ because of its role in the solution of linear systems, $A\{1,2\}$ because of the symmetry inherent in the relation

$$X \in A\{1,2\} \iff A \in X\{1,2\},$$

and $A\{1,3\}$ and $A\{1,4\}$ because of their minimization properties, which will be studied in the next chapter.

As in (4), let $A^{(1)}$ be a fixed, but arbitrary element of $A\{1\}$, where $A \in \mathbb{C}_r^{m\times n}$. Also, let $F \in \mathbb{C}_{n-r}^{n\times(n-r)}$, $K^* \in \mathbb{C}_{m-r}^{m\times(m-r)}$, $B \in \mathbb{C}_r^{n\times r}$ be given matrices whose columns are bases for $N(A)$, $N(A^*)$, and $R(A^{(1)}A)$, respectively. We shall show that the general solution of

$$AXA = A \tag{1.1}$$

is

$$X = A^{(1)} + FY + BZK, \tag{105}$$

where $Y \in \mathbb{C}^{(n-r)\times m}$ and $Z \in \mathbb{C}^{r\times(m-r)}$ are arbitrary.

Clearly $AF = O$ and $KA = O$. Therefore RHS(105) satisfies (1.1). Since $R(I_n - A^{(1)}A) = N(A)$ and $R((I_m - AA^{(1)})^*) = N(A^*)$ by (28) and Lemma 1(g), there exist uniquely defined matrices G, H, D such that

$$FG = I_n - A^{(1)}A, \quad HK = I_m - AA^{(1)}, \quad BD = A^{(1)}A. \tag{106}$$

Since these products are idempotent, we have, by Lemma 2,

$$GF = DB = I_n, \quad KH = I_m. \tag{107}$$

Moreover, it is easily verified that

$$GB = O, \quad DF = O. \tag{108}$$

Using (107) and (108), we obtain easily from (105)

$$Y = G(X - A^{(1)}), \quad Z = D(X - A^{(1)})H. \tag{109}$$

Now, let X be an arbitrary element of $A\{1\}$. Upon substituting in (105) the expression (109) for Y and Z, it is found that (105) is satisfied. We have shown, therefore, that (105) does indeed give the general solution of (1.1).

We recall that $A^{(1)}, F, G, H, K, B, D$ are fixed matrices. Therefore, not only does (105) give X uniquely in terms of Y and Z, but also (109) gives Y and Z uniquely in terms of X. Therefore, different choices of Y and Z in (105) must yield different $\{1\}$-inverses X. Thus, the characterization (105) is completely efficient, and contains the smallest possible number of arbitrary parameters.

It is interesting to compare the number of arbitrary elements in the characterizations (4) and (105). In (4) this is mn, the number of elements of Z. In (105) it is $mn - r^2$, the total number of elements in Y and Z. Clearly, (105) contains fewer arbitrary elements, except in the trivial case $r = 0$, as previously noted.

The case of $A\{1,3\}$ is easier. If, as before, the columns of F are a basis for $N(A)$, it is readily seen that (17) can be written in the alternative form

$$A\{1,3\} = \{A^{(1,3)} + FY : Y \in \mathbb{C}^{(n-r) \times m}\}. \tag{110}$$

This is easily shown to be an efficient characterization. Here the number of arbitrary parameters is $m(m - r)$. Evidently, this is less than the number in the efficient characterization (105) of $A\{1\}$, unless $r = m$, in which case every $\{1\}$-inverse is a $\{1,3\}$-inverse, since $AA^1 = I_m$ by Lemma 1.2.

Similarly, if the columns of K^* are a basis for $N(A^*)$,

$$A\{1,4\} = \{A^{(1,4)} + YK : Y \in \mathbb{C}^{n \times (m-r)}\}, \tag{111}$$

where $A^{(1,4)}$ is a fixed, but arbitrary element of $A\{1,4\}$.

Efficient characterization of $A\{1,2\}$ is somewhat more difficult. Let $A^{(1,2)}$ be a fixed, but arbitrary element of $A\{1,2\}$, and let

$$A^{(1,2)} = Y_0 Z_0$$

be a full-rank factorization. As before, let the columns of F and K^* form bases for the null spaces of A and A^*, respectively. Then we shall show that

$$A\{1,2\} = \{(Y_0 + FU)(Z_0 + VK) : U \in \mathbb{C}^{(n-r) \times r}, V \in \mathbb{C}^{r \times (m-r)}\}. \tag{112}$$

Indeed, it is easily seen that (1.1) and (1.2) are satisfied if X is taken as the product expression in RHS(112). Moreover, if

$$FG = I_n - A^{(1,2)}A, \quad HK = I_m - AA^{(1,2)},$$

it can be shown that

$$U = GXAY_0, \quad V = Z_0AXH. \tag{113}$$

It is found that the product in RHS(112) reduces to X if the expressions in (113) are substituted for U and V.

Relation (112) contains $r(m + n - 2r)$ arbitrary parameters. This is less than the number in the efficient characterization (105) of $A\{1\}$ by $(m - r)(n - r)$, which vanishes only if A is of full (row or column) rank, in which case every $\{1\}$-inverse is a $\{1,2\}$-inverse.

Exercises

Ex. 80. In (106) obtain explicit formulas for G, H, and D in terms of A, $A^{(1)}$, F, K, B, and $\{1\}$-inverses of the latter three matrices.

Ex. 81. Consider the problem of obtaining all $\{1\}$-inverses of the matrix A of (0.80). Note that the parametric representation of Ex. 1.6 does not give all $\{1\}$-inverses. (In this connection see Ex. 1.11.) Obtain in two ways parametric representation that do in fact give all $\{1\}$-inverses: first by (4) and then by (105). Note that a very simple $\{1\}$-inverse (in fact, a $\{1,2\}$-inverse) is obtained by taking all the arbitrary parameters equal to zero in the representation of Ex. 1.6. Verify that possible choices of F and K are

$$F = \begin{bmatrix} 1 & 0 & 0 & 0 \\ 0 & 1 & -1+2i & 0 \\ 0 & -2 & 0 & i \\ 0 & 0 & -2 & -1-i \\ 0 & 0 & 1 & 0 \\ 0 & 0 & 0 & 1 \end{bmatrix}, \quad K = [3i \quad 1 \quad 3].$$

Compare the number of arbitrary parameters in the two representations.

Ex. 82. Under the hypotheses of Theorem 12, let F and K^* be matrices whose columns are bases for $N(A)$ and $N(A^*)$, respectively. Then, (59) can be written in the alternative form

$$X = A_{T,S}^{(1,2)} + FZK, \tag{114}$$

where Z is an arbitrary element of $\mathbb{C}^{(n-r) \times (m-r)}$. Moreover,

$$\operatorname{rank} X = r + \operatorname{rank} Z. \tag{115}$$

PROOF. Clearly the right member of (114) satisfies (58). On the other hand, substituting in (59) the first two equations (106) gives (114) with $Z = GYH$.

Moreover, (114) and Theorem 12(c) give

$$XP_{L,S} = A_{T,S}^{(1,2)},$$

and, therefore,

$$X(I_m - P_{L,S}) = FZK.$$

Consequently, $R(X)$ contains the range of each of the two terms of RHS(114). Furthermore, the intersection of the latter two ranges is $\{0\}$, since $R(F) = N(A) = M$, which is a subspace complementary to $T = R(A_{T,S}^{(1,2)})$. Therefore, $R(X)$ is the direct sum of the two ranges mentioned and, by statement (c) of Ex. 0.1, $\operatorname{rank} X$ is the sum of the ranks of the two terms in RHS(114).

Now, the first term is a $\{1,2\}$-inverse of A and its rank is therefore r by Theorem 1.2, while the rank of the second term is $\operatorname{rank} Z$ by Ex. 1.7. This establishes (115).   □

Ex. 83. Exercise 82 gives

$$A\{1\}_{T,S} = \{A_{T,S}^{(1,2)} + FZK : Z \in \mathbb{C}^{(n-r)\times(m-r)}\},$$

where $A\{1\}_{T,S}$ is defined in Corollary 9. Show that this characterization is efficient.

Ex. 84. Show that if $A \in \mathbb{C}^{m\times n}$, $A\{1\}_{T,S}$ contains matrices of all ranks from r to $\min\{m,n\}$.

Ex. 85. Let $A = ST$ be a full-rank factorization of $A \in \mathbb{C}_r^{m\times n}$, let Y_0 and Z_0 be particular $\{1\}$-inverses of T and S, respectively, and let F and K be defined as in Ex. 82. Then, show that:

$$S\{1\} = \{Z_0 + VK : V \in \mathbb{C}^{r\times(m-r)}\},$$
$$T\{1\} = \{Y_0 + FU : U \in \mathbb{C}^{(n-r)\times r}\},$$
$$AA\{1\} = SS\{1\} = \{S(Z_0 + VK) : V \in \mathbb{C}^{r\times(m-r)}\},$$
$$A\{1\}A = T\{1\}T = \{(Y_0 + FU)T : U \in \mathbb{C}^{(n-r)\times r}\},$$
$$A\{1\} = \{Y_0Z_0 + Y_0VK + FUZ_0 + FWK :$$
$$U \in \mathbb{C}^{(n-r)\times r}, V \in \mathbb{C}^{r\times(m-r)}, W \in \mathbb{C}^{(n-r)\times(m-r)}\},$$
$$= A\{1,2\} + \{FXK : X \in \mathbb{C}^{(n-r)\times(m-r)}\}.$$

Show that all the preceding characterizations are efficient.

Ex. 86. For the matrix A of (0.80), obtain all the characterizations of Ex. 85. *Hint*: Use the full-rank factorization of A given in (0.85), and take

$$Z_0 = \begin{bmatrix} -\frac{1}{2}i & 0 & 0 \\ 0 & -\frac{1}{3} & 0 \end{bmatrix}.$$

9. Restricted Generalized Inverses

In a linear equation

$$A\mathbf{x} = \mathbf{b},$$

with given $A \in \mathbb{C}^{m\times n}$ and $\mathbf{b} \in \mathbb{C}^m$, the points $\mathbf{x}$ are sometimes constrained to lie in a given subspace S of $\mathbb{C}^n$, resulting in a "constrained" linear equation

$$A\mathbf{x} = \mathbf{b} \quad \text{and} \quad \mathbf{x} \in S. \tag{116}$$

In principle, this situation presents no difficulty since (116) is equivalent to the following, "unconstrained" but larger, linear system

$$\begin{bmatrix} A \\ P_{S^\perp} \end{bmatrix} \mathbf{x} = \begin{bmatrix} \mathbf{b} \\ \mathbf{0} \end{bmatrix}, \quad \text{where } P_{S^\perp} = I - P_S.$$

Another approach to the solution of (116) that does not increase the size of the problem is to interpret A as representing an element of $\mathcal{L}(S, \mathbb{C}^m)$, the space of linear transformations from S to $\mathbb{C}^m$, instead of an element of $\mathcal{L}(\mathbb{C}^n, \mathbb{C}^m)$, see, e.g., Sections 4 and 6.1. This interpretation calls for the following definitions.

Let $A \in \mathcal{L}(\mathbb{C}^n, \mathbb{C}^m)$ and let S be a subspace of $\mathbb{C}^n$. The *restriction* of A to S, denoted by $A_{[S]}$, is a linear transformation from S to $\mathbb{C}^m$ defined by

$$A_{[S]}\mathbf{x} = A\mathbf{x}, \quad \mathbf{x} \in S. \tag{117}$$

Conversely, let $B \in \mathcal{L}(S, \mathbb{C}^m)$. The *extension* of B to $\mathbb{C}^n$, denoted by ext B, is the linear transformation from $\mathbb{C}^n$ to $\mathbb{C}^m$ defined by

$$(\text{ext } B)\mathbf{x} = \begin{cases} B\mathbf{x}, & \text{if } \mathbf{x} \in S, \\ \mathbf{0}, & \text{if } \mathbf{x} \in S^{\perp}. \end{cases} \tag{118}$$

Restricting an $A \in \mathcal{L}(\mathbb{C}^n, \mathbb{C}^m)$ to S and then extending to $\mathbb{C}^n$ results in $\text{ext}(A_{[S]}) \in \mathcal{L}(\mathbb{C}^n, \mathbb{C}^m)$ given by

$$\text{ext}(A_{[S]})\mathbf{x} = \begin{cases} A\mathbf{x}, & \text{if } \mathbf{x} \in S, \\ \mathbf{0}, & \text{if } \mathbf{x} \in S^{\perp}. \end{cases} \tag{119}$$

From (119) it should be clear that if $A \in \mathcal{L}(\mathbb{C}^n, \mathbb{C}^m)$ is represented by the matrix $A \in \mathbb{C}^{m \times n}$, then $\text{ext}(A_{[S]})$ is represented by AP_S. The following lemma is then obvious:

LEMMA 4. *Let $A \in \mathbb{C}^{m \times n}$, $\mathbf{b} \in \mathbb{C}^m$, and let S be a subspace of $\mathbb{C}^n$. The system*

$$A\mathbf{x} = \mathbf{b}, \quad \mathbf{x} \in S, \tag{116}$$

is consistent if and only if the system

$$AP_S\mathbf{z} = \mathbf{b} \tag{120}$$

is consistent, in which case $\mathbf{x}$ is a solution of (116) if and only if

$$\mathbf{x} = P_S\mathbf{z},$$

where $\mathbf{z}$ is a solution of (120). $\square$

From Lemma 4 and Corollary 2 it follows that the general solution of (116) is

$$\mathbf{x} = P_S(AP_S)^{(1)}\mathbf{b} + P_S(I - (AP_S)^{(1)}AP_S)\mathbf{y}, \tag{121}$$

$$\text{for arbitrary } (AP_S)^{(1)} \in (AP_S)\{1\} \text{ and } \mathbf{y} \in \mathbb{C}^n.$$

We are thus led to study generalized inverses of $\text{ext}(A_{[S]}) = AP_S$, and from (121) it appears that $P_S(AP_S)^{(1)}$, rather than $A^{(1)}$, plays the role of a $\{1\}$-inverse in solving the linear system (116); hence the following definition:

DEFINITION 2. Let $A \in \mathbb{C}^{m \times n}$ and let S be a subspace of $\mathbb{C}^n$. A matrix $X \in \mathbb{C}^{n \times m}$ is an *S-restricted $\{i, j, \ldots, k\}$-inverse* of A if

$$X = P_S(AP_S)^{(i,j,\ldots,k)} \tag{122}$$

for any $(AP_S)^{(i,j,\ldots,k)} \in (AP_S)\{i, j, \ldots, k\}$.

The role that S-restricted generalized inverses play in constrained problems is completely analogous to the role played by the corresponding generalized inverse in the unconstrained situation. Thus, for example, the following result is the constrained analog of Corollary 2.

COROLLARY 11. *Let* $A \in \mathbb{C}^{m \times n}$ *and let* S *be a subspace of* $\mathbb{C}^n$. *Then the equation*

$$A\mathbf{x} = \mathbf{b}, \quad \mathbf{x} \in S, \tag{116}$$

is consistent if and only if

$$AX\mathbf{b} = \mathbf{b},$$

where X *is any* S-*restricted* $\{1\}$-*inverse of* A. *If consistent, the general solution of* (116) *is*

$$\mathbf{x} = X\mathbf{b} + (I - XA)\mathbf{y}$$

with X *as above, and arbitrary* $\mathbf{y} \in S$. $\qquad\square$

Exercises

Ex. 87. Let I be the identity transformation in $\mathcal{L}(\mathbb{C}^n, \mathbb{C}^n)$ and let S be a subspace of $\mathbb{C}^n$. Show that

$$\text{ext}(I_{[S]}) = P_S.$$

Ex. 88. Let $A \in \mathcal{L}(\mathbb{C}^n, \mathbb{C}^m)$. Show that $A_{[R(A^*)]}$, the restriction of A to $R(A^*)$, is a one-to-one mapping of $R(A^*)$ onto $R(A)$.

SOLUTION. We show first that $A_{[R(A^*)]}$ is one-to-one on $R(A^*)$. Clearly it suffices to show that A is one-to-one on $R(A^*)$. Let $\mathbf{u}, \mathbf{v} \in R(A^*)$ and suppose that $A\mathbf{u} = A\mathbf{v}$, i.e., $\mathbf{u}$ and $\mathbf{v}$ are mapped to the same point. Then $A(\mathbf{u} - \mathbf{v}) = \mathbf{0}$, i.e.,

$$\mathbf{u} - \mathbf{v} \in N(A).$$

But we also have

$$\mathbf{u} - \mathbf{v} \in R(A^*),$$

since $\mathbf{u}$ and $\mathbf{v}$ are in $R(A^*)$. Therefore,

$$\mathbf{u} - \mathbf{v} \in N(A) \cap R(A^*)$$

and by (0.26), $\mathbf{u} = \mathbf{v}$, proving the A is one-to-one on $R(A^*)$.

We show next that $A_{[R(A^*)]}$ is a mapping onto $R(A)$, i.e., that

$$R(A_{[R(A^*)]}) = R(A).$$

This follows since, for any $\mathbf{x} \in \mathbb{C}^n$,

$$A\mathbf{x} = AA^\dagger A\mathbf{x} = AP_{R(A^*)}\mathbf{x} = A_{[R(A^*)]}\mathbf{x}.$$

Ex. 89. Let $A \in \mathbb{C}^{m \times n}$. Show that

$$\text{ext}(A_{[R(A^*)]}) = A. \tag{123}$$

Ex. 90. From Ex. 88 it follows that the linear transformation

$$A_{[R(A^*)]} \in \mathcal{L}(R(A^*), R(A))$$

has an inverse

$$(A_{[R(A^*)]})^{-1} \in \mathcal{L}(R(A), R(A^*)).$$

Show that this inverse is the restriction of $A^\dagger$ to $R(A)$, namely

$$(A^\dagger)_{[R(A)]} = (A_{[R(A^*)]})^{-1}. \tag{124}$$

SOLUTION. From Exs. 88, 38, and 58 it follows that, for any $\mathbf{y} \in R(A)$, $A^\dagger \mathbf{y}$ is the unique element of $R(A^*)$ satisfying

$$A\mathbf{x} = \mathbf{y}.$$

Therefore

$$A^\dagger\mathbf{y} = (A_{[R(A^*)]})^{-1}\mathbf{y}, \quad \text{for all } \mathbf{y} \in R(A).$$

Ex. 91. Show that the extension of $(A_{[R(A^*)]})^{-1}$ to $\mathbb{C}^m$ is the Moore–Penrose inverse of A,

$$\text{ext}((A_{[R(A^*)]})^{-1}) = A^\dagger. \tag{125}$$

Compare with (123).

Ex. 92. Let each of the following two linear equations be consistent

$$A_1\mathbf{x} = \mathbf{b}_1, \tag{126a}$$

$$A_2\mathbf{x} = \mathbf{b}_2. \tag{126b}$$

Show that (126a) and (126b) have a common solution if and only if the linear equation

$$A_2 P_{N(A_1)}\mathbf{y} = \mathbf{b}_2 - A_2 A_1^{(1)}\mathbf{b}_1$$

is consistent, in which case the general common solution of (126a) and (126b) is

$$\mathbf{x} = A_1^{(1)}\mathbf{b}_1 + P_{N(A_1)}(A_2 P_{N(A_1)})^{(1)}(\mathbf{b}_2 - A_2 A_1^{(1)}\mathbf{b}_1) + N(A_1) \cap N(A_2)$$

or, equivalently,

$$\mathbf{x} = A_2^{(1)}\mathbf{b}_2 + P_{N(A_2)}(A_1 P_{N(A_2)})^{(1)}(\mathbf{b}_1 - A_1 A_2^{(1)}\mathbf{b}_2) + N(A_1) \cap N(A_2).$$

Hint. Substitute the general solution of (126a),

$$\mathbf{x} = A_1^{(1)}\mathbf{b}_1 + P_{N(A_1)}\mathbf{y}, \quad \mathbf{y} \text{ arbitrary},$$

in (126b).

Ex. 93. Exercise 92 illustrates the need for $P_{N(A_1)}(A_2 P_{N(A_1)})^{(1)}$, an $N(A_1)$-restricted $\{1\}$-inverse of A_2. Other applications call for other, similarly restricted, generalized inverses. The $N(A_1)$-restricted $\{1,2,3,4\}$-inverse of A_2 was studied for certain Hilbert space operators by Minamide and Nakamura [**556**] and [**557**], who characterized it as the unique solution X of the five equations

$$A_1 X = O,$$
$$A_2 X A_2 = A_2 \text{ on } N(A_1),$$
$$X A_2 X = X,$$
$$(A_2 X)^* = A_2 X,$$

and

$$P_{N(A_1)}(X A_2)^* = X A_2 \quad \text{on } N(A_1).$$

Show that $P_{N(A_1)}(A_2 P_{N(A_1)})^\dagger$ is the unique solution of these five equations.

10. The Bott–Duffin Inverse

Consider the constrained system

$$A\mathbf{x} + \mathbf{y} = \mathbf{b}, \quad x \in L, \ \mathbf{y} \in L^\perp, \tag{127}$$

with given $A \in \mathbb{C}^{n \times n}$, $\mathbf{b} \in \mathbb{C}^n$, and a subspace L of $\mathbb{C}^n$. Such systems arise in electrical network theory; see, e.g., Bott and Duffin [**120**] and Section 13 below. As in Section 9 we conclude that the consistency of (127) is equivalent to the consistency of the following system:

$$(AP_L + P_{L^\perp})\mathbf{z} = \mathbf{b} \tag{128}$$

and that $\begin{bmatrix} \mathbf{x} \\ \mathbf{y} \end{bmatrix}$ is a solution of (127) if and only if

$$\mathbf{x} = P_L \mathbf{z}, \quad \mathbf{y} = P_{L\perp} \mathbf{z} = \mathbf{b} - AP_L \mathbf{z}, \tag{129}$$

where $\mathbf{z}$ is a solution of (128).

If the matrix $(AP_L + P_{L\perp})$ is nonsingular, then (127) is consistent for all $\mathbf{b} \in \mathbb{C}^m$ and the solution

$$\mathbf{x} = P_L(AP_L + P_{L\perp})^{-1}\mathbf{b}, \quad \mathbf{y} = \mathbf{b} - A\mathbf{x},$$

is unique. The transformation

$$P_L(AP_L + P_{L\perp})^{-1}$$

was introduced and studied by Bott and Duffin [**120**], who called it the *constrained inverse* of A. Since it exists only when $(AP_L + P_{L\perp})$ is nonsingular, one may be tempted to introduce generalized inverses of this form, namely

$$P_L(AP_L + P_{L\perp})^{(i,j,\dots,k)} \quad (1 \leq i,j,\dots,k \leq 4),$$

which do exist for all A and L. This section, however, is restricted to the Bott–Duffin inverse.

DEFINITION 3. Let $A \in \mathbb{C}^{n\times n}$ and let L be a subspace of $\mathbb{C}^n$. If $(AP_L + P_{L\perp})$ is nonsingular, the *Bott–Duffin inverse of A with respect to L*, denoted by $A_{(L)}^{(-1)}$, is defined by

$$A_{(L)}^{(-1)} = P_L(AP_L + P_{L\perp})^{-1}. \tag{130}$$

Some properties of $A_{(L)}^{(-1)}$ are collected in

THEOREM 16 (Bott and Duffin [**120**]). *Let $(AP_L + P_{L\perp})$ be nonsingular. Then:*

(a) *The equation*

$$A\mathbf{x} + \mathbf{y} = \mathbf{b}, \quad x \in L, \quad y \in L^\perp, \tag{127}$$

has for every $\mathbf{b}$, *the unique solution*

$$\mathbf{x} = A_{(L)}^{(-1)}\mathbf{b}, \tag{131a}$$

$$\mathbf{y} = (I - AA_{(L)}^{(-1)})\mathbf{b}. \tag{131b}$$

(b) A, P_L, *and* $A_{(L)}^{(-1)}$ *satisfy*

$$P_L = A_{(L)}^{(-1)}AP_L = P_L AA_{(L)}^{(-1)}, \tag{132a}$$

$$A_{(L)}^{(-1)} = P_L A_{(L)}^{(-1)} = A_{(L)}^{(-1)}P_L. \tag{132b}$$

PROOF. (a) This follows from the equivalence of (127) and (128)–(129).

(b) From (130), $P_L A_{(L)}^{(-1)} = A_{(L)}^{(-1)}$. Postmultiplying $A_{(L)}^{(-1)}(AP_L + P_{L\perp}) = P_L$ by P_L gives $A_{(L)}^{(-1)}AP_L = P_L$. Therefore $A_{(L)}^{(-1)}P_{L\perp} = O$ and $A_{(L)}^{(-1)}P_L = A_{(L)}^{(-1)}$. Multiplying (131b) by P_L gives $(P_L - P_L AA_{(L)}^{(-1)})\mathbf{b} = \mathbf{0}$ for all $\mathbf{b}$, thus $P_L = P_L AA_{(L)}^{(-1)}$). $\qquad\square$

From these results it follows that the Bott–Duffin inverse $A_{(L)}^{(-1)}$, whenever it exists, is the $\{1,2\}$-inverse of $(P_L A P_L)$ having range L and null space $L^{\perp}$.

COROLLARY 12. *If $A P_L + P_{L^{\perp}}$ is nonsingular, then:*

(a) $A_{(L)}^{(-1)} = (A P_L)_{L, L^{\perp}}^{(1,2)} = (P_L A)_{L, L^{\perp}}^{(1,2)} = (P_L A P_L)_{L, L^{\perp}}^{(1,2)}$; *and*

(b) $(A_{(L)}^{(-1)})_{(L)}^{(-1)} = P_L A P_L$.

PROOF. (a) From (132a), $\dim L = \operatorname{rank} P_L \leq \operatorname{rank} A_{(L)}^{(-1)}$. Similarly, from (132b), $\operatorname{rank} A_{(L)}^{(-1)} \leq \dim L$, $R(A_{(L)}^{(-1)}) \subset R(P_L) = L$, and $N(A_{(L)}^{(-1)}) \supset N(P_L) = L^{\perp}$. Therefore,

$$\operatorname{rank} A_{(L)}^{(-1)} = \dim L \tag{133}$$

and

$$R(A_{(L)}^{(-1)}) = L, \quad N(A_{(L)}^{(-1)}) = L^{\perp}. \tag{134}$$

Now $A_{(L)}^{(-1)}$ is a $\{1,2\}$-inverse of $A P_L$:

$$A P_L A_{(L)}^{(-1)} A P_L = A P_L, \quad \text{by (132a)},$$

and

$$A_{(L)}^{(-1)} A P_L A_{(L)}^{(-1)} = A_{(L)}^{(-1)}, \quad \text{by (132a) and (132b)}.$$

That $A_{(L)}^{(-1)}$ is a $\{1,2\}$-inverse of $P_L A$ and of $P_L A P_L$ is similarly proved.

(b) We show first that $(A_{(L)}^{(-1)})_{(L)}^{(-1)}$ is defined, i.e., that $(A_{(L)}^{(-1)} P_L + P_{L^{\perp}})$ is nonsingular. From (132b), $A_{(L)}^{(-1)} P_L + P_{L^{\perp}} = A_{(L)}^{(-1)} + P_{L^{\perp}}$, which is a nonsingular matrix since its columns span $L + L^{\perp} = \mathbb{C}^n$, by (134). Now $P_L A P_L$ is a $\{1,2\}$-inverse of $A_{(L)}^{(-1)}$, by (a), and therefore by Theorem 1.2 and (133),

$$\operatorname{rank} P_L A P_L = \operatorname{rank} A_{(L)}^{(-1)} = \dim L.$$

This result, together with

$$R(P_L A P_L) \subset R(P_L) = L, \quad N(P_L A P_L) \supset N(P_L) = L^{\perp},$$

shows that

$$R(P_L A P_L) = L, \quad N(P_L A P_L) = L^{\perp},$$

proving that

$$P_L A P_L = (A_{(L)}^{(-1)})_{L, L^{\perp}}^{(1,2)}$$
$$= (A_{(L)}^{(-1)})_{(L)}^{(-1)}. \qquad \square$$

Exercises

EX. 94. Show that the following statements are equivalent, for any $A \in \mathbb{C}^{n \times n}$ and a subspace $L \subset \mathbb{C}^n$:

(a) $A P_L + P_{L^{\perp}}$ is nonsingular.

(b) $\mathbb{C}^n = AL \oplus L^{\perp}$, i.e., $AL = \{A\mathbf{x} : \mathbf{x} \in L\}$ and $L^{\perp}$ are complementary subspaces of $\mathbb{C}^n$.

(c) $\mathbb{C}^n = P_L R(A) \oplus L^{\perp}$.

(d) $\mathbb{C}^n = P_L AL \oplus L^{\perp}$.

(e) $\operatorname{rank} P_L A P_L = \dim L$.

Thus, each of the above conditions is necessary and sufficient for the existence of $A_{(L)}^{(-1)}$, the Bott–Duffin inverse of A with respect to L.

Ex. 95. *A converse to Corollary* 12. If any one of the following three $\{1,2\}$-inverses exist

$$(AP_L)_{L,L^\perp}^{(1,2)}, \quad (P_L A)_{L,L^\perp}^{(1,2)}, \quad (P_L A P_L)_{L,L^\perp}^{(1,2)},$$

then all three exist, $AP_L + P_{L^\perp}$ is nonsingular, and

$$(AP_L)_{L,L^\perp}^{(1,2)} = (P_L A)_{L,L^\perp}^{(1,2)} = (P_L A P_L)_{L,L^\perp}^{(1,2)} = A_{(L)}^{(-1)}.$$

Hint. Condition (b) in Ex. 94 is equivalent to the existence of $(AP_L)_{L,L^\perp}^{(1,2)}$.

Ex. 96. Let K be a matrix whose columns form a basis for L. Then $A_{(L)}^{(-1)}$ exists if and only if $K^* A K$ is nonsingular, in which case

$$A_{(L)}^{(-1)} = K(K^* A K)^{-1} K^* \quad \text{(Bott and Duffin [120])}.$$

PROOF. Follows from Corollary 12 and Theorem 13(d). $\qquad\qquad\square$

Ex. 97. If A is Hermitian and $A_{(L)}^{(-1)}$ exists, then $A_{(L)}^{(-1)}$ is Hermitian.

Ex. 98. Using the notation

$$A = [a_{ij}] \quad (i,j \in \overline{1,n})$$
$$A_{(L)}^{(-1)} = [t_{ij}] \quad (i,j \in \overline{1,n})$$
$$d_{A,L} = \det(AP_L + L^\perp), \tag{135}$$
$$\psi_{A,L} = \log d_{A,L} \tag{136}$$

show that:

(a) $\dfrac{\partial \psi_{A,L}}{\partial a_{ij}} = t_{ji} \ (i,j \in \overline{1,n})$.

(b) $\dfrac{\partial t_{kl}}{\partial a_{ij}} = t_{ki} t_{jl} \ (i,j,k,l \in \overline{1,n})$.   (Bott and Duffin [120, Theorem 3]).

Bott and Duffin called $d_{A,L}$ the *discriminant* of A and $\psi_{A,L}$ the *potential* of $A_{(L)}^{(-1)}$.

Ex. 99. Let $A \in \mathbb{C}^{n \times n}$ be nonsingular and let L be a subspace of $\mathbb{C}^n$. Then $A_{(L)}^{(-1)}$ exists if and only if $A_{(L^\perp)}^{(-1)}$ exists.

Hint. Use $A^{-1} P_{L^\perp} + P_L = A^{-1}(AP_L + P_{L^\perp})$ to show that $(A^{-1} P_{L^\perp} + P_L)^{-1} = (AP_L + P_{L^\perp})^{-1} A$.

Ex. 100. Let $A \in \mathbb{C}^{n \times n}$ be nonsingular, let L be a subspace of $\mathbb{C}^n$, let $d_{A,L}$ and $\psi_{A,L}$ be given by (128) and (136), respectively, and similarly define

$$d_{A^{-1},L^\perp} = \det(A^{-1} P_{L^\perp} + P_L),$$
$$\psi_{A^{-1},L^\perp} = \log d_{A^{-1},L^\perp}.$$

Then:

(a) $d_{A^{-1},L^\perp} = \dfrac{d_{A,L}}{\det A}$.

(b) $(A^{-1})_{(L^\perp)}^{(-1)} = A - A A_{(L)}^{(-1)} A$.   (Bott and Duffin [120, Theorem 4]).

Ex. 101. If $\Re\langle Au, u \rangle > 0$ for every nonzero vector u, then

$$d_{A,L} \neq 0, \quad \Re\langle A_{(L)}^{(-1)} u, u \rangle \geq 0$$

for every vector u and $\Re(t_{ii}) \geq 0$, where $A_{(L)}^{(-1)} = [t_{ij}]$ (Bott and Duffin, [120, Theorem 6]).

Ex. 102. Let $A, B \in \mathbb{C}^{n \times n}$ and let L be a subspace of $\mathbb{C}^n$ such that both $A_{(L)}^{(-1)}$ and $B_{(L)}^{(-1)}$ exist. Then

$$B_{(L)}^{(-1)} A_{(L)}^{(-1)} = (AP_L B)_{(L)}^{-1}.$$

11. An Application of {1}-Inverses in Interval Linear Programming

For two vectors $\mathbf{u} = (u_i), \mathbf{v} = (v_i) \in \mathbb{R}^m$, let

$$\mathbf{u} \leq \mathbf{v}$$

denote the fact that $u_i \leq v_i$ for $i = 1, \ldots, m$. A linear programming problem of the form

$$\text{maximize}\{\mathbf{c}^T \mathbf{x} : \mathbf{a} \leq Ax \leq \mathbf{b}\}, \tag{137}$$

with given $\mathbf{a}, \mathbf{b} \in \mathbb{R}^m$, $\mathbf{c} \in \mathbb{R}^n$, $A \in \mathbb{R}^{m \times n}$, is called an *interval linear program* (also a *linear program with two-sided constraints*) and denoted by IP$(\mathbf{a}, \mathbf{b}, \mathbf{c}, A)$ or simply by IP. Any linear programming problem with bounded constraint set can be written as an IP, see, e.g., Robers and Ben-Israel [694].

In this section, which is based on the work of Ben-Israel and Charnes [79], the optimal solutions of (137) are obtained by using {1}-inverses of A, in the special case where A is of full row rank. More general cases were studied by Zlobec and Ben-Israel [893], [894] (see also Exs. 103 and 104), and an iterative method for solving the general IP appears in Robers and Ben-Israel [694]. Applications of interval programming are given in Ben-Israel, Charnes, and Robers [80], and Robers and Ben-Israel [693]. References for other applications of generalized inverses in linear programming are Pyle [661] and Cline and Pyle [203].

The IP (137) is called *consistent* (also *feasible*) if the set

$$F = \{\mathbf{x} \in \mathbb{R}^n : \mathbf{a} \leq Ax \leq \mathbf{b}\} \tag{138}$$

is nonempty, in which case the elements of F are called the feasible solutions of IP$(\mathbf{a}, \mathbf{b}, \mathbf{c}, A)$. A consistent IP$(\mathbf{a}, \mathbf{b}, \mathbf{c}, A)$ is called *bounded* if

$$\max\{\mathbf{c}^T \mathbf{x} : x \in F\}$$

is finite, in which case the optimal solutions of IP$(\mathbf{a}, \mathbf{b}, \mathbf{c}, A)$ are its feasible solutions $\mathbf{x}_0$ which satisfy

$$\mathbf{c}^T \mathbf{x}_0 = \max\{\mathbf{c}^T \mathbf{x} : \mathbf{x} \in F\}.$$

Boundedness is equivalent to $\mathbf{c} \in R(A^T)$ as the following lemma shows:

LEMMA 5. *Let* $\mathbf{a}, \mathbf{b} \in \mathbb{R}^m$, $\mathbf{c} \in \mathbb{R}^n$, $A \in \mathbb{R}^{m \times n}$ *be such that* IP$(\mathbf{a}, \mathbf{b}, \mathbf{c}, A)$ *is consistent. Then* IP$(\mathbf{a}, \mathbf{b}, \mathbf{c}, A)$ *is bounded if and only if*

$$\mathbf{c} \in N(A)^{\perp}. \tag{139}$$

PROOF. From (138), $F = F + N(A)$. Therefore,

$$\max\{\mathbf{c}^T \mathbf{x} : \mathbf{x} \in F\} = \max\{\mathbf{c}^T \mathbf{x} : \mathbf{x} \in F + N(A)\}$$
$$= \max\{(P_{R(A^T)}\mathbf{c} + P_{N(A)}\mathbf{c})^T \mathbf{x} : \mathbf{x} \in F + N(A)\}, \text{ by } (0.26),$$
$$= \max\{\mathbf{c}^T P_{R(A^T)}\mathbf{x} : \mathbf{x} \in F\} + \max\{\mathbf{c}^T \mathbf{x} : \mathbf{x} \in N(A)\},$$

where the first term

$$\max\{\mathbf{c}^T P_{R(A^T)}\mathbf{x} : \mathbf{x} \in F\} = \max\{\mathbf{c}^T A^\dagger A\mathbf{x} : \mathbf{a} \leq A\mathbf{x} \leq \mathbf{b}\}$$

is finite and the second term

$$\max\{\mathbf{c}^T\mathbf{x} : \mathbf{x} \in N(A)\}$$

is finite if and only if $\mathbf{c} \in N(A)^\perp$. $\square$

We introduce now a function $\eta\colon \mathbb{R}^m \times \mathbb{R}^m \times \mathbb{R}^m \to \mathbb{R}^m$, defined for $\mathbf{u}, \mathbf{v}, \mathbf{w} \in \mathbb{R}^m$ by

$$\eta(\mathbf{u}, \mathbf{v}, \mathbf{w}) = [\eta_i] \quad (i \in \overline{1, m}),$$

where

$$\eta_i = \begin{cases} u_i, & \text{if } w_i < 0, \\ v_i, & \text{if } w_i > 0, \\ \lambda_i u_i + (1 - \lambda_i)v_i, & \text{where } 0 \leq \lambda_i \leq 1, \text{ if } w_i = 0. \end{cases} \tag{140}$$

A component of $\eta(\mathbf{u}, \mathbf{v}, \mathbf{w})$ is equal to the corresponding component of $\mathbf{u}$ or $\mathbf{v}$, if the corresponding component of $\mathbf{w}$ is negative or positive, respectively. If a component of $\mathbf{w}$ is zero, then the corresponding component of $\eta(\mathbf{u}, \mathbf{v}, \mathbf{w})$ is the closed interval with the corresponding components of $\mathbf{u}$ and $\mathbf{v}$ as endpoints. Thus η maps points in $\mathbb{R}^m \times \mathbb{R}^m \times \mathbb{R}^m$ into sets in $\mathbb{R}^m$, and any statement below about $\eta(\mathbf{u}, \mathbf{v}, \mathbf{w})$ is meant for all values of $\eta(\mathbf{u}, \mathbf{v}, \mathbf{w})$, unless otherwise specified.

The next result gives all the optimal solutions of IP$(\mathbf{a}, \mathbf{b}, \mathbf{c}, A)$ with A of full row rank.

THEOREM 17 (Ben-Israel and Charnes [**79**]). *Let* $\mathbf{a}, \mathbf{b} \in \mathbb{R}^m$, $\mathbf{c} \in \mathbb{R}^n$, $A \in \mathbb{R}^{m \times n}$ *be such that* IP$(\mathbf{a}, \mathbf{b}, \mathbf{c}, A)$ *is consistent and bounded and let* $A^{(1)}$ *be any* $\{1\}$-*inverse of* A. *Then the general optimal solution of* IP$(\mathbf{a}, \mathbf{b}, \mathbf{c}, A)$ *is*

$$\mathbf{x} = A^{(1)}\eta(\mathbf{a}, \mathbf{b}, A^{(1)T}\mathbf{c}) + \mathbf{y}, \quad \mathbf{y} \in N(A). \tag{141}$$

PROOF. From $A \in \mathbb{R}^{m \times n}_m$ it follows that $R(A) = \mathbb{R}^m$, so that any $\mathbf{u} \in \mathbb{R}^m$ can be written as

$$\mathbf{u} = A\mathbf{x} \tag{142}$$

where

$$\mathbf{x} = A^{(1)}\mathbf{u} + \mathbf{y}, \quad \mathbf{y} \in N(A), \quad \text{by Corollary 2.} \tag{143}$$

Substituting (142) and (143) in (137) we get, by using (139), the equivalent IP

$$\max\{\mathbf{c}^T A^{(1)}\mathbf{u} : \mathbf{a} \leq \mathbf{x} \leq \mathbf{b}\}$$

whose general optimal solution is, by the definition (140) of η,

$$\mathbf{u} = \eta(\mathbf{a}, \mathbf{b}, A^{(1)T}\mathbf{c})$$

which gives (141) by using (143). $\square$

Exercises

Ex. 103. Let $\mathbf{a}, \mathbf{b} \in \mathbb{R}^m, \mathbf{c} \in \mathbb{R}^n, A \in \mathbb{R}^{m \times n}$ be such that $\mathrm{IP}(\mathbf{a}, \mathbf{b}, \mathbf{c}, A)$ is consistent and bounded. Let $A^{(1)} \in A\{1\}$ and let $\mathbf{z}_0 \in N(A^T)$ satisfy

$$\mathbf{z}^T \boldsymbol{\eta}_0 \leq 0$$

for some $\boldsymbol{\eta}_0 \in \eta(\mathbf{a}, \mathbf{b}, (A^{(1)} P_{R(A)})^T \mathbf{c} + \mathbf{z}_0)$. Then

$$\mathbf{x}_0 = A^{(1)} P_{R(A)} \boldsymbol{\eta}_0 + \mathbf{y}, \quad \mathbf{y} \in N(A)$$

is an optimal solution of $\mathrm{IP}(\mathbf{a}, \mathbf{b}, \mathbf{c}, A)$ if and only if it is a feasible solution (Zlobec and Ben-Israel [**894**]).

Ex. 104. Let $\mathbf{b} \in \mathbb{R}^m, \mathbf{c} \in \mathbb{R}^n, A \in \mathbb{R}^{m \times n}$ and let $\mathbf{u} \in \mathbb{R}^n$ be a positive vector such that the problem

$$\min\{\mathbf{c}^T \mathbf{x} : A\mathbf{x} = \mathbf{b}, 0 \leq \mathbf{x} \leq \mathbf{u}\} \tag{144}$$

is consistent. Let $\mathbf{z}_0 \in R(A^T)$ satisfy

$$\mathbf{z}^T \boldsymbol{\eta}_0 \leq \mathbf{z}^T A^\dagger \mathbf{b}$$

for some $\boldsymbol{\eta}_0 \in \eta(\mathbf{0}, \mathbf{u}, P_{N(A)} \mathbf{c} + \mathbf{z}_0)$. Then

$$\mathbf{x}_0 = A^\dagger \mathbf{b} + P_{N(A)} \boldsymbol{\eta}_0$$

is an optimal solution of (144) if and only if it is a feasible solution (Zlobec and Ben-Israel [**894**]).

12. A $\{1, 2\}$-Inverse for the Integral Solution of Linear Equations

We use the notation $\mathbb{Z}, \mathbb{Z}^m, \mathbb{Z}^{m \times n}, \mathbb{Z}_r^{m \times n}$ of Section 0.8.

Any vector in $\mathbb{Z}^m$ will be called an *integral vector*. Similarly, any element of $\mathbb{Z}^{m \times n}$ will be called an *integral matrix*.

Let $A \in \mathbb{Z}^{m \times n}, \mathbf{b} \in \mathbb{Z}^m$ and let the linear equation

$$A\mathbf{x} = \mathbf{b} \tag{5}$$

be consistent. In many applications one has to determine if (5) has integral solutions, in which case one has to find some or all of them. If A is a *unit matrix* (i.e., A is nonsingular and its inverse is also integral) then (5) has the unique integral solution $\mathbf{x} = A^{-1} \mathbf{b}$ for any integral $\mathbf{b}$.

In this section, which is based on the work of Hurt and Waid [**434**], we study the integral solution of (5) for any $A \in \mathbb{Z}^{m \times n}$ and $\mathbf{b} \in \mathbb{Z}^m$. Using the *Smith normal form* of A (Theorem 0.4), a $\{1, 2\}$-inverse is found (Corollary 13) which can be used to determine the existence of integral solutions, and to list all of them if they exist (Corollaries 14 and 15).

COROLLARY 13 (Hurt and Waid [**434**]). *Let $A \in \mathbb{Z}^{m \times n}$. Then there is an $n \times m$ matrix X satisfying*

$$AXA = A, \tag{1.1}$$

$$XAX = X, \tag{1.2}$$

$$AX \in \mathbb{Z}^{m \times m}, \quad XA \in \mathbb{Z}^{n \times n}. \tag{145}$$

PROOF. Let

$$PAQ = S \tag{0.132}$$

be the Smith normal form of A and let

$$\widehat{A} = QS^{\dagger}P. \tag{146}$$

Then

$$PAQ = S = SS^{\dagger}S = PAQS^{\dagger}PAQ = PA\widehat{A}AQ,$$

proving $A = A\widehat{A}A$. $\widehat{A}A\widehat{A} = \widehat{A}$ is similarly proved. The integrality of $A\widehat{A}$ and $\widehat{A}A$ follows from that of $PA\widehat{A} = SS^{\dagger}P$ and $\widehat{A}AQ = QS^{\dagger}S$, respectively.   □

In the rest of this section we denote by $\widehat{A}, \widehat{B}$ the $\{1,2\}$-inverses of A, B as given in Corollary 13.

COROLLARY 14 (Hurt and Waid [**434**]). *Let A, B, D be integral matrices and let the matrix equation*

$$AXB = D \tag{1}$$

be consistent. Then (1) has an integral solution if and only if the matrix

$$\widehat{A}D\widehat{B}$$

is integral, in which case the general integral solution of (1) is

$$X = \widehat{A}D\widehat{B} + Y - \widehat{A}AYB\widehat{B}, \quad Y \in \mathbb{Z}^{n \times m}.$$

PROOF. Follows from Corollary 13 and Theorem 1.   □

COROLLARY 15 (Hurt and Waid [**434**]). *Let A and $\mathbf{b}$ be integral, and let the vector equation*

$$A\mathbf{x} = \mathbf{b} \tag{5}$$

be consistent. Then (5) has an integral solution if and only if the vector

$$\widehat{A}\mathbf{b}$$

is integral, in which case the general integral solution of (5) is

$$\mathbf{x} = \widehat{A}\mathbf{b} + (I - \widehat{A}A)\mathbf{y}, \quad \mathbf{y} \in \mathbb{Z}^{n}.$$

Exercises

EX. 105. Use the results of Sections 11 and 12 to find the integral optimal solutions of the interval program

$$\max\{\mathbf{c}^T\mathbf{x} : \mathbf{a} \le \mathbf{x} \le \mathbf{b}\}$$

where $\mathbf{a}, \mathbf{b}, \mathbf{c}$, and A are integral.

EX. 106. If $\mathbb{Z}$ is the *ring of polynomials with real coefficients*, or the *ring of polynomials with complex coefficients*, the results of this section hold; see, e.g., Marcus and Minc [**534**, p. 40]. Interpret Corollaries 13 and 15 in these two cases.

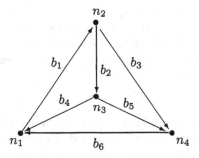

FIGURE 1. An example of a network.

13. An Application of the Bott–Duffin Inverse to Electrical Networks

In this section which is based on Bott and Duffin [**120**], we keep the discussion of electrical networks at the minimum sufficient to illustrate the application of the Bott–Duffin inverse studied in Section 10. The reader is referred to the original work of Bott and Duffin for further information.

An *electrical network* is described topologically in terms of its *graph* consisting of *nodes* (also *vertices*, *junctions*, etc.) and *branches* (also *edges*), and electrically in terms of its (branch) *currents* and *voltages*.

Let the *graph* consist of m elements called *nodes* denoted by n_i, $i \in \overline{1,m}$ (which, in the present limited discussion, can be represented by m points in the plane), and n ordered pairs of nodes called *branches* denoted by b_j, $j \in \overline{1,n}$ (represented here by directed segments joining the paired nodes).

For example, the network represented by Figure 1 has four nodes n_1, n_2, n_3, and n_4, and six branches $b_1 = \{n_1, n_2\}$, $b_2 = \{n_2, n_3\}$, $b_3 = \{n_2, n_4\}$, $b_4 = \{n_3, n_1\}$, $b_5 = \{n_3, n_4\}$, and $b_6 = \{n_4, n_1\}$.

A graph with m nodes and n branches can be represented by an $m \times n$ matrix called the *(node–branch) incidence matrix*, denoted by $M = [m_{ij}]$ and defined as follows:

(i) The ith row of M corresponds to the node n_i, $i \in \overline{1,m}$.

(ii) The jth column of M corresponds to the branch b_j, $j \in \overline{1,n}$.

(iii) If $b_j = \{n_k, n_l\}$, then

$$m_{ij} = \begin{cases} 1, & i = k, \\ -1, & i = l, \\ 0, & i \neq k, l. \end{cases}$$

For example, the incidence matrix of the graph of Fig 1 is

$$M = \begin{bmatrix} 1 & 0 & 0 & -1 & 0 & -1 \\ -1 & 1 & 1 & 0 & 0 & 0 \\ 0 & -1 & 0 & 1 & 1 & 0 \\ 0 & 0 & -1 & 0 & -1 & 1 \end{bmatrix}.$$

Two nodes n_k and n_l (or the corresponding rows of M) are called *directly connected* if either $\{n_k, n_l\}$ or $\{n_l, n_k\}$ is a branch, i.e., if there is a column in M having its nonzero entries in rows k and l. Two nodes n_k and n_l (or the corresponding rows of M) are called *connected* if there is a sequence of nodes

$$\{n_k, n_p, \ldots, n_q, n_l\}$$

in which every two adjacent nodes are directly connected. Finally, a graph (or its incidence matrix) is called *connected* if every two nodes are connected.

In this section we consider only *direct current* (DC) *networks*, referring the reader to Bott and Duffin [**120**] and to Ex. 108 below, for alternating current (AC) networks. A DC network is described electrically in terms of two real valued functions, the *current* and the *potential*, defined on the sets of branches and nodes respectively.

For $j = 1, \ldots, m$, the *current in branch* b_j, denoted by y_j, is the current (measured in amperes) flowing in b_j. The sign of y_j is *positive* if it flows in the direction of b_j and is *negative* if it flows in the opposite direction.

For $i = 1, \ldots, m$, the *potential* at node n_i, denoted by p_i, is the voltage difference (measured in volts) between n_i and some reference point, which can be taken as one of the nodes. A related function which is more often used, is the *voltage*, defined on the set of branches. For $j = 1, \ldots, n$, the *voltage across branch* $b_j = \{n_k, n_l\}$, denoted by x_j, is defined as the potential difference

$$x_k = p_k = p_l.$$

From the definition of the incidence matrix M it is clear that the vector of branch voltages $\mathbf{x} = [x_j]$ and the vector of node potentials $\mathbf{p} = [p_i]$ are related by

$$\mathbf{x} = M^T \mathbf{p}. \tag{147}$$

The currents and voltages are assumed to satisfy Kirchhoff laws. The *Kirchhoff current law* is a conservation theorem for the currents (or electrical charges), stating that for each node, the net current entering the node is zero, i.e., the sum of incoming currents equals the sum of outgoing currents. From the definition of the incidence matrix M it follows that the Kirchhoff current law can be written as

$$M\mathbf{y} = \mathbf{0}. \tag{148}$$

The *Kirchhoff voltage law* states that the potential function is single valued. This statement usually assumes the equivalent form that the sum of the branch voltages directed around any closed circuit is zero.

From (147), (148), and (0.26), it follows that the Kirchhoff current and voltage laws define two complementary orthogonal subspaces:

$N(M)$, the currents satisfying Kirchhoff current law; and

$R(M^T)$, the voltages satisfying Kirchhoff voltage law.

Each branch b_j, $j \in \overline{1, n}$, of the network will be regarded as having a *series voltage generator* of v_j volts and a *parallel current generator* of w_j amperes.

These are related to the branch currents and voltages by *Ohm's law*

$$a_j(x_j - v_j) + (y_j - w_j) = 0, \quad j \in \overline{1,n}, \tag{149}$$

where $a_j > 0$ is the conductivity of the branch b_j, measured in mhos.[3]

Thus the branch currents $\mathbf{y}$ and voltages $\mathbf{x}$ are found by solving the following constrained system:

$$A\mathbf{x} + \mathbf{y} = A\mathbf{v} + \mathbf{w}, \quad \mathbf{x} \in R(M^T), \quad \mathbf{y} \in N(M), \tag{150}$$

where $A = \text{diag}\,(a_j)$ is the diagonal matrix of branch conductivities, $\mathbf{v}$ and $\mathbf{w}$ are the given vectors of generated voltages and currents, respectively, and M is the incidence matrix. It can be shown that the Bott–Duffin inverse of A with respect to $R(M^T)$, $A^{(-1)}_{(R(M^T))}$, exists; see, e.g., Ex. 107 below. Therefore, by Theorem 16, the unique solution of (150) is

$$\mathbf{x} = A^{(-1)}_{(R(M^T))}(A\mathbf{v} + \mathbf{w}), \tag{151a}$$

$$\mathbf{y} = (I - AA^{(-1)}_{(R(M^T))})(A\mathbf{v} + \mathbf{w}). \tag{151b}$$

The physical significance of the matrix $A^{(-1)}_{(R(M^T))}$ should be clear from (151a). The $(i,j)^{\text{th}}$ entry of $A^{(-1)}_{(R(M^T))}$ is the voltage across branch b_i as a result of inserting a current source of 1 ampere in branch b_j; $i,j \in \overline{1,n}$. Because of this property, $A^{(-1)}_{(R(M^T))}$ is called the *transfer matrix* of the network.

Since the conductivity matrix A is nonsingular, the network equations (150) can be rewritten as

$$A^{-1}\mathbf{y} + \mathbf{x} = A^{-1}\mathbf{w} + \mathbf{v}, \quad \mathbf{y} \in N(M), \quad \mathbf{x} \in R(M^T). \tag{152}$$

By Exs. 107 and 99, the unique solution of (152) is

$$\mathbf{y} = (A^{-1})^{(-1)}_{(N(M))}(A^{-1}\mathbf{w} + \mathbf{v}), \tag{153a}$$

$$\mathbf{x} = (I - A^{-1}(A^{-1})^{(-1)}_{(N(M))})(A^{-1}\mathbf{w} + \mathbf{v}A^{-1}\mathbf{w} + \mathbf{v}). \tag{153b}$$

The matrix $(A^{-1})^{(-1)}_{(N(M))}$ is called the *dual transfer matrix*, its $(i,j)^{\text{th}}$ entry being the current in branch b_i as a result of inserting a 1 volt generator parallel to branch b_j. Comparing the corresponding equations in (151) and in (153), we prove that the transfer matrices $A^{(-1)}_{(R(M^T))}$ and $(A^{-1})^{(-1)}_{(N(M))}$ satisfy

$$A^{-1}(A^{-1})^{(-1)}_{(N(M))} + A^{(-1)}_{(R(M^T))}A = I, \tag{154}$$

which can also be proved directly from Ex. 100(b).

The correspondence between results like (151) and (153) is called *electrical duality*; see, e.g., the discussion in Bott and Duffin [120], Duffin [241], and Sharpe and Styan [745], [746], [747], for further results on duality and on applications of generalized inverses in electrical networks.

---

[3]*mho*, the unit of conductance, is the reciprocal of *ohm*, the unit of resistance.

Exercises

Ex. 107. Let $A \in \mathbb{C}^{n \times n}$ be such that $\langle A\mathbf{x}, \mathbf{x} \rangle \neq 0$ for every nonzero vector $\mathbf{x}$ in L, a subspace of $\mathbb{C}^n$. Then $A_{(L)}^{(-1)}$ exists, i.e., $(AP_L + P_{L^\perp})$ is nonsingular.

PROOF. If $A\mathbf{x} + \mathbf{y} = \mathbf{0}$ for some $\mathbf{x} \in L$ and $\mathbf{y} \in L^\perp$, then $A\mathbf{x} \in L^\perp$ and therefore $\langle A\mathbf{x}, \mathbf{x} \rangle = 0$. $\qquad\square$

See also Exs. 101 and 94(b) above.

Ex. 108. In AC networks without mutual coupling, equations (149) still hold for the branches, by using complex, instead of real, constants and variables. The complex a_j is then the admittance of branch b_j. AC networks with mutual coupling due to transformers, are still represented by (150), where the *admittance matrix A* is symmetric, its off-diagonal elements giving the mutual couplings; see, e.g., Bott and Duffin [**120**].

Ex. 109. *Incidence matrix.* Let M be a connected $m \times n$ incidence matrix. Then, for any $M^{(1,3)} \in M\{1,3\}$,

$$I - MM^{(1,3)} = \frac{1}{m}\,\mathbf{e}\mathbf{e}^T,$$

where $\mathbf{e}\mathbf{e}^T$ is the $m \times m$ matrix whose elements are all 1. See also Ijiri [**436**].

PROOF. From $(I - MM^{(1,3)})M = O$ it follows for any two directly connected nodes n_i and n_j (i.e., for any column of M having its $+1$ and -1 in rows i and j), that the ith and jth columns of $I - MM^{(1,3)}$ are identical. Since M is connected, all columns of $I - MM^{(1,3)}$ are identical. Since $I - MM^{(1,3)}$ is symmetric, all its rows are also identical. Therefore, all elements of $I - MM^{(1,3)}$ are equal, say

$$I - MM^{(1,3)} = \alpha\mathbf{e}\mathbf{e}^T,$$

for some real α. Now $I - MM^{(1,3)}$ is idempotent, proving that $\alpha = 1/m$. $\qquad\square$

Ex. 110. Let M be a connected $m \times n$ incidence matrix. Then rank $M = m - 1$.
PROOF.

$$P_{N(M^T)} = I - P_{R(M)}, \quad \text{by (0.27)},$$

$$= I - MM^{(1,3)}, \quad \text{by Ex. 1.9 and Lemma 3,}$$

$$= \frac{1}{m}\,\mathbf{e}\mathbf{e}^T, \quad \text{by Ex. 109,}$$

proving that $\dim N(M^T) = \operatorname{rank} P_{N(M^T)} = 1$ and, therefore,

$$\operatorname{rank} M = \dim R(M) = m - \dim N(M^T) = m - 1. \qquad\square$$

Ex. 111. *Set inclusion matrices*, Bapat [**45**]. Let $0 \le r \le c \le n$ be integers. The *set inclusion matrix W_{rc}* and the *set intersection matrix $\overline{W}_{rc}$* are $\binom{n}{r} \times \binom{n}{c}$ matrices, with rows indexed by r–element sets, and columns indexed by c–element sets, and the $(R, C)^{\text{th}}$ element

$$W_{rc}[R, C] = \begin{cases} 1, & \text{if } R \subset C, \\ 0, & \text{otherwise}, \end{cases}$$

$$\overline{W}_{rc}[R, C] = \begin{cases} 1, & \text{if } R \cap C = \emptyset, \\ 0, & \text{otherwise}. \end{cases}$$

When $r = 0$, both W_{0c} and $\overline{W}_{0c}$ are the $1 \times \binom{n}{c}$ vector of all ones.

Let $0 \le r \le c \le n - r$. Then

$$W_{rc}^\dagger = \sum_{i=0}^{r} \frac{(-1)^i}{\binom{n-i-r}{c-r}}\, \overline{W}_{ic}^T W_{ir}.$$

Ex. 112. *Trees.* Let a connected network consist of m nodes and n branches, and let M be its incidence matrix. A *tree* is defined as consisting of the m nodes, and any $m-1$ branches which correspond to linearly independent columns of M. Show that:

(a) A tree is a connected network which contains no closed circuit.

(b) Any column of M not among the $m-1$ columns corresponding to a given tree, can be expressed uniquely as a linear combination of those $m-1$ columns, using only the coefficients $0, +1$, and -1.

(c) Any branch not in a given tree, lies in a unique closed circuit whose other branches, or the branches obtained from them by reversing their directions, belong to the tree.

Ex. 113. Let $A = \operatorname{diag}(a_j)$, $a_j \neq 0$, $j \in \overline{1, n}$, and let M be a connected $m \times n$ incidence matrix. Show that the discriminant (see Ex. 98)

$$d_{A, R(M^T)} = \det(A P_{R(M^T)} + P_{N(M)})$$

is the sum, over all trees $\{b_{j_1}, b_{j_2}, \ldots, b_{j_{m-1}}\}$ in the network, of the products

$$a_{j_1} a_{j_2} \cdots a_{j_{m-1}} \quad \text{(Bott and Duffin } [\mathbf{120}]).$$

Suggested Further Reading

SECTION 1. Bjerhammar [**103**], Hearon [**409**], Jones [**450**], Morris and Odell [**583**], Sheffield [**749**].

SECTION 4. Afriat [**3**], Chipman and Rao [**191**], Graybill and Marsaglia [**319**], Greville [**330**], [**347**], Groß and Trenkler [**355**], [**356**], Li and Wei [**516**], Mizel and Rao [**571**], Takane and Yanai [**793**], Wedderburn [**853**].

SECTION 5. Bhatia [**96**], Dunford and Schwartz [**246**, pp. 556–565], Frame [**285**], [**286**], [**287**], [**288**], Gantmacher [**296**], Lancaster [**494**], Lancaster and Tismenetsky [**495**, Chapter 9], Rinehart [**691**], Robinson [**696**].

SECTION 6. Ward, Boullion, and Lewis [**851**].

SECTION 7. Afriat [**3**], Anderson and Duffin [**21**], Ben-Israel [**65**], Chipman and Rao [**191**], Glazman and Ljubich [**299**], Greville [**330**], Petryshyn [**641**], Stewart [**780**], Trenkler [**815**].

PARTIAL ORDERS FOR MATRICES. Baksalary [**31**], Baksalary and Hauke [**32**], Baksalary and Mitra [**35**], Baksalary and Pukelsheim [**36**], Baksalary, Pukelsheim and Styan [**37**], Bapat, Jain, and Snyder [**48**], Carlson [**167**], Drazin [**235**], Groß [**342**], [**346**], [**348**], Groß, Hauke, and Markiewicz [**349**], [**350**], Groß and Troschke [**357**], Guterman [**362**], [**361**], Hartwig [**384**], Hartwig and Drazin [**389**], Hartwig and Styan [**395**], Hauke and Markiewicz [**400**], [**399**], Jain, Mitra, and Werner [**443**], Liski [**518**], Liski and Wang [**519**], Markiewicz [**535**], Mitra [**561**], [**562**], [**563**], [**564**], Mitra and Hartwig [**565**], Nordström [**617**].

SECTION 10. Chen [**180**], Rao and Mitra [**679**].

SECTION 11. For applications of generalized inverses in mathematical programming see also Beltrami [**58**], Ben-Israel [**64**], [**68**], Ben-Israel and Kirby [**82**], Charnes and Cooper [**175**], Charnes, Cooper, and Thompson [**176**], Charnes and Kirby [**178**], Kirby [**475**], Nelson, Lewis, and Boullion [**604**], Rosen [**710**], [**711**], Zlobec [**890**].

SECTION 12. Batigne [**52**], Batigne, Hall, and Katz [**53**], Bowman and Burdet [**127**], Charnes and Granot [**177**].

$\{1, 2\}$-INVERSES OF POLYNOMIAL MATRICES. Bose and Mitra [**118**], Sontag [**768**].

SECTION 13. Ben-Israel and Charnes [**78**], Sharpe and Styan [**745**], [**746**], [**747**].

CHAPTER 3

Minimal Properties of Generalized Inverses

1. Least-Squares Solutions of Inconsistent Linear Systems

For given $A \in \mathbb{C}^{m \times n}$ and $\mathbf{b} \in \mathbb{C}^m$, the linear system

$$Ax = b \tag{1}$$

is *consistent*, i.e., has a solution for $\mathbf{x}$, if and only if $\mathbf{b} \in R(A)$. Otherwise, the *residual* vector

$$r = b - Ax \tag{2}$$

is nonzero for all $\mathbf{x} \in \mathbb{C}^n$, and it may be desired to find an *approximate solution* of (1), by which is meant a vector $\mathbf{x}$ making the residual vector (2) "closest" to zero in some sense, i.e., minimizing some norm of (2). An approximate solution that is often used, especially in statistical applications, is the *least-squares solution* of (1), defined as a vector $\mathbf{x}$ minimizing the Euclidean norm of the residual vector, i.e., minimizing the sum of squares of moduli of the residuals

$$\sum_{i=1}^{m} |r_i|^2 = \sum_{i=1}^{m} |b_i - \sum_{j=1}^{n} a_{ij} x_j|^2 = \|b - Ax\|^2. \tag{3}$$

In this section the Euclidean vector norm – see, e.g., Ex. 0.10 – is denoted simply by $\|\ \|$.

The following theorem shows that $\|Ax - b\|$ is minimized by choosing $\mathbf{x} = X\mathbf{b}$, where $X \in A\{1,3\}$, thus establishing a relation between the $\{1,3\}$-inverses and the least-squares solutions of $Ax = b$, characterizing each of these two concepts in terms of the other.

THEOREM 1. *Let* $A \in \mathbb{C}^{m \times n}$, $\mathbf{b} \in \mathbb{C}^m$. *Then* $\|Ax - b\|$ *is smallest when* $\mathbf{x} = A^{(1,3)}\mathbf{b}$, *where* $A^{(1,3)} \in A\{1,3\}$. *Conversely, if* $X \in \mathbb{C}^{n \times m}$ *has the property that, for all* $\mathbf{b}$, $\|Ax - b\|$ *is smallest when* $\mathbf{x} = X\mathbf{b}$, *then* $X \in A\{1,3\}$.

PROOF. From (0.26)

$$b = (P_{R(A)} + P_{R(A)^{\perp}})b. \tag{4}$$

$$\therefore\ b - Ax = (P_{R(A)}b - Ax) + P_{N(A^*)}b.$$

$$\therefore\ \|Ax - b\|^2 = \|Ax - P_{R(A)}b\|^2 + \|P_{N(A^*)}b\|^2, \quad \text{by Ex. 0.46.} \tag{5}$$

Evidently, (5) assumes its minimum value if and only if

$$Ax = P_{R(A)}b, \tag{6}$$

which holds if $\mathbf{x} = A^{(1,3)}\mathbf{b}$ for any $A^{(1,3)} \in A\{1,3\}$, since, by Theorem 2.8, (2.28), and Lemma 2.3,

$$AA^{(1,3)} = P_{R(A)}. \tag{7}$$

Conversely, if X is such that for all $\mathbf{b}$, $\|A\mathbf{x} - \mathbf{b}\|$ is smallest when $\mathbf{x} = X\mathbf{b}$, (6) gives $AX\mathbf{b} = P_{R(A)}\mathbf{b}$ for all $\mathbf{b}$, and therefore

$$AX = P_{R(A)}.$$

Thus, by Theorem 2.3, $X \in A\{1,3\}$. $\qquad\square$

COROLLARY 1. *A vector* $\mathbf{x}$ *is a least-squares solution of* $A\mathbf{x} = \mathbf{b}$ *if and only if*

$$A\mathbf{x} = P_{R(A)}\mathbf{b} = AA^{(1,3)}\mathbf{b}.$$

Thus, the general least-squares solution is

$$\mathbf{x} = A^{(1,3)}\mathbf{b} + (I_n - A^{(1,3)}A)\mathbf{y}, \tag{8}$$

with $A^{(1,3)} \in A\{1,3\}$ *and arbitrary* $\mathbf{y} \in \mathbb{C}^n$.

It will be noted that the least-squares solution is unique only when A is of full column rank (the most frequent case in statistical applications). Otherwise, (8) is an infinite set of such solutions.

Exercises and Supplementary Notes

EX. 1. *Normal equation.* Show that a vector $\mathbf{x}$ is a least-squares solution of $A\mathbf{x} = \mathbf{b}$ if and only if $\mathbf{x}$ is a solution of

$$A^*A\mathbf{x} = A^*\mathbf{b}, \tag{9}$$

often called the *normal equation* of $A\mathbf{x} = \mathbf{b}$.

SOLUTION. It follows from (4) and (6) that $\mathbf{x}$ is a least-squares solution if and only if

$$A\mathbf{x} - \mathbf{b} \in N(A^*),$$

which is (9). $\qquad\square$

ALTERNATIVE SOLUTION. A necessary condition for the vector $\mathbf{x}^0$ to be a least-squares solution of $A\mathbf{x} = \mathbf{b}$ is that the partial derivatives $\partial f/\partial x_j$ of the function

$$f(\mathbf{x}) = \|A\mathbf{x} - \mathbf{b}\|^2 = \sum_{i=1}^{m}(\sum_{j=1}^{n} a_{ij}x_j - b_i)^*(\sum_{j=1}^{n} a_{ij}x_j - b_i) \tag{10}$$

vanish at $\mathbf{x}^0$, i.e., that $\nabla f(\mathbf{x}^0) = \mathbf{0}$, where

$$\nabla f(\mathbf{x}^0) = \left(\frac{\partial f}{\partial x_j}(\mathbf{x}^0)\right),$$

is the *gradient* of f at $\mathbf{x}^0$. Now it can be shown that the gradient of (8) at $\mathbf{x}^0$ is

$$\nabla f(\mathbf{x}^0) = 2A^*(A\mathbf{x} - \mathbf{b}),$$

proving the necessity of (9). The sufficiency follows from the identity

$$(A\mathbf{x} - \mathbf{b})^*(A\mathbf{x} - \mathbf{b}) - (A\mathbf{x}^0 - \mathbf{b})^*(A\mathbf{x}^0 - \mathbf{b})$$
$$= (\mathbf{x} - x^0)^*A^*A(\mathbf{x} - \mathbf{x}^0) + 2\Re\{(\mathbf{x} - x^0)^*A^*(\mathbf{x} - x^0)\},$$

which holds for all $\mathbf{x}, \mathbf{x}^0 \in \mathbb{C}^n$. $\qquad\square$

EX. 2. For any $A \in \mathbb{C}^{m \times n}$ and $\mathbf{b} \in \mathbb{C}^m$, the normal equation (9) is consistent.

EX. 3. The general least-squares solution of $A\mathbf{x} = \mathbf{b}$,

$$\mathbf{x} = A^{(1,3)}\mathbf{b} + (I_n - A^{(1,3)}A)\mathbf{y}, \quad \mathbf{y} \in \mathbb{C}^n, \tag{8}$$

$$= (A^*A)^{(1)}A^*\mathbf{b} + (I_n - (A^*A)^{(1)}A^*A)\mathbf{y}, \quad \mathbf{y} \in \mathbb{C}^n, \tag{11}$$

where $(A^*A)^{(1)}$ is any $\{1\}$-inverse of A^*A.

PROOF. By (1.12a), $(A^*A)^{(1)}A^*$ is a $\{1,2,3\}$-inverse of A (little more than needed here). $\qquad\square$

EX. 4. *Ill-conditioning.* The linear equation $A\mathbf{x} = \mathbf{b}$ and the matrix A are said to be *ill-conditioned* (or badly conditioned) if the solutions are very sensitive to small changes in the data, see, e.g., [615, Chapter 8] and [873].

If the matrix A is ill-conditioned, then A^*A is even worse conditioned, see Ex. 6.11 below. The following example illustrates the ill-conditioning of the normal equation. Let

$$A = \begin{bmatrix} 1 & 1 \\ \epsilon & 0 \\ 0 & \epsilon \end{bmatrix} \quad \text{and let the elements of} \quad A^TA = \begin{bmatrix} 1+\epsilon^2 & 1 \\ 1 & 1+\epsilon^2 \end{bmatrix}$$

be computed using double-precision and then rounded to single-precision with t binary digits. If $|\epsilon| < \sqrt{2^{-t}}$, then the rounded A^TA is

$$\mathrm{fl}(A^TA) = \begin{bmatrix} 1 & 1 \\ 1 & 1 \end{bmatrix} \quad \text{(fl denotes *floaating point*)}$$

which is of rank 1, whereas A is of rank 2. Thus for any $\mathbf{b} \in \mathbb{R}^3$, the computed normal equation

$$\mathrm{fl}(A^TA)\mathbf{x} = \mathrm{fl}(A^T\mathbf{b})$$

may be inconsistent, or may have solutions which are not least-squares solutions of $A\mathbf{x} = \mathbf{b}$.

EX. 5. *Noble's method.* Let again $A \in \mathbb{C}_n^{m \times n}$ and assume that A is partitioned as

$$A = \begin{bmatrix} A_1 \\ A_2 \end{bmatrix} \quad \text{where } A_1 \in \mathbb{C}_n^{n \times n}.$$

Then A may be factorized as

$$A = \begin{bmatrix} I \\ S \end{bmatrix} A_1 \quad \text{where } S = A_2 A_1^{-1} \in \mathbb{C}^{(m-n) \times n}. \tag{12}$$

Now let $\mathbf{b} \in \mathbb{C}^m$ be partitioned as $\mathbf{b} = \begin{bmatrix} \mathbf{b}_1 \\ \mathbf{b}_2 \end{bmatrix}$, $\mathbf{b}_1 \in \mathbb{C}^n$. Then the normal equation reduces to

$$(I + S^*S)A_1\mathbf{x} = \mathbf{b}_1 + S^*\mathbf{b}_2 \tag{13}$$

(which reduces further to $A_1\mathbf{x} = \mathbf{b}_1$ if and only if $A\mathbf{x} = \mathbf{b}$ is consistent).

The matrix S can be obtained by applying Gauss–Jordan elimination to the matrix

$$\begin{bmatrix} A_1 & \mathbf{b}_1 & I \\ A_2 & \mathbf{b}_2 & O \end{bmatrix}$$

transforming it into

$$\begin{bmatrix} I & A_1^{-1}\mathbf{b}_1 & A_1^{-1} \\ O & \mathbf{b}_2 - S\mathbf{b}_1 & -S \end{bmatrix}$$

from which S can be read. (See Noble [615, pp. 262–265].)

Ex. 6. *Iterative refinement of solutions.* Let $\mathbf{x}^{(0)}$ be an approximate solution of the consistent equation $A\mathbf{x} = \mathbf{b}$ and let $\widehat{\mathbf{x}}$ be an exact solution. Then the error $\delta\mathbf{x} = \widehat{\mathbf{x}} - \mathbf{x}^{(0)}$ satisfies

$$A\delta\mathbf{x} = A\widehat{\mathbf{x}} - A\mathbf{x}^{(0)}$$
$$= \mathbf{b} - A\mathbf{x}^{(0)}$$
$$= \mathbf{r}^{(0)}, \quad \text{the } residual \text{ corresponding to } \mathbf{x}^{(0)}.$$

This suggests the following *iterative refinement of solutions*, due to Wilkinson [**872**] (see also Moler [**572**]):

The initial approximation: $\mathbf{x}^{(0)}$, given.
The kth residual: $\mathbf{r}^{(k)} = \mathbf{b} - A\mathbf{x}^{(k)}$.
The kth correction, $\delta\mathbf{x}^{(k)}$, is obtained by solving $A\delta\mathbf{x}^{(k)} = \mathbf{r}^{(k)}$.
The $(k+1)^{\text{st}}$ approximation: $\mathbf{x}^{(k+1)} = \mathbf{x}^{(k)} + \delta\mathbf{x}^{(k)}$.
Double precision is used in computing the residuals, but not elsewhere.
The iteration is stopped if $\|\delta\mathbf{x}^{(k+1)}\|/\|\delta\mathbf{x}^{(k)}\|$ falls below a prescribed number.
If the sequence $\{\mathbf{x}^{(k)} : k = 0, 1, \ldots\}$ converges, it converges to a solution of $A\mathbf{x} = \mathbf{b}$.

The use of this method to solve linear equations which are equivalent to the normal equation, such as (66) or (13), has been successful in finding, or improving, least-squares solutions. The reader is referred to Golub and Wilkinson [**312**], Björck [**105**], [**107**], and Björck and Golub [**111**].

Ex. 7. Show that the vector $\mathbf{x}$ is a least-squares solution of $A\mathbf{x} = \mathbf{b}$ if and only if there is a vector $\mathbf{r}$ such that the vector $\begin{bmatrix} \mathbf{r} \\ \mathbf{x} \end{bmatrix}$ is a solution of

$$\begin{bmatrix} I & A \\ A^* & O \end{bmatrix} \begin{bmatrix} \mathbf{r} \\ \mathbf{x} \end{bmatrix} = \begin{bmatrix} \mathbf{b} \\ \mathbf{0} \end{bmatrix}. \tag{14}$$

Ex. 8. Let $A \in \mathbb{C}^{m \times n}$ and let $\mathbf{b}_1, \mathbf{b}_2, \ldots, \mathbf{b}_k \in \mathbb{C}^m$. Show that a vector $\mathbf{x}$ minimizes

$$\sum_{i=1}^{k} \|A\mathbf{x} - \mathbf{b}_i\|^2$$

if and only if $\mathbf{x}$ is a least-squares solution of

$$A\mathbf{x} = \frac{1}{k} \sum_{i=1}^{k} \mathbf{b}_i.$$

Ex. 9. Let $A_i \in \mathbb{C}^{m \times n}$, $\mathbf{b}_i \in \mathbb{C}^m$ $(i = 1, \ldots, k)$. Show that a vector $\mathbf{x}$ minimizes

$$\sum_{i=1}^{k} \|A_i\mathbf{x} - \mathbf{b}_i\|^2 \tag{15}$$

if and only if $\mathbf{x}$ is a solution of

$$\left(\sum_{i=1}^{k} A_i^* A_i\right)\mathbf{x} = \sum_{i=1}^{k} A_i^* \mathbf{b}_i. \tag{16}$$

SOLUTION. $\mathbf{x}$ minimizes (15) if and only if $\mathbf{x}$ is a least-squares solution of the system

$$\begin{bmatrix} A_1 \\ A_2 \\ \vdots \\ A_k \end{bmatrix} \mathbf{x} = \begin{bmatrix} \mathbf{b}_1 \\ \mathbf{b}_2 \\ \vdots \\ \mathbf{b}_k \end{bmatrix},$$

whose normal equation is (16). $\square$

Constrained Least-Squares Solutions

EX. 10. A vector $\mathbf{x}$ is said to be a *constrained least-squares solution* if $\mathbf{x}$ is a solution of the constrained minimization problem: Minimize $\|A\mathbf{x} - \mathbf{b}\|$ subject to the given constraints. Let $A_1 \in \mathbb{C}^{m_1 \times n}$, $\mathbf{b}_1 \in \mathbb{C}^{m_1}$, $A_2 \in \mathbb{C}^{m_2 \times n}$, $\mathbf{b}_2 \in R(A_2)$. Characterize the solutions of the problem:

$$\text{minimize } \|A_1\mathbf{x} - \mathbf{b}_1\|^2 \tag{17a}$$

$$\text{subject to } A_2\mathbf{x} = \mathbf{b}_2. \tag{17b}$$

SOLUTION. The general solution of (17b) is

$$\mathbf{x} = A_2^{(1)}\mathbf{b}_2 + (I - A_2^{(1)}A_2)\mathbf{y}, \tag{18}$$

where $A_2^{(1)} \in A_2\{1\}$ and $\mathbf{y}$ ranges over $\mathbb{C}^n$. Substituting (18) in $A_1\mathbf{x} = \mathbf{b}_1$ gives the equation

$$A_1(I - A_2^{(1)}A_2)\mathbf{y} = \mathbf{b}_1 - A_1A_2^{(1)}\mathbf{b}_2, \tag{19a}$$

$$\text{abbreviated as } \ \overline{A}\mathbf{y} = \overline{\mathbf{b}}. \tag{19b}$$

Therefore $\mathbf{x}$ is a constrained least-squares solution of (17) if and only if $\mathbf{x}$ is given by (18) with $\mathbf{y}$ a least-squares solution of (19),

$$\mathbf{x} = A_2^{(1)}\mathbf{b}_2 + (I - A_2^{(1)}A_2)[\overline{A}^{(1,3)}\overline{\mathbf{b}} + (I - \overline{A}^{(1,3)}\overline{A})\mathbf{z}], \quad \mathbf{z} \in \mathbb{C}^n, \tag{20}$$

where $\overline{A}^{(1,3)}$ is an arbitrary $\{1,3\}$-inverse of $\overline{A}$. $\square$

EX. 11. Show that a vector $\mathbf{x} \in \mathbb{C}^n$ is a solution of (17) if and only if there is a vector $\mathbf{y} \in \mathbb{C}^{m_2}$ such that the vector $\begin{bmatrix} \mathbf{x} \\ \mathbf{y} \end{bmatrix}$ is a solution of

$$\begin{bmatrix} A_1^*A_1 & A_2^* \\ A_2 & O \end{bmatrix} \begin{bmatrix} \mathbf{x} \\ \mathbf{y} \end{bmatrix} = \begin{bmatrix} A_1^*\mathbf{b}_1 \\ \mathbf{b}_2 \end{bmatrix}. \tag{21}$$

Compare this with Ex. 1. Similarly, find a characterization analogous to that given in Ex. 7. See also Björck and Golub [**111**].

2. Solutions of Minimum Norm

When the system (1) has a multiplicity of solutions for $\mathbf{x}$, there is a unique solution of minimum-norm. This follows from Ex. 2.88, restated here as,

LEMMA 1. *Let $A \in \mathbb{C}^{m \times n}$. Then A is a one-to-one mapping of $R(A^*)$ onto $R(A)$.* $\square$

COROLLARY 2. *Let $A \in \mathbb{C}^{m \times n}$, $\mathbf{b} \in R(A)$. Then there is a unique solution of*

$$A\mathbf{x} = \mathbf{b} \tag{1}$$

given as the unique solution of (1) which lies in $R(A^*)$.

PROOF. By Lemma 1, Eq. (1) has a unique solution $\mathbf{x}_0$ in $R(A^*)$. Now the general solution is given as

$$\mathbf{x} = \mathbf{x}_0 + \mathbf{y}, \quad \mathbf{y} \in N(A),$$

and, by Ex. 2.46,

$$\|\mathbf{x}\|^2 = \|\mathbf{x}_0\|^2 + \|\mathbf{y}\|^2$$

proving that $\|\mathbf{x}\| > \|\mathbf{x}_0\|$ unless $\mathbf{x} = \mathbf{x}_0$. $\qquad\square$

The following theorem relates minimum-norm solutions of $A\mathbf{x} = \mathbf{b}$ and $\{1,4\}$-inverses of A, characterizing each of these two concepts in terms of the other.

THEOREM 2. Let $A \in \mathbb{C}^{m \times n}$, $\mathbf{b} \in \mathbb{C}^m$. If $A\mathbf{x} = \mathbf{b}$ has a solution for $\mathbf{x}$, the unique solution for which $\|\mathbf{x}\|$ is smallest is given by

$$\mathbf{x} = A^{(1,4)}\mathbf{b},$$

where $A^{(1,4)} \in A\{1,4\}$. Conversely, if $X \in \mathbb{C}^{n \times m}$ is such that, whenever $A\mathbf{x} = \mathbf{b}$ has a solution, $\mathbf{x} = X\mathbf{b}$ is the solution of minimum-norm, then $X \in A\{1,4\}$.

PROOF. If $A\mathbf{x} = \mathbf{b}$ is consistent, then for any $A^{(1,4)} \in A\{1,4\}$, $\mathbf{x} = A^{(1,4)}\mathbf{b}$ is a solution (by Corollary 2.2), lies in $R(A^*)$ (by Ex. 1.9) and thus, by Lemma 1, is the unique solution in $R(A^*)$, and thus the unique minimum-norm solution by Corollary 2.

Conversely, let X be such that, for all $\mathbf{b} \in R(A)$, $\mathbf{x} = X\mathbf{b}$ is the solution of $A\mathbf{x} = \mathbf{b}$ of minimum-norm. Setting $\mathbf{b}$ equal to each column of A, in turn, we conclude that

$$XA = A^{(1,4)}A$$

and $X \in A\{1,4\}$ by Theorem 2.4. $\qquad\square$

The unique minimum-norm least-squares solution of $A\mathbf{x} = \mathbf{b}$, and the generalized inverse $A^\dagger$ of A, are related as follows:

COROLLARY 3 (Penrose [636]). Let $A \in \mathbb{C}^{m \times n}$, $\mathbf{b} \in \mathbb{C}^m$. Then, among the least-squares solutions of $A\mathbf{x} = \mathbf{b}$, $A^\dagger\mathbf{b}$ is the one of minimum-norm. Conversely, if $X \in \mathbb{C}^{n \times m}$ has the property that, for all $\mathbf{b}$, $X\mathbf{b}$ is the minimum-norm least-squares solution of $A\mathbf{x} = \mathbf{b}$, then $X = A^\dagger$.

PROOF. By Corollary 1, the least-squares solutions of $A\mathbf{x} = \mathbf{b}$ coincide with the solutions of

$$A\mathbf{x} = AA^{(1,3)}\mathbf{b}. \tag{6}$$

Thus the minimum-norm least-squares solution of $A\mathbf{x} = \mathbf{b}$ is the minimum-norm solution of (6). But, by Theorem 2, the latter is

$$\mathbf{x} = A^{(1,4)}AA^{(1,3)}\mathbf{b}$$

$$= A^\dagger\mathbf{b}$$

by Theorem 1.4.

A matrix X having the properties stated in the last sentence of the theorem must satisfy $X\mathbf{b} = A^\dagger\mathbf{b}$ for all $\mathbf{b} \in \mathbb{C}^m$ and, therefore, $X = A^\dagger$. $\qquad\square$

The minimum-norm least-squares solution, $\mathbf{x}_0 = A^\dagger\mathbf{b}$ (also called the approximate solution; e.g., Penrose [636]) of $A\mathbf{x} = \mathbf{b}$, can thus be characterized by the following two inequalities:

$$\|A\mathbf{x}_0 - \mathbf{b}\| \le \|A\mathbf{x} - \mathbf{b}\|, \quad \text{for all } \mathbf{x} \tag{22a}$$

and

$$\|\mathbf{x}_0\| < \|\mathbf{x}\| \tag{22b}$$

for any $\mathbf{x} \neq \mathbf{x}_0$ which gives equality in (22a).

Exercises and Supplementary Notes

Ex. 12. Let A be given by (0.80) and let

$$\mathbf{b} = \begin{bmatrix} -i \\ 1 \\ 1 \end{bmatrix}.$$

Show that the general least-squares solution of $A\mathbf{x} = \mathbf{b}$ is

$$\mathbf{x} = \frac{1}{19}\begin{bmatrix} 0 \\ 1 \\ 0 \\ -4 \\ 0 \\ 0 \end{bmatrix} + \begin{bmatrix} 1 & 0 & 0 & 0 & 0 & 0 \\ 0 & 0 & -\frac{1}{2} & 0 & -1+2i & \frac{1}{2}i \\ 0 & 0 & 1 & 0 & 0 & 0 \\ 0 & 0 & 0 & 0 & -2 & -1-i \\ 0 & 0 & 0 & 0 & 1 & 0 \\ 0 & 0 & 0 & 0 & 0 & 1 \end{bmatrix}\begin{bmatrix} y_1 \\ y_2 \\ y_3 \\ y_4 \\ y_5 \\ y_6 \end{bmatrix},$$

where $y_1, y_2, \dots, y_6$ are arbitrary, while the residual vector for the least-squares solution is

$$\frac{1}{19}\begin{bmatrix} 2i \\ 12 \\ -2 \end{bmatrix}.$$

Ex. 13. In Ex. 12 show that the minimum-norm least-squares solution is

$$\mathbf{x} = \frac{1}{874}\begin{bmatrix} 0 \\ 26 - 36i \\ 13 - 18i \\ -55 - 9i \\ -12 - 2i \\ -46 + 59i \end{bmatrix}.$$

Ex. 14. Let $A \in \mathbb{C}^{m \times n}$, $\mathbf{b} \in \mathbb{C}^m$, and $\mathbf{a} \in \mathbb{C}^n$. Show that if $A\mathbf{x} = \mathbf{b}$ has a solution for $\mathbf{x}$, then the unique solution for which $\|\mathbf{x} - \mathbf{a}\|$ is smallest is given by

$$\mathbf{x} = A^{(1,4)}\mathbf{b} + (I - A^{(1,4)}A)\mathbf{a}$$
$$= A^{(1,4)}\mathbf{b} + P_{N(A)}\mathbf{a}.$$

Ex. 15. *Matrix spaces.* For any $A, B \in \mathbb{C}^{m \times n}$ define

$$R(A, B) = \{Y = AXB \in \mathbb{C}^{m \times n} : X \in \mathbb{C}^{n \times m}\} \tag{23}$$

and

$$N(A, B) = \{X \in \mathbb{C}^{n \times m} : AXB = O\} \tag{24}$$

which we shall call the *range* and *null space* of (A, B), respectively. Let $\mathbb{C}^{m \times n}$ be endowed with the inner product

$$\langle X, Y \rangle = \text{trace } Y^*X = \sum_{i=1}^{m} \sum_{j=1}^{n} x_{ij}\,\overline{y_{ij}}, \tag{25}$$

for $X = [x_{ij}]$, $Y = [y_{ij}] \in \mathbb{C}^{m \times n}$. Then, for every $A, B \in \mathbb{C}^{m \times n}$, the sets $R(A, B)$ and $N(A^*, B^*)$ are complementary orthogonal subspaces of $\mathbb{C}^{m \times n}$.

SOLUTION. As in Ex. 2.2 we use the one-to-one correspondence

$$v_{n(i-1)+j} = x_{ij} \quad (i \in \overline{1, m}, \, j \in \overline{1, n}) \tag{26}$$

between the matrices $X = [x_{ij}] \in \mathbb{C}^{m \times n}$ and the vectors $\mathbf{v} = \text{vec}(X) = [v_k] \in \mathbb{C}^{mn}$. The correspondence (26) is a nonsingular linear transformation mapping $\mathbb{C}^{m \times n}$ onto C^{mn}. Linear subspaces of $\mathbb{C}^{m \times n}$ and C^{mn} thus correspond under (26).

It follows from (26) that the inner product (25) is equal to the standard inner product of the corresponding vectors $\text{vec}(X)$ and $\text{vec}(Y)$. Thus $\langle X, Y \rangle = \langle \text{vec}(X), \text{vec}(Y) \rangle = \text{vec}(Y)^* \text{vec}(X)$. Also, from (2.10) we deduce that under (26), $R(A, B)$ and $N(A^*, B^*)$ correspond to $R(A \otimes B^T)$ and $N(A^* \otimes B^{*T})$, respectively. By (2.8), the latter is the same as $N((A \otimes B^T)^*)$, which by (0.27) is the orthogonal complement of $R(A \otimes B^T)$ in C^{mn}. Therefore, $R(A, B)$ and $N(A^*, B^*)$ are orthogonal complements in $\mathbb{C}^{m \times n}$. $\qquad\square$

EX. 16. Let A, B, C be matrices of compatible dimensions. Then the following statements are equivalent:

(a) $R(C) \subset R(A)$ and $R(B^*) \subset R(A^*)$,
(b) $BA^{(1)}C$ is invariant under the choice of $A^{(1)} \in A\{1\}$,
(c) $N(A, A) \subset N(B, C)$.

PROOF. (a) $\Longrightarrow$ (b) (this part is due to Rao and Mitra [**678**, Lemma 2.2.4(iii)]). Let $BA\{1\}C$ denote $\{BXC : X \in A\{1\}\}$. Then

$BA\{1\}C = BA^\dagger C + \{BZC - BA^\dagger AZAA^\dagger C : Z \text{ arbitrary}\}, \quad \text{by Corollary 2.1,}$

$\qquad = \{BA^\dagger C\} \quad \text{if} \quad B = BA^\dagger A, C = AA^\dagger C.$

(b) $\Longleftrightarrow$ (c) Let $Z = Z_1 + Z_2$, $AZ_1 A = O$, $Z_2 = A^* X A^*$. Then

$BA\{1\}C = BA^\dagger C + \{BZ_1 C + BA^* X A^* C - BA^\dagger AA^* X A^* AA^\dagger C : AZ_1 A = O\}$

$\qquad = BA^\dagger C + \{BZ_1 C : AZ_1 A = O\}$

$\qquad = \{BA^\dagger C\} \quad \text{if and only if (c).}$

(c) $\Longrightarrow$ (a) Suppose $R(B^*) \not\subset R(A^*)$, i.e., there is a vector $\mathbf{x} \in R(B^*)$ with $\mathbf{x} = \mathbf{x}_1 + \mathbf{x}_2, \mathbf{x}_1 \in R(A^*), 0 \neq \mathbf{x}_2 \in N(A)$. Let the vector $\mathbf{y}$ satisfy $C^* \mathbf{y} \neq \mathbf{0}$ ($C \neq O$ can be assumed), and let $X = \mathbf{x}_2 \mathbf{y}^*$. Then

$$AXA = O, \quad BXC \neq O,$$

contradicting (c). An analogous proof applies to the case $R(C) \not\subset R(A)$. $\qquad\square$

EX. 17. *Characterization of* $\{1, 3\}$-, $\{1, 4\}$-, *and* $\{1, 2, 3, 4\}$-*inverses.* Let the norm used in $\mathbb{C}^{m \times n}$ be the Frobenius norm

$$\|X\|_F = \sqrt{\text{trace } X^* X}. \tag{0.50}$$

Show that for every $A \in \mathbb{C}^{m \times n}$:

(a) $X \in A\{1, 3\}$ if and only if X is a least-squares solution of

$$AX = I_m, \tag{27}$$

i.e., minimizing $\|AX - I\|_F$.

(b) $X \in A\{1, 4\}$ if and only if X is a least-squares solution of

$$XA = I_n. \tag{28}$$

(c) $A^\dagger$ is the minimum-norm least-squares solution of both (27) and (28).

SOLUTION. These results are based on the fact that the norm $\|X\|_F$ defined by (0.50) is merely the Euclidean norm of the corresponding vector $\mathrm{vec}(X)$.

(a) Writing (27) as

$$(A \otimes I)\mathrm{vec}(X) = \mathrm{vec}(I), \qquad (29)$$

it follows from Corollary 1 that the general least-squares solution of (29) is

$$\mathrm{vec}(X) = (A \otimes I)^{(1,3)} \mathrm{vec}(I) + (I - (A \otimes I)^{(1,3)}(A \otimes I))\mathbf{y}, \qquad (30)$$

where $\mathbf{y}$ is an arbitrary element of $\mathbb{C}^{mn}$. From (2.8) and (2.9) it follows that for every $A^{(1,3)} \in A\{1,3\}$, $(A^{(1,3)} \otimes I)$ is a $\{1,3\}$-inverse of $(A \otimes I)$. Therefore the general least-squares solution of (27) is the matrix corresponding to (30), namely

$$X = A^{(1,3)} + (I - A^{(1,3)}A)Y, \quad Y \in \mathbb{C}^{n \times m},$$

which is the general $\{1,3\}$-inverse of A by Corollary 2.3.

(b) Taking the conjugate transpose of (28), we get

$$A^* X^* = I_n.$$

The set of least-squares solutions of the last equation is, by (a),

$$A^*\{1,3\},$$

which coincides with $A\{1,4\}$.

(c) This is left to the reader.   □

EX. 18. Let A, B, D be complex matrices having dimensions consistent with the matrix equation

$$AXB = D.$$

Show that the minimum-norm least-squares solution of the last equation is

$$X = A^\dagger D B^\dagger \quad \text{(Penrose [636])}.$$

EX. 19. Let $A \in \mathbb{C}^{m \times n}$ and let X be a $\{1\}$-inverse of A; i.e., let X satisfy

$$AXA = A. \qquad (1.1)$$

Then the following are equivalent:

(a) $X = A^\dagger$;

(b) $X \in R(A^*, A^*)$; and

(c) X is the minimum-norm solution of (1.1) (Ben-Israel [67]).

PROOF. The general solution of (1.1) is, by Theorem 2.1,

$$X = A^\dagger A A^\dagger + Y - A^\dagger A Y A A^\dagger, \quad Y \in \mathbb{C}^{n \times m},$$
$$= A^\dagger + Y - A^\dagger A Y A A^\dagger. \qquad (31)$$

Now it is easy to verify that

$$A^\dagger \in R(A^*, A^*), \quad Y - A^\dagger A Y A A^\dagger \in N(A, A),$$

and using the Frobenius norm (0.50) it follows from Ex. 15 that X of (31) satisfies

$$\|X\|_F^2 = \|A^\dagger\|_F^2 + \|Y - A^\dagger A Y A A^\dagger\|_F^2,$$

and the equivalence of (a), (b), and (c) is obvious.   □

EX. 20. *Restricted generalized inverses.* Let the matrix $A \in \mathbb{C}^{m \times n}$ and let the subspace $S \subset \mathbb{C}^n$ be given. Then, for any $\mathbf{b} \in \mathbb{C}^m$, the point $X\mathbf{b} \in S$ minimizes $\|A\mathbf{x} - \mathbf{b}\|$ in S if and only if $X = P_S(AP_S)^{(1,3)}$ is any S-restricted $\{1.3\}$-inverse of A.

PROOF. Follows from Section 2.9 and Theorem 1.   □

EX. 21. Let A, S be as in Ex. 20. Then for any $\mathbf{b} \in \mathbb{C}^m$ for which the system

$$A\mathbf{x} = \mathbf{b}, \quad \mathbf{x} \in S, \tag{2.116}$$

is consistent, $X\mathbf{b}$ is the minimum-norm solution of (2.116) if and only if $X = P_S(AP_S)^{(1,4)}$ is any S-restricted $\{1,4\}$-inverse of A.

PROOF. Follows from Section 2.9 and Theorem 2. $\qquad\square$

EX. 22. Let A, S be as above. Then, for any $\mathbf{b} \in \mathbb{C}^m$, $X\mathbf{b}$ is the minimum-norm least-squares solution of (2.116) if and only if $X = P_S(AP_S)^{\dagger}$, the S-restricted Moore-Penrose inverse of A (Minamide and Nakamura [**556**]).

EX. 23. *Constrained least-squares solutions.* Let $A_1 \in \mathbb{C}^{m_1 \times n}$, $\mathbf{b}_1 \in \mathbb{C}^{m_1}$, $A_2 \in \mathbb{C}^{m_2 \times n}$, $\mathbf{b}_2 \in R(A_2)$. The minimum-norm solution of

$$\text{minimize } \|A_1\mathbf{x} - \mathbf{b}_1\|^2, \tag{17a}$$

$$\text{subject to } \quad A_2\mathbf{x} = \mathbf{b}_2, \tag{17b}$$

is

$$\mathbf{x} = A_2^{\dagger}\mathbf{b}_2 + P_{N(A_2)}\left(A_1 P_{N(A_2)}\right)^{\dagger}(\mathbf{b}_1 - A_1 A_2^{\dagger}\mathbf{b}_2). \tag{32}$$

PROOF. Follows as in Ex. 10, with $\mathbf{z} = \mathbf{0}$ in (20). $\qquad\square$

See also Ex. 6.92 below.

EX. 24. (Albert [**10**, Lemma, p. 185]). Let $V \in \mathbb{C}^{n \times n}$ be Hermitian and nonsingular and let $X \in \mathbb{C}^{n \times k}$. Then

$$(V^{-1}X)^{\dagger} = X^{\dagger}V(I - (VP_{N(X^*)})^{\dagger}(VP_{N(X^*)})). \tag{33}$$

PROOF. Denote $P_{N(X^*)}$ by Q. The unique minimum-norm least-squares solution of

$$(V^{-1}X)^*\mathbf{x} = \mathbf{y}$$

is $\widehat{\mathbf{x}} = (V^{-1}X)^{*\dagger}\mathbf{y}$. If $\mathbf{u}^*$ minimizes $\|X^*\mathbf{u} - \mathbf{y}\|^2$, then $\mathbf{x}^* := V\mathbf{u}^*$ minimizes $\|(V^{-1}X)^*\mathbf{x} - \mathbf{y}\|^2$. Moreover,

$$\|\mathbf{x}^*\| = \|V\mathbf{u}^*\| > \|\widehat{\mathbf{x}}\| \quad \text{unless} \quad \widehat{x} = V\mathbf{u}^*.$$

The general least-squares solution of $X^*\mathbf{u} = \mathbf{y}$ is

$$\mathbf{u}(\mathbf{w}) = X^{*\dagger}\mathbf{y} - Q\mathbf{w}, \quad \mathbf{w} \text{ arbitrary.} \tag{a}$$

The square

$$\|V\mathbf{u}(\mathbf{w})\|^2 = \|VX^{*\dagger}\mathbf{y} - VQ\mathbf{w}\|^2$$

is minimized when

$$\widehat{\mathbf{w}} = (VQ)^{\dagger}VX^{*\dagger}\mathbf{y}. \tag{b}$$

Moreover,

$$V\mathbf{u}(\widehat{\mathbf{w}}) = \widehat{\mathbf{x}}, \tag{c}$$

for otherwise $\|V\mathbf{u}(\widehat{\mathbf{w}})\| > \|\widehat{\mathbf{x}}\|$ and, for $\widehat{\mathbf{u}} := V^{-1}\widehat{\mathbf{x}}$,

$$\|V\widehat{\mathbf{u}}\| = \|\widehat{\mathbf{x}}\| < \|V\mathbf{u}(\widehat{\mathbf{w}})\|$$

a contradiction to $\mathbf{u}(\widehat{\mathbf{w}})$ minimizing $\|V\mathbf{u}\|$. Combining (a), (b), and (c) we get

$$(V^{-1}X)^{*\dagger}\mathbf{y} = (I - (VQ)^{\dagger}(VQ))VX^{*\dagger}\mathbf{y}$$

for all $\mathbf{y}$, proving (33).   □

3. Tikhonov Regularization

Let $A \in \mathbb{C}^{m \times n}$, $\mathbf{b} \in \mathbb{C}^m$. The minimum-norm least-squares solution of

$$A\mathbf{x} = \mathbf{b} \tag{1}$$

is $\mathbf{x} = A^\dagger \mathbf{b}$. It is literally the solution of a *two-stage minimization problem*:

   Stage 1:

$$\text{minimize} \quad \|A\mathbf{x} - \mathbf{b}\|. \tag{34}$$

   Stage 2:

$$\text{minimize} \quad \{\|\mathbf{x}\| \text{ among all solutions of Stage 1}\}. \tag{35}$$

The idea of *Tikhonov regularization* (Tikhonov [806], [805]) is to replace these two stages by one problem,

$$\min_{\mathbf{x} \in \mathbb{C}^n} \ f_{\alpha^2}(\mathbf{x}) \tag{36}$$

where the function

$$f_{\alpha^2}(\mathbf{x}) = \|A\mathbf{x} - \mathbf{b}\|^2 + \alpha^2 \|\mathbf{x}\|^2 \tag{37}$$

depends on a positive real parameter α^2. Let $\mathbf{x}_{\alpha^2}$ be the miminizer of f_{α^2}. Then

$$f_{\alpha^2} \to A^\dagger \mathbf{b} \quad \text{as} \quad \alpha \to 0, \quad \text{see Ex. 25 below,}$$

and it may seem that the limit $\alpha \to 0$ is desirable, for then (36) tends to coincide with the two-stage problem (34)–(35). There are, however, applications and contexts where the minimization problem (36), with positive α^2, is preferred. Some examples:

   (a) The constrained least squares problem

$$\text{minimize} \ \|A\mathbf{x} - \mathbf{b}\| \quad \text{subject to} \quad \|\mathbf{x}\| = p \tag{38}$$

has the function (37) as Lagrangian and α^2 as the Lagrange multiplier, see Ex. 29.

   (b) The norm $\|\mathbf{x}_{\alpha^2}\|$ is a monotone decreasing function of α, see Theorem 3. Positive values of α are required, if it is necessary to control the norm of the solution. An example is *ridge estimation*, see Section 8.4, where a trade-off between *bias* (which increases with α) and *variance* (or norm, decreases with α), may determine an optimal α.

   (c) If A is ill-conditioned, the solution $\mathbf{x}_{\alpha^2}$ of (36) is more stable, numerically, than the minimal-norm least-squares solution, see Ex. 6.13.

   The dependence of the solution $\mathbf{x}_{\alpha^2}$ is described in the following:

   THEOREM 3. *The function $f_{\alpha^2}(\mathbf{x})$ has a unique minimizer $\mathbf{x}_{\alpha^2}$ given by*

$$\mathbf{x}_{\alpha^2} = (A^*A + \alpha^2 I)^{-1} A^* \mathbf{b} \tag{39}$$

whose norm $\|\mathbf{x}_{\alpha^2}\|$ is a monotone decreasing function of α^2.

PROOF. The function (37) is a special case of (15) with $k = 2, A_1 = A, A_2 = \alpha I, \mathbf{b}_1 = \mathbf{b}$, and $\mathbf{b}_2 = \mathbf{0}$. Substituting these values in (16) we get

$$(A^*A + \alpha^2 I)\mathbf{x} = A^*\mathbf{b}, \tag{40}$$

which has the unique solution (39), since $(A^*A + \alpha^2 I)$ is nonsingular.

Using (0.26) or Lemma 1, it is possible to write $\mathbf{b}$ (uniquely) as

$$\mathbf{b} = A\mathbf{v} + \mathbf{u}, \quad \mathbf{v} \in R(A^*), \quad \mathbf{u} \in N(A^*). \tag{41}$$

Substituting this in (39) gives

$$\mathbf{x}_{\alpha^2} = (A^*A + \alpha^2 I)^{-1}A^*A\mathbf{v}. \tag{42}$$

Now let $\{\mathbf{v}_1, \mathbf{v}_2, \dots, \mathbf{v}_r\}$ be an o.n. basis of $R(A^*)$ consisting of eigenvectors of A^*A corresponding to nonzero eigenvalues, say

$$A^*A\mathbf{v}_j = \sigma_j^2 \mathbf{v}_j \quad (\sigma_j > 0, \ j \in \overline{1,r}). \tag{0.35a}$$

If $\mathbf{v} = \sum_{j=1}^r \beta_j \mathbf{v}_j$ is the representation of $\mathbf{v}$ in terms of the above basis, then (42) gives

$$\mathbf{x}_{\alpha^2} = \sum_{j=1}^r \frac{\sigma_j^2 \beta_j}{\sigma_j^2 + \alpha^2} \mathbf{v}_j$$

whose norm squared is

$$\|\mathbf{x}_{\alpha^2}\|^2 = \sum_{j=1}^r \left(\frac{\sigma_j^2}{\sigma_j^2 + \alpha^2}\right)^2 |\beta_j|^2,$$

a monotone decreasing function of α^2. $\qquad\square$

Problems of minimizing expressions like (37) in infinite-dimensional spaces and subject to linear constraints often arise in control theory. The reader is referred to [645], especially to Section 4.4 and pp. 353–354, where additional references are given. Tikhonov regularization originated, and is still mainly used, for solving linear operator equations, see references on page 151.

Exercises

Ex. 25. (den Broeder and Charnes [136]). For any $A \in \mathbb{C}^{m \times n}$, as $\lambda \to 0$ through any neighborhood of 0 in $\mathbb{C}$, the following limit exists and

$$\lim_{\lambda \to 0}(A^*A + \lambda I)^{-1}A^* = A^\dagger. \tag{43}$$

PROOF. We must show that

$$\lim_{\lambda \to 0}(A^*A + \lambda I)^{-1}A^*\mathbf{y} = A^\dagger\mathbf{y} \tag{44}$$

for all $\mathbf{y} \in \mathbb{C}^m$. Since $N(A^*) = N(A^\dagger)$, by Ex. 2.38, (44) holds trivially for $\mathbf{y} \in N(A^*)$. Therefore it suffices to prove (44) for $\mathbf{y} \in N(A^*)^\perp = R(A)$. By Lemma 1, for any $\mathbf{y} \in R(A)$, there is a unique $\mathbf{x} \in R(A^*)$ such that $\mathbf{y} = A\mathbf{x}$. Proving (44) thus amounts to proving, for all $\mathbf{x} \in R(A^*)$,

$$\lim_{\lambda \to 0}(A^*A + \lambda I)^{-1}A^*A\mathbf{x} = A^\dagger A\mathbf{x} \tag{45}$$

$$= \mathbf{x}, \quad \text{since } A^\dagger A = P_{R(A^*)}.$$

It thus suffices to show that

$$\lim_{\lambda \to 0}(A^*A + \lambda I_n)^{-1}A^*A = P_{R(A^*)}.$$

Now let $A^*A = FF^*$, $F \in \mathbb{C}_r^{n \times r}$ be a full-rank factorization. Then

$$(A^*A + \lambda I_n)^{-1}A^*A = (FF^* + \lambda I_n)^{-1}FF^*$$

for any λ for which the inverses exist. We now use the identity

$$(FF^* + \lambda I_n)^{-1}FF^* = F(F^*F + \lambda I_r)^{-1}F^*$$

and note that F^*F is nonsingular so that $\lim_{\lambda \to 0}(F^*F + \lambda I_r)^{-1} = (F^*F)^{-1}$. Collecting these facts we conclude that

$$\lim_{\lambda \to 0}(A^*A + \lambda I_n)^{-1}A^*A = F(F^*F)^{-1}F^*$$
$$= FF^\dagger = P_{R(A^*)}$$

since the columns of F are a basis for $R(A^*A) = R(A^*)$.   □

See also Exs. 4.21 and 6.13 below, Chernoff [185], Foster [283], and Ben-Israel [70].

Ex. 26. An alternative proof of (43).
PROOF. It suffices to prove

$$\lim_{\lambda \to 0}(A^*A + \lambda I)^{-1}A^*A\mathbf{x} = \mathbf{x} \tag{3.45}$$

for all $\mathbf{x} \in R(A^*)$. Let $\{\mathbf{v}_1, \dots, \mathbf{v}_r\}$ be a basis for $R(A^*)$ consisting of eigenvectors of A^*A, say

$$A^*A\mathbf{v}_j = \sigma_j^2 \mathbf{v}_j \quad (\sigma_j > 0, \ j \in \overline{1,r}). \tag{0.35a}$$

Writing $\mathbf{x} \in R(A^*)$ in terms of this basis

$$\mathbf{x} = \sum_{j=1}^{r} \xi_j \mathbf{v}_j,$$

we verify that, for all $\lambda \neq -\sigma_1^2, -\sigma_2^2, \dots, -\sigma_r^2$,

$$(A^*A + \lambda I)^{-1}A^*A\mathbf{x} = \sum_{j=1}^{r} \frac{\sigma_j^2 \xi_j}{\sigma_j^2 + \lambda}\mathbf{v}_j,$$

which tends, as $\lambda \to 0$, to $\sum_{j=1}^{r} \xi_j \mathbf{v}_j = \mathbf{x}$.   □

Ex. 27. (Boyarintsev [130, Theorem 1.2.3]). The approximation error of (43) for real positive λ is

$$\|A^\dagger - (A^*A + \lambda I)^{-1}A^*\|_2 \leq \lambda \|A^\dagger\|_2^3 \tag{46}$$

where $\|\cdot\|_2$ is the spectral norm (0.56.2).
PROOF.

$$A^\dagger - (A^*A + \lambda I)^{-1}A^* = (A^*A + \lambda I)^{-1}((A^*A + \lambda I)A^\dagger - A^*)$$
$$= \lambda(A^*A + \lambda I)^{-1}A^\dagger, \quad \text{since } A^* = A^*AA^\dagger,$$
$$= \lambda(A^*A + \lambda I)^{-1}A^*AA^\dagger A^{\dagger *}A^\dagger, \quad \text{since } A^*A^{\dagger *} = A^\dagger A.$$

If $\lambda > 0$, then $\|(A^*A + \lambda I)^{-1}A^*A\|_2 \leq 1$ and, therefore,

$$\|A^\dagger - (A^*A + \lambda I)^{-1}A^*\|_2 \leq \lambda \|A^\dagger A^{\dagger *}A^\dagger\|_2$$

and (46) follows since $\|\cdot\|_2$ is multiplicative.   □

Ex. 28. Use Theorem 3 and Ex. 25 to conclude that the solutions $\{\mathbf{x}_{\alpha 2}\}$ of the minimization problems:

$$\text{minimize } \{\|A\mathbf{x} - \mathbf{b}\|^2 + \alpha^2\|\mathbf{x}\|^2\}$$

converge to $A^\dagger \mathbf{b}$ as $\alpha \to 0$. Explain this result in view of Corollary 3.

Ex. 29. Let $A \in \mathbb{C}_r^{m \times n}$, $\mathbf{b} \in \mathbb{C}^m$, and let $0 < p < \sqrt{r}\|\mathbf{u}\|$ where $\mathbf{u}$ is given by (41). Show that the problem

$$\text{minimize } \|A\mathbf{x} - \mathbf{b}\| \quad \text{subject to} \quad \|\mathbf{x}\| = p \tag{38}$$

has the unique solution

$$\mathbf{x} = (A^*A + \alpha^2 I)^{-1} A^* \mathbf{b}$$

where α is (uniquely) determined by

$$\|(A^*A + \alpha^2 I)^{-1} A^* \mathbf{b}\| = p.$$

Hint. Use Theorem 3.

See also Forsythe and Golub [**282**, Section 7] and Forsythe [**281**].

Ex. 30. For a given $A \in \mathbb{C}^{m \times n}$, $\mathbf{b} \in \mathbb{C}^m$, and a positive real number p, solve the problem

$$\text{minimize } \|A\mathbf{x} - \mathbf{b}\| \quad \text{subject to} \quad \|\mathbf{x}\| \leq p. \tag{47}$$

SOLUTION. If

$$\|A^\dagger \mathbf{b}\| \leq p, \tag{48}$$

then $\mathbf{x} = A^\dagger \mathbf{b}$ is a solution of (47) and is the unique solution if and only if (48) is an equality.

If (48) does not hold, then (47) has the unique solution given in Ex. 29. □

See also Balakrishnan [**40**, Theorem 2.3].

4. Weighted Generalized Inverses

It may be desired to give different weights to the different squared residuals of the linear system $A\mathbf{x} = \mathbf{b}$. This is a more general problem than the one solved by the $\{1, 3\}$-inverse. A still further generalization which, however, presents no greater mathematical difficulty, is the minimizing of a given PD quadratic form in the residuals or, in other words, the minimizing of

$$\|A\mathbf{x} - \mathbf{b}\|_W^2 = (A\mathbf{x} - \mathbf{b})^* W (A\mathbf{x} - \mathbf{b}), \tag{49}$$

where W is a given PD matrix, see Ex. 0.4.

When A is not of full column rank, this problem does not have a unique solution for $\mathbf{x}$ and we may choose from the class of "generalized least-squares solutions" the one for which

$$\|\mathbf{x}\|_Q^2 = \mathbf{x}^* Q \mathbf{x} \tag{50}$$

is smallest, where Q is a second PD matrix. If $A \in \mathbb{C}^{m \times n}$, W is of order m and Q of order n.

Since every inner product in $\mathbb{C}^n$ can be represented as $\mathbf{x}^* Q \mathbf{y}$ for some PD matrix Q (see Ex. 0.4), it follows that the problem of minimizing (49), and the problem of minimizing (50) among all the minimizers of (49), differ from the problems treated in Sections 1 and 2 only in the different choices of inner products and their associated norms in $\mathbb{C}^m$ and $\mathbb{C}^n$. These seemingly more general problems can be reduced by a simple transformation to the "unweighted" problems considered in Sections 1 and 2. Every PD matrix H has a unique PD *square root*: that is a PD matrix K such that $K^2 = H$ (see, e.g., Ex. 31 and Ex. 6.37 below). Let us denote this K by $H^{1/2}$ and its inverse by $H^{-1/2}$.

We shall now introduce the transformations

$$\widetilde{A} = W^{1/2}AQ^{-1/2}, \quad \widetilde{\mathbf{x}} = Q^{1/2}\mathbf{x}, \quad \widetilde{\mathbf{b}} = W^{1/2}\mathbf{b}, \tag{51}$$

and it is easily verified that

$$\|A\mathbf{x} - \mathbf{b}\|_W = \|\widetilde{A}\widetilde{\mathbf{x}} - \widetilde{\mathbf{b}}\| \tag{52}$$

and

$$\|\mathbf{x}\|_Q = \|\widetilde{\mathbf{x}}\|, \tag{53}$$

expressing the norms $\|\ \|_W$ and $\|\ \|_Q$ in terms of the Euclidean norms of the transformed vectors.

Similarly, the relations

$$X = Q^{-1/2}YW^{1/2}, \quad \text{or} \quad Y = Q^{1/2}XW^{-1/2}, \tag{54}$$

result in

$$\widetilde{\mathbf{x}} = Y\widetilde{\mathbf{b}} \quad \Longleftrightarrow \quad \mathbf{x} = X\mathbf{b}, \tag{55}$$

$$\widetilde{A}Y\widetilde{A} = \widetilde{A} \quad \Longleftrightarrow \quad AXA = A, \tag{56}$$

$$(\widetilde{A}Y)^* = \widetilde{A}Y \quad \Longleftrightarrow \quad (WAX)^* = WAX, \tag{57}$$

$$(Y\widetilde{A})^* = Y\widetilde{A} \quad \Longleftrightarrow \quad (QXA)^* = QXA. \tag{58}$$

These observations lead to the following two theorems.

THEOREM 4. *Let $A \in \mathbb{C}^{m \times n}$, $\mathbf{b} \in \mathbb{C}^m$, and let $W \in \mathbb{C}^{m \times m}$ be positive definite. Then $\|A\mathbf{x} - \mathbf{b}\|_W$ is smallest when $\mathbf{x} = X\mathbf{b}$, where X satisfies*

$$AXA = A, \quad (WAX)^* = WAX. \tag{59}$$

Conversely, if $X \in \mathbb{C}^{n \times m}$ has the property that, for all $\mathbf{b}$, $\|A\mathbf{x} - \mathbf{b}\|_W$ is smallest when $\mathbf{x} = X\mathbf{b}$, then X satisfies (59).

PROOF. In view of (52), it follows from Theorem 1 that $\|A\mathbf{x} - \mathbf{b}\|_W$ is smallest when $\widetilde{\mathbf{x}} = Y\widetilde{\mathbf{b}}$, where Y satisfies

$$\widetilde{A}Y\widetilde{A} = \widetilde{A}, \quad (\widetilde{A}Y)^* = \widetilde{A}Y, \tag{60}$$

and also if $Y \in \mathbb{C}^{n \times m}$ has the property that, for all $\widetilde{\mathbf{b}}$, $\|\widetilde{A}\widetilde{\mathbf{x}} - \widetilde{\mathbf{b}}\|$ is smallest when $\widetilde{\mathbf{x}} = Y\widetilde{\mathbf{b}}$, then Y satisfies (60).

Now let X, Y be related by (54) with $Q = I$. The proof then follows from (55), (56), and (57). □

See also Ex. 33.

THEOREM 5. *Let $A \in \mathbb{C}^{m \times n}$, $\mathbf{b} \in \mathbb{C}^m$, and let $Q \in \mathbb{C}^{n \times n}$ be positive definite. If $A\mathbf{x} = \mathbf{b}$ has a solution for $\mathbf{x}$, the unique solution for which $\|\mathbf{x}\|_Q$ is smallest is given by $\mathbf{x} = X\mathbf{b}$, where X satisfies*

$$AXA = A, \quad (QXA)^* = QXA. \tag{61}$$

Conversely, if $X \in \mathbb{C}^{n \times m}$ is such that, whenever $A\mathbf{x} = \mathbf{b}$ has a solution, $\mathbf{x} = X\mathbf{b}$ is the solution for which $\|\mathbf{x}\|_Q$ is smallest, then X satisfies (61).

PROOF. In view of (51),

$$Ax = b \quad \Longleftrightarrow \quad \widetilde{A}\widetilde{x} = \widetilde{b}.$$

Then it follows from (53) and Theorem 2 that, if $Ax = b$ has a solution for x, the unique solution for which $\|x\|_Q$ is smallest is given by $\widetilde{x} = Y\widetilde{b}$, where Y satisfies

$$\widetilde{A}Y\widetilde{A} = \widetilde{A}, \cdot (Y\widetilde{A})^* = Y\widetilde{A}, \tag{62}$$

and, furthermore, if $Y \in \mathbb{C}^{n \times m}$ has the property that, whenever $Ax = b$ has a solution, $\|x\|_Q$ is smallest when $\widetilde{x} = Y\widetilde{b}$, then Y satisfies (62).

As in the proof of Theorem 4, let X, Y be related by (54) with $W = I$. The proof is completed by (55), (56), and (58). $\qquad\square$

See also Ex. 35.

From Theorems 4 and 5 and Corollary 3, we can easily deduce:

COROLLARY 4. *Let $A \in \mathbb{C}^{m \times n}$, $b \in \mathbb{C}^m$, and let $W \in \mathbb{C}^{m \times m}$ and $Q \in \mathbb{C}^{n \times n}$ be positive definite. Then, there is a unique matrix*

$$X = A_{(W,Q)}^{(1,2)} \in A\{1, 2\}$$

satisfying

$$(WAX)^* = WAX, \quad (QXA)^* = QXA. \tag{63}$$

Moreover, $\|Ax - b\|_W$ assumes its minimum value for $x = Xb$, and in the set of vectors x for which this minimum value is assumed, $x = Xb$ is the one for which $\|x\|_Q$ is smallest.

If $Y \in \mathbb{C}^{n \times m}$ has the property that, for all b, $x = Yb$ is the vector of $\mathbb{C}^m$ for which $\|x\|_Q$ is smallest among those for which $\|Ax - b\|_W$ assumes its minimum value, then $Y = A_{(W,Q)}^{(1,2)}$. $\qquad\square$

$A_{(W,Q)}^{(1,2)}$ is called the $\{W, Q\}$-*weighted* $\{1, 2\}$-*inverse of* A. See also Exs. 36–43.

Exercises

EX. 31. *Square root.* Let H be Hermitian PD with the spectral decomposition

$$H = \sum_{i=1}^{k} \lambda_i E_i. \tag{2.90}$$

Then

$$H^{1/2} = \sum_{i=1}^{k} \lambda_i^{1/2} E_i.$$

EX. 32. *Cholesky factorization.* Let H be Hermitian PD. Then it can be factorized as

$$H = R_H^* R_H, \tag{64}$$

where R_H is an upper triangular matrix. Equation (64) is called the *Cholesky factorization* of H; see, e.g., Wilkinson [**872**].

Show that the results of Section 4 can be derived by using the Cholesky factorization

$$Q = R_Q^* R_Q \quad \text{and} \quad W = R_W^* R_W \tag{65}$$

of Q and W, respectively, instead of their square-root factorizations.
Hint: Instead of (51) use

$$\tilde{A} = R_W A R_Q^{-1}, \quad \tilde{\mathbf{x}} = R_Q \mathbf{x}, \quad \tilde{\mathbf{b}} = R_W \mathbf{b}.$$

Ex. 33. Let A, $\mathbf{b}$, and W be as in Theorem 4. Show that a vector $\mathbf{x} \in \mathbb{C}^n$ minimizes $\|A\mathbf{x} - \mathbf{b}\|_W$ if and only if $\mathbf{x}$ is a solution of

$$A^* W A\mathbf{x} = A^* W \mathbf{b},$$

and compare with Ex. 1.

Ex. 34. Let $A_1 \in \mathbb{C}^{m_1 \times n}$, $\mathbf{b}_1 \in \mathbb{C}^{m_1}$, $A_2 \in \mathbb{C}^{m_2 \times n}$, $\mathbf{b}_2 \in R(A_2)$, and let $W \in \mathbb{C}^{m_1 \times m_1}$ be PD. Consider the problem

$$\text{minimize } \|A_1 \mathbf{x} - \mathbf{b}_1\|_W \quad \text{subject to} \quad A_2 \mathbf{x} = \mathbf{b}_2. \tag{66}$$

Show that a vector $\mathbf{x} \in \mathbb{C}^n$ is a minimizer of (66) if and only if there is a vector $\mathbf{y} \in \mathbb{C}^{m_2}$ such that the vector $\begin{bmatrix} \mathbf{x} \\ \mathbf{y} \end{bmatrix}$ is a solution of

$$\begin{bmatrix} A_1^* W A_1 & A_2^* \\ A_2 & O \end{bmatrix} \begin{bmatrix} \mathbf{x} \\ \mathbf{y} \end{bmatrix} = \begin{bmatrix} A_1^* W \mathbf{b}_1 \\ \mathbf{b}_2 \end{bmatrix}.$$

Compare with Ex. 11.

Ex. 35. Let $A \in \mathbb{C}^{m \times n}$, $\mathbf{b} \in R(A)$, and let $Q \in \mathbb{C}^{n \times n}$ be PD. Show that the problem

$$\text{minimize } \|\mathbf{x}\|_Q \quad \text{subject to} \quad A\mathbf{x} = \mathbf{b} \tag{67}$$

has the unique minimizer

$$\mathbf{x} = Q^{-1} A^* (A Q^{-1} A^*)^{(1)} \mathbf{b}$$

and the minimum value

$$\mathbf{b}^* (A Q^{-1} A^*)^{(1)} \mathbf{b}$$

where $(A Q^{-1} A^*)^{(1)}$ is any $\{1\}$-inverse of $A Q^{-1} A^*$ (Rao [**671**, p. 49]).

OUTLINE OF SOLUTION. Problem (67) is equivalent to the problem

$$\text{minimize } \|\tilde{\mathbf{x}}\| \quad \text{subject to} \quad \tilde{A} \tilde{\mathbf{x}} = \tilde{\mathbf{b}},$$

where $\tilde{\mathbf{x}} = Q^{1/2} \tilde{\mathbf{b}}$, $\tilde{A} = A Q^{-1/2}$, $\tilde{\mathbf{b}} = \mathbf{b}$. The unique minimizer of the last problem is, by Theorem 2,

$$\tilde{\mathbf{x}} = Y \tilde{\mathbf{b}}, \quad \text{for any } Y \in \tilde{A}\{1,4\}.$$

Therefore the unique minimizer of (67) is

$$\mathbf{x} = Q^{-1/2} X \mathbf{b}, \quad \text{for any } X \in (A Q^{-1/2})\{1,4\}.$$

Complete the proof by choosing

$$X = Q^{-1/2} A^* (A Q^{-1} A^*)^{(1)}$$

which by Theorem 1.3 is a $\{1,2,4\}$-inverse of $A Q^{-1/2}$.

Ex. 36. *The weighted inverse* $A_{(W,Q)}^{(1,2)}$. Chipman [**187**] first called attention to the unique $\{1,2\}$-inverse given by Corollary 4. However, instead of the second equation of (63) he used

$$(X A V)^* = X A V.$$

Show that these two relations are equivalent. How are Q and V related?

Ex. 37. Use Theorems 4 and 5 to show that

$$A_{(W,Q)}^{(1,2)} = Q^{-1/2}(W^{1/2}AQ^{-1/2})^\dagger W^{1/2}, \tag{68a}$$

or, equivalently, using (65),

$$A_{(W,Q)}^{(1,2)} = R_Q^{-1}(R_W A R_Q^{-1})^\dagger R_W. \tag{68b}$$

See also Ex. 6.93.

Ex. 38. Use Exs. 33 and 35 to show that

$$A_{(W,Q)}^{(1,2)} = Q^{-1}A^*WA(A^*WAQ^{-1}A^*WA)^{(1)}A^*W.$$

Ex. 39. For a given A and an arbitrary $X \in A\{1,2\}$, do there exist PD matrices W and Q such that $X = A_{(W,Q)}^{(1,2)}$? Show that this question reduces to the following simpler one. Given an idempotent E, is there a PD matrix V, such that VE is Hermitian? Show that such a V is given by

$$V = E^*HE + (I - E^*)K(I - E),$$

where H and K are arbitrary PD matrices. (This slightly generalizes a result of Ward, Boullion, and Lewis [851], who took $H = K = I$.)
SOLUTION. Since H and K are PD, $\mathbf{x}^*V\mathbf{x} = 0$ only if both the equations

$$E\mathbf{x} = \mathbf{0}, \quad (I - E)\mathbf{x} = \mathbf{0}, \tag{69}$$

hold. But addition of the two equations (69) gives $\mathbf{x} = \mathbf{0}$. Therefore V is PD. Moreover,

$$VE = E^*HE$$

is clearly Hermitian.   □

Ex. 40. As a particular illustration, let $E = \left(\begin{smallmatrix} 1 & 1 \\ 0 & 0 \end{smallmatrix}\right)$ and show that V can be taken as any matrix of the form

$$V = \begin{bmatrix} a & a \\ a & b \end{bmatrix} \tag{70}$$

where $b > a > 0$. Show that (70) can be written in the form

$$V = aE^*E + c(I - E^*)(I - E),$$

where a and c are arbitrary positive scalars.

Ex. 41. Use Ex. 39 to prove that if X is an arbitrary $\{1,2\}$-inverse of A, there exist PD matrices W and Q such that $X = A_{(W,Q)}^{(1,2)}$ (Ward, Boullion, and Lewis [851]).

Ex. 42. Show that

$$A_{(W,Q)}^{(1,2)} = A_{T,S}^{(1,2)}$$

(see Theorem 2.12(c)), where the subspaces T, S and the PD matrices W, Q are related by

$$T = Q^{-1}N(A)^\perp \tag{71}$$

and

$$S = W^{-1}R(A) \tag{72}$$

or, equivalently, by

$$Q = P_{N(A),T}^* Q_1 P_{N(A),T} + P_{T,N(A)}^* Q_2 P_{T,N(A)} \tag{73}$$

and

$$W = P^*_{R(A),S} W_1 P_{R(A),S} + P^*_{S,R(A)} W_2 P_{S,R(A)} \qquad (74)$$

where Q_1, Q_2, W_1, and W_2 are arbitrary PD matrices of appropriate dimensions. SOLUTION. From (63), we have

$$XA = Q^{-1}A^*X^*Q,$$

and therefore

$$R(X) = R(XA) = Q^{-1}R(A^*) = Q^{-1}N(A)^{\perp}$$

by Corollary 2.7 and (0.26). Also,

$$AX = W^{-1}XAW,$$

and therefore

$$N(X) = N(AX) = N(A^*W) = W^{-1}N(A^*) = W^{-1}R(A)^{\perp}$$

by Corollary 2.7 and (0.27). Finally, from Exs. 39 and 2.23 it follows that the general PD matrix Q mapping T onto $N(A)^{\perp}$ is given by (73). Equation (74) is similarly proved. $\qquad\square$

EX. 43. Let $A = FG$ be a full-rank factorization. Use Ex. 42 and Theorem 2.13(d) to show that

$$A^{(1,2)}_{(W,Q)} = Q^{-1}G^*(F^*WAQ^{-1}G^*)^{-1}F^*W.$$

Compare with Ex. 37.

5. Least-Squares Solutions and Basic Solutions

Berg [86] showed that theMoore–Penrose inverse $A^\dagger$ is a convex combination of ordinary inverses $\{A^{-1}_{IJ} : (I,J) \in \mathcal{N}(A)\}$,

$$A^\dagger = \sum_{(I,J)\in\mathcal{N}(A)} \lambda_{IJ} \widehat{A^{-1}_{IJ}}, \qquad (75)$$

where $\widehat{X}$ denotes that X is padded with the right number of zeros in the right places.

Equivalently, for any $\mathbf{b} \in \mathbb{R}^m$, the *minimum-norm least-squares solution*[1] of the linear equations

$$A\mathbf{x} = \mathbf{b} \qquad (1)$$

is

$$A^\dagger \mathbf{b} = \sum_{(I,J)\in\mathcal{N}(A)} \lambda_{IJ} \widehat{A^{-1}_{IJ}\mathbf{b}_I}, \qquad (76)$$

a convex combination of *basic solutions* $A^{-1}_{IJ}\mathbf{b}_I$, where $\mathbf{b}_I$ is the Ith subvector of $\mathbf{b}$. The representation (76) was given by Ben-Tal and Teboulle [84] for A of full column rank, from which the general case follows easily.

What is curious about these convex combinations is that the weights are proportional to the squares of the determinants of the A_{IJ}'s,

$$\lambda_{IJ} = \frac{\det^2 A_{IJ}}{\sum_{(K,L)\in\mathcal{N}(A)} \det^2 A_{KL}}, \quad (I,J) \in \mathcal{N}(A). \qquad (77)$$

---

[1]Here *norm* means Euclidean norm.

We recall that the sum of squares in the denominator of (77) is the *volume* of A, see Section 0.5.

For the sake of motivation, consider a trivial system with one variable x,

$$\begin{bmatrix} a_1 \\ \vdots \\ a_m \end{bmatrix} x = \begin{bmatrix} b_1 \\ \vdots \\ b_m \end{bmatrix}.$$

The least-squares solution is

$$x = \frac{\sum\limits_{i=1}^{m} a_i b_i}{\sum\limits_{k=1}^{m} a_k^2} = \sum_{i=1}^{m} \lambda_i x_i,$$

a convex combination of the *basic solutions* $\{x_i = a_i^{-1} b_i : a_i \neq 0\}$ with *weights*

$$\lambda_i = \frac{a_i^2}{\sum\limits_{k=1}^{m} a_k^2},$$

which explains (77) for the case $n = 1$. This explanation also works for the general case, since by taking exterior products the system of equations (1) reduces to a system with one column, whose nonzero coefficients are the $r \times r$ determinants $\{\det A_{IJ} : (I, J) \in \mathcal{N}(A)\}$.

LEMMA 2 (Solution of Full-Rank Systems).

(a) *Let* $C \in \mathbb{R}_r^{m \times r}$, $\mathbf{b} \in \mathbb{R}^m$. *Then the (unique) least-squares solution* $\mathbf{y}$ *of*

$$C\mathbf{y} = \mathbf{b}, \tag{78}$$

is

$$\mathbf{y} = \sum_{I \in \mathcal{I}(C)} \mu_{I*} C_{I*}^{-1} \mathbf{b}_I, \tag{79}$$

where μ_{I*} *is given by*

$$\mu_{I*} = \frac{\text{vol}^2 \, C_{I*}}{\text{vol}^2 \, C}. \tag{80}$$

(b) *Let* $R \in \mathbb{R}_r^{r \times n}$, $\mathbf{y} \in \mathbb{R}^r$. *Then the minimum-norm solution of*

$$R\mathbf{x} = \mathbf{y}, \tag{81}$$

is

$$\mathbf{x} = \sum_{J \in \mathcal{J}(R)} \nu_{*J} R_{*J}^{-1} \mathbf{y}, \tag{82}$$

where ν_{*J} *is given by*

$$\nu_{*J} = \frac{\text{vol}^2 \, R_{*J}}{\text{vol}^2 \, R}. \tag{83}$$

PROOF. (a) The coefficients y_i satisfy the *normal equation*

$$C^T C \mathbf{y} = C^T \mathbf{b},$$

rewritten as,

$$C^T \mathbf{c}^{(1)} \wedge \cdots \wedge C^T \mathbf{c}^{(i-1)} \wedge C^T \mathbf{b} \wedge C^T \mathbf{c}^{(i+1)} \wedge \cdots \wedge C^T \mathbf{c}^{(r)}$$
$$= y_i \, (C^T \mathbf{c}^{(1)} \wedge \cdots \wedge C^T \mathbf{c}^{(r)}). \quad (84)$$

The left-hand side of (84) is

$$C_r(C^T)(\mathbf{c}^{(1)} \wedge \cdots \wedge \mathbf{c}^{(i-1)} \wedge \mathbf{b} \wedge \mathbf{c}^{(i+1)} \wedge \cdots \wedge \mathbf{c}^{(r)})$$

which simplifies to

$$\text{LHS}(84) = \sum_{I \in \mathcal{I}(C)} \det C_{I*} \, \det C_{I*}[i \leftarrow \mathbf{b}_I]$$
$$= \sum_{I \in \mathcal{I}(C)} \det{}^2 C_{I*} \, (C_{I*}^{-1} \mathbf{b}_I)_i,$$

and RHS(84) is y_i times (0.110b). The Cramer rule for the least-squares solution is therefore

$$y_i = \sum_{I \in \mathcal{I}(C)} \mu_{I*}(C_{I*}^{-1} \mathbf{b}_I)_i, \quad (85)$$

with μ_{I*} given by (80), and $(C_{I*}^{-1} \mathbf{b}_I)_i$ is the ith component of the solution $C_{I*}^{-1} \mathbf{b}_I$ of the $r \times r$ system

$$C_{I*} \, \mathbf{y} = \mathbf{b}_I. \quad (86)$$

Combining (85) for $i = 1, \ldots, r$, we obtain the least squares solution $\mathbf{y}$ as the convex combination (79) of "basic solutions".[2] $\square$

Lemma 2(a) is due to Jacobi [**442**] and has been rediscovered by Subrahmanyan [**785**], Ben-Tal and Teboulle [**84**], and others. See Farebrother [**269**], Sheynin [**750**], and [**85**] for further details and references.

REMARK 1. Lemma 2 suffices for computing $A^\dagger \mathbf{b}$, the minimum-norm least-squares solution of a linear equation

$$A\mathbf{x} = \mathbf{b}, \quad (1)$$

for general $A \in \mathbb{R}_r^{m \times n}$. Indeed, $A^\dagger \mathbf{b}$ is literally the solution of a *two-stage minimization problem*:

Stage 1:

$$\text{minimize} \quad \|A\mathbf{x} - \mathbf{b}\|. \quad (34)$$

Stage 2:

$$\text{minimize} \quad \{\|\mathbf{x}\| \text{ among all solutions of Stage 1}\}. \quad (35)$$

Stage 1 (least squares) has a unique solution only if $r = n$. Stage 2 has the (unique) solution $\mathbf{x} = A^\dagger \mathbf{b}$, see also Ex. 45 below.

For any full-rank factorization $A = CR$ the above two stages can be separated:

Stage 1:

$$\text{minimize} \quad \|C\mathbf{y} - \mathbf{b}\|. \quad (87)$$

---

[2]This derivation follows that of Marcus [**531**, § 3.1, Example 1.5(c)] and Ben-Tal and Teboulle [**84**].

Stage 2:

$$\text{minimize} \quad \{\|\mathbf{x}\| \text{ among all solutions of } R\mathbf{x} = \mathbf{y}\}. \tag{88}$$

Stage 1 now has the unique solution $\mathbf{y} = C^\dagger \mathbf{b}$. This is an implementation of the fact that

$$A^\dagger = R^\dagger C^\dagger \tag{1.20}$$

is a full-rank factorization of $A^\dagger$.

Combining Remark 1 and Lemma 2, we prove Berg's theorem in geometric form.

THEOREM 6 (Berg [86]). *Let $A \in \mathbb{R}_r^{m \times n}$, $\mathbf{b} \in \mathbb{R}^m$. Then the minimum-norm least-squares solution of*

$$A\mathbf{x} = \mathbf{b}, \tag{1}$$

is the convex combination

$$\mathbf{x} = \sum_{(I,J) \in \mathcal{N}(A)} \lambda_{IJ} \widehat{A_{IJ}^{-1}} \mathbf{b}_I, \tag{89}$$

with weights given by (77).

PROOF. Follows by substituting (79) in RHS(81). Then (89) follows from (82) with weights

$$\lambda_{IJ} = \mu_{I*} \nu_{*J}$$

which, by (80) and (0.83), are (77). □

Since (89) holds for all $\mathbf{b}$, we proved Berg's representation (75) of the Moore–Penrose inverse as a convex combination of ordinary inverses of $r \times r$ submatrices,

$$A^\dagger = \sum_{(I,J) \in \mathcal{N}} \lambda_{IJ} \widehat{A_{IJ}^{-1}}, \tag{75}$$

where $\widehat{A_{IJ}^{-1}}$ is an $n \times m$ matrix with the inverse of A_{IJ} in position (J, I) and zeros elsewhere.

Consider next a *weighted* (or *scaled*) least-squares problem

$$\text{minimize } \|D^{1/2}(A\mathbf{x} - \mathbf{b})\|, \tag{90}$$

where $D = \text{diag}(d_i)$ is a given diagonal matrix with all (weights) $d_i > 0$.

THEOREM 7 (Ben-Tal and Teboulle [84]). *The solutions of* (90), *i.e., the least-squares solutions of*

$$D^{1/2} A\mathbf{x} = D^{1/2}\mathbf{b}, \tag{91}$$

satisfy the normal equation

$$A^T D A\mathbf{x} = A^T D\mathbf{b}. \tag{92}$$

The minimum-norm (weighted) least-squares solution of (91) *is*

$$\mathbf{x}(D) = \sum_{(I,J) \in \mathcal{N}(A)} \lambda_{IJ}(D) \widehat{A_{IJ}^{-1}} \mathbf{b}_I, \tag{93}$$

with weights

$$\lambda_{IJ}(D) = \frac{(\prod_{i\in I} d_i) \det^2 A_{IJ}}{\sum_{(K,L)\in\mathcal{N}(A)} (\prod_{i\in K} d_i) \det^2 A_{KL}}.\tag{94}$$

PROOF. If $A = CR$ is a full-rank factorization of A, then $D^{1/2}A = (D^{1/2}C)R$ is a full-rank factorization of $D^{1/2}A$. The first stage (87) for the problem (91) is

$$\text{minimize}\quad \|D^{1/2}C\mathbf{y} - D^{1/2}\mathbf{b}\|,\tag{95}$$

whose solution, using Lemma 2(a), is

$$\mathbf{y} = \sum_{I\in\mathcal{I}(A)} \mu_{I*}(D^{1/2})C_{I*}^{-1}\mathbf{b}_I,\tag{96}$$

with

$$\mu_{I*}(D^{1/2}) = \frac{(\prod_{i\in I} d_i) \det^2 C_{I*}}{\sum_{K\in\mathcal{I}(A)} (\prod_{i\in K} d_i) \det^2 C_{K*}}.\tag{97}$$

The second stage is still (88),

$$\text{minimize}\quad \{\|\mathbf{x}\| : R\mathbf{x} = \mathbf{y}\},$$

with $\mathbf{y}$ from (96). Therefore the minimum-norm (weighted) least-squares solution of (91) is (93) with weights (94). □

Theorem 7 was proved, in the full-rank case, by Ben-Tal and Teboulle in [84], together with extensions from least squares to minimizing isotone functions of $|A\mathbf{x} - \mathbf{b}|$, the vector of absolute values of $A\mathbf{x} - \mathbf{b}$.

Note that the scaling matrix D appears only in the convex weights λ_{IJ}. Therefore, for any scaling matrix D, the solution $\mathbf{x}(D)$ is in the convex hull of the basic solutions $\{\widehat{A_{IJ}^{-1}\mathbf{b}_I} : (I,J) \in \mathcal{N}(A)\}$, a compact set that does not depend on D. This fact is important for the convergence of interior point methods,[3] in particular, the Dikin method [231], see also Vanderbei and Lagarias [833].

Put differently, let $\mathcal{D}_+$ denote the *positive diagonal matrices* and, for any $D \in \mathcal{D}_+$, consider the operators

$$\boldsymbol{\xi}_D : \mathbb{R}^m \to \mathbb{R}^n, \qquad \text{defined by}\quad \boldsymbol{\xi}_D(\mathbf{b}) = (A^T DA)^\dagger A^T D\mathbf{b},\tag{98}$$

$$\boldsymbol{\eta}_D : \mathbb{R}^{m\times n} \to \mathbb{R}^{m\times m}, \qquad \text{defined by}\quad \boldsymbol{\eta}_D(A) = A(A^T DA)^\dagger A^T D,\tag{99}$$

mapping $\mathbf{b}$ into the solution (93) and the matrix A into the oblique projector $A(A^T DA)^\dagger A^T D$, respectively. The above results imply the uniform boundedness of these operators over $\mathcal{D}_+$.

The uniform boundedness is lost if the weight matrix D is nondiagonal, as shown by following example:

EXAMPLE 1. (Forsgren and Sporre [280, p. 43]).

$$A = \begin{bmatrix} 0 \\ 1 \end{bmatrix}, \quad W(\epsilon) = \begin{bmatrix} \frac{2}{\epsilon} & 1 \\ 1 & \epsilon \end{bmatrix}, \quad \text{with}\quad (A^T W(\epsilon)A)^{-1}A^T W(\epsilon) = \begin{bmatrix} \frac{1}{\epsilon} & 1 \end{bmatrix}.$$

---

[3]Such methods solve, in each iteration, a weighted least-squares problem with fixed $A, \mathbf{b}$ and a different scaling matrix.

Exercises

Ex. 44. Other representations of $A^\dagger$, e.g., [**138**], can be obtained by summing (75) in special ways. Summing (75) over $I \in \mathcal{I}(A)$ we obtain, using Ex. 0.64 and (0.103),

$$A^\dagger = \sum_{J \in \mathcal{J}(A)} \lambda_{*J} \widehat{A^\dagger_{*J}}, \tag{100}$$

a convex combination of the Moore–Penrose inverses of maximal full column rank submatrices A_{*J}, with weights

$$\lambda_{*J} = \frac{\mathrm{vol}^2 A_{*J}}{\mathrm{vol}^2 A}, \tag{101}$$

and $\widehat{A^\dagger_{*J}}$ is the $n \times m$ matrix with $A^\dagger_{*J}$ in rows J and zeros elsewhere. Similarly, summing (75) over $J \in \mathcal{J}(A)$ gives

$$A^\dagger = \sum_{I \in \mathcal{I}(A)} \lambda_{I*} \widehat{A^\dagger_{I*}}, \tag{102}$$

where

$$\lambda_{I*} = \frac{\mathrm{vol}^2 A_{I*}}{\mathrm{vol}^2 A}, \tag{103}$$

and $\widehat{A^\dagger_{I*}}$ is an $n \times m$ matrix with $A^\dagger_{I*}$ in columns I and zeros elsewhere.

Ex. 45. The two stages (34)–(35) can be combined (in the limit):

$$\text{minimize } \|Ax - b\|^2 + \alpha^2 \|x\|^2, \quad \text{where } \alpha \to 0.$$

Ex. 46. *Corresponding $r \times r$ submatrices of A and $A^\dagger$.* Let $A \in \mathbb{R}_r^{m \times n}$, $r > 0$. Then the determinants of the corresponding (in transposed position) $r \times r$ submatrices of A and $A^\dagger$ are proportional,

$$\det(A^\dagger)_{JI} = \frac{\det A_{IJ}}{\mathrm{vol}^2 A}, \quad \forall\, (I, J) \in \mathcal{N}(A). \tag{104}$$

PROOF. From (0.110a)–(0.110c) we calculate

$$C_r(A)^\dagger = \frac{1}{\mathrm{vol}^2 A} (\mathbf{r}_{(1)} \wedge \cdots \wedge \mathbf{r}_{(r)})(\mathbf{c}^{(1)} \wedge \cdots \wedge \mathbf{c}^{(r)}). \tag{105}$$

We conclude that $\mathcal{N}(A^\dagger) = \mathcal{J}(A) \times \mathcal{I}(A)$ and (104) follows from (0.103). $\square$

6. Minors of the Moore–Penrose Inverse

This section, based on Miao and Ben-Israel [**550**], is a continuation of the previous section, but otherwise does not belong in this chapter on minimal properties of generalized inverses.

If the matrix $A \in \mathbb{R}^{n \times n}$ is nonsingular, then the adjoint formula for its inverse

$$A^{-1} = \frac{1}{\det A} \,\mathrm{adj}\, A, \tag{106}$$

has a well-known generalization, the *Jacobi identity*, which relates the *minors* of A^{-1} to those of A. First, some notation: for index sets $\alpha, \beta \in Q_{k,n}$ denote by:

$A[\alpha, \beta]$, the submatrix of A having row indices α and column indices β; and

$A[\alpha', \beta']$, the submatrix obtained from A by deleting rows indexed by α and columns indexed by β.

Then the Jacobi identity is: For any α, $\beta \in Q_{k,n}$,

$$\det A^{-1}[\beta, \alpha] = (-1)^{s(\alpha)+s(\beta)} \frac{\det A[\alpha', \beta']}{\det A}, \tag{107}$$

where $s(\alpha)$ is the sum of the integers in α, see, e.g., Brualdi and Schneider [137]. By convention, $\det A[\emptyset, \emptyset] = 1$.

Moore [575] gave, for any $A \in \mathbb{R}_r^{m \times n}$, a determinantal formula for the entries of the Moore–Penrose inverse $A^\dagger$, see Appendix A, Section A.2. This formula was rediscovered by Berg [86] and was further generalized to matrices defined over an integral domain by Bapat, Bhaskara Rao, and Prasad [49].

A similar result holds for the minors of $A^\dagger$, for $A \in \mathbb{R}_r^{m \times n}$. Theorem 8 expresses them in terms of the minors of the maximal nonsingular submatrices A_{IJ} of A. This requires the notation of §0.5.3. In addition, for $\alpha \in Q_{k,m}$, $\beta \in Q_{k,n}$, let

$$\mathcal{I}(\alpha) = \{I \in \mathcal{I}(A) : \alpha \subseteq I\},$$
$$\mathcal{J}(\beta) = \{J \in \mathcal{J}(A) : \beta \subseteq J\},$$
$$\mathcal{N}(\alpha, \beta) = \{(I, J) \in \mathcal{N}(A) : \alpha \subseteq I, \ \beta \subseteq J\}.$$

Then, by Ex. 0.64(c),

$$\mathcal{N}(\alpha, \beta) = \mathcal{I}(\alpha) \times \mathcal{J}(\beta).$$

For $\alpha = (\alpha_1, \ldots, \alpha_k)$ and $\beta = (\beta_1, \ldots, \beta_k)$ in $Q_{k,n}$, we denote by:

$A[\beta \leftarrow I_\alpha]$ the matrix obtained from A by replacing the β_i^{th} column with the unit vector $\mathbf{e}_{\alpha_i}$, for all $i \in \overline{1,k}$.

Finally, the coefficient $(-1)^{s(\alpha)+s(\beta)} \det A[\alpha', \beta']$, of $\det A[\alpha, \beta]$ in the Laplace expansion of $\det A$, is denoted by

$$\frac{\partial}{\partial |A_{\alpha\beta}|} |A|. \tag{108}$$

Using the above notation we rewrite (108) as

$$\frac{\partial}{\partial |A_{\alpha\beta}|} |A| = (-1)^{s(\alpha)+s(\beta)} \det A[\alpha', \ \beta'] = \det A[\beta \leftarrow I_\alpha], \tag{109}$$

and the Jacobi identity as

$$\det A^{-1}[\beta, \alpha] = \frac{\det A[\beta \leftarrow I_\alpha]}{\det A}, \tag{110a}$$

$$= \frac{1}{\det A^T A} \det A^T A[\beta \leftarrow I_\alpha]. \tag{110b}$$

THEOREM 8. *Let* $A \in \mathbb{R}_r^{m \times n}$ *and* $1 \le k \le r$. *Then, for any* $\alpha \in Q_{k,m}$, $\beta \in Q_{k,n}$,

$$\det A^\dagger[\beta, \alpha] = \begin{cases} 0, & \text{if } \mathcal{N}(\alpha, \beta) = \emptyset, \\ \dfrac{1}{\text{vol}^2 A} \displaystyle\sum_{(I,J) \in \mathcal{N}(\alpha,\beta)} \det A_{IJ} \dfrac{\partial}{\partial |A_{\alpha\beta}|} |A_{IJ}|, & \text{otherwise.} \end{cases} \tag{111}$$

PROOF. See proof of [**550**, Theorem 1]. □

As a special case, if $\alpha = I \in \mathcal{I}(A)$, $\beta = J \in \mathcal{J}(A)$, then $\mathcal{N}(\alpha,\beta)$ contains only one element, i.e., (I,J). Now Theorem 8 gives the identity,

$$\det(A^\dagger)_{JI} = \frac{\det A_{IJ}}{\mathrm{vol}^2\, A}, \quad \forall\, (I,J) \in \mathcal{N}(A) . \tag{3.104}$$

We saw in Section 5 that the Moore–Penrose inverse is a convex combination of ordinary inverses of $r \times r$ submatrices

$$A^\dagger = \sum_{(I,J) \in \mathcal{N}} \lambda_{IJ} \widehat{A_{IJ}^{-1}}, \tag{75}$$

where each $\widehat{A_{IJ}^{-1}}$ is an $n \times m$ matrix with the inverse of A_{IJ} in position (J,I) and zeros elsewhere, and

$$\lambda_{IJ} = \frac{\det^2 A_{IJ}}{\mathrm{vol}^2\, A}, \quad (I,J) \in \mathcal{N}(A). \tag{77}$$

Theorem 8 allows a stronger claim than (3.75), i.e., every minor of $A^\dagger$ in position (β,α) is the same convex combination of the minors of $\widehat{A_{IJ}^{-1}}$'s in the corresponding position.

THEOREM 9. *Let* $A \in \mathbb{R}^{m \times n}_r$ *and* $1 \leq k \leq r$. *Then, for any* $\alpha \in Q_{k,m}$, $\beta \in Q_{k,n}$,

$$\det A^\dagger[\beta,\alpha] = \sum_{(I,J) \in \mathcal{N}(A)} \lambda_{IJ}\, \det \widehat{A_{IJ}^{-1}}[\beta,\alpha]. \tag{112}$$

PROOF. From Theorem 8 it follows that

$$\det A^\dagger[\beta,\alpha] = \sum_{(I,J) \in \mathcal{N}(\alpha,\beta)} \frac{\det^2 A_{IJ}}{\mathrm{vol}^2\, A} \frac{\det A_{IJ}[\beta \leftarrow I_\alpha]}{\det A_{IJ}},$$

$$= \sum_{(I,J) \in \mathcal{N}(\alpha,\beta)} \lambda_{IJ}\, \det \widehat{A_{IJ}^{-1}}[\beta,\alpha],$$

by (110a). We prove (112) by showing that the sum over $\mathcal{N}(\alpha,\beta)$ is the same as the sum over the larger set $\mathcal{N}(A)$. Indeed, if $(I,J) \in \mathcal{N}(A)$, and either $I \notin \mathcal{I}(\alpha)$ or $J \notin \mathcal{J}(\beta)$, then there is at least one column, or row, of zeros in $\widehat{A_{IJ}^{-1}}[\beta,\alpha]$, thus $\det \widehat{A_{IJ}^{-1}}[\beta,\alpha] = 0$. □

By applying Berg's formula to $A^\dagger$, it follows from (104) that the same weights appear in the convex decomposition of A into ordinary inverses of the submatrices $(A^\dagger)_{JI}$,

$$A = \sum_{(I,J) \in \mathcal{N}(A)} \lambda_{IJ}\, \widehat{(A^\dagger)_{JI}^{-1}}, \tag{113}$$

where $\widehat{(A^\dagger)_{JI}^{-1}}$ is the $m \times n$ matrix with the inverse of the $(J,I)^{\text{th}}$ submatrix of $A^\dagger$ in position (I,J) and zeros elsewhere.

Finally applying (112) to $A^\dagger$, we establish a remarkable property of the convex decomposition (113) of A: Every minor of A is the same convex combination of the minors of $\widehat{(A^\dagger)_{JI}^{-1}}$'s.

THEOREM 10. *Let $A \in \mathbb{R}_r^{m \times n}$, $r > 0$. Then there is a convex decomposition of A,*

$$A = \sum_{(I,J) \in \mathcal{N}(A)} \lambda_{IJ} B_{IJ} \tag{114}$$

such that, for all $k = 1, \ldots, r$ and for every $\alpha \in Q_{k,m}$, $\beta \in Q_{k,n}$,

$$\det A[\alpha, \beta] = \sum_{(I,J) \in \mathcal{N}(A)} \lambda_{IJ} \det B_{IJ}[\alpha, \beta], \tag{115}$$

where B_{IJ} is an $m \times n$ matrix with an $r \times r$ nonsingular matrix in position (I, J), zeros elsewhere.

Exercises

Ex. 47. (Miao and Ben-Israel [**550**, Corollary 1]). Reasoning as in Theorem 9, it can be shown that summing (112) over $I \in \mathcal{I}(\alpha)$ is equivalent to summing over $I \in \mathcal{I}(A)$. Similarly, summing over $J \in \mathcal{J}(\beta)$ or over $J \in \mathcal{J}(A)$ give the same result. We summarize these observations as follows: Let $A \in \mathbb{R}_r^{m \times n}$ and $1 \leq k \leq r$. Then, for any $\alpha \in Q_{k,m}$, $\beta \in Q_{k,n}$,

$$\det A^\dagger[\beta, \alpha] = 0 \quad \text{if } \mathcal{J}(\beta) = \emptyset \text{ or } \mathcal{I}(\alpha) = \emptyset,$$

and, otherwise,

$$\det A^\dagger[\beta, \alpha] = \sum_{J \in \mathcal{J}(A)} \lambda_{*J} \det \widehat{A_{*J}^\dagger}[\beta, \alpha] = \sum_{J \in \mathcal{J}(\beta)} \lambda_{*J} \det A_{*J}^\dagger[\beta, \alpha],$$

$$= \sum_{I \in \mathcal{I}(A)} \lambda_{I*} \det \widehat{A_{I*}^\dagger}[\beta, \alpha] = \sum_{I \in \mathcal{I}(\alpha)} \lambda_{I*} \det A_{I*}^\dagger[\beta, \alpha].$$

7. Essentially Strictly Convex Norms and the Associated Projectors and Generalized Inverses

(This section requires familiarity with the basic properties of convex functions and convex sets in finite-dimensional spaces; see, e.g., Rockafellar [**703**].)

In the previous sections various generalized inverses were characterized and studied in terms of their minimization properties with respect to the class of ellipsoidal (or weighted Euclidean) norms

$$\|\mathbf{x}\|_U = (\mathbf{x}^* U \mathbf{x})^{1/2}, \tag{50}$$

where U is positive definite.

Given any two ellipsoidal norms $\|\ \|_W$ and $\|\ \|_U$ on $\mathbb{C}^m$ and $\mathbb{C}^n$, respectively (defined by (50) and two given PD matrices $W \in \mathbb{C}^{m \times m}$ and $U \in \mathbb{C}^{n \times n}$), it was shown in Corollary 4 (page 119) that every $A \in \mathbb{C}^{m \times n}$ has a unique $\{1, 2\}$-inverse $A_{(W,U)}^{(1,2)}$ with the following minimization property:

For any $\mathbf{b} \in \mathbb{C}^m$, the vector $A_{(W,U)}^{(1,2)} \mathbf{b}$ satisfies

$$\|A A_{(W,U)}^{(1,2)} \mathbf{b} - \mathbf{b}\|_W \leq \|A\mathbf{x} - \mathbf{b}\|_W, \quad \text{for all } \mathbf{x} \in \mathbb{C}^n, \tag{116a}$$

and

$$\|A_{(W,U)}^{(1,2)} \mathbf{b}\|_U < \|\mathbf{x}\|_U \tag{116b}$$

for any $A_{(W,U)}^{(1,2)} \mathbf{b} \neq \mathbf{x} \in \mathbb{C}^n$ which gives equality in (116a). In particular, for $W = I_m$ and $U = I_n$, the inverse mentioned above is the Moore–Penrose inverse

$$A_{(I_m,I_n)}^{(1,2)} = A^\dagger, \quad \text{for every } A \in \mathbb{C}^{m \times n}.$$

In this section, which is based on Erdelsky [**255**], Newman and Odell [**609**], and Holmes [**427**], similar minimizations are attempted for norms in the more general class of essentially strictly convex norms. The resulting projectors and generalized inverses are, in general, not even linear transformations, but they still retain many useful properties that justify their study.

We denote by $\alpha, \beta, \phi, \ldots$ various vector norms on finite-dimensional spaces; see, e.g., Ex. 0.8.

Let ϕ be a norm on $\mathbb{C}^n$ and let L be a subspace of $\mathbb{C}^n$. Then for any point $\mathbf{x} \in \mathbb{C}^n$ there is a point $\mathbf{y} \in L$ which is "closest" to $\mathbf{x}$ in the norm ϕ, i.e., a point $\mathbf{y} \in L$ satisfying

$$\phi(\mathbf{y} - \mathbf{x}) = \inf\{\phi(\boldsymbol{\ell} - \mathbf{x}) : \boldsymbol{\ell} \in L\}, \tag{117}$$

see Ex. 48 below. Generally, the closest point is not unique; see, e.g., Ex. 49. However, Lemma 1 below guarantees the uniqueness of closest points, for the special class of essentially strictly convex norms.

From the definition of a vector norm (see § 0.1.5), it is obvious that every norm ϕ on $\mathbb{C}^n$ is a *convex function*, i.e., for every $\mathbf{x}, \mathbf{y} \in \mathbb{C}^n$ and $0 \leq \lambda \leq 1$,

$$\phi(\lambda \mathbf{x} + (1 - \lambda)\mathbf{y}) \leq \lambda \phi(\mathbf{x}) + (1 - \lambda)\phi(\mathbf{y}).$$

A function $\phi : \mathbb{C}^n \to \mathbb{R}$ is called *strictly convex* if, for all $\mathbf{x} \neq \mathbf{y} \in \mathbb{C}^n$ and $0 < \lambda < 1$,

$$\phi(\lambda \mathbf{x} + (1 - \lambda)\mathbf{y}) < \lambda \phi(\mathbf{x}) + (1 - \lambda)\phi(\mathbf{y}). \tag{118}$$

If $\phi : \mathbb{C}^n \to \mathbb{R}$ is a norm, then (118) is clearly violated for $\mathbf{y} = \mu\mathbf{x}$, $\mu \geq 0$. Thus a norm ϕ on $\mathbb{C}^n$ is not strictly convex. Following Holmes [**427**], a norm ϕ on $\mathbb{C}^n$ is called *essentially strictly convex* (abbreviated *e.s.c.*) if ϕ satisfies (118) for all $\mathbf{x} \neq \mathbf{0}$ and $\mathbf{y} \notin \{\mu\mathbf{x} : \mu \geq 0\}$. Equivalently, a norm ϕ on $\mathbb{C}^n$ is e.s.c. if

$$\left. \begin{array}{l} \mathbf{x} \neq \mathbf{y} \in \mathbb{C}^n \\ \phi(\mathbf{x}) = \phi(\mathbf{y}) \\ 0 < \lambda < 1 \end{array} \right\} \implies \phi(\lambda \mathbf{x} + (1 - \lambda)\mathbf{y}) < \lambda \phi(\mathbf{x}) + (1 - \lambda)\phi(\mathbf{y}).$$

$$\tag{119}$$

The following lemma is a special case of a result in Clarkson [**199**].

LEMMA 3. *Let ϕ be any e.s.c. norm on $\mathbb{C}^n$. Then for any subspace $L \subset \mathbb{C}^n$ and any point $\mathbf{x} \in \mathbb{C}^n$, there is a unique point $\mathbf{y} \in L$ closest to $\mathbf{x}$, i.e.,*

$$\phi(\mathbf{y} - \mathbf{x}) = \inf\{\phi(\boldsymbol{\ell} - \mathbf{x}) : \boldsymbol{\ell} \in L\}. \tag{117}$$

PROOF. To prove uniqueness, let $\mathbf{y}_1, \mathbf{y}_2 \in L$ satisfy (117) and $\mathbf{y}_1 \neq \mathbf{y}_2$. Then, for any $0 < \lambda < 1$,

$$\phi(\lambda \mathbf{y}_1 + (1 - \lambda)\mathbf{y}_2 - \mathbf{x}) < \phi(\mathbf{y}_1 - \mathbf{x}), \quad \text{by (119)},$$

showing that the point $\lambda \mathbf{y}_1 + (1 - \lambda)\mathbf{y}_2$, which is in L, is closer to $\mathbf{x}$ than $\mathbf{y}_1$, a contradiction. $\square$

DEFINITION 1. Let ϕ be an e.s.c. norm on $\mathbb{C}^n$ and let L be a subspace of $\mathbb{C}^n$. Then the ϕ-*metric projector on* L, denoted by $P_{L,\phi}$ is the mapping $P_{L,\phi} : \mathbb{C}^n \to L$ assigning to each point in $\mathbb{C}^n$ its (unique) closest point in L, i.e.,

$$P_{L,\phi}(\mathbf{x}) \in L$$

and

$$\phi(P_{L,\phi}(\mathbf{x}) - \mathbf{x}) \leq \phi(\boldsymbol{\ell} - \mathbf{x}), \quad \text{for all } \mathbf{x} \in \mathbb{C}^n, \ \boldsymbol{\ell} \in L. \tag{120}$$

If ϕ is a general norm, then the projector $P_{L,\phi}$ defined as above is a point-to-set mapping,[4] since the closest point $P_{L,\phi}(\mathbf{x})$ need not be unique for all $\mathbf{x} \in \mathbb{C}^n$ and $L \subset \mathbb{C}^n$.

Some properties of $P_{L,\phi}$ in the e.s.c. case are collected in the following theorem, a special case of results by Aronszajn and Smith, and Hirschfeld; see also Singer [**763**, p. 140, Theorem 6.1].

THEOREM 11. *Let ϕ be an e.s.c. norm on $\mathbb{C}^n$. Then, for any subspace L of $\mathbb{C}^n$ and every point $\mathbf{x} \in \mathbb{C}^n$:*

(a) $P_{L,\phi}(\mathbf{x}) = \mathbf{x}$ *if and only if* $\mathbf{x} \in L$;
(b) $P_{L,\phi}^2(\mathbf{x}) = P_{L,\phi}(\mathbf{x})$;
(c) $P_{L,\phi}(\lambda \mathbf{x}) = \lambda P_{L,\phi}(\mathbf{x})$ *for all* $\lambda \in \mathbb{C}$;
(d) $P_{L,\phi}(\mathbf{x} + \mathbf{y}) = P_{L,\phi}(\mathbf{x}) + \mathbf{y}$ *for all* $\mathbf{y} \in L$;
(e) $P_{L,\phi}(\mathbf{x} - P_{L,\phi}(\mathbf{x})) = \mathbf{0}$;
(f) $|\phi(\mathbf{x} - P_{L,\phi}(\mathbf{x})) - \phi(\mathbf{y} - P_{L,\phi}(\mathbf{y}))| \leq \phi(\mathbf{x} - \mathbf{y})$ *for all* $\mathbf{y} \in \mathbb{C}^n$;
(g) $\phi(\mathbf{x} - P_{L,\phi}(\mathbf{x})) \leq \phi(\mathbf{x})$;
(h) $\phi(P_{L,\phi}(\mathbf{x})) \leq 2\phi(\mathbf{x})$; *and*
(i) $P_{L,\phi}$ *is continuous on* $\mathbb{C}^n$.

PROOF. (a) Follows from (117) and (120) since the infimum in (117) is zero if and only if $\mathbf{x} \in L$.

(b)
$$P_{L,\phi}^2(\mathbf{x}) = P_{L,\phi}(P_{L,\phi}(\mathbf{x}))$$
$$= P_{L,\phi}(\mathbf{x}), \quad \text{by (a), since } P_{L,\phi}(\mathbf{x}) \in L.$$

(c) For any $\mathbf{z} \in L$ and $\lambda \neq 0$,

$$\phi(\lambda \mathbf{x} - \mathbf{z}) = \phi\left(\lambda \mathbf{x} - \lambda \frac{\mathbf{z}}{\lambda}\right)$$
$$= |\lambda|\phi\left(\mathbf{x} - \frac{\mathbf{z}}{\lambda}\right)$$
$$\geq |\lambda|\phi\left(\mathbf{x} - P_{L,\phi}(\mathbf{x})\right), \quad \text{by (120)},$$
$$= \phi(\lambda \mathbf{x} - \lambda P_{L,\phi}(\mathbf{x})),$$

which proves (c) for $\lambda \neq 0$. For $\lambda = 0$, (c) is obvious.

(d) From (120) it follows that, for all $\mathbf{z} \in L$,

$$\phi(P_{L,\phi}(\mathbf{x}) + \mathbf{y} - (\mathbf{x} + \mathbf{y})) \leq \phi(\mathbf{z} + \mathbf{y} - (\mathbf{x} + \mathbf{y})),$$

proving (d).

(e) Follows from (d).

---

[4]Excellent surveys of metric projectors in normed linear spaces are given in Deutsch [**230**], and Holmes [**427**, Section 32]; see also Exs. 68–76 below.

(f) For all $\mathbf{x}, \mathbf{y} \in \mathbb{C}^n$,

$$\phi(\mathbf{x} - P_{L,\phi}(\mathbf{x})) \leq \phi(\mathbf{x} - P_{L,\phi}(\mathbf{y})) \leq \phi(\mathbf{x} - \mathbf{y}) + \phi(\mathbf{y} - P_{L,\phi}(\mathbf{y}))$$

and, thus,

$$\phi(\mathbf{x} - P_{L,\phi}(\mathbf{x})) - \phi(\mathbf{y} - P_{L,\phi}(\mathbf{y})) \leq \phi(\mathbf{x} - \mathbf{y}),$$

from which (f) follows by interchanging $\mathbf{x}$ and $\mathbf{y}$.

(g) Follows from (f) by taking $\mathbf{y} = \mathbf{0}$.

(h)
$$\phi(P_{L,\phi}(\mathbf{x})) \leq \phi(P_{L,\phi}(\mathbf{x}) - \mathbf{x}) + \phi(\mathbf{x})$$
$$\leq 2\,\phi(\mathbf{x}), \quad \text{by (g)}.$$

(i) Let $\{\mathbf{x}_k\} \subset \mathbb{C}^n$ be a sequence converging to $\mathbf{x}$,

$$\lim_{k \to \infty} \mathbf{x}_k = \mathbf{x}.$$

Then the sequence $\{P_{L,\phi}(\mathbf{x}_k)\}$ is bounded, by (h), and hence contains a convergent subsequence, also denoted by $\{P_{L,\phi}(\mathbf{x}_k)\}$. Let

$$\lim_{k \to \infty} P_{L,\phi}(\mathbf{x}_k) = \mathbf{y}.$$

Then

$$\phi(P_{L,\phi}(\mathbf{x}_k) - \mathbf{x}_k) \leq \phi(P_{L,\phi}(\mathbf{x}) - \mathbf{x}_k)$$

for $k = 1, 2, \ldots$ and in the limit,

$$\phi(\mathbf{y} - \mathbf{x}) \leq \phi(P_{L,\phi}(\mathbf{x}) - \mathbf{x})$$

proving the $\mathbf{y} = P_{L,\phi}(\mathbf{x})$. $\qquad\square$

The function $P_{L,\phi}$ is homogeneous by Theorem 11(c) but, in general, it is not additive; i.e., it does not necessarily satisfy

$$P_{L,\phi}(\mathbf{x} + \mathbf{y}) = P_{L,\phi}(\mathbf{x}) + P_{L,\phi}(\mathbf{y}), \quad \text{for all } \mathbf{x}, \mathbf{y} \in \mathbb{C}^n.$$

Thus, in general, $P_{L,\phi}$ is not a linear transformation. The following three corollaries deal with cases where $P_{L,\phi}$ is linear.

For any $\ell \in L$ we define the *inverse image of ℓ under $P_{L,\phi}$*, denoted by $P_{L,\phi}^{-1}(\ell)$, as

$$P_{L,\phi}^{-1}(\ell) = \{\mathbf{x} \in \mathbb{C}^n : P_{L,\phi}(\mathbf{x}) = \ell\}.$$

We recall that a *linear manifold* (also *affine set, flat, linear variety*) in $\mathbb{C}^n$ is a set of the form

$$\mathbf{x} + L = \{\mathbf{x} + \ell : \ell \in L\},$$

where $\mathbf{x}$ and L are a given point and subspace, respectively, in $\mathbb{C}^n$.

The following result is a special case of Theorem 6.4 in Singer [763].

COROLLARY 5. *Let ϕ be an e.s.c. norm on $\mathbb{C}^n$ and let L be a subspace of $\mathbb{C}^n$. Then the following statements are equivalent:*

(a) *$P_{L,\phi}$ is additive;*

(b) *$P_{L,\phi}^{-1}(\mathbf{0})$ is a linear subspace; and*

(c) *$P_{L,\phi}^{-1}(\ell)$ is a linear manifold for any $\ell \in L$.*

PROOF. First we show that

$$P_{L,\phi}^{-1}(\mathbf{0}) = \{\mathbf{x} - P_{L,\phi}(\mathbf{x}) : \mathbf{x} \in \mathbb{C}^n\}. \tag{121}$$

From Theorem 11(f) it follows that

$$P_{L,\phi}^{-1}(\mathbf{0}) \supset \{\mathbf{x} - P_{L,\phi}(\mathbf{x}) : \mathbf{x} \in \mathbb{C}^n\}.$$

The reverse containment follows by writing each $\mathbf{x} \in P_{L,\phi}^{-1}(\mathbf{0})$ as

$$\mathbf{x} = \mathbf{x} - P_{L,\phi}(\mathbf{x}).$$

The equivalence of (a) and (b) is obvious from (121). The equivalence of (b) and (c) follows from

$$P_{L,\phi}^{-1}(\boldsymbol{\ell}) = \boldsymbol{\ell} + P_{L,\phi}^{-1}(\mathbf{0}), \quad \text{for all } \boldsymbol{\ell} \in L, \tag{122}$$

which is a result of Theorem 11(d) and (e). $\qquad\square$

COROLLARY 6. *Let L be a hyperplane of $\mathbb{C}^n$, i.e., an $(n-1)$-dimensional subspace of $\mathbb{C}^n$. Then $P_{L,\phi}$ is additive for any e.s.c. norm ϕ on $\mathbb{C}^n$.*

PROOF. Let $\mathbf{u}$ be a vector not contained in L. Then any $\mathbf{x} \in \mathbb{C}^n$ is uniquely represented as

$$\mathbf{x} = \lambda\mathbf{u} + \boldsymbol{\ell}, \quad \text{where } \lambda \in \mathbb{C}, \boldsymbol{\ell} \in L,$$

Therefore, by (121),

$$\begin{aligned}
P_{L,\phi}^{-1}(\mathbf{0}) &= \{\lambda\mathbf{u} + (\boldsymbol{\ell} - P_{L,\phi}(\lambda\mathbf{u} + \boldsymbol{\ell})) : \lambda \in \mathbb{C}, \boldsymbol{\ell} \in L\} \\
&= \{\lambda\mathbf{u} + P_{L,\phi}(-\lambda\mathbf{u}) : \lambda \in \mathbb{C}\}, \quad \text{by Theorem 11(d)}, \\
&= \{\lambda(\mathbf{u} - P_{L,\phi}(\mathbf{u})) : \lambda \in \mathbb{C}\}, \quad \text{by Theorem 11(c)},
\end{aligned}$$

is a line, proving that $P_{L,\phi}$ is additive, by Corollary 5. $\qquad\square$

COROLLARY 7 (Erdelsky [**255**]). *Let ϕ be an e.s.c. norm on $\mathbb{C}^n$ and let r be an integer, $1 \leq r < n$. If $P_{L,\phi}$ is additive for all r-dimensional subspaces of $\mathbb{C}^n$, then it is additive for all subspaces of higher dimension.*

PROOF. Let L be a subspace with $\dim L > r$ and assume that $P_{L,\phi}$ is not additive. Then by Corollary 5, $P_{L,\phi}^{-1}(\mathbf{0})$ is not a subspace, i.e., there exist $\mathbf{x}_1, \mathbf{x}_2 \in P_{L,\phi}^{-1}(\mathbf{0})$ such that $P_{L,\phi}(\mathbf{x}_1 + \mathbf{x}_2) = \mathbf{y} \neq \mathbf{0}$. Now let M be an r-dimensional subspace of L which contains $\mathbf{y}$. Then $\mathbf{x}_1, \mathbf{x}_2 \in P_{M,\phi}^{-1}(\mathbf{0})$, but $P_{M,\phi}(\mathbf{x}_1 + \mathbf{x}_2) = \mathbf{y} \neq \mathbf{0}$, a contradiction of the hypothesis that $P_{M,\phi}$ is additive. $\qquad\square$

See also Exs. 71–74 for additional results on the linearity of the projectors $P_{L,\phi}$.

Following Boullion and Odell [**123**, pp. 43–44] we define generalized inverses associated with pairs of e.s.c. norms as follows.

DEFINITION 2. Let α and β be e.s.c. norms on $\mathbb{C}^m$ and $\mathbb{C}^n$, respectively. For any $A \in \mathbb{C}^{m \times n}$ we define the *generalized inverse associated with α and β* (also called the *α-β generalized inverse*, see, e.g, Boullion and Odell [**123**, p. 44]), denoted by $A_{\alpha,\beta}^{(-1)}$, as

$$A_{\alpha,\beta}^{(-1)} = (I - P_{N(A),\beta})A^{(1)}P_{R(A),\alpha}, \tag{123}$$

where $A^{(1)}$ is any $\{1\}$-inverse of A.

RHS(123) means that the three transformations

$$P_{R(A),\alpha} : \mathbb{C}^m \to R(A),$$

$$A^{(1)} : \mathbb{C}^m \to \mathbb{C}^n,$$

and

$$(I - P_{N(A),\beta}) : \mathbb{C}^n \to P_{N(a),\beta}^{-1}(\mathbf{0}),$$

see, e.g., (121), are performed in this order. We now show that $A_{\alpha,\beta}^{(-1)}$ is a single-valued transformation which does not depend on the particular {1}-inverse used in its definition. For any $\mathbf{y} \in \mathbb{C}^m$, the set

$$\{A^{(1)} P_{R(A),\alpha}(\mathbf{y}) : A^{(1)} \in A\{1\}\}$$

obtained as $A^{(1)}$ ranges over $A\{1\}$ is, by Theorem 1.2, the set of solutions of the linear equation

$$A\mathbf{x} = P_{R(A),\alpha}(\mathbf{y}),$$

a set which can be written as

$$A^\dagger P_{R(A),\alpha}(\mathbf{y}) + \{\mathbf{z} : \mathbf{z} \in N(A)\}.$$

Now, for any $\mathbf{z} \in N(A)$, it follows from Theorem 11(a) and (d) that

$$(I - P_{N(A),\beta})(A^\dagger P_{R(A),\alpha}(\mathbf{y}) + \mathbf{z}) = (I - P_{N(A),\beta})A^\dagger P_{R(A),\alpha}(\mathbf{y})$$

proving that

$$A_{\alpha,\beta}^{(-1)}(\mathbf{y}) = (I - P_{N(A),\beta})A^\dagger P_{R(A),\alpha}(\mathbf{y}), \quad \text{for all } \mathbf{y} \in \mathbb{C}^n, \qquad (124)$$

independently of the {1}-inverse $A^{(1)}$ used in the definition (123).

If the norms α and β are Euclidean, then $P_{R(A),\alpha}$ and $P_{N(A),\beta}$ reduce to the orthogonal projectors $P_{R(A)}$ and $P_{N(A)}$, respectively, and $A_{\alpha,\beta}^{(-1)}$ is, by (124), just the Moore–Penrose inverse $A^\dagger$; see also Exs. 69–72 and 76 below. Thus many properties of $A^\dagger$ are specializations of the corresponding properties of $A_{\alpha,\beta}^{(-1)}$, some of which are collected in the following theorem. In particular, the minimization properties of $A^\dagger$ are special cases of statements (i) and (j) below.

As in the case of linear transformations, we denote

$$N(A_{\alpha,\beta}^{(-1)}) = \{\mathbf{y} \in \mathbb{C}^m : A_{\alpha,\beta}^{(-1)}(\mathbf{y}) = \mathbf{0}\},$$
$$R(A_{\alpha,\beta}^{(-1)}) = \{A_{\alpha,\beta}^{(-1)}(\mathbf{y}) : \mathbf{y} \in \mathbb{C}^m\}.$$

THEOREM 12 (Erdelsky [**255**], Newman and Odell [**609**]). *Let α and β be e.s.c. norms on $\mathbb{C}^m$ and $\mathbb{C}^n$, respectively. Then, for any $A \in \mathbb{C}^{m \times n}$:*

(a) $A_{\alpha,\beta}^{(-1)} : \mathbb{C}^m \to \mathbb{C}^n$ *is a homogeneous transformation.*

(b) $A_{\alpha,\beta}^{(-1)}$ *is additive (hence linear) if $P_{R(A),\alpha}$ and $P_{N(A),\beta}$ are additive.*

(c) $N(A_{\alpha,\beta}^{(-1)}) = P_{R(A),\alpha}^{-1}(\mathbf{0}).$

(d) $R(A_{\alpha,\beta}^{(-1)}) = P_{N(A),\beta}^{-1}(\mathbf{0}).$

(e) $AA_{\alpha,\beta}^{(-1)} = P_{R(A),\alpha}.$

(f) $A_{\alpha,\beta}^{(-1)}A = I - P_{N(A),\beta}.$

(g) $AA_{\alpha,\beta}^{(-1)}A = A.$

(h) $A_{\alpha,\beta}^{(-1)}AAA_{\alpha,\beta}^{(-1)} = A_{\alpha,\beta}^{(-1)}.$

(i) *For any* $\mathbf{b} \in \mathbb{C}^m$, *an* α-*approximate solution of* $A\mathbf{x} = \mathbf{b}$ *is defined as any vector* $\mathbf{x} \in \mathbb{C}^n$ *minimizing* $\alpha(A\mathbf{x} - \mathbf{b})$. *Then* $\mathbf{x}$ *is an* α-*approximate solution of* (1) *if and only if*

$$A\mathbf{x} = AA^{(-1)}_{\alpha,\beta}(\mathbf{b}). \tag{125}$$

(j) *For any* $\mathbf{b} \in \mathbb{C}^m$, *the equation*

$$A\mathbf{x} = \mathbf{b} \tag{1}$$

has a unique α-*approximate solution of minimal* β-*norm, given by* $A^{(-1)}_{\alpha,\beta}(\mathbf{b})$; *that is, for every* $\mathbf{b} \in \mathbb{C}^m$,

$$\alpha(AA^{(-1)}_{\alpha,\beta}(\mathbf{b}) - \mathbf{b}) \leq \alpha(A\mathbf{x} - \mathbf{b}), \quad \text{for all } \mathbf{x} \in \mathbb{C}^n, \tag{126a}$$

and

$$\beta(A^{(-1)}_{\alpha,\beta}(\mathbf{b})) \leq \beta(\mathbf{x}) \tag{126b}$$

for any $\mathbf{x} \neq A^{(-1)}_{\alpha,\beta}(\mathbf{b})$ *with equality in* (126a).

PROOF. (a) Follows from the definition and Theorem 11(c).

(b) Obvious from definition (123).

(c) From (123) it is obvious that

$$N(A^{(-1)}_{\alpha,\beta}) \supset P^{-1}_{R(A),\alpha}(\mathbf{0}).$$

Conversely, if $\mathbf{y} \neq P^{-1}_{R(A),\alpha}(\mathbf{0})$; i.e., if $P^{-1}_{R(A),\alpha}(\mathbf{y}) \neq \mathbf{0}$, then $A^{\dagger} P^{-1}_{R(A),\alpha}(\mathbf{y}) \neq \mathbf{0}$ since $(A^{\dagger})_{[R(A)]}$ is nonsingular (see Ex. 2.90) and, consequently,

$$(I - P_{N(A),\beta})A^{\dagger} P^{-1}_{R(A),\alpha}(\mathbf{y}) \neq \mathbf{0}, \quad \text{by Theorem 11(a).}$$

(d) From (121) and the definition (123) it is obvious that

$$R(A^{(-1)}_{\alpha,\beta}) \subset P^{-1}_{N(A),\beta}(\mathbf{0}).$$

Conversely, let $\mathbf{x} \in P^{-1}_{N(A),\beta}(\mathbf{0})$. Then, by (121),

$$
\begin{aligned}
\mathbf{x} &= (I - P_{N(A),\beta})\mathbf{z}, \quad \text{for some } \mathbf{z} \in \mathbb{C}^n, \\
&= (I - P_{N(A),\beta})P_{R(A^*)}\mathbf{z}, \quad \text{by Theorem 11(d),} \\
&= (I - P_{N(A),\beta})A^{\dagger} A\mathbf{z} \\
&= (I - P_{N(A),\beta})A^{\dagger} P_{R(A),\alpha}(A\mathbf{z}) \\
&= A^{(-1)}_{\alpha,\beta}(A\mathbf{z}). \tag{127}
\end{aligned}
$$

(e) Obvious from (124).

(f) For any $\mathbf{z} \in \mathbb{C}^n$ it follows from (127) that $(I - P_{N(A),\beta})\mathbf{z} = A^{(-1)}_{\alpha,\beta}(A\mathbf{z})$.

(g) Obvious from (e) and Theorem 11(a).

(h) Obvious from (f) and (d).

(i) A vector $\mathbf{x} \in \mathbb{C}^n$ is an α-approximate solution of (1) if and only if

$$\alpha(A\mathbf{x} - \mathbf{b}) \leq \alpha(\mathbf{y} - \mathbf{b}), \quad \text{for all } \mathbf{y} \in R(A),$$

or, equivalently,

$$
\begin{aligned}
A\mathbf{x} &= P_{R(A),\alpha}(\mathbf{b}), \quad \text{by (120),} \\
&= AA^{(-1)}_{\alpha,\beta}(\mathbf{b}), \quad \text{by (e).}
\end{aligned}
$$

(j) From (125) it follows that $\mathbf{x}$ is an α-approximate solution of (1) if and only if

$$\mathbf{x} = A^\dagger AA_{\alpha,\beta}^{(-1)}(\mathbf{b}) + \mathbf{z}, \quad \mathbf{z} \in N(A), \tag{128}$$

$$= A^\dagger P_{R(A),\alpha}(\mathbf{b}) + \mathbf{z}, \quad \mathbf{z} \in N(A), \quad \text{by (e).} \tag{129}$$

Now, by Lemma 3 and Definition 1, the β-norm of

$$A^\dagger P_{R(A),\alpha}(\mathbf{b}) + \mathbf{z}, \quad \mathbf{z} \in N(A),$$

is minimized uniquely at

$$\mathbf{z} = -P_{N(A),\beta} A^\dagger P_{R(A),\alpha}(\mathbf{b}),$$

which substituted in (128) gives

$$\mathbf{x} = (I - P_{N(A),\beta}) A^\dagger P_{R(A),\alpha}(\mathbf{b})$$

$$= A_{\alpha,\beta}^{(-1)}(\mathbf{b}). \qquad \square$$

See Exs. 76–79 for additional results on the generalized inverse $A_{\alpha,\beta}^{(-1)}$.

Exercises

EX. 48. *Closest points.* Let ϕ be a norm on $\mathbb{C}^n$ and let L be a nonempty closed set in $\mathbb{C}^n$. Then, for any $\mathbf{x} \in \mathbb{C}^n$, the infimum

$$\inf\{\phi(\boldsymbol{\ell} - \mathbf{x}) : \boldsymbol{\ell} \in L\}$$

is attained at some point $\mathbf{y} \in L$ called ϕ-closest to $\mathbf{x}$ in L.

PROOF. Let $\mathbf{z} \in L$. Then the set

$$K = L \cap \{\boldsymbol{\ell} \in \mathbb{C}^n : \phi(\boldsymbol{\ell} - \mathbf{x}) \leq \phi(\mathbf{z} - \mathbf{x})\}$$

is closed (being the intersection of two closed sets) and bounded, hence compact. The continuous function $\phi(\boldsymbol{\ell} - \mathbf{x})$ attains its minimum at some $\boldsymbol{\ell} \in K$ but, by definition of K,

$$\inf\{\phi(\boldsymbol{\ell} - \mathbf{x}) : \boldsymbol{\ell} \in K\} = \inf\{\phi(\boldsymbol{\ell} - \mathbf{x}) : \boldsymbol{\ell} \in L\}. \qquad \square$$

EX. 49. Let ϕ be the ℓ_1-norm on $\mathbb{R}^2$,

$$\phi(\mathbf{x}) = \phi\left(\begin{bmatrix} x_1 \\ x_2 \end{bmatrix}\right) = |x_1| + |x_2|,$$

see, e.g., Ex. 0.10, and let $L = \{\mathbf{x} \in \mathbb{R}^2 : x_1 + x_2 = 1\}$. Then the set of ϕ-closest points in L to $\binom{1}{1}$ is $\{\binom{\alpha}{-\alpha} : -1 \leq \alpha \leq 1\}$.

EX. 50. Let $\| \ \|$ be the Euclidean norm on $\mathbb{C}^n$, let $S \subset \mathbb{C}^n$ be a convex set, and let $\mathbf{x}, \mathbf{y}$ be two points in $\mathbb{C}^n$: $\mathbf{x} \notin S$ and $\mathbf{y} \in S$. Then the following statements are equivalent:

(a) $\mathbf{y}$ is $\| \ \|$-closest to $\mathbf{x}$ in S.
(b) $\mathbf{s} \in S \implies \Re\langle \mathbf{y} - \mathbf{x}, \mathbf{s} - \mathbf{y}\rangle \geq 0$.

PROOF. (Adapted from Goldstein [302, p. 99]).

(a) $\implies$ (b) For any $0 \leq \lambda \leq 1$ and $\mathbf{s} \in S$,

$$\mathbf{y} + \lambda(\mathbf{s} - \mathbf{y}) \in S.$$

Now

$$0 \leq \|\mathbf{x} - \mathbf{y} - \lambda(\mathbf{s} - \mathbf{y})\|^2 - \|\mathbf{x} - \mathbf{y}\|^2$$

$$= 2\lambda\Re\langle \mathbf{y} - \mathbf{x}, \mathbf{s} - \mathbf{y}\rangle + \lambda^2\|\mathbf{s} - \mathbf{x}\|^2$$

$$< 0, \quad \text{if} \quad \Re\langle \mathbf{y} - \mathbf{x}, \mathbf{s} - \mathbf{y}\rangle < 0 \text{ and } 0 < \lambda < -\frac{2\Re\langle \mathbf{y} - \mathbf{x}, \mathbf{s} - \mathbf{y}\rangle}{\|\mathbf{s} - \mathbf{y}\|^2},$$

a contradiction to (a).

(b) $\implies$ (a) For any $\mathbf{s} \in S$,

$$\|\mathbf{x} - \mathbf{s}\|^2 - \|\mathbf{x} - \mathbf{y}\|^2 = \|\mathbf{s}\|^2 - 2\Re\langle\mathbf{s}, \mathbf{x}\rangle + 2\Re\langle\mathbf{y}, \mathbf{x}\rangle - \|\mathbf{y}\|^2$$
$$= \|\mathbf{s} - \mathbf{y}\|^2 + 2\Re\langle\mathbf{y} - \mathbf{x}, \mathbf{s} - \mathbf{y}\rangle$$
$$\geq 0, \quad \text{if (b).} \qquad \square$$

EX. 51. *A hyperplane separation theorem.* Let S be a nonempty closed convex set in $\mathbb{C}^n$, $\mathbf{x}$ a point not in S. Then there is a real hyperplane

$$\{\mathbf{z} \in \mathbb{C}^n : \Re\langle\mathbf{u}, \mathbf{z}\rangle = \alpha\}, \quad \text{for some } \mathbf{0} \neq \mathbf{u} \in \mathbb{C}^n, \ \alpha \in \mathbb{R},$$

which separates S and $\mathbf{x}$, in the sense that

$$\Re\langle\mathbf{u}, \mathbf{x}\rangle < \alpha \quad \text{and} \quad \Re\langle\mathbf{u}, \mathbf{s}\rangle \geq \alpha \text{ for all } \mathbf{s} \in S.$$

PROOF. Let $\mathbf{x}_S$ be the $\|\ \|$-closest point to $\mathbf{x}$ in S, where $\|\ \|$ is the Euclidean norm. The point $\mathbf{x}_S$ is unique, by the same proof as in Lemma 3, since $\|\ \|$ is e.s.c. Then, for any $\mathbf{s} \in S$,

$$\Re\langle\mathbf{x}_S - \mathbf{x}, \mathbf{s}\rangle \geq \Re\langle\mathbf{x}_S - \mathbf{x}, \mathbf{x}_S\rangle, \quad \text{by Ex. 50,}$$
$$> \Re\langle\mathbf{x}_S - \mathbf{x}, \mathbf{x}\rangle,$$

since

$$\Re\langle\mathbf{x}_S - \mathbf{x}, \mathbf{x}_S - \mathbf{x}\rangle = \|\mathbf{x}_S - \mathbf{x}\|^2 > 0.$$

The proof is completed by choosing

$$\mathbf{u} = \mathbf{x}_S - \mathbf{x}, \quad \alpha = \Re\langle\mathbf{x}_S - \mathbf{x}, \mathbf{x}_S\rangle. \qquad \square$$

Gauge Functions and Their Duals

EX. 52. A function $\phi : \mathbb{C}^n \to \mathbb{R}$ is called a *gauge function* (also a *Minkowski functional*) if:

(G1) ϕ is continuous, and for all $\mathbf{x}, \mathbf{y} \in \mathbb{C}^n$;

(G2) $\phi(\mathbf{x}) \geq 0$ and $\phi(\mathbf{x}) = 0$ only if $\mathbf{x} = \mathbf{0}$;

(G3) $\phi(\alpha\mathbf{x}) = \alpha\phi(\mathbf{x})$ for all $\alpha \geq 0$; and

(G4) $\phi(\mathbf{x} + \mathbf{y}) \leq \phi(\mathbf{x}) + \phi(\mathbf{y})$.

A gauge function ϕ is called *symmetric* if, for all $\mathbf{x} = (x_1, x_2, \ldots, x_n)^T \in \mathbb{C}^n$:

(G5) $\phi(\mathbf{x}) = \phi(x_1, x_2, \ldots, x_n) = \phi(x_{\pi(1)}, x_{\pi(2)}, \ldots, x_{\pi(n)})$, for every permutation $\{\pi(1), \pi(2), \ldots, \pi(n)\}$ of $\{1, 2, \ldots, n\}$; and

(G6) $\phi(\mathbf{x}) = \phi(x_1, x_2, \ldots, x_n) = \phi(\lambda_1 x_1, \lambda_2 x_2, \ldots, \lambda_n x_n)$, for every scalar sequence $\{\lambda_1, \lambda_2, \ldots, \lambda_n\}$ satisfying

$$\begin{cases} |\lambda_i| = 1, & \text{if } \phi : \mathbb{C}^n \to \mathbb{R}, \\ \lambda_i = \pm 1, & \text{if } \phi : \mathbb{R}^n \to \mathbb{R}, \end{cases} \quad i \in \overline{1, n}.$$

Let $\phi : \mathbb{C}^n \to \mathbb{R}$ satisfy (G1)–(G3). The *dual function*[5] of ϕ is the function $\phi_D : \mathbb{C}^n \to \mathbb{R}$ defined by

$$\phi_D(\mathbf{y}) = \sup_{\mathbf{x} \neq \mathbf{0}} \frac{\Re\langle\mathbf{y}, \mathbf{x}\rangle}{\phi(\mathbf{x})}. \tag{130}$$

Then:

---

[5]Originally, ϕ_D was called the *conjugate* of ϕ by Bonnesen and Fenchel [**116**] and von Neumann [**839**]. However, in the modern convexity literature, the word *conjugate function* has a different meaning, see, e.g., Rockafellar [**703**].

(a) The supremum in (130) is attained, and

$$\phi_D(\mathbf{y}) = \max_{\mathbf{x} \in S_i} \frac{\Re\langle \mathbf{y}, \mathbf{x} \rangle}{\phi(\mathbf{x})}, \quad i = 1 \text{ or } \phi, \tag{131a}$$

where

$$S_1 = \{\mathbf{x} \in \mathbb{C}^n : \|\mathbf{x}\|_1 = \sum_{i=1}^n |x_i| = 1\} \tag{131b}$$

and

$$S_\phi = \{\mathbf{x} \in \mathbb{C}^n : \phi(\mathbf{x}) = 1\}. \tag{131c}$$

(b) ϕ_D is a gauge function.

(c) ϕ_D satisfies (G5) [(G6)] if ϕ does.

(d) If ϕ is a gauge function (i.e., if ϕ also satisfies (G4)), then ϕ is the conjugate of ϕ_D (Bonnesen and Fenchel [**116**], von Neumann [**839**]).

PROOF. (a) From (G3) it follows that the constraint $\mathbf{x} \neq \mathbf{0}$ in (130) can be replaced by $\mathbf{x} \in S_1$ or, alternatively, by $\mathbf{x} \in S_\phi$. The supremum is attained since S_1 is compact.

(b),(c) The continuity of ϕ_D follows from (G1), (131a) and the compactness of S_1. It is easy to show that ϕ shares with ϕ_D each of the properties (G2), (G3), (G5), and (G6), while (G4) holds for ϕ_D, by definition (130), without requiring that it hold for ϕ.

(d) From (130) it follows that

$$\Re\langle \mathbf{y}, \mathbf{x} \rangle \leq \phi(\mathbf{x})\phi_D(\mathbf{y}), \quad \text{for all } \mathbf{x}, \mathbf{y} \in \mathbb{C}^n, \tag{132}$$

and, hence,

$$\phi(\mathbf{x}) \geq \sup_{\mathbf{y} \neq \mathbf{0}} \frac{\Re\langle \mathbf{y}, \mathbf{x} \rangle}{\phi_D(\mathbf{y})}. \tag{133}$$

To show equality in (133) we note that the set

$$B = \{\mathbf{z} : \phi(\mathbf{z}) \leq 1\}$$

is a closed convex set in $\mathbb{C}^n$, an easy consequence of the definition of a gauge function. From the hyperplane separation theorem (see, e.g., Ex. 51 above) we conclude:

If a point $\mathbf{x}$ is contained in every closed half-space $\{\mathbf{z} : \Re\langle \mathbf{u}, \mathbf{z} \rangle \leq 1\}$ which contains B, then $\mathbf{x} \in B$, i.e., $\phi(\mathbf{x}) \leq 1$. $\qquad$ (134)

From (131a) and (131c) it follows that

$$B \subset \{\mathbf{z} : \Re\langle \mathbf{y}, \mathbf{z} \rangle \leq 1\}$$

is equivalent to

$$\phi_D(\mathbf{y}) \leq 1.$$

Statement (134) is thus equivalent to

$$\{\phi_D(\mathbf{y}) \leq 1 \implies \Re\langle \mathbf{y}, \mathbf{x} \rangle \leq 1\} \implies \phi(\mathbf{x}) \leq 1,$$

which proves equality in (133). $\qquad\qquad\qquad\qquad\qquad\qquad\qquad\qquad$ $\square$

Convex Bodies and Gauge Functions

EX. 53. A *convex body* in $\mathbb{C}^n$ is a closed bounded convex set with nonempty interior.

Let $B \subset \mathbb{C}^n$ be a convex body and let $\mathbf{0} \in \operatorname{int} B$ where $\operatorname{int} B$ denotes the *interior* of B. The *gauge function* (or *Minkowski functional*) *of* B is the function $\phi^B : \mathbb{C}^n \to \mathbb{R}$ defined by

$$\phi^B(\mathbf{x}) = \inf\{\lambda > 0 : \mathbf{x} \in \lambda B\}. \tag{135}$$

Then:

(a) ϕ^B is a gauge function, i.e., it satisfies (G1)–(G4) of Ex. 52.
(b) $B = \{\mathbf{x} \in \mathbb{C}^n : \phi^B(\mathbf{x}) \le 1\}$.
(c) $\operatorname{int} B = \{\mathbf{x} \in \mathbb{C}^n : \phi^B(\mathbf{x}) < 1\}$.

Conversely, if $\phi : \mathbb{C}^n \to \mathbb{R}$ is any gauge function, then ϕ is the gauge function ϕ^B of a convex body B defined by

$$B = \{\mathbf{x} \in \mathbb{C}^n : \phi^B(\mathbf{x}) \le 1\}, \tag{136}$$

which has $\mathbf{0}$ as an interior point.

Thus (135) and (136) establish a one-to-one correspondence between all gauge functions $\phi : \mathbb{C}^n \to \mathbb{R}$ and all convex bodies $B \subset \mathbb{C}^n$ with $\mathbf{0} \in \operatorname{int} B$.

EX. 54. A set $B \in \mathbb{C}^n$ is called *equilibrated* if

$$\mathbf{x} \in B, \quad |\lambda| \le 1 \quad \Longrightarrow \quad \lambda\mathbf{x} \in B.$$

Clearly, $\mathbf{0}$ is an interior point of any equilibrated convex body.

Let B be a convex body, $\mathbf{0} \in \operatorname{int} B$. Then B is equilibrated if and only if its gauge function ϕ^B satisfies

$$\phi^B(\lambda\mathbf{x}) = |\lambda|\phi^B(\mathbf{x}), \quad \text{for all } \lambda \in \mathbb{C}, \mathbf{x} \in \mathbb{C}^n. \tag{137}$$

EX. 55. *Vector norms.* From the definition of a *vector norm* (§0.1.5) and a gauge function (Ex. 52) it follows that a function $\phi : \mathbb{C}^n \to \mathbb{R}$ is a norm if and only if ϕ is a gauge function satisfying (137).

Thus (135) and (136) establish a one-to-one correspondence between all norms $\phi : \mathbb{C}^n \to \mathbb{R}$ and all equilibrated convex bodies $B \in \mathbb{C}^n$ (Householder [**432**, Chapter 2]).

EX. 56. If a norm $\phi : \mathbb{C}^n \to \mathbb{R}$ is *unitarily invariant* (i.e., if $\phi(U\mathbf{x}) = \phi(\mathbf{x})$ for all $\mathbf{x} \in \mathbb{C}^n$ and any unitary matrix $U \in \mathbb{C}^{n \times n}$), then ϕ is a symmetric gauge function (see Ex. 52). Is the converse true?

Dual Norms

EX. 57. The *dual* (also *polar*) of a nonempty set $B \subset \mathbb{C}^n$ is the set B_D defined by

$$B_D = \{\mathbf{y} \in \mathbb{C}^n : \mathbf{x} \in B \implies \Re\langle \mathbf{y}, \mathbf{x} \rangle \le 1\}. \tag{138}$$

Let $B \subset \mathbb{C}^n$ be an equilibrated convex body. Then:

(a) B_D is an equilibrated convex body.
(b) $(B_D)_D = B$, i.e., B is the dual of its dual.
(c) Let ϕ^B be the norm corresponding to B via (135). Then the dual of ϕ^B, computed by (130),

$$\phi_D^B(\mathbf{y}) = \sup_{\mathbf{x} \neq 0} \frac{\Re\langle \mathbf{y}, \mathbf{x} \rangle}{\phi^B(\mathbf{x})}, \tag{139}$$

is the norm corresponding to B_D. The norm ϕ_D^B, defined by (139), is called the *dual* of ϕ^B.

(d) $\left(\phi_D^B\right)_D = \phi^B$, i.e., ϕ^B is the dual of its dual. Such pairs $\{\phi^B, \phi_D^B\}$ are called *dual norms* (Householder [**432**, Chapter 2]).

Ex. 58. ℓ_p*-norms.* If ϕ is an ℓ_p-norm, $p \geq 1$ (see Exs. 0.9–10), then its dual is an ℓ_q-norm where q is determined by

$$\tfrac{1}{p} + \tfrac{1}{q} = 1.$$

In particular, the ℓ_1- and ℓ_∞-norms are dual, while the Euclidean norm (the ℓ_2-norm) is self-dual.

Ex. 59. *The generalized Cauchy–Schwartz inequality.* Let $\{\phi, \phi_D\}$ be dual norms on $\mathbb{C}^n$. Then

$$\Re\langle \mathbf{y}, \mathbf{x}\rangle \leq \phi(\mathbf{x})\phi_D(\mathbf{y}), \quad \text{for all } \mathbf{x}, \mathbf{y} \in \mathbb{C}^n, \tag{132}$$

and for any $\mathbf{x} \neq \mathbf{0}\,[\mathbf{y} \neq \mathbf{0}]$ there exists a $\mathbf{y} \neq \mathbf{0}\,[\mathbf{x} \neq \mathbf{0}]$ giving equality in (132). Such pairs $\{\mathbf{x}, \mathbf{y}\}$ are called *dual vectors* (with respect to the norm ϕ).

If ϕ is the Euclidean norm, then (132) reduces to the classical Cauchy–Schwartz inequality (0.4) (Householder [**432**]).

Ex. 60. *A Tchebycheff solution of* $A\mathbf{x} = \mathbf{b}$, $A \in \mathbb{C}_n^{(n+1)\times n}$. *A Tchebycheff approximate solution* of the system

$$A\mathbf{x} = \mathbf{b} \tag{1}$$

is, by the definition in Theorem 12(i), a vector $\mathbf{x}$ minimizing the Tchebycheff norm

$$\|\mathbf{r}\|_\infty = \max_{i=1,\dots,m} \{|r_i|\}$$

of the residual vector

$$\mathbf{r} = \mathbf{b} - A\mathbf{x}. \tag{2}$$

Let $A \in \mathbb{C}_n^{(n+1)\times n}$ and $\mathbf{b} \in \mathbb{C}^{n+1}$ be such that (1) is inconsistent. Then (1) has a unique Tchebycheff approximate solution given by

$$\mathbf{x} = A^\dagger(\mathbf{b} + \mathbf{r}), \tag{140}$$

where the residual $\mathbf{r} = [r_i]$ is

$$r_i = \frac{\sum_{j=1}^{n+1} |(P_{N(A^*)}\mathbf{b})_j|^2}{\sum_{j=1}^{n+1} |(P_{N(A^*)}\mathbf{b})_j|} \frac{(P_{N(A^*)}\mathbf{b})_i}{|(P_{N(A^*)}\mathbf{b})_i|}, \quad i \in \overline{1, n+1}. \tag{141}$$

(The real case appeared in Cheney [**184**, p. 41] and Meicler [**540**].)
PROOF. From

$$\mathbf{r}(\mathbf{x}) - \mathbf{b} = -A\mathbf{x} \in R(A)$$

it follows that any residual $\mathbf{r}$ satisfies

$$P_{N(A^*)}\mathbf{r} = P_{N(A^*)}\mathbf{b}$$

or, equivalently,

$$\langle P_{N(A^*)}\mathbf{b}, \mathbf{r}\rangle = \langle \mathbf{b}, P_{N(A^*)}\mathbf{b}\rangle, \tag{142}$$

since $\dim N(A^*) = 1$ and $\mathbf{b} \notin R(A)$. (Equation (142) represents the hyperplane of residuals; see, e.g., Cheney [**184**, Lemma, p. 40]). A routine computation now shows, that among all vectors $\mathbf{r}$ satisfying (142) there is a unique vector of minimum Tchebycheff norm given by (141), from which (140) follows since $N(A) = \{\mathbf{0}\}$. $\square$

EX. 61. Let $A \in \mathbb{C}_n^{(n+1)\times n}$ and $\mathbf{b} \in \mathbb{C}^{n+1}$ be such that (1) is inconsistent. Then, for any norm ϕ on $\mathbb{C}^n$, a ϕ-approximate solution of (1) is given by

$$\mathbf{x} = A^\dagger(\mathbf{b} + \mathbf{r}),$$

where the residual $\mathbf{r}$ is a dual vector of $P_{N(A^*)}\mathbf{b}$ with respect to the norm ϕ, and the error of approximation is

$$\phi(\mathbf{r}) = \frac{\langle \mathbf{b}, P_{N(A^*)}\mathbf{b}\rangle}{\phi_D(P_{N(A^*)}\mathbf{b})}.$$

PROOF. Follows from (142) and Ex. 59.   □

EX. 62. Let $\{\phi, \phi_D\}$ be dual norms with unit balls $B = \{\mathbf{x} : \phi(\mathbf{x}) \leq 1\}$ and $B_D = \{\mathbf{y} : \phi_D(\mathbf{y}) \leq 1\}$, respectively, and let $\{\mathbf{x}_0, \mathbf{y}_0\}$ be dual vectors of norm one, i.e., $\phi(\mathbf{x}_0) = 1$, $\phi_D(\mathbf{y}) = 1$, and

$$\Re\langle \mathbf{x}_0, \mathbf{y}_0\rangle = \phi(\mathbf{x}_0)\phi_D(\mathbf{y}_0).$$

Then:

(a) The hyperplane

$$H = \{\mathbf{x} : \Re\langle \mathbf{y}_0, \mathbf{x}\rangle = \phi(\mathbf{x}_0)\phi_D(\mathbf{y}_0)\}$$

supports B at $\mathbf{x}_0$, that is, $\mathbf{x}_0 \in H$ and B lies on one side of H, i.e.,

$$\mathbf{x} \in B \implies \Re\langle \mathbf{y}_0, \mathbf{x}\rangle \leq \Re\langle \mathbf{y}_0, \mathbf{x}_0\rangle = \phi(\mathbf{x}_0)\phi_D(\mathbf{y}_0).$$

(b) The hyperplane

$$\{\mathbf{y} : \Re\langle \mathbf{x}_0, \mathbf{y}\rangle = \phi(\mathbf{x}_0)\phi_D(\mathbf{y}_0)\}$$

supports B_D at $\mathbf{y}_0$.

PROOF. Follows from (132).   □

EX. 63. A closed convex set B is called *rotund* if its boundary contains no line segments or, equivalently, if each one of its boundary points is an extreme point.

A closed convex set is called *smooth* if it has, at each boundary point, a unique supporting hyperplane.

Show that an equilibrated convex body B is rotund if and only if its dual set B_D is smooth.

PROOF. *If*: If B is not rotund, then its boundary contains two points $\mathbf{x}_0 \neq \mathbf{x}_1$ and the line segment $\{\lambda\mathbf{x}_1 + (1 - \lambda)\mathbf{x}_0 : 0 \leq \lambda \leq 1\}$ joining them; that is

$$\phi(\lambda\mathbf{x}_1 + (1 - \lambda)\mathbf{x}_0) = 1, \quad 0 \leq \lambda \leq 1,$$

where ϕ is the gauge function of B.

For any $0 < \lambda < 1$ let $\mathbf{y}_\lambda$ be a dual vector of $\mathbf{x}_\lambda = \lambda\mathbf{x}_1 + (1 - \lambda)\mathbf{x}_0$ with $\phi_D(\mathbf{y}_\lambda) = 1$. Then

$$\Re\langle \mathbf{x}_\lambda, \mathbf{y}_\lambda\rangle = 1$$

and, by (132),

$$\Re\langle \mathbf{x}_0, \mathbf{y}_\lambda\rangle = \Re\langle \mathbf{x}_1, \mathbf{y}_\lambda\rangle = 1,$$

showing that $\mathbf{y}_\lambda$ is a dual vector of both $\mathbf{x}_0$ and $\mathbf{x}_1$ and, by Ex. 62(b), both hyperplanes

$$\{\mathbf{y} : \Re\langle \mathbf{x}_\lambda, \mathbf{y}\rangle = 1\}, \quad \lambda = 0, 1,$$

support B_D at $\mathbf{y}_\lambda$.

Only if: Follows by reversing the above steps.   □

For additional results on rotundity see the survey of Cudia [211].

Ex. 64. Let ϕ be a norm on $\mathbb{C}^n$ and let B be its unit ball,

$$B = \{\mathbf{x} : \phi(\mathbf{x}) \leq 1\}.$$

Then:

(a) ϕ is e.s.c. if and only if B is rotund.

(b) ϕ is Gateaux differentiable; that is, the limit

$$\phi'(\mathbf{x}; \mathbf{y}) = \lim_{t \to 0} \frac{\phi(\mathbf{x} + t\mathbf{y}) - \phi(\mathbf{x})}{t}$$

exists for all $\mathbf{x}, \mathbf{y} \in \mathbb{C}^n$ if and only if B is smooth.

Ex. 65. Give an example of dual norms $\{\phi, \phi_D\}$ such that ϕ is e.s.c. but ϕ_D is not.

SOLUTION. Let

$$B = \left\{ \begin{bmatrix} x_1 \\ x_2 \end{bmatrix} \in \mathbb{R}^2 : x_1 \geq \tfrac{1}{2}(x_2 + 1)^2 - 1, \; x_2 \geq \tfrac{1}{2}(x_1 + 1)^2 - 1 \right\}.$$

Then B is an equilibrated convex body. B is rotund but not smooth (the points $\begin{pmatrix} 1 \\ 1 \end{pmatrix}$ and $\begin{pmatrix} -1 \\ -1 \end{pmatrix}$ are "corners" of B), so, by Ex. 61, the dual set B_D is not rotund. Hence, by Ex. 64(a), the gauge function ϕ^B is an e.s.c. norm, but its dual ϕ_D^B is not. $\square$

Norms of Homogeneous Transformations

Ex. 66. (Bauer [54], Householder [432]). Let α and β be norms on $\mathbb{C}^n$ and $\mathbb{C}^m$, respectively. Let $A : \mathbb{C}^n \to \mathbb{C}^m$ be a continuous transformation that is homogeneous; that is,

$$A(\lambda \mathbf{x}) = \lambda A(\mathbf{x}), \quad \text{for all } \lambda \in \mathbb{C}, \, \mathbf{x} \in \mathbb{C}^n.$$

The *norm* (also *least upper bound*) *of* A *corresponding to* $\{\alpha, \beta\}$, denoted by $\|A\|_{\alpha, \beta}$ (also by $\mathrm{lub}_{\alpha,\beta}(A)$), is defined as

$$
\begin{aligned}
\|A\|_{\alpha, \beta} &= \sup_{\mathbf{x} \neq 0} \frac{\beta(A\mathbf{x})}{\alpha(\mathbf{x})} \\
&= \max_{\alpha(\mathbf{x}) = 1} \beta(A\mathbf{x}),
\end{aligned}
\tag{143}
$$

since A is continuous and homogeneous. Then for any A, A_1, A_2 as above:

(a) $\|A\|_{\alpha, \beta} \geq 0$ with equality if and only if A is the zero transformation.

(b) $\|\lambda A\|_{\alpha, \beta} = |\lambda| \, \|A\|_{\alpha, \beta}$ for all $\lambda \in \mathbb{C}$.

(c) $\|A_1 + A_2\|_{\alpha, \beta} \leq \|A_1\|_{\alpha, \beta} + \|A_2\|_{\alpha, \beta}$.

(d) If B_α, B_β are the unit balls of α, β, respectively, then

$$\|A\|_{\alpha, \beta} = \inf\{\lambda > 0 : AB_\alpha \subset \lambda B_\beta\}.$$

(e) If $A_1 : \mathbb{C}^n \to \mathbb{C}^m$ and $A_2 : \mathbb{C}^m \to \mathbb{C}^p$ are continuous homogeneous transformations, and if α, β, and γ are norms on $\mathbb{C}^n, \mathbb{C}^m$, and $\mathbb{C}^p$, respectively, then

$$\|A_2 A_1\|_{\alpha, \gamma} \leq \|A_1\|_{\alpha, \beta} \, \|A_2\|_{\beta, \gamma}.$$

(f) If $A : \mathbb{C}^n \to \mathbb{C}^m$ is a linear transformation, and if $\alpha = \beta$, i.e., the same norm is used in $\mathbb{C}^n$ and $\mathbb{C}^m$, then definition (143) reduces to that given in Ex. 0.35.

Ex. 67. Let α and β be norms on $\mathbb{C}^n$ and $\mathbb{C}^m$, respectively. Then, for any $A \in \mathbb{C}^{m \times n}$,

$$\|A\|_{\alpha, \beta} = \|A^*\|_{\beta_D, \alpha_D}.
\tag{144}$$

PROOF. From (132) and (143) it follows that, for all $\mathbf{x} \in \mathbb{C}^n$, $\mathbf{y} \in \mathbb{C}^m$,

$$\Re\langle A\mathbf{x}, \mathbf{y} \rangle \leq \beta(A\mathbf{x})\beta_D(\mathbf{y}) \leq \|A\|_{\alpha,\beta}\, \alpha(\mathbf{x})\, \beta_D(\mathbf{y}),$$

with equality for at least one pair $\mathbf{x} \neq \mathbf{0}$, $\mathbf{y} \neq \mathbf{0}$. The dual inequalities

$$\Re\langle \mathbf{x}, A^*\mathbf{y} \rangle \leq \alpha(\mathbf{x})\alpha_D(A^*\mathbf{y}) \leq \|A^*\|_{\beta_D,\alpha_D}\, \beta_D(\mathbf{y})\alpha(\mathbf{x})$$

then show that

$$\|A\|_{\alpha,\beta} \leq \|A^*\|_{\beta_D,\alpha_D},$$

from which (144) follows by reversing the roles of A and A^* and by using Ex. 57(d).
$\square$

Projective Bounds and Norms

Ex. 68. (Erdelsky [255]). Let α be an e.s.c. norm on $\mathbb{C}^n$. The *projective bound* of α, denoted by $Q(\alpha)$, is defined as

$$Q(\alpha) = \sup_L \|P_{L,\alpha}\|_{\alpha,\alpha} \tag{145}$$

where the supremum is taken over all subspaces L with dimension $1 \leq \dim L \leq n-1$. (The α-metric projector $P_{L,\alpha}$ is continuous and homogeneous, by Theorem 11(c) and (i), allowing the use of (143) to define $\|P_{L,\alpha}\|_{\alpha,\alpha}$.) Then:

(a) The supremum in (145) is finite and is attained for a k-dimensional subspace for each $k = 1, 2, \ldots, n-1$.

(b) The projective bound satisfies

$$1 \leq Q(\alpha) < 2 \tag{146}$$

and the upper limit is approached arbitrarily closely by e.s.c. norms.

Ex. 69. (Erdelsky [255]). An e.s.c. norm α on $\mathbb{C}^n$ for which the projective bound

$$Q(\alpha) = 1$$

is called a *projective norm*. All ellipsoidal norms,

$$\|\mathbf{x}\|_U = (\mathbf{x}^*U\mathbf{x})^{1/2}, \quad U \text{ positive definite}, \tag{50}$$

are projective.

Conversely, for spaces of dimension ≥ 3, all projective norms are ellipsoidal, both in the real case (Kakutani [455]) and in the complex case (Bohnenblust [115]). An example of a nonellipsoidal projective norm on $\mathbb{R}^2$ is

$$\alpha\left(\begin{bmatrix} x_1 \\ x_2 \end{bmatrix}\right) = \begin{cases} (|x_1|^p + |x_2|^p)^{1/p}, & \text{if } x_1 x_2 \geq 0, \\ (|x_1|^q + |x_2|^q)^{1/q}, & \text{if } x_1 x_2 < 0, \end{cases}$$

where $(1/p) + (1/q) = 1$, $1 < p \neq 2$.

Ex. 70. (Erdelsky [255]). If α is a projective norm, L is a subspace for which the α-metric projector $P_{L,\alpha}$ is linear, and N denotes

$$N = P_{L,\alpha}^{-1}(\mathbf{0}), \tag{147}$$

then

$$L = P_{N,\alpha}^{-1}(\mathbf{0}). \tag{148}$$

PROOF. $L \subset P_{N,\alpha}^{-1}(\mathbf{0})$: If $\mathbf{x} \in L$ and $\mathbf{y} \in N$, then

$$P_{L,\alpha}^{-1}(\mathbf{x} + \mathbf{y}) = \mathbf{x},$$

by Theorem 11(a) and, consequently,

$$\alpha(\mathbf{x}) \leq \|P_{L,\alpha}\|_{\alpha,\alpha}\, \alpha(\mathbf{x} + \mathbf{y})$$
$$\leq Q(\alpha)\, \alpha(\mathbf{x} + \mathbf{y})$$
$$= \alpha(\mathbf{x} + \mathbf{y})$$

for all $\mathbf{y} \in N$, proving that $P_{N,\alpha}(\mathbf{x}) = \mathbf{0}$.

$P_{N,\alpha}^{-1}(\mathbf{0}) \subset L$: If $\mathbf{x} \in P_{N,\alpha}^{-1}(\mathbf{0})$, then, by (121), it can be written as

$$\mathbf{x} = \mathbf{x}_1 + \mathbf{x}_2, \quad \mathbf{x}_1 \in L, \ \mathbf{x}_2 \in N.$$

Therefore,

$$\mathbf{0} = P_{N,\alpha}(\mathbf{x}) = P_{N,\alpha}(\mathbf{x}_1) + \mathbf{x}_2, \quad \text{by Theorem 11(d),}$$
$$= \mathbf{x}_2, \quad \text{since } L \subset P_{N,\alpha}^{-1}(\mathbf{0}),$$

proving that

$$\mathbf{x} = \mathbf{x}_1 \in L. \qquad \qquad \square$$

Projective Norms and the Linearity of Metric Projectors

The following four exercises probe the relations between the linearity of the α-metric projector $P_{L,\alpha}$ and the projectivity of the norm α. Exercise 71 shows that

$$\alpha \text{ projective} \implies P_{L,\alpha} \text{ linear for all } L,$$

and a partial converse is proved in Ex. 73.

EX. 71. (Erdelsky [255]). If α is a projective norm on $\mathbb{C}^n$, then $P_{L,\alpha}$ is linear for all subspaces L of $\mathbb{C}^n$.

PROOF. By Corollary 7 it suffices to prove the linearity of $P_{L,\alpha}$ for all one-dimensional subspaces L.

Let $\dim L = 1$, $\boldsymbol{\ell} \in L$, $\alpha(\boldsymbol{\ell}) = 1$, and let $\boldsymbol{\ell} + N$ be a supporting hyperplane of $B_\alpha = \{\mathbf{x} : \alpha(\mathbf{x}) \leq 1\}$ at $\boldsymbol{\ell}$. Since

$$\alpha(\boldsymbol{\ell}) \leq \alpha(\mathbf{x}), \quad \text{for all } \mathbf{x} \in \boldsymbol{\ell} + N,$$

it follows from Definition 1 that

$$P_{N,\alpha}(\boldsymbol{\ell}) = \mathbf{0}$$

and hence

$$L \subset P_{N,\alpha}^{-1}(\mathbf{0}).$$

Now $P_{N,\alpha}$ is linear by Corollary 6, since $\dim N = n - 1$, which also shows that $P_{N,\alpha}^{-1}(\mathbf{0})$ is a one-dimensional subspace, by (121), and hence

$$L = P_{N,\alpha}^{-1}(\mathbf{0}).$$

From Ex. 70 it follows then that

$$N = P_{L,\alpha}^{-1}(\mathbf{0}),$$

and the linearity of $P_{L,\alpha}$ is established by Corollary 5(b). $\qquad \square$

EX. 72. (Erdelsky [255]). If α is an e.s.c. norm on $\mathbb{C}^n$, L is a subspace for which $P_{L,\alpha}$ is linear, and N denotes

$$N = P_{L,\alpha}^{-1}(\mathbf{0}), \tag{147}$$

then

$$L = P_{N,\alpha}^{-1}(\mathbf{0}) \tag{148}$$

if, and only if,

$$P_{L,\alpha} + P_{N,\alpha} = I.$$

PROOF. Follows from (121). □

EX. 73. (Erdelsky [255]). Let α be an e.s.c. norm on $\mathbb{C}^n$ and let $1 \leq k \leq n-1$ be an integer such that, for every k-dimensional subspace L of $\mathbb{C}^n$,

$$P_{L,\alpha} \text{ is linear}$$

and

$$L = P_{N,\alpha}^{-1}(0), \tag{148}$$

where N is given by (147). Then α is projective.

PROOF. Let α be nonprojective; i.e., let $Q(\alpha) > 1$. Then there is a k-dimensional subspace L and two points $\mathbf{x}, \mathbf{y}$ in $\mathbb{C}^n$ such that

$$\mathbf{y} = P_{L,\alpha}(\mathbf{x}) \tag{149a}$$

and

$$\alpha(\mathbf{y}) = \|P_{L,\alpha}\|_{\alpha,\alpha} \, \alpha(\mathbf{x}) = Q(\alpha) \, \alpha(\mathbf{x}) > \alpha(\mathbf{x}). \tag{149b}$$

Let $N = P_{L,\alpha}^{-1}(0)$. Then

$$\mathbf{0} \neq \mathbf{y} - \mathbf{x} \in N, \quad \text{by (149b), (149a), and (121),} \tag{149c}$$

and

$$\alpha(\mathbf{x}) = \alpha(\mathbf{y} - (\mathbf{y} - \mathbf{x})) < \alpha(\mathbf{y}). \tag{149d}$$

Now

$$\mathbf{y} = P_{L,\alpha}(\mathbf{y}) + P_{N,\alpha}(\mathbf{y}), \quad \text{by (148) and Ex. 72}$$
$$= \mathbf{y} + P_{N,\alpha}(\mathbf{y}), \quad \text{by (149a) and Theorem 11(a),} \tag{149e}$$

proving that

$$P_{N,\alpha}(\mathbf{y}) = \mathbf{0}, \tag{149f}$$

which, by (149c) and (120), contradicts (149d). □

EX. 74. (Newman and Odell [609]). Let ϕ_p be the ℓ_p-norm, $1 < p < \infty$, on $\mathbb{C}^n$. The P_{L,ϕ_p} is linear for every subspace L if and only if $p = 2$.

Essentially Strictly Convex Norms

EX. 75. (Erdelsky [255]). Let α be an e.s.c. norm on $\mathbb{C}^n$, $\mathbf{0} \neq \mathbf{x} \in \mathbb{C}^n$ and let L be a subspace of $\mathbb{C}^n$. Then

$$\mathbf{x} \in P_{L,\alpha}^{-1}(\mathbf{0})$$

if, and only if, there is a dual $\mathbf{y}$ of $\mathbf{x}$ with respect to α (i.e., a vector $\mathbf{y} \neq \mathbf{0}$ satisfying $\langle \mathbf{y}, \mathbf{x} \rangle = \alpha(\mathbf{x}) \, \alpha_D(\mathbf{y})$), such that $\mathbf{y} \in L^\perp$.

EX. 76. (Erdelsky [255]). If α and α_D are both e.s.c. norms on $\mathbb{C}^n$, L is a subspace of $\mathbb{C}^n$ for which $P_{L,\alpha}$ is linear, and $N = P_{L,\alpha}^{-1}(\mathbf{0})$, then:

(a) $L^\perp = P_{N^\perp, \alpha_D}^{-1}(\mathbf{0})$.

(b) $P_{N^\perp, \alpha_D} = (P_{L,\alpha})^*$.

PROOF. (a) Since both α and α_D are e.s.c., it follows from Exs. 64(a), 63, and 62 that every $\mathbf{0} \neq \mathbf{x}$ has a dual $\mathbf{0} \neq \mathbf{y}$ with respect to α and $\mathbf{x}$ is a dual of $\mathbf{y}$. Now

$$\mathbf{y} \in P_{N^\perp,\alpha_D}^{-1}(\mathbf{0}) \quad \Longleftrightarrow \quad \mathbf{x} \in N^{\perp\perp} = N,$$

by Ex. 75, which also shows that

$$\mathbf{x} \in N \quad \Longleftrightarrow \quad \mathbf{y} \in L^\perp,$$

proving (a).

(b) By (a) and Corollary 5(b), $P_{N^\perp,\alpha_D}$ is linear. Let $\mathbf{x}$ and $\mathbf{y}$ be arbitrary vectors, written as

$$\mathbf{x} = \mathbf{x}_1 + \mathbf{x}_2, \quad \mathbf{x}_1 \in L, \ \mathbf{x}_2 \in N, \quad \text{by (121)},$$

and

$$\mathbf{y} = \mathbf{y}_1 + \mathbf{y}_2, \quad \mathbf{y}_1 \in N^\perp, \ \mathbf{y}_2 \in L^\perp, \quad \text{by (a) and (121)}.$$

Then

$$\langle P_{L,\alpha}(\mathbf{x}), \mathbf{y} \rangle = \langle \mathbf{x}_1, \mathbf{y}_1 \rangle = \langle \mathbf{x}, P_{N^\perp,\alpha_D}(\mathbf{y}) \rangle. \qquad \square$$

EX. 77. (Erdelsky [**255**]). *Dual norms.* Let α and α_D be dual norms on $\mathbb{C}^n$. Then:

(a) If α and α_D are both e.s.c., then $Q(\alpha) = Q(\alpha_D)$.
(b) If α is projective, then α_D is e.s.c.
(c) If α is projective, then so is α_D.

α-β Generalized Inverses

EX. 78. (Erdelsky [**255**]). Let α and β be e.s.c. norms on $\mathbb{C}^m$ and $\mathbb{C}^n$, respectively, and let $A \in \mathbb{C}^{m \times n}$.

If $B \in \mathbb{C}^{n \times m}$ satisfies

$$AB = P_{R(A),\alpha}, \tag{150a}$$
$$BA = I - P_{N(A),\beta}, \tag{150b}$$
$$\operatorname{rank} B = \operatorname{rank} A, \tag{150c}$$

then

$$B = A_{\alpha,\beta}^{(-1)}.$$

Thus, if the α-β generalized inverse of A is linear, it can be defined by (150a)–(150c).

EX. 79. (Erdelsky [**255**]). Let α and β be e.s.c. norms on $\mathbb{C}^m$ and $\mathbb{C}^n$, respectively. Then

$$(A_{\alpha,\beta}^{(-1)})_{\beta,\alpha}^{(-1)} = A, \quad \text{for all } A \in \mathbb{C}^{m \times n}, \tag{151}$$

if and only if α and β are projective norms.

PROOF. *If:* If α and β are projective, then $A_{\alpha,\beta}^{(-1)}$ is linear for any A, by Theorem 12(b) and Ex. 71. Let $\widehat{R} = R(A_{\alpha,\beta}^{(-1)})$ and $\widehat{N} = N(A_{\alpha,\beta}^{(-1)})$. Then, by Exs. 70, 71, 75, and Theorem 12(c), (d), (e), (g), and (h),

$$A_{\alpha,\beta}^{(-1)} A = I - P_{N(A),\beta} = P_{\widehat{R},\beta},$$
$$A A_{\alpha,\beta}^{(-1)} = P_{R(A),\alpha} = I - P_{\widehat{N},\alpha},$$
$$\operatorname{rank} A_{\alpha,\beta}^{(-1)} = \operatorname{rank} A,$$

and (151) follows from Ex. 78.

Only if: If (151) holds for all $A \in \mathbb{C}^{m \times n}$, then

$$I - P_{N(A),\beta} = A_{\alpha,\beta}^{(-1)} A = A_{\alpha,\beta}^{(-1)} (A_{\alpha,\beta}^{(-1)})_{\beta,\alpha}^{(-1)} = P_{\widehat{R},\beta},$$
$$P_{R(A),\alpha} = A A_{\alpha,\beta}^{(-1)} = (A_{\alpha,\beta}^{(-1)})_{\beta,\alpha}^{(-1)} A_{\alpha,\beta}^{(-1)} = I - P_{\widehat{N},\alpha},$$

and α and β are projective by Ex. 73. $\square$

Ex. 80. (Erdelsky [**255**]). If α and β are e.s.c. norms on $\mathbb{C}^m$ and $\mathbb{C}^n$, respectively, then

$$(A_{\alpha,\beta}^{(-1)})^* = (A^*)_{\beta_D,\alpha_D}^{(-1)}, \quad \text{for all } A \in \mathbb{C}^{m \times n}. \tag{152}$$

PROOF. From Theorem 12(d) and (f) and Exs. 70, 71, and 72,

$$A A_{\alpha,\beta}^{(-1)} = P_{R(A),\alpha} = I - P_{N,\alpha}, \quad N = P_{R(A),\alpha}^{-1}(\mathbf{0}),$$
$$A_{\alpha,\beta}^{(-1)} A = I - P_{N(A),\beta} = P_{M,\beta}, \quad M = P_{N(A),\beta}^{-1}(\mathbf{0}),$$

and

$$R(A) = P_{N,\alpha}^{-1}(\mathbf{0}),$$
$$N(A) = P_{M,\beta}^{-1}(\mathbf{0}).$$

Since α_D and β_D are e.s..c. norms, by Ex. 77(b), it follows from Ex. 76(b) that

$$A A_{\alpha,\beta}^{(-1)} = I - (P_{R(A)^\perp,\alpha_D})^* = I - (P_{N(A^*),\alpha_D})^*,$$
$$A_{\alpha,\beta}^{(-1)} A = (P_{N(A)^\perp,\beta_D})^* = (P_{R(A^*),\beta_D})^*,$$

and hence

$$(A_{\alpha,\beta}^{(-1)})^* A^* = I - P_{N(A^*),\alpha_D},$$
$$A^* (A_{\alpha,\beta}^{(-1)})^* = P_{R(A^*),\beta_D},$$

from which (152) follows by using Ex. 78. $\square$

Ex. 81. (Erdelsky [**255**]). If α and β are e.s.c. norms on $\mathbb{C}^m$ and $\mathbb{C}^n$, respectively, then, for any $O \neq A \in \mathbb{C}^{m \times n}$,

$$\frac{1}{\|A_{\alpha,\beta}^{(-1)}\|_{\beta,\alpha}} \leq \inf \{\|X\|_{\alpha,\beta} : X \in \mathbb{C}^{m \times n}, \ \text{rank}(A + X) < \text{rank } A\}$$

$$\leq \frac{q}{\|A_{\alpha,\beta}^{(-1)}\|_{\beta,\alpha}}, \tag{153}$$

where

$$q = \begin{cases} 1, & \text{if rank } A = m, \\ Q(\alpha), & \text{otherwise.} \end{cases}$$

In particular, if α is projective,

$$\frac{1}{\|A_{\alpha,\beta}^{(-1)}\|_{\beta,\alpha}} = \inf\{\|X\|_{\alpha,\beta} : X \in \mathbb{C}^{m \times n}, \ \text{rank}(A + X) < \text{rank } A\}. \tag{154}$$

A special case of (154) is given in Ex. 6.30 below.

8. An Extremal Property of the Bott–Duffin Inverse with Application to Electrical Networks

An important extremal property of the Bott–Duffin inverse, studied in Sections 2.10 and 2.13, is stated in the following theorem:

THEOREM 13 (Bott and Duffin [**120**]). *Let $A \in \mathbb{C}^{n \times n}$ be Hermitian and let L be a subspace of $\mathbb{C}^n$ such that $A_{(L)}^{(-1)}$ exists.*[6] *Then, for any two vectors $\mathbf{v}, \mathbf{w} \in \mathbb{C}^n$, the quadratic function*

$$q(\mathbf{x}) = \tfrac{1}{2} (\mathbf{x} - \mathbf{v})^* A (\mathbf{x} - \mathbf{v}) - \mathbf{w}^* \mathbf{x} \tag{155}$$

has a unique stationary value in L, when

$$\mathbf{x} = A_{(L)}^{(-1)} (A\mathbf{v} + \mathbf{w}). \tag{156}$$

Conversely, if the Hermitian matrix A and the subspace L are such that for any two vectors $\mathbf{v}, \mathbf{w} \in \mathbb{C}^n$, the quadratic function (155) *has a stationary value in L, then $A_{(L)}^{(-1)}$ exists and the stationary point is unique for any $\mathbf{v}, \mathbf{w}$ given by* (156).

PROOF. A *stationary point* of q in L is a point $\mathbf{x} \in L$ at which the gradient $\nabla q(\mathbf{x}) = \left[\frac{\partial}{\partial x_j} q(\mathbf{x}) \right]$ $(j \in \overline{1,n})$, is orthogonal to L, i.e.,

$$\nabla q(\mathbf{x}) \in L^{\perp}. \tag{157}$$

The value of q at a stationary point is called a *stationary value* of q.

Differentiating (155) we get from (157),

$$\nabla q(\mathbf{x}) = A(\mathbf{x} - \mathbf{v}) - \mathbf{w} \in L^{\perp},$$

and, by taking $\mathbf{y} = -\nabla q(\mathbf{x})$, we conclude that $\mathbf{x}$ is a stationary point of q in L if and only if $\mathbf{x}$ is a solution of

$$A\mathbf{x} + \mathbf{y} = A\mathbf{v} + \mathbf{w}, \quad \mathbf{x} \in L, \ \mathbf{y} \in L^{\perp}. \tag{158}$$

Thus the existence of a stationary value of q for any $\mathbf{v}, \mathbf{w}$ is equivalent to the consistency of (158) for any $\mathbf{v}, \mathbf{w}$, i.e., to the existence of $A_{(L)}^{(-1)}$, in which case (156) is the unique stationary point in L. $\qquad\square$

COROLLARY 8. *Let $A \in \mathbb{C}^{n \times n}$ be Hermitian positive definite and let L be a subspace of $\mathbb{C}^n$. Then, for any $\mathbf{v}, \mathbf{w} \in \mathbb{C}^n$, the function*

$$q(\mathbf{x}) = \tfrac{1}{2} (\mathbf{x} - \mathbf{v})^* A (\mathbf{x} - \mathbf{v}) - \mathbf{w}^* \mathbf{x} \tag{155}$$

has a unique minimum in L when

$$\mathbf{x} = A_{(L)}^{(-1)} (A\mathbf{v} + \mathbf{w}). \tag{156}$$

PROOF. Follows from Theorem 13, since $A_{(L)}^{(-1)}$ exists, by Ex. 2.107, and the stationary value of q is actually a minimum since A is PD. $\qquad\square$

We return now to the direct current electrical network of Section 2.13, consisting of m nodes $\{n_i : i \in \overline{1,m}\}$ and n branches $\{b_j : j \in \overline{1,n}\}$, with

$a_j > 0$, the *conductance* of b_j;
$A = \mathrm{diag}\,(a_j)$, the *conductance matrix*;
x_j, the *voltage across b_j*;
y_j, the *current in b_j*;
v_j, the *voltage generated by the sources in series with b_j*;

---

[6]See Ex. 2.94 for conditions equivalent to the existence of $A_{(L)}^{(-1)}$.

w_j, the *current generated by the sources in parallel with* b_j; and M, the (node-branch) *incidence matrix*.

We recall that the branch voltages $\mathbf{x}$ and currents $\mathbf{y}$ are uniquely determined by the following three physical laws:

$$A\mathbf{x} + \mathbf{y} = A\mathbf{v} + \mathbf{w} \quad (Ohm's \ law), \tag{159}$$

$$\mathbf{y} \in N(M) \quad (Kirchhoff's \ current \ law), \tag{160}$$

$$\mathbf{x} \in R(M^T) \quad (Kirchhoff's \ voltage \ law), \tag{161}$$

and that $\mathbf{x}$, $\mathbf{y}$ are related by

$$\mathbf{x} = A^{(-1)}_{(R(M^T))}(A\mathbf{v} + \mathbf{w}), \tag{2.151a}$$

$$\mathbf{y} = (I - AA^{(-1)}_{(R(M^T))})(A\mathbf{v} + \mathbf{w}), \tag{2.151b}$$

or, dually, by (2.153b) and (2.153a).

A classical variational principle of Kelvin [**804**] and Maxwell [**537**, pp. 903–908], states that the voltages $\mathbf{x}$ and the currents $\mathbf{y}$ are such that the rate of energy dissipation is minimized. This variational principle is given in the following corollary:

COROLLARY 9. *Let* A, M, $\mathbf{x}$, $\mathbf{y}$, $\mathbf{v}$, $\mathbf{w}$ *be as above. Then:*

(a) *The vector* $\mathbf{x}_0$ *of branch voltages is the unique minimizer of*

$$q(\mathbf{x}) = \tfrac{1}{2}(\mathbf{x} - \mathbf{v})^* A(\mathbf{x} - \mathbf{v}) - \mathbf{w}^* \mathbf{x} \tag{155}$$

in $R(M^T)$, *and the vector* $\mathbf{y}_0$ *of branch currents is*

$$\mathbf{y}_0 = -\nabla q(\mathbf{x}_0) = -A(\mathbf{x}_0 - \mathbf{v}) + \mathbf{w} \in R(M^T)^\perp = N(M). \tag{162}$$

(b) *The vector* $\mathbf{y}_0$ *is the unique minimizer of*

$$p(\mathbf{y}) = \tfrac{1}{2}(\mathbf{y} - \mathbf{w})^* A^{-1}(\mathbf{y} - \mathbf{w}) - \mathbf{v}^* \mathbf{y} \tag{163}$$

in $N(M)$, *and the vector* $\mathbf{x}_0$ *is*

$$\mathbf{x}_0 = -\nabla p(\mathbf{y}_0) = -A^{-1}(\mathbf{y}_0 - \mathbf{w}) + \mathbf{v} \in N(M)^\perp = R(M^T). \tag{164}$$

PROOF. Since the conductance matrix A is PD, it follows by comparing (156) and (2.151a) that $\mathbf{x}_0$ is the unique minimizer of (155) in $R(M^T)$, and the argument used in the proof of Theorem 13 shows that $\mathbf{y}_0 = -\nabla q(\mathbf{x}_0)$ as given in (162). Part (b) follows from the dual derivation (2.153a) and (2.153b) of $\mathbf{y}_0$ and $\mathbf{x}_0$, respectively, as solutions of the dual network equations (2.152). □

Corollary 9 shows that the voltage $\mathbf{x}$ is uniquely determined by the function (155) to be minimized subject to Kirchhoff's voltage law (161). Kirchhoff's current law (160) and Ohm's law (159) are then consequences of (162).

Dually, the current $\mathbf{y}$ is uniquely determined by the function (163) to be minimized subject to Kirchhoff's current law (160), and the other two laws (159) and (161) then follow from (164).

Corollary 9 is a special case of the Duality Theory of Convex Programming; see, e.g., Rockafellar [**703**].

Exercises

EX. 82. Let $A \in \mathbb{C}^{n \times n}$ be Hermitian PD, and let the subspace $L \subset \mathbb{C}^n$ and the vector $\mathbf{w} \in \mathbb{C}^n$ be given. Then the quadratic function

$$\tfrac{1}{2}\mathbf{x}^* A \mathbf{x} - \mathbf{w}^* \mathbf{x} \qquad (165)$$

has a minimum in L if and only if the system

$$A\mathbf{x} - \mathbf{w} \in L^{\perp}, \quad \mathbf{x} \in L, \qquad (166)$$

is consistent, in which case the solutions $\mathbf{x}$ of (166) are the minimizers of (165) in L.

EX. 83. Show that the consistency of (166) is equivalent to the condition

$$\mathbf{x} \in L, \quad A\mathbf{x} = \mathbf{0} \quad \Longrightarrow \quad \mathbf{w}^* \mathbf{x} = 0,$$

which is obviously equivalent to the boundedness from below of (165) in L, hence to the existence of a minimizer in L.

EX. 84. Show that $A_{(L)}^{(-1)}$ exists if and only if the system (166) has a unique solution for any $\mathbf{w} \in \mathbb{C}^n$, in which case this solution is

$$\mathbf{x} = A_{(L)}^{(-1)} \mathbf{w}.$$

EX. 85. Give the general solution of (166) in case it is consistent but $A_{(L)}^{(-1)}$ does not exist.

Suggested Further Reading

SECTION 1. Desoer and Whalen [**226**], Eldén [**249**], [**250**], Erdélyi and Ben-Israel [**265**], Leringe and Wedin [**508**], Osborne [**621**], Peters and Wilkinson [**640**], Robinson [**698**], and the references on applications to statistics on p. 328 and system theory on p. 329.

SECTION 2. Erdélyi and Ben-Israel [**265**], Osborne [**622**], Rosen [**712**].

SECTION 3. See the references in p. 369.

SECTION 5. The geometry of approximate solutions: Hanke and Neumann [**372**], Miao and Ben-Israel [**551**], [**552**], Zietak [**888**].

The bounds on $\|\boldsymbol{\xi}_D\|$ and $\|\boldsymbol{\eta}_D\|$, and associated condition numbers, were studied by Stewart [**782**] and Todd [**810**], and studied further, applied, and extended, by Bobrovnikova and Vavasis [**114**], Forsgren [**279**], Forsgren and Sporre [**280**], Gonzaga and Lara [**313**], Hanke and Neumann [**372**], Ikramov [**437**], O'Leary [**619**], Vavasis and Ye [**834**], M. Wei [**857**], and others.

SECTION 6. Bapat [**46**].

SECTION 7. Bhatia [**96**].

SECTION 8. Further references on the extremal properties of the network functions and solutions are Dennis [**225**], Stern [**777**], [**778**], Guillemin [**359**].

CHAPTER 4

Spectral Generalized Inverses

1. Introduction

In this chapter we shall study generalized inverses having some of the spectral properties (i.e., properties relating to eigenvalues and eigenvectors) of the inverse of a nonsingular matrix. Only square matrices are considered, since only they have eigenvalues and eigenvectors.

If A is nonsingular it is easy to see that every eigenvector of A associated with the eigenvalue λ is also an eigenvector of A^{-1} associated with the eigenvalue λ^{-1}. (A nonsingular matrix does not have 0 as an eigenvalue.)

A matrix $A \in \mathbb{C}^{n \times n}$ that is not diagonable does not have n linearly independent eigenvectors (see Ex. 2.22). However, it does have n linearly independent principal vectors, see § 0.7.

It is not difficult to show (see Ex. 2) that, if A is nonsingular, a vector $\mathbf{x}$ is a λ^{-1}-vector of A^{-1} of grade p if and only if it is a λ-vector of A of grade p. In the remainder of this chapter, we shall explore the extent to which singular matrices have generalized inverses with comparable spectral properties.

The four Penrose equations of Chapter 1,

$$AXA = A, \tag{1}$$

$$XAX = X, \tag{2}$$

$$(AX)^* = AX, \tag{3}$$

$$(XA)^* = XA, \tag{4}$$

will now be supplemented further by the following equations applicable only to square matrices

$$A^k XA = A^k, \tag{1^k}$$

$$AX = XA, \tag{5}$$

$$A^k X = XA^k, \tag{5^k}$$

$$AX^k = X^k A, \tag{6^k}$$

In these equations k is a given positive integer.

This chapter deals mostly with the $\{1^k, 2, 5\}$-*inverse* of A, where k is the *index* of A. This inverse, called the *Drazin inverse*, has important spectral properties that make it extremely useful in many applications. We also study a special case of the Drazin inverse, the *group inverse*.

Exercises

Ex. 1. A square matrix A is diagonable if and only if all its principal vectors are eigenvectors.

Ex. 2. If A is nonsingular, $\mathbf{x}$ is a λ^{-1}-vector of A^{-1} of grade p if and only if it is a λ-vector of A of grade p. (*Hint*: Show that $A^{-p}(A - \lambda I)^p = (-\lambda)^p (A^{-1} - \lambda^{-1}I)^p$. Using this and the analogous relation obtained by replacing A by A^{-1}, show that $(A - \lambda I)^r \mathbf{x} = \mathbf{0}$ if and only if $(A^{-1} - \lambda^{-1}I)^r \mathbf{x} = \mathbf{0}$ for $r = 0, 1, \dots$)

Ex. 3. If A is nonsingular and diagonable, A^{-1} is the *only* matrix related to A by the property stated in Ex. 2.

Ex. 4. If A is nonsingular and not diagonable, there are matrices other than A^{-1} having the spectral relationship to A described in Ex. 2. For example, consider

$$A = \begin{bmatrix} \lambda & 1 \\ 0 & \lambda \end{bmatrix}, \quad X = \begin{bmatrix} \lambda^{-1} & c \\ 0 & \lambda^{-1} \end{bmatrix} \quad (\lambda, c \neq 0).$$

Show that for $p = 1, 2$, $\mathbf{x}$ is a λ^{-1}-vector of X of grade p if and only if it is a λ-vector of A of grade p. (Note that $X = A^{-1}$ for $c = -\lambda^{-2}$.)

2. The Matrix Index

It is readily seen that the set of three equations (1^k), (2), and (5) is equivalent to the set

$$AX = XA, \tag{5}$$

$$A^{k+1}X = A^k, \tag{7}$$

$$AX^2 = X. \tag{8}$$

It is evident also that if (7) holds for some positive integer k, then it holds for every integer $\ell > k$. It follows also from (7) that

$$\operatorname{rank} A^k = \operatorname{rank} A^{k+1}. \tag{9}$$

Therefore, a solution X for (7) (and, consequently, of the set (5), (7), (8)) exists only if (9) holds. In this connection, the following definition is useful.

Definition 1. The smallest positive integer k for which (9) holds, is called the *index*[1] of A, and denoted $\operatorname{Ind} A$.

The next lemma collects properties of the matrix index that are needed below.

Lemma 1. *Let $A \in \mathbb{C}^{n \times n}$ and $\operatorname{Ind} A = k$. Then:*

(a) *All matrices $\{A^\ell : \ell \geq k\}$ have the same rank, the same range and the same null space.*

(b) *Their transposes $\{(A^\ell)^T : \ell \geq k\}$ all have the same rank, the same range and the same null space.*

(c) *Their conjugate transposes $\{(A^\ell)^* : \ell \geq k\}$ all have the same rank, the same range and the same null space.*

(d) *Moreover, for no ℓ less than k do A^ℓ and a higher power of A (or their transposes or conjugate transposes) have the same range or the same null space.*

---

[1] Some writers (e.g., MacDuffee [**528**]) define the index as the degree of the minimal polynomial.

PROOF. It may be well to point out first that (9) necessarily holds for *some* positive integer k (see Ex. 7).

(a)   It follows from (9) and Ex. 1.10 that

$$R(A^{k+1}) = R(A^k). \tag{10}$$

Therefore (7) holds for some X, and multiplication on the left by $A^{\ell-k}$ gives

$$A^\ell = A^{\ell+1}X \quad (\ell \geq k). \tag{11}$$

It follows from (11) that all the matrices $\{A^\ell : \ell \geq k\}$ have the same range and the same rank. From Ex. 1.10 and the fact that A^k and A^ℓ have the same rank, it follows that they have the same null space. (See Ex. 6 for an alternative proof of $R(A^\ell) = R(A^{\ell+1})$ for all $\ell \geq k$.).

(b) and (c)   The statements about the transposes and conjugate transposes are obtained by applying (a) to A^T and A^* and noting that $(A^\ell)^T = (A^T)^\ell$ and $(A^\ell)^* = (A^*)^\ell$.

(d)   If an equality of ranges of the kind ruled out in part (d) should occur, there must be some $\ell < k$ such that A^ℓ or its transpose or conjugate transpose have the same range as the corresponding matrix with exponent $\ell + 1$. But this would imply rank $A^\ell = $ rank $A^{\ell+1}$, and k would not be the index of A. Similarly, equality of null spaces would imply that A^ℓ and $A^{\ell+1}$ have the same nullity, and therefore the same rank. $\square$

THEOREM 1. *Let $A \in \mathbb{C}^{n \times n}$. Then the following statements are equivalent:*

(a) Ind $A = k$.
(b) *The smallest positive exponent for which* (7) *holds is k.*
(c) *If A is singular and $m(\lambda)$ is its minimal polynomial, k is the multiplicity of $\lambda = 0$ as a zero of $m(\lambda)$.*
(d) *If A is singular, k is the maximal grade of 0-vectors of A.*

PROOF. (a) $\Longleftrightarrow$ (b)   Clearly (11) implies

$$\text{rank } A^{\ell+1} = \text{rank } A^\ell, \tag{12}$$

and by Ex. 1.10, (12) implies

$$R(A^{\ell+1}) = R(A^\ell),$$

so that (11) holds. Thus (12) and (11) are equivalent, proving (a).

(b) $\Longleftrightarrow$ (c)   Let

$$m(\lambda) = \lambda^\ell p(\lambda)$$

where $p(0) \neq 0$. Let k be defined by (b), and we must now show that $k = \ell$. We have

$$p(A)A^\ell = O.$$

If $\ell > k$, then

$$O = p(A)A^\ell X = p(A)A^{\ell-1},$$

where $\lambda^{\ell-1}p(\lambda)$ is of lower degree than $m(\lambda)$, contrary to the definition of the minimal polynomial.

Since $p(0) \neq 0$, we can write[2]

$$m(\lambda) = c\lambda^\ell(1 - \lambda q(\lambda)), \tag{13}$$

---

[2]For this device we are indebted to M.R. Hestenes (see [**77**, p. 687, footnote 56]).

where $c \neq 0$ and q is a polynomial. It follows that

$$A^{\ell+1}q(A) = A^{\ell}. \tag{14}$$

If $\ell < k$, this would contradict (b).

(a) $\Longleftrightarrow$ (d)   Let A have index k and let h be the maximal grade of the 0-vectors of A. We must show that $h = k$. The definition of h implies that $N(A^{\ell}) = N(A^h)$ for all $\ell \geq h$, but $N(A^{h-1})$ is a proper subspace of $A(A^h)$. It follows from Lemma 1 that $h = k$. $\square$

Exercises

Ex. 5.   Let $A \in \mathbb{C}^{n \times n}$ have index k, and let ℓ be a positive integer. Then $R(A^{\ell})$ and $N(A^{\ell})$ are complementary subspaces if and only if $\ell \geq k$.

PROOF.   Let $A \in \mathbb{C}^{n \times n}$. If A is nonsingular, $R(A^{\ell}) = \mathbb{C}^n$ and $N(A^{\ell}) = \{\mathbf{0}\}$ for all $\ell = 1, 2, \dots$. Thus $R(A^{\ell})$ and $N(A^{\ell})$ are trivially complementary. Since a nonsingular matrix has index 1, it remains to prove the theorem for singular matrices. Now, for any positive integer ℓ,

$$\dim R(A^{\ell}) + \dim N(A^{\ell}) = \operatorname{rank} A^{\ell} + \operatorname{nullity} A^{\ell} = n.$$

It therefore follows from statement (c) of Ex. 0.1 that $R(A^{\ell})$ and $N(A^{\ell})$ are complementary if and only if

$$R(A^{\ell}) \cap N(A^{\ell}) = \{\mathbf{0}\}. \tag{15}$$

Since, for any positive integer ℓ,

$$R(A^{\ell+1}) \subset R(A^{\ell}),$$
and
$$N(A^{\ell}) \subset N(A^{\ell+1}),$$

it follows that (15) is equivalent to

$$\dim R(A^{\ell}) = \dim R(A^{\ell+1}), \tag{16}$$

or $\ell \geq k$, by Definition 1. $\square$

Ex. 6.   Let $A \in \mathbb{C}^{n \times n}$. If, for some positive integer k,

$$R(A^{k+1}) = R(A^k), \tag{10}$$

then, for all integers $\ell > k$,

$$R(A^{\ell+1}) = R(A^{\ell}).$$

[*Hint*: $R(A^{k+1}) = AR(A^k)$ and $R(A^{\ell}) = A^{\ell-k}R(A^k)$.]

Ex. 7.   Let $A \in \mathbb{C}^{n \times n}$. Show that (9) holds for some k between 1 and n, inclusive.

PROOF.   Since $n \geq \operatorname{rank}(A^k) \geq \operatorname{rank}(A^{k+1}) \geq 0$, eventually $\operatorname{rank} A^k = \operatorname{rank} A^{k+1}$ for some $k \in \overline{1, n}$. $\square$

3. Spectral Inverse of a Diagonable Matrix

In investigating the existence of generalized inverses of a singular square matrix, we shall begin with diagonable matrices because they are the easiest to deal with. Evidently some extension must be made of the spectral property enjoyed by nonsingular matrices, because a singular matrix has 0 as one of its eigenvalues. Given a diagonable matrix $A \in \mathbb{C}^{n \times n}$, let us seek a matrix X such that every eigenvector of A associated with the eigenvalue λ (for every λ in the spectrum of A) is also an eigenvector of X associated with the eigenvalue $\lambda^{\dagger}$, where $\lambda^{\dagger}$ is defined in (1.8), page 43.

Since A has n linearly independent eigenvectors, there is a nonsingular matrix P, having such a set of eigenvectors as columns, such that

$$AP = PJ \tag{17}$$

where

$$J = \operatorname{diag}(\lambda_1, \lambda_2, \dots, \lambda_n)$$

is a Jordan form of A. We shall need the diagonal matrix obtained from J by replacing each diagonal element λ_i by $\lambda_i^\dagger$. By Ex. 1.22, this is, in fact, the Moore–Penrose inverse of J; that is,

$$J^\dagger = \operatorname{diag}(\lambda_1^\dagger, \lambda_2^\dagger, \dots, \lambda_n^\dagger).$$

Because of the spectral requirement imposed on X, we must have

$$XP = PJ^\dagger. \tag{18}$$

Solving (17) and (18) for A and X givs

$$A = PJP^{-1}, \quad X = PJ^\dagger P^{-1}. \tag{19}$$

Since J and $J^\dagger$ are both diagonal, they commute with each other. As a result, it follows from (19) that $X \in A\{1, 2, 5\}$.

We do not wish to limit our consideration to diagonable matrices. We began with them because they are easier to work with. The result just obtained suggests that we should examine the existence and properties (especially spectral properties) of $\{1, 2, 5\}$-inverses for square matrices in general.

4. The Group Inverse

It follows from (5) and from Corollary 2.7 that a $\{1, 2, 5\}$-inverse of A, if it exists, is a $\{1, 2\}$-inverse X such that $R(X) = R(A)$ and $N(X) = N(A)$. By Theorem 2.12, there is at most one such inverse.

This unique $\{1, 2, 5\}$-inverse is called the *group inverse* of A, and is denoted by $A^\#$. The name "group inverse" was given by I. Erdélyi [258], because the positive and negative powers of a given matrix A (the latter being interpreted as powers of $A^\#$), together with the projector $AA^\#$ as the unit element, constitute an Abelian group; see Ex. 10. Both he and Englefield [254] (who called it the "commuting reciprocal inverse") drew attention to the spectral properties of the group inverse. As we shall see later, however, the group inverse is a particular case of the *Drazin inverse* [233], or $\{1^k, 2, 5\}$-inverse, which predates it [258] and [254].

The group inverse is not restricted to diagonable matrices; however, it does not exist for all square matrices. By Section 2.6 and Theorem 2.12, such an inverse exists if and only if $R(A)$ and $N(A)$ are complementary subspaces. This is equivalent, by Ex. 5, to A having index 1. We have, therefore, the following theorem.

THEOREM 2. *A square matrix A has a group inverse if and only if* Ind $A = 1$, *i.e.*,

$$\operatorname{rank} A = \operatorname{rank} A^2. \tag{20}$$

When the group inverse exists, it is unique.   □

The following theorem gives an alternative condition for the existence of $A^{\#}$, and an explicit formula for its computation.

THEOREM 3 (Cline [201]). *Let a square matrix* A *have the full-rank factorization*

$$A = FG. \tag{21}$$

Then A *has a group inverse if and only if* GF *is nonsingular, in which case*

$$A^{\#} = F(GF)^{-2}G. \tag{22}$$

PROOF. Let $r = \operatorname{rank} A$. Then $GF \in \mathbb{C}^{r \times r}$. Now

$$A^2 = FGFG,$$

and so

$$\operatorname{rank} A^2 = \operatorname{rank} GF$$

by Ex. 1.7. Therefore (20) holds if and only if GF is nonsingular, and the first part of the theorem is established. It is easily verified that (1), (2), and (5) hold with A given by (21) and X by RHS(22). Formula (22) then follows from the uniqueness of the group inverse. $\qquad\square$

For an important class of matrices, the group inverse and the Moore–Penrose inverse are the same. We shall call a square matrix A *range-Hermitian* (such a matrix is also called an EP_r or EP matrix, e.g., Schwerdtfeger [735], Pearl [631] and other writers) if

$$R(A^*) = R(A), \tag{23}$$

or, equivalently, if

$$N(A^*) = N(A), \tag{24}$$

the equivalence follows from (0.26).

Using the notation of Theorem 2.12, the preceding discussion shows that

$$A^{\#} = A^{(1,2)}_{R(A),N(A)},$$

while Ex. 2.38 establishes that

$$A^{\dagger} = A^{(1,2)}_{R(A^*),N(A^*)}.$$

The two inverses are equal, therefore, if and only if $R(A) = R(A^*)$ and $N(A) = N(A^*)$. Thus we have proved:

THEOREM 4. $A^{\#} = A^{\dagger}$ *if and only if* A *is range-Hermitian.* $\qquad\square$

The approach of (19) can be extended from diagonable matrices to all square matrices of index 1. To do this we shall need the following lemma:

LEMMA 2. *Let* J *be a square matrix in Jordan form. Then* J *is range-Hermitian if and only if it has index* 1.

PROOF. *Only if*: Follows from Ex. 12.

If: If J is nonsingular, $\operatorname{rank} J = \operatorname{rank} J^2$ and J is range-Hermitian by Ex. 14. If J has only 0 as an eigenvalue, it is nilpotent. In this case, it follows easily from the structure of the Jordan form that $\operatorname{rank} J^2 < \operatorname{rank} J$ unless J is the null matrix O, in which case it is trivially range-Hermitian.

If J has both zero and nonzero eigenvalues, it can be partitioned in the form

$$J = \begin{bmatrix} J_1 & O \\ O & J_0 \end{bmatrix}, \tag{25}$$

where J_1 is nonsingular and has as eigenvalues the nonzero eigenvalues of J, while J_0 is nilpotent. By the same reasoning employed in the preceding paragraph, rank $J = $ rank J^2 implies $J_0 = O$. It then follows from Ex. 14 that J is range-Hermitian. $\qquad\square$

THEOREM 5 (Erdélyi [258]). *Let A have index 1 and let*

$$A = PJP^{-1},$$

where P is nonsingular and J is a Jordan normal form of A. Then

$$A^{\#} = PJ^{\dagger}P^{-1}. \tag{26}$$

PROOF. It is easily verified that relations (1), (2), (5), and (20) are similarity invariants. Therefore

$$J^{\#} = P^{-1}A^{\#}P \tag{27}$$

and also rank $J = $ rank J^2. It then follows from Lemma 2 and Theorem 4 that

$$J^{\#} = J^{\dagger}, \tag{28}$$

and (26) follows from (27) and (28). $\qquad\square$

Exercises

EX. 8. An alternative proof of uniqueness of the group inverse in Theorem 2 is as follows. Let $X, Y \in A\{1, 2, 5\}$, $E = AX = XA$, and $F = AY = YA$. Then $E = F$ since

$$E = AX = AYAX = FE,$$
$$F = YA = YAXA = FE.$$

Therefore,

$$X = EX = FX = YE = YF = Y.$$

EX. 9. (Properties of the group inverse).
   (a) If A is nonsingular, $A^{\#} = A^{-1}$.
   (b) $A^{\#\#} = A$.
   (c) $A^{*\#} = A^{\#*}$.
   (d) $A^{T\#} = A^{\#T}$.
   (e) $(A^{\ell})^{\#} = (A^{\#})^{\ell}$ for every positive integer ℓ.

EX. 10. Let A have index 1 and denote $(A^{\#})^j$ by A^{-j} for $j = 1, 2, \ldots$. Also denote $AA^{\#}$ by A^0. Then show that

$$A^{\ell}A^m = A^{\ell+m}$$

for all integers ℓ and m. (Thus, the "powers" of A, positive, negative and zero, constitute an Abelian group under matrix multiplication.)

EX. 11. Show that

$$A^{\#} = A(A^3)^{(1)}A, \tag{29}$$

where $(A^3)^{(1)}$ is an arbitrary $\{1\}$-inverse of A^3.

EX. 12. Every range-Hermitian matrix has index 1.

PROOF. If A is range-Hermitian, then by (0.26), $N(A) = R(A)^{\perp}$. Thus $R(A)$ and $N(A)$ are complementary subspaces.   □

EX. 13. Let A be range-Hermitian, and let $P = P_{R(A)} = P_{R(A^*)}$. Then $A^{\dagger}$ is the unique solution of the system

$$AX = XA = P, \tag{30a}$$

$$PX = XP = X. \tag{30b}$$

EX. 14. A nonsingular matrix is range-Hermitian.

EX. 15. A normal matrix is range-Hermitian. [*Hint*: Use Corollary 1.2.]

REMARK. It follows from Exs. 12 and 15 that

{Hermitian matrices} $\subset$ {normal matrices}

$\subset$ {range-Hermitian matrices} $\subset$ {matrices of index 1}.

EX. 16. A square matrix A is range-Hermitian if and only if A commutes with $A^{\dagger}$.

EX. 17. (Katz [**471**]). A square matrix A is range-Hermitian if and only if there is a matrix Y such that $A^* = YA$.

EX. 18. (Katz and Pearl [**473**]). A matrix in $\mathbb{C}^{n \times n}$ is range-Hermitian if and only if it is similar to a matrix of the form $\begin{pmatrix} B & O \\ O & O \end{pmatrix}$, where B is nonsingular.

PROOF. See Lemma 2.   □

EX. 19. (Ben-Israel [**70**]). Let $A \in \mathbb{C}^{n \times n}$. Then A has index 1 if and only if the limit

$$\lim_{\lambda \to 0} (\lambda I_n + A)^{-1} A$$

exists, in which case

$$\lim_{\lambda \to 0} (\lambda I_n + A)^{-1} A = AA^{\#}.$$

REMARK. Here $\lambda \to 0$ means $\lambda \to 0$ through any neighborhood of 0 in $\mathbb{C}$ which excludes the nonzero eigenvalues of A.

PROOF. Let rank $A = r$ and let $A = FG$ be a full-rank factorization. Then the identity

$$(\lambda I_n + A)^{-1} A = F(\lambda I_r + GF)^{-1} G$$

holds whenever the inverse in question exists. Therefore the existence of

$$\lim_{\lambda \to 0} (\lambda I_n + A)^{-1} A$$

is equivalent to the existence of

$$\lim_{\lambda \to 0} (\lambda I_r + GF)^{-1}$$

which, in turn, is equivalent to the nonsingularity of GF. The proof is completed by using Theorems 2 and 3.   □

EX. 20. Let $A \in \mathbb{C}^{n \times n}$. Then A is range-Hermitian if and only if

$$\lim_{\lambda \to 0} (\lambda I_n + A)^{-1} P_{R(A)} = A^{\dagger}.$$

PROOF. Follows from Ex. 19 and Theorem 4.   □

Ex. 21. Let $O \neq A \in \mathbb{C}^{m \times n}$. Then

$$\lim_{\lambda \to 0} (\lambda I_n + A^* A)^{-1} A^* = A^\dagger \quad \text{(den Broeder and Charnes [136])}. \tag{3.43}$$

Proof.

$$\lim_{\lambda \to 0} (\lambda I_n + A^* A)^{-1} A^* = \lim_{\lambda \to 0} (\lambda I_n + A^* A)^{-1} P_{R(A^* A)} A^*$$
$$\text{(since } R(A^*) = R(A^* A))$$
$$= (A^* A)^\dagger A^* \quad \text{(by Ex. 20 since } A^* A \text{ is range-Hermitian)}$$
$$= A^\dagger \quad \text{(by Ex. 1.18(d))}. \qquad \square$$

Ex. 22. *The "reverse-order" property for the Moore–Penrose inverse.* For some pairs of matrices A, B the relation

$$(AB)^\dagger = B^\dagger A^\dagger \tag{31}$$

holds, and for others it does not. There does not seem to be a simple criterion for distinguishing the cases in which (31) holds. The following result is due to Greville [327].

For matrices A, B such that AB exists, the reverse-order relation (31) holds if and only if

$$R(A^* AB) \subset R(B) \quad \text{and} \quad R(BB^* A^*) \subset R(A^*). \tag{32}$$

Proof. *If*: We have

$$BB^\dagger A^* AB = A^* AB \tag{33}$$

and

$$A^\dagger ABB^* A^* = BB^* A^*. \tag{34}$$

Taking conjugate transposes of both sides of (33) gives

$$B^* A^* ABB^\dagger = B^* A^* A, \tag{35}$$

and then multiplying on the right by $A^\dagger$ and on the left by $(AB)^{*\dagger}$ yields

$$ABB^\dagger A^\dagger = AB(AB)^\dagger. \tag{36}$$

Multiplying (34) on the left by $B^\dagger$ and on the right by $(AB)^{*\dagger}$ gives

$$B^\dagger A^\dagger AB = (AB)^\dagger AB. \tag{37}$$

It follows from (36) and (37) that $B^\dagger A^\dagger \in (AB)\{1, 3, 4\}$.

Finally, the equations,

$$B^* A^* = B^* BB^\dagger A^\dagger AA^*, \quad B^\dagger A^\dagger = B^\dagger B^{*\dagger} B^* A^* A^{*\dagger} A^\dagger$$

show that

$$\operatorname{rank} B^\dagger A^\dagger = \operatorname{rank} B^* A^* = \operatorname{rank} AB,$$

and therefore $B^\dagger A^\dagger \in (AB)\{2\}$ by Theorem 1.2, and so (31) holds.

Only if: We have

$$B^* A^* = B^\dagger A^\dagger ABB^* A^*,$$

and multiplying on the left by $ABB^* B$ gives

$$ABB^* (I - A^\dagger A) BB^* A^* = O.$$

Since the left member is Hermitian and $I - A^\dagger A$ is idempotent, it follows that

$$(I - A^\dagger A)BB^*A^* = O,$$

which is equivalent to (34). In an analogous manner, (33) is obtained.   □

EX. 23. (Arghiriade [24]). For matrices A, B such that AB exists, (31) holds if and only if A^*ABB^* is range-Hermitian.

PROOF. We shall show that the condition that A^*ABB^* be range-Hermitian is equivalent to (32), and the result will then follow from Ex. 22. Let C denote A^*ABB^*, and observe that

$$R(A^*AB) = R(C), \qquad R(BB^*A^*) = R(C^*)$$

because

$$CB^{*\dagger} = A^*AB, \qquad C^*A^\dagger = BB^*A^*.$$

Therefore it is sufficient to prove that $R(C) = R(C^*)$ if and only if $R(C) \subset R(B)$ and $R(C^*) \subset R(A^*)$.

If: A^*A and BB^* are Hermitian, and therefore of index 1 by Ex. 12. Since $R(BB^*) = R(B)$ by Corollary 1.2, it follows from Ex. 54 with $F = A^*A$, $G = BB^*$ that

$$R(C) = R(A^*) \cap R(B).$$

Reversing the assignments of F and G gives

$$R(C^*) = R(A^*) \cap R(B).$$

Thus $R(C) = R(C^*)$.

Only if: Obvious.   □

5. Spectral Properties of the Group Inverse

Even when A is not diagonable, the group inverse has spectral properties comparable to those of the inverse of a nonsingular matrix. However, in this case, $A^\#$ is not the only matrix having such properties. This has already been illustrated in the case of a nonsingular matrix (see Ex. 4).

We note that if a square matrix A has index 1, its 0-vectors are all of grade 1, i.e., null vectors of A. This follows from the fact that (20) implies $N(A^2) = N(A)$ by Ex. 1.10.

The following two lemmas are needed in order to establish the spectral properties of the group inverse. The second is stated in greater generality than is required for the immediate purpose because it will be used in connection with spectral generalized inverses other than the group inverse.

LEMMA 3. Let $\mathbf{x}$ be a λ-vector of A with $\lambda \neq 0$. Then $\mathbf{x} \in R(A^\ell)$ where ℓ is an arbitrary positive integer.

PROOF. We have

$$(A - \lambda I)^p \mathbf{x} = \mathbf{0}$$

for some positive integer p. Expanding the left member by the binomial theorem, transposing the last term, and dividing by its coefficient $(-\lambda)^{p-1} \neq 0$ gives

$$\mathbf{x} = c_1 A\mathbf{x} + c_2 A^2\mathbf{x} + \cdots + c_p A^p\mathbf{x}, \tag{38}$$

where

$$c_i = (-1)^{i-1}\lambda^{-i}\binom{p}{i}.$$

Successive multiplication of (38) by A gives

$$Ax = c_1 A^2 x + c_2 A^3 x + \cdots + c_p A^{p+1} x,$$
$$A^2 x = c_1 A^3 x + c_2 A^4 x + \cdots + c_p A^{p+2} x,$$
$$\cdots = \cdots \cdots \cdots \cdots \tag{39}$$
$$A^{\ell-1} x = c_1 A^\ell x + c_2 A^{\ell+1} x + \cdots + c_p A^{p+\ell-1} x,$$

and successive substitution of equations (39) in RHS(38) gives eventually

$$x = A^\ell q(A) x,$$

where q is some polynomial. □

LEMMA 4. *Let A be a square matrix and let*

$$X A^{\ell+1} = A^\ell \tag{40}$$

for some positive integer ℓ. Then every λ-vector of A of grade p for $\lambda \neq 0$ is a λ^{-1}-vector of X of grade p.

PROOF. The proof will be by induction on the grade p. Let $\lambda \neq 0$ and $Ax = \lambda x$. Then $A^{\ell+1} x = \lambda^{\ell+1} x$, and therefore $x = \lambda^{-\ell-1} A^{\ell+1} x$. Accordingly,

$$X x = \lambda^{-\ell-1} X A^{\ell+1} x = \lambda^{-1} x.$$

proving the lemma for $p = 1$.

Suppose the lemma is true for $p = 1, 2, \ldots, r$, and let x be a λ-vector of A of grade $r + 1$. Then, by Lemma 3,

$$x = A^\ell y$$

for some y. Thus

$$(X - \lambda^{-1} I) x = (X - \lambda^{-1} I) A^\ell y = X(A^\ell - \lambda^{-1} A^{\ell+1}) y$$
$$= X(I - \lambda^{-1} A) A^\ell y = -\lambda^{-1} X(A - \lambda I) x.$$

By the induction hypothesis, $(A - \lambda I) x$ is a λ^{-1}-vector of X of grade r. Consequently

$$(X - \lambda^{-1} I)^r (A - \lambda I) x = 0,$$
$$z = (X - \lambda^{-1} I)^{r-1} (A - \lambda I) x \neq 0,$$
$$X z = \lambda^{-1} z.$$

Therefore

$$(X - \lambda^{-1} I)^{r+1} x = -\lambda^{-1} X(X - \lambda^{-1} I)^r (A - \lambda I) x = 0,$$
$$(X - \lambda^{-1} I)^r x = -\lambda^{-1} X z = -\lambda^{-2} z \neq 0.$$

This completes the induction. □

The following theorem shows that for every matrix A of index 1, the group inverse is the only matrix in $A\{1\}$ or $A\{2\}$ having spectral properties comparable to those of the inverse of a nonsingular matrix. For convenience, we introduce:

DEFINITION 2. X is an *S-inverse* of A (or A and X *S-inverses* of each other) if they share the property that, for every $\lambda \in \mathbb{C}$ and every vector x, x is a λ-vector of A of grade p if and only if it is a $\lambda^\dagger$-vector of X of grade p.

THEOREM 6. *Let $A \in \mathbb{C}^{n \times n}$ have index 1. Then $A^{\#}$ is the unique S-inverse of A in $A\{1\} \cup A\{2\}$. If A is diagonable, $A^{\#}$ is the only S-inverse of A.*

PROOF. First we shall show that $A^{\#}$ is an S-inverse of A. Since $X = A^{\#}$ satisfies (40) with $\ell = 1$, it follows from Lemma 4 that $A^{\#}$ satisfies the "if" part of the definition of S-inverse for $\lambda \neq 0$. Replacing A by $A^{\#}$ establishes the "only if" part for $\lambda \neq 0$, since $A^{\#\#} = A$ (see Ex. 9(b)).

Since both A and $A^{\#}$ have index 1, all their 0-vectors are null vectors as pointed out in the second paragraph of this section. Thus, in order to prove that $A^{\#}$ satisfies the definition of S-inverse for $\lambda = 0$, we need only show that $N(A) = N(A^{\#})$. But this follows from the commutativity of A and $A^{\#}$ and Ex. 1.10.

Let $r = \operatorname{rank} A$ and consider the equation

$$AP = PJ$$

where P is nonsingular and J is a Jordan form of A. The columns of P are λ-vectors of A. Since A has index 1, those columns which are not null vectors are associated with nonzero eigenvalues, and are therefore in $R(A)$ by Lemma 3. Since there are r of them and they are linearly independent, they span $R(A)$. But, by hypothesis, these columns are also λ^{-1}-vectors of X and therefore in $R(X)$. Since $\operatorname{rank} X = r$, these r vectors span $R(X)$, and so $R(X) = R(A)$. Thus X is a $\{1,2\}$-inverse of A such that $R(X) = R(A)$ and $N(X) = N(A)$. But $A^{\#}$ is the only such matrix, and so $X = A^{\#}$.

It was shown in Section 3 that if A is diagonable, an S-inverse of A must be a $\{1,2,5\}$-inverse. Since $A^{\#}$ is the only such inverse, this completes the proof. $\square$

6. The Drazin Inverse

We have seen that the group inverse does not exist for all square matrices, but only for those of index 1. However, we shall show in this section that every square matrix has a unique $\{1^k, 2, 5\}$-inverse, where k is its index. This inverse, called the *Drazin inverse*, was first studied by Drazin [233] (though in the more general context of rings and semigroups without specific reference to matrices).

We start informally. Let $A \in \mathbb{C}^{n \times n}$ have index k. Then $R(A^k)$ and $N(A^k)$ are complementary subspace, see Ex. 5, and the restriction $A_{[R(A^k)]}$ of A to $R(A^k)$ is invertible (being a one-to-one mapping of $R(A^k)$ onto itself.) Let $X \in \mathbb{C}^{n \times n}$ be defined by

$$X\mathbf{u} = \begin{cases} A_{[R(A^k)]}^{-1}\mathbf{u}, & \text{if } \mathbf{u} \in R(A^k), \\ \mathbf{0}, & \text{if } \mathbf{u} \in N(A^k). \end{cases} \tag{41}$$

It follows from this definition that the relations $AX\mathbf{u} = XA\mathbf{u}$ and $XAX\mathbf{u} = X\mathbf{u}$ hold for $\mathbf{u} \in R(A^k)$ and for $\mathbf{u} \in N(A^k)$, and therefore in all of $\mathbb{C}^n$. The matrix X is thus a $\{2,5\}$-inverse of A.

Definition (41) says that AX is the identity in $R(A^k)$, i.e., $AXA^k\mathbf{x} = A^k\mathbf{x}$ for all $\mathbf{x} \in \mathbb{C}^n$, allowing for zero in both sides (if $\mathbf{x} \in N(A^k)$.) Therefore, X is also a $\{1^k\}$-inverse of A.

The matrix X is thus a $\{1^k, 2, 5\}$-inverse of A, properties that define it uniquely, as shown in Theorem 7 below. X is called the *Drazin inverse* of A, and is denoted by A^D.

The following lemma is needed for proving the existence and uniqueness of the Drazin inverse.

LEMMA 5. *If Y is a $\{1^\ell, 5\}$-inverse of square matrix A, then*

$$X = A^\ell Y^{\ell+1}$$

is a $\{1^\ell, 2, 5\}$-inverse.

PROOF. We have

$$A^{\ell+1} Y = A^\ell, \quad AY = YA.$$

Clearly X satisfies (5). We have then

$$A^\ell X A = A^{2\ell+1} Y^{\ell+1} = A^{2\ell} Y^\ell = A^{2\ell-1} Y^{\ell-1} = \cdots = A^\ell,$$

and

$$XAX = A^{2\ell+1} Y^{2\ell+2} = A^{2\ell} Y^{2\ell+1} = \cdots = A^\ell Y^{\ell+1} = X. \qquad \square$$

THEOREM 7. *Let $A \in \mathbb{C}^{n \times n}$ have index k. Then A has a unique $\{1^k, 2, 5\}$-inverse, which is expressible as a polynomial in A, and is also the unique $\{1^\ell, 2, 5\}$-inverse for every $\ell \geq k$.*

PROOF. The matrix $q(A)$ of (14) is a $\{1^k, 5\}$-inverse of A. Therefore, by Lemma 5,

$$X = A^k (q(A))^{k+1} \tag{42}$$

is a $\{1^k, 2, 5\}$-inverse.[3] This proves the existence of such an inverse.

A matrix X that satisfies (7) clearly satisfies (11) for all $\ell \geq k$. Therefore, a $\{1^k, 2, 5\}$-inverse of A is a $\{1^\ell, 2, 5\}$-inverse for all $\ell \geq k$.

Uniqueness will be proved by adapting the proof of uniqueness of the group inverse given in Ex. 8. Let $X, Y \in A\{1^\ell, 2, 5\}$, $E = AX = XA$, and $F = AY = YA$. Note that E and F are idempotent. Then $E = F$, since

$$E = AX = A^\ell X^\ell = AYA^\ell X^\ell = FAX = FE,$$

$$F = YA = Y^\ell A^\ell = Y^\ell A^\ell XA = YAE = FE.$$

The proof is then completed exactly as in the case of the group inverse. $\qquad \square$

This unique $\{1^k, 2, 5\}$-inverse is the *Drazin inverse*, and we shall denote it by A^D. The group inverse is the particular case of the Drazin inverse for matrices of index 1.

The Drazin inverse has a simple representation in terms of the Jordan form:

THEOREM 8. *Let $A \in \mathbb{C}^{n \times n}$ have the Jordan form*

$$A = XJX^{-1} = X \begin{bmatrix} J_1 & O \\ O & J_0 \end{bmatrix} X^{-1} \tag{43}$$

where J_0 and J_1 are the parts of J corresponding to zero and nonzero eigenvalues. Then

$$A^D = X \begin{bmatrix} J_1^{-1} & O \\ O & O \end{bmatrix} X^{-1}. \tag{44}$$

PROOF. Let A be singular of index k (i.e., the biggest block in the submatrix J_0 is $k \times k$). Then the matrix given by (44) is a $\{1^k, 2, 5\}$-inverse of A. $\qquad \square$

---

[3]See also Ex. 34 below.

EXAMPLE 1. The matrix

$$A = \begin{bmatrix} 0 & -2 & 0 & -5 & 2 \\ 0 & -1 & 0 & -2 & 0 \\ 0 & 0 & 0 & 2 & 0 \\ 0 & 1 & 0 & 2 & 0 \\ -2 & -2 & 1 & -8 & 4 \end{bmatrix}$$

has the Jordan form

$$A = X \begin{bmatrix} J_1(1) & & \\ & J_2(2) & \\ & & J_2(0) \end{bmatrix} X^{-1} = X \begin{bmatrix} 1 & & & \\ & 2 & 1 & \\ & 0 & 2 & \\ & & & 0 & 1 \\ & & & 0 & 0 \end{bmatrix} X^{-1}$$

see Example 2.1, p. 66. The Drazin inverse A^D is, by Theorem 8,

$$A^D = X \begin{bmatrix} (J_1(1))^{-1} & & \\ & (J_2(2))^{-1} & \\ & & O \end{bmatrix} X^{-1} = X \begin{bmatrix} 1 & & & \\ & \frac{1}{2} & -\frac{1}{4} & \\ & 0 & \frac{1}{2} & \\ & & & 0 & 0 \\ & & & 0 & 0 \end{bmatrix} X^{-1}$$

$$= \begin{bmatrix} 1 & \frac{3}{2} & -\frac{1}{2} & \frac{7}{2} & -\frac{1}{2} \\ 0 & -1 & 0 & -2 & 0 \\ 0 & 2 & 0 & 4 & 0 \\ 0 & 1 & 0 & 2 & 0 \\ \frac{1}{2} & 2 & -\frac{1}{4} & 4 & 0 \end{bmatrix}.$$

For $A \in \mathbb{C}^{n \times n}$ with spectrum $\lambda(A)$, and a scalar function $f \in \mathcal{F}(A)$, the corresponding matrix function $f(A)$ is

$$f(A) = \sum_{\lambda \in \lambda(A)} E_\lambda \sum_{k=0}^{\nu(\lambda)-1} \frac{f^{(k)}(\lambda)}{k!} (A - \lambda I_n)^k \qquad (2.50)$$

(see Definition 2.1). The analogous result for the Drazin inverse is:

COROLLARY 1. *Let $A \in \mathbb{C}^{n \times n}$ have spectrum $\lambda(A)$. Then*

$$A^D = \sum_{0 \neq \lambda \in \lambda(A)} E_\lambda \sum_{k=0}^{\nu(\lambda)-1} \frac{(-1)^k}{\lambda^{k+1}} (A - \lambda I_n)^k. \qquad (45)$$

PROOF. Theorem 8 shows that the Drazin inverse is the matrix function corresponding to $f(z) = 1/z$, defined on nonzero eigenvalues. $\square$

EXAMPLE 2. An alternative expression of the Drazin inverse of A in Example 1, using (45) and the projectors E_1, E_2 of Example 2.1, is

$$A^D = E_1 + \tfrac{1}{2} E_2 - \tfrac{1}{4} E_2 (A - 2I).$$

In the computation of A^D when the index of A exceeds 1, it is not easy to avoid raising A to a power. When ill-conditioning of A is serious, perhaps the best method is the sequential procedure of Cline [201], which involves full-rank factorization of matrices of successively smaller order, until a nonsingular matrix is reached. Thus, we take

$$A = B_1 G_1, \tag{46a}$$

$$G_i B_i = B_{i+1} G_{i+1} \quad (i = 1, 2, \ldots, k-1), \tag{46b}$$

where k is the index of A. Then

$$A^D = B_1 B_2 \cdots B_k (G_k B_k)^{-k-1} G_k G_{k-1} \cdots G_1. \tag{46c}$$

We saw in Exs. 0.79–0.80 that the inverse of a nonsingular matrix A is a polynomial in A. An analogous result is:

COROLLARY 2 (Englefield [**254**]). *Let $A \in \mathbb{C}^{n \times n}$. Then there is a $\{1, 2\}$-inverse of A expressible as a polynomial in A if and only if $\operatorname{Ind} A = 1$, in which case the only such inverse is the group inverse, which is given by*

$$A^{\#} = A(q(A))^2, \tag{47}$$

where q is defined by (13).

PROOF. *Only if*: A $\{1, 2\}$-inverse of A that is a polynomial in A necessarily commutes with A, and is therefore a $\{1, 2, 5\}$-inverse. The group inverse $A^{\#}$ is the only such inverse, and A has a group inverse if and only if its index is 1.

If: If $\operatorname{Ind} A = 1$, then A has a group inverse, which is a $\{1, 2\}$-inverse, and in this case coincides with the Drazin inverse. It is therefore expressible as a polynomial in A by Theorem 7. Formula (47) is merely the specialization of (42) for $k = 1$. □

COROLLARY 3 (Pearl [**633**]). *Let $A \in \mathbb{C}^{n \times n}$. Then $A^{\dagger}$ is expressible as polynomial in A if and only if A is range-Hermitian.*

Exercises

EX. 24. Show that, for a matrix A of index k that is not nilpotent, with B_i and G_i defined by (46a) and (46b), $G_k B_k$ is nonsingular. (*Hints*: Express A^k and A^{k+1} in terms of B_i and G_i ($i = 1, 2, \ldots, k$), and let r_k denote the number of columns of B_k, which is also the number of rows of G_k. Show that rank $A^k = r_k$, while rank $A^{k+1} = \operatorname{rank} G_k B_k$. Therefore rank $A^{k+1} = \operatorname{rank} A^k$ implies that $G_k B_k$ is nonsingular.)

EX. 25. Use Theorem 7 to verify (46c).

EX. 26. The Drazin inverse preserves similarity: If X is nonsingular, then

$$A = XBX^{-1} \quad \Longrightarrow \quad A^D = XB^D X^{-1}.$$

EX. 27. *Properties of the Drazin inverse.*
  (a) $(A^*)^D = (A^D)^*$.
  (b) $(A^T)^D = (A^D)^T$.
  (c) $(A^\ell)^D = (A^D)^\ell$ for $\ell = 1, 2, \ldots$.
  (d) If A has index k, A^ℓ has index 1 and $(A^\ell)^{\#} = (A^D)^\ell$ for $\ell \geq k$.
  (e) $(A^D)^D = A$ if and only if A has index 1.
  (f) A^D has index 1, and $(A^D)^{\#} = A^2 A^D$.
  (g) $((A^D)^D)^D = A^D$.
  (h) If A has index k, $R(A^D) = R(A^\ell)$ and $N(A^D) = N(A^\ell)$ for all $\ell \geq k$.

EX. 28. Let A be a square singular matrix, $\text{Ind}\,A = k$. If the system

$$A\mathbf{x} = \mathbf{b}, \quad \mathbf{x} \in R(A^k),$$

has a solution, it is uniquely given by

$$\mathbf{x} = A^D \mathbf{b}.$$

PROOF. Let a solution $\mathbf{x}$ be written as $\mathbf{x} = A^k\mathbf{y}$ for some $\mathbf{y}$. Then $A^{k+1}\mathbf{y} = \mathbf{b}$, and

$$\begin{aligned} \mathbf{x} = A^k\mathbf{y} &= A^{k+1}A^D\mathbf{y} \\ &= A^D A^{k+1}\mathbf{y} \\ &= A^D\mathbf{b}. \end{aligned}$$

Uniqueness follows from $R(A^k) \cap N(A^k) = \{\mathbf{0}\}$, see Ex. 5. $\square$

EX. 29. $R(A^D)$ is the subspace spanned by all the λ-vectors of A for all nonzero eigenvalues λ, and $N(A^D)$ is the subspace spanned by all the 0-vectors of A, and these are complementary subspaces.

EX. 30. $AA^D = A^D A$ is idempotent and is the projector on $R(A^D)$ along $N(A^D)$. Alternatively, if A has index k, it is the projector on $R(A^\ell)$ along $N(A^\ell)$ for all $\ell \geq k$.

EX. 31. If A and X are S-inverses of each other, they have the same index.

EX. 32. $A^D(A^D)^\# = AA^D$.

EX. 33. (Campbell and Meyer [**159**, Theorem 7.8.4]). Let $A, B \in \mathbb{C}^{n \times n}$. Then:
  (a) $(AB)^D = A[(BA)^2]^D B$.
If $AB = BA$ then:
  (b) $(AB)^D = B^D A^D = A^D B^D$; and
  (c) $A^D B = BA^D$ and $AB^D = B^D A$.

PROOF. (a) Let $Y = A[(BA)^2]^D B$. Then it is easy to verify that Y is a $\{2,5\}$-inverse of AB. Let $k = \max\{\text{Ind}(AB), \text{Ind}(BA)\}$. Then

$$\begin{aligned} (AB)^{k+2}Y = (AB)^{k+2}A[(BA)^2]^D B &= (AB)^{k+1}ABA[(BA)^2]^D B \\ &= (AB)^{k+1}A(BA)^D B = A(BA)^{k+1}(BA)^D B = A(BA)^k B \\ &= (AB)^{k+1}, \end{aligned}$$

showing that Y is a $\{1^k\}$-inverse of AB.

(b) and (c) follow from the fact that A^D, B^D are polynomials in A, B, respectively; see Theorem 7. $\square$

EX. 34. Let $A \in \mathbb{C}^{n \times n}$ have index k. Then, for all $\ell \geq k$,

$$A^D = A^\ell (q(A))^{\ell+1},$$

where q is defined by (13). See also (42).

EX. 35. If A is nilpotent, $A^D = O$.

EX. 36. If $\ell > m > 0$, $A^m(A^D)^\ell = (A^D)^{\ell-m}$.

EX. 37. If $m > 0$ and $\ell - m \geq k$, $A^\ell(A^D)^m = A^{\ell-m}$.

Ex. 38. Let A have index k, and define as follows a set of matrices B_j where j ranges over all the integers:

$$B_j = \begin{cases} A^j & \text{for } j \geq k, \\ A^k (A^D)^{k-j} & \text{for } 0 \leq j < k, \\ (A^D)^{-j} & \text{for } j < 0. \end{cases}$$

Is the set of matrices $\{B_j\}$ an Abelian group under matrix multiplication with unit element B_0 and multiplication rule $B_\ell B_m = B_{\ell+m}$? Is there an equivalent, but easier way of defining the matrices B_j?

Ex. 39. (Greville [328]). If ℓ is any integer not less than the index of A,

$$A^D = A^\ell (A^{2\ell+1})^{(1)} A^\ell, \tag{48}$$

where $(A^{2\ell+1})^{(1)}$ is an arbitrary $\{1\}$-inverse of $A^{2\ell+1}$. Note that (29) is a particular case of (48).

Ex. 40. (Cline [201]). If ℓ is any integer not less than the index of A,

$$(A^D)^\dagger = (A^\ell)^\dagger A^{2\ell+1} (A^\ell)^\dagger.$$

HINT: Use Ex. 2.61, noting that $\mathbb{R}(A^D) = R(A^\ell)$ and $N(A^D) = N(A^\ell)$.

Ex. 41. (Meyer [543], Boyarintsev [130, Theorem 1.8.6]). If $A \in \mathbb{C}^{n \times n}$ has index k then

$$A^D = \lim_{\alpha \to 0} (A^{k+1} + \alpha^2 I)^{-1} A^k, \quad \alpha \text{ real},$$

and the approximation error is

$$\| A^D - (A^{k+1} + \alpha^2 I)^{-1} A^k \| \leq \frac{\alpha^2 \| A^D \|^{k+2}}{1 - \alpha^2 \| A^D \|^{k+1}},$$

where $\| \cdot \|$ is the spectral norm (0.56.2).

7. Spectral Properties of the Drazin Inverse

The spectral properties of the Drazin inverse have been studied by Cline [201] and Greville [328]; some of them will be mentioned here.

The spectral properties of the Drazin inverse are the same as those of the group inverse with regard to nonzero eigenvalues and the associated eigenvectors, but weaker for 0-vectors. The necessity for such weakening is apparent from the following theorem.

THEOREM 9. *Let $A \in \mathbb{C}^{n \times n}$ and let $X \in A\{1\} \cup A\{2\}$ be an S-inverse of A. Then both A and X have index 1.*

PROOF. First, let $X \in A\{1\}$, and suppose that $\mathbf{x}$ is a 0-vector of A of grade 2. Then, $A\mathbf{x}$ is a null-vector of A. Since X is an S-inverse of A, $A\mathbf{x}$ is also a null-vector of X. Thus,

$$0 = XA\mathbf{x} = AXA\mathbf{x} = A\mathbf{x},$$

which contradicts the assumption that $\mathbf{x}$ is a 0-vector of A of grade 2. Hence, A has no 0-vectors of grade 2, and therefore has index 1, by Theorem 1(d). By Ex. 31, X has also index 1.

If $X \in A\{2\}$, we reverse the roles of A and X. $\qquad \square$

Accordingly, we relax the definition of the S-inverse (Definition 2, p. 162) as follows.

DEFINITION 3. X is an S'-*inverse* of A if, for all $\lambda \neq 0$, a vector $\mathbf{x}$ is a λ^{-1}-vector of X of grade p if and only if it is a λ-vector of A of grade p, and $\mathbf{x}$ is a 0-vector of X if and only if it is a 0-vector of A (without regard to grade).

THEOREM 10. *For every square matrix A, A and A^D are S'-inverses of each other.*

PROOF. Since A^D satisfies
$$A^D A^{k+1} = A^k, \quad A(A^D)^2 = A^D,$$
the part of Definition 3 relating to nonzero eigenvalues follows from Lemma 4. Since A^D has index 1 by Ex. 27(f), all its 0-vectors are null vectors. Thus the part of Definition 3 relating to 0-vectors follows from Ex. 29.   □

8. Index 1-Nilpotent Decomposition of a Square Matrix

The following theorem plays an important role in the study of spectral generalized inverses of matrices of index greater than 1. It is implicit in Wedderburn's [**853**] results on idempotent and nilpotent parts, but is not stated by him in this form.

THEOREM 11. *A square matrix A has a unique decomposition*
$$A = B + N, \tag{49}$$
such that B has index 1, N is nilpotent, and
$$NB = BN = O. \tag{50}$$
Moreover,
$$B = (A^D)^\#. \tag{51}$$
See Ex. 42.

PROOF. Suppose A has a decomposition (49) such that B has index 1, N is nilpotent and (50) holds. We shall first show that this implies (51), and therefore the decomposition is unique if it exists.

Since
$$B^\# = B(B^\#)^2 = (B^\#)^2 B,$$
we have
$$B^\# N = N B^\# = O.$$
Consequently,
$$AB^\# = BB^\# = B^\# A. \tag{52}$$
Moreover,
$$A(B^\#)^2 = B(B^\#)^2 = B^\#. \tag{53}$$
Because of (50), we have
$$A^\ell = (B+N)^\ell = B^\ell + N^\ell \quad (\ell = 1, 2, \dots). \tag{54}$$
If ℓ is sufficiently large so that $N^\ell = O$,
$$A^\ell = B^\ell,$$

and for such ℓ,

$$A^{\ell+1}B^{\#} = B^{\ell+1}B^{\#} = B^{\ell}. \tag{55}$$

It follows from (52), (53), and (55) that $X = B^{\#}$ satisfies (5), (7), and (8), and therefore

$$B^{\#} = A^{D},$$

which is equivalent to (51).

It remains to show that this decomposition has the required properties. Clearly B has index 1. By taking

$$N = A - (A^{D})^{\#} \tag{56}$$

and noting that

$$(A^{D})^{\#} = A^{2}A^{D}$$

by Ex. 27(f), it is easily verified that (50) holds. Therefore (54) follows, and if k is the index of A,

$$A^{k} = B^{k} + N^{k} = A^{2k}(A^{D})^{k} + N^{k} = A^{k} + N^{k},$$

and therefore $N^{k} = O$. $\qquad\square$

We shall call the matrix N given by (56) the *nilpotent part* of A and shall denote it by $A^{(N)}$.

THEOREM 12. *Let $A \in \mathbb{C}^{n\times n}$. Then A and X are S'-inverses of each other if*

$$X^{D} = (A^{D})^{\#}. \tag{57}$$

Moreover, if $X \in A\{1\} \cup A\{2\}$, it is an S'-inverse of A only if (57) holds.

PROOF. If (57), A and X have the same range and the same null space, and consequently the projectors XX^{D} and $AA^{D} = A^{D}(A^{D})^{\#}$ are equal. Thus, if ℓ is the maximum of the indices of A and X,

$$XA^{\ell+1} = X(A^{D})^{\#}A^{D}A^{\ell+1} = XX^{D}A^{\ell} = A^{\ell} \tag{58}$$

by Ex. 30. By interchanging the roles of A and X we obtain also

$$AX^{\ell+1} = X^{\ell}. \tag{59}$$

From (58) and (59), Lemma 4, Ex. 29 and the fact that A^{D} and X^{D} have the same null space, we deduce that A and X are S'-inverses of each other.

On the other hand, let A and X be S'-inverses of each other, and let $X \in A\{1\}$. Then, by Ex. 29,

$$N(A^{D}) = N(X^{D}),$$

and so,

$$(A^{D})^{\#}X^{(N)} = (X^{D})^{\#}A^{(N)} = O.$$

Similarly, since

$$R(A^{D}) = R(X^{D}),$$

(2.59) gives

$$N(A^{D*}) = N(X^{D*}),$$

and therefore

$$X^{(N)}(A^D)^\# = (X^D)^\# A^{(N)} = O.$$

Consequently

$$A = AXA = (A^D)^\#(X^D)^\#(A^D)^\# + A^{(N)}X^{(N)}A^{(N)},$$

and therefore

$$A^D = A^D A A^D = A A^D (X^D)^\# A A^D = (X^D)^\#, \tag{60}$$

since AA^D is the projector on the range of $(X^D)^\#$ along its null space. But (60) is equivalent to (57).

If $X \in A\{2\}$, we reverse the roles of A and X. $\qquad\square$

Referring back to the proof of Theorem 6, we note that if A has index 1, a matrix X that is an S-inverse of A and also either a $\{1\}$-inverse or a $\{2\}$-inverse, is automatically a $\{1,2\}$-inverse. However, a similar remark does not apply when the index of A is greater than 1 and X is an S'-inverse of A. This is because $A^{(N)}$ is no longer a null matrix (as it is when A has index 1) and its properties must be taken into account. (For details see Ex. 50.)

Exercises

Ex. 42. Let $A \in \mathbb{C}^{n \times n}$ be given in Jordan form

$$A = XJX^{-1} = X \begin{bmatrix} J_1 & O \\ O & J_0 \end{bmatrix} X^{-1} \tag{43}$$

where J_0 and J_1 are the parts of J corresponding to zero and nonzero eigenvalues. Then

$$B = X \begin{bmatrix} J_1 & O \\ O & O \end{bmatrix} X^{-1}, \quad N = X \begin{bmatrix} O & O \\ O & J_0 \end{bmatrix} X^{-1} \tag{61}$$

give the Wedderburn decomposition (49) of A, see Theorem 11. If A is nonsingular, (61) reads: $B = A$, $N = O$.

9. Quasi-Commuting Inverses

Erdélyi [259] calls A and X *quasi-commuting inverses* of each other if they are $\{1, 2, 5^k, 6^k\}$-inverses of each other for some positive integer k. He noted that for such pairs of matrices the spectrum of X is obtained by replacing each eigenvalue λ of A by $\lambda^\dagger$. The following theorem shows that quasi-commuting inverses have much more extensive spectral properties.

THEOREM 13. *If A and X are quasi-commuting inverses, they are S'-inverses.*

PROOF. If A and X are $\{1, 2, 5^\ell, 6^\ell\}$-inverses of each other, then

$$XA^{\ell+1} = A^\ell XA = A^\ell,$$

and similarly,

$$AX^{\ell+1} = X^\ell. \tag{59}$$

In view of Lemma 4 and Ex. 29, all that remains in order to prove that A and X are S'-inverses of each other is to show that A^D and X^D have the same null space. Now,

$$A^D\mathbf{x} = 0 \quad \Longrightarrow \quad 0 = A^{\ell+1}A^D\mathbf{x} = A^\ell\mathbf{x}$$
$$\Longrightarrow \quad 0 = X^{2\ell}A^\ell\mathbf{x} = A^\ell X^{2\ell}\mathbf{x} = X^\ell\mathbf{x} \quad \text{(by (59))}$$
$$\Longrightarrow \quad 0 = (X^D)^{\ell+1}X^\ell\mathbf{x} = X^D\mathbf{x}.$$

Since the roles of A and X are symmetrical, the reverse implication follows by interchanging them. □

COROLLARY 4. *A and X are quasi-commuting inverses of each other if and only if* (57) *holds and* $A^{(N)}$ *and* $X^{(N)}$ *are* $\{1,2\}$-*inverses of each other.*
PROOF. *If*: A and X are $\{1,2\}$-inverses of each other by Ex. 47. Choose ℓ sufficiently large so that $(A^{(N)})^\ell = O$. Then

$$XA^\ell = ((X^D)^\# + X^{(N)})((A^D)^\#)^\ell$$
$$= ((X^D)^\# + X^{(N)})((X^D)^\#)^\ell = ((X^D)^\#)^{\ell-1} = A^\ell X.$$

By interchanging A and X, it follows also that A commutes with X^ℓ.

Only if: By Theorem 13, A and X are S'-inverses of each other. Then, by Theorem 12, (57) holds, and by Ex. 50, $A^{(N)}$ and $X^{(N)}$ are $\{1,2\}$-inverses of each other. □

10. Other Spectral Generalized Inverses

Greville [**328**] calls X a *strong spectral inverse* if equations (19) are satisfied. Although this is not quite obvious, the relationship is a reciprocal one, and they can be called strong spectral inverses of each other. If A has index 1, Theorem 5 shows that $A^\#$ is the only strong spectral inverse. Greville has shown that strong spectral inverses are quasi-commuting, but, for a matrix A with index greater than 1, the set of strong spectral inverses is a proper subset of the set of quasi-commuting inverses. Strong spectral inverses have some remarkable and, in some respects, complicated properties, and there are a number of open questions concerning them. As these properties relate to matrices of index greater than 1, which are not for most purposes a very important class, they will not be discussed further here. The interested reader may consult Greville [**328**].

Cline [**201**] has pointed out that a square matrix A of index 1 has a $\{1,2,3\}$-inverse whose range is $R(A)$. This is, therefore, a "least-squares" inverse and also has spectral properties (see Exs. 50 and 51). Greville [**329**] has extended this notion to square matrices of arbitrary index, but his extension raises some questions that have not been answered (see the conclusion of [**329**]).

Exercises

Ex. 43. If A has index 1, $A^{(N)} = O$.

Ex. 44. If A is nilpotent, $\operatorname{rank} A^{\ell+1} < \operatorname{rank} A^\ell$ unless $A^\ell = O$,

Ex. 45. If A is nilpotent, the smallest positive integer ℓ such that $A^\ell = O$ is called the *index of nilpotency* of A. Show that this is the same as the index of A (see Definition 1).

Ex. 46. A and $A^{(N)}$ have the same index.

Ex. 47. $\operatorname{rank} A = \operatorname{rank} A^D + \operatorname{rank} A^{(N)}$.

Ex. 48. $A^D A^{(N)} = A^{(N)} A^D = O$.

Ex. 49. Every 0-vector of A of grade p is a 0-vector of $A^{(N)}$ of grade p.

Ex. 50. Let A and X satisfy (57). Then $X \in A\{1\}$ if and only if $A^{(N)} \in A^{(N)}\{1\}$. Similar statements with $\{1\}$ replaced by $\{2\}$ and by $\{1,2\}$ are also true.

Ex. 51. If A has index 1, show that $X = A^{\#} A A^{\dagger} \in A\{1,2,3\}$ (Cline). Show that this X has the properties of an S-inverse of A with respect to nonzero eigenvalues (but, in general, not with respect to 0-vectors). What is the condition on A that this X be an S-inverse of A?

Ex. 52. For square A with arbitrary index, Greville has suggested as an extension of Cline's inverse

$$X = A^D A A^{\dagger} + A^{(1)} A^{(N)} A^{\dagger},$$

where $A^{(1)}$ is an arbitrary element of $A\{1\}$. Show that $X \in A\{1,2,3\}$ and has some spectral properties. Describe its spectral properties precisely.

Ex. 53. Can a matrix A of index greater than 1 have an S-inverse? It can if we are willing to accept an "inverse" that is neither a $\{1\}$-inverse nor a $\{2\}$-inverse. Let

$$A^{(S)} = A^D + A^{(N)}.$$

Show that $A^{(S)}$ is an S-inverse of A and that $X = A^{(S)}$ is the unique solution of the four equations

$$AX = XA, \qquad A^{\ell+1} X = A^{\ell},$$
$$AX^{\ell+1} = X^{\ell}, \qquad A - X = A^{\ell} X^{\ell} (A - X),$$

for every positive integer ℓ not less than the index of A. Show also that $A^{(S)} = A^{\#}$ if A has index 1 and $(A^{(S)})^{(S)} = A$. In your opinion, can $A^{(S)}$ properly be called a generalized inverse of A?

Ex. 54. Let F be a square matrix of index 1, and let G be such that $R(FG) \subset R(G)$. Then,

$$R(FG) = R(F) \cap R(G).$$

PROOF. Evidently, $R(FG) \subset R(F)$ and therefore

$$R(FG) \subset R(F) \cap R(G).$$

Now let $\mathbf{x} \in R(F) \cap R(G)$, and we must show that $\mathbf{x} \in R(FG)$. Since F has index 1, it has a group inverse $F^{\#}$, which, by Corollary 2, can be expressed as a polynomial in F, say $p(F)$. We have

$$\mathbf{x} = F\mathbf{y} = G\mathbf{z}$$

for some $\mathbf{y}, \mathbf{z}$, and therefore

$$\mathbf{x} = FF^{\#}\mathbf{x} = FF^{\#}G\mathbf{z} = Fp(F)G\mathbf{z}.$$

Since $R(FG) \subset R(G)$,

$$FG = GH$$

for some H, and, consequently,

$$F^\ell G = GH^\ell$$

for every nonnegative integer ℓ. Thus

$$\mathbf{x} = Fp(F)G\mathbf{z} = FGp(H)\mathbf{z} \subset R(FG).$$ □

(This is a slight extension of a result of Arghiriade [24].)

Suggested Further Reading

SECTION 4. For range-Hermitian matrices see Arghiriade [24], Katz [471], Katz and Pearl [473], Pearl [631], [632], [633]. For matrices of index 1 see Ben-Israel [70].

GROUP INVERSES. Haynsworth and Wall [404], [405], Heinig [412], Puystjens and Hartwig [658], Robert [695].

REVERSE–ORDER PROPERTY. Djordjević [232], Erdélyi [257], Groß [343], Hartwig [387], Shinozaki and Sibuya [751], [752], Sun and Wei [788], Wei [858], Wibker, Howe, and Gilbert [871].

SECTION 6. Hartwig ([383], [385]), Hartwig, Wang, and Wei [396], Prasad, Bhaskara Rao, and Bapat [649], Sibuya [758], Stanimirović and Djordjević [775], Wei [859], Wei and Wu [861].

DRAZIN INVERSES OF INFINITE MATRICES AND LINEAR OPERATORS. See references in p. 368.

APPLICATIONS OF THE DRAZIN INVERSE. Campbell [146], [147], [149], [148], [152], [155], [153], [156], [158], Campbell and Meyer [159], Campbell, Meyer, and Rose [160], Campbell and Rose [162], [163], Hartwig and Hall [390], Hartwig and Levine [511], [512], [391], [392], Meyer ([544], [545]), Meyer and Shoaf [548].

SECTION 10. Poole and Boullion [644], Ward, Boullion, and Lewis [852], Scroggs and Odell [736].

CHAPTER 5

Generalized Inverses of Partitioned Matrices

1. Introduction

In this chapter we study linear equations and matrices in partitioned form. For example, in computing a (generalized or ordinary) inverse of a matrix $A \in \mathbb{C}^{m \times n}$, the size or difficulty of the problem may be reduced if A is partitioned as

$$A = \begin{bmatrix} A_{11} & A_{12} \\ A_{21} & A_{22} \end{bmatrix}.$$

The typical result here is the sought for inverse expressed in terms of the submatrices A_{ij}.

Partitioning by columns and by rows is used in Section 2 to solve linear equations and to compute generalized inverses and related items.

Intersections of linear manifolds are studied in Section 3 and used in Section 4 to obtain common solutions of pairs of linear equations and to invert matrices partitioned by rows.

Greville's method for computing $A^\dagger$ for $A \in \mathbb{C}^{m \times n}$, $n \geq 2$, is based on partitioning A as

$$A = [A_{n-1} \quad \mathbf{a}_n]$$

where $\mathbf{a}_n$ is the nth column of A. $A^\dagger$ is then expressed in terms of $\mathbf{a}_n$ and $A_{n-1}^\dagger$, which is computed in the same way, using the partition

$$A_{n-1} = [A_{n-2} \quad \mathbf{a}_{n-1}], \quad \text{etc.}$$

Greville's method and some of its consequences are studied in Section 5.

Bordered matrices, the subject of Section 6, are matrices of the form

$$\begin{bmatrix} A & U \\ V^* & O \end{bmatrix},$$

where $A \in \mathbb{C}^{m \times n}$ is given and U and V are chosen so that the resulting bordered matrix is nonsingular. Moreover,

$$\begin{bmatrix} A & U \\ V^* & O \end{bmatrix}^{-1} = \begin{bmatrix} A^\dagger & V^{*\dagger} \\ U^\dagger & O \end{bmatrix}$$

expressing generalized inverses in terms of an ordinary matrix.

2. Partitioned Matrices and Linear Equations

Consider the linear equation

$$A\mathbf{x} = \mathbf{b} \tag{1}$$

with given matrix A and vector $\mathbf{b}$, in the following three cases.

Case 1. $A \in \mathbb{C}_r^{r \times n}$, i.e., A **is of full row rank.** Let the columns of A be rearranged, if necessary, so that the first r columns are linearly independent. A rearrangement of columns may be interpreted as postmultiplication by a suitable permutation matrix; thus,

$$AQ = [A_1 \quad A_2] \quad \text{or} \quad A = [A_1 \quad A_2]Q^T, \tag{2}$$

where Q is an $n \times n$ permutation matrix (hence $Q^{-1} = Q^T$) and A_1 consists of r linearly independent columns, so that $A_1 \in \mathbb{C}_r^{r \times r}$, i.e., A_1 is nonsingular.

The matrix A_2 is in $\mathbb{C}^{r \times (n-r)}$ and if $n = r$, this matrix and other items indexed by the subscript 2 are to be interpreted as absent.

Corresponding to (2), let the vector $\mathbf{x}$ be partitioned

$$\mathbf{x} = \begin{bmatrix} \mathbf{x}_1 \\ \mathbf{x}_2 \end{bmatrix}, \quad \mathbf{x}_1 \in \mathbb{C}^r. \tag{3}$$

Using (2) and (3) we rewrite (1) as

$$[A_1 \quad A_2]Q^T \begin{bmatrix} \mathbf{x}_1 \\ \mathbf{x}_2 \end{bmatrix} = \mathbf{b} \tag{4}$$

easily shown to be satisfied by the vector

$$\begin{bmatrix} \mathbf{x}_1 \\ \mathbf{x}_2 \end{bmatrix} = Q \begin{bmatrix} A_1^{-1}\mathbf{b} \\ O \end{bmatrix}, \tag{5}$$

which is thus a particular solution of (1).

The general solution of (1) is obtained by adding to (5) the general element of $N(A)$, i.e., the general solution of

$$A\mathbf{x} = \mathbf{0}. \tag{6}$$

In (2), the columns of A_2 are linear combinations of the columns of A_1, say,

$$A_2 = A_1 T \quad \text{or} \quad T = A_1^{-1}A_2 \in \mathbb{C}^{r \times (n-r)}, \tag{7}$$

where the matrix T is called the *multiplier* corresponding to the partition (2), a name suggested by T being the "ratio" of the last $n - r$ columns of AQ to its first r columns.

Using (2), (3), and (7) permits writing (6) as

$$A_1[I_r \quad T]Q^T \begin{bmatrix} \mathbf{x}_1 \\ \mathbf{x}_2 \end{bmatrix} = \mathbf{0}, \tag{8}$$

whose general solution is clearly

$$\begin{bmatrix} \mathbf{x}_1 \\ \mathbf{x}_2 \end{bmatrix} = Q \begin{bmatrix} -T \\ I_{n-r} \end{bmatrix} \mathbf{y}, \tag{9}$$

where $\mathbf{y} \in \mathbb{C}^{n-r}$ is arbitrary.

Adding (5) and (9) we obtain the general solution of (1),

$$\begin{bmatrix} \mathbf{x}_1 \\ \mathbf{x}_2 \end{bmatrix} = Q \begin{bmatrix} A_1^{-1}\mathbf{b} \\ O \end{bmatrix} + Q \begin{bmatrix} -T \\ I_{n-r} \end{bmatrix} \mathbf{y}, \quad \mathbf{y} \text{ arbitrary.} \tag{10}$$

Thus an advantage of partitioning A as in (2) is that it permits solving (1) by working with matrices smaller or more convenient than A. We also note that the null space of A is completely determined by the multiplier T

and the permutation matrix Q, indeed (9) shows that the columns of the $n \times (n - r)$ matrix

$$Q \begin{bmatrix} -T \\ I_{n-r} \end{bmatrix} \tag{11}$$

form a basis for $N(A)$.

Case 2. $A \in \mathbb{C}_r^{m \times r}$, i.e., **$A$ is of full column rank.** Unlike Case 1, here the linear equation (1) may be inconsistent. If, however, (1) is consistent, then it has a unique solution. Partitioning the rows of A is useful for both checking the consistency of (1) and for computing its solution, if consistent.

Let the rows of A be rearranged, if necessary, so that the first r rows are linearly independent. This is written, analogously to (2), as

$$PA = \begin{bmatrix} A_1 \\ A_2 \end{bmatrix} \quad \text{or} \quad A = P^T \begin{bmatrix} A_1 \\ A_2 \end{bmatrix}, \tag{12}$$

where P is an $m \times m$ permutation matrix, and $A_1 \in \mathbb{C}_r^{r \times r}$.

If $m = r$, the matrix A_2 and other items with the subscript 2 are to be interpreted as absent.

In (12) the rows of A_2 are linear combinations of the rows of A_1, say,

$$A_2 = SA_1 \quad \text{or} \quad S = A_2 A_1^{-1} \in \mathbb{C}^{(m-r) \times r}, \tag{13}$$

where again S is called the *multiplier* corresponding to the partition (12), giving the "ratio" of the last $(m - r)$ rows of PA to its first r rows.

Corresponding to (12), let the permutation matrix P be partitioned as

$$P = \begin{bmatrix} P_1 \\ P_2 \end{bmatrix}, \quad P_1 \in \mathbb{C}^{r \times m}. \tag{14}$$

Equation (1) can now be written, using (12), (13), and (14), as

$$\begin{bmatrix} I_r \\ S \end{bmatrix} A_1 \mathbf{x} = \begin{bmatrix} P_1 \\ P_2 \end{bmatrix} \mathbf{b}, \tag{15}$$

from which the conclusions below easily follow:

(a) Equation (1) is consistent if and only if

$$P_2 \mathbf{b} = S P_1 \mathbf{b} \tag{16}$$

i.e., the "ratio" of the last $m - r$ components of the vector $P\mathbf{b}$ to its first r components is the multiplier S of (13).

(b) If (16) holds, then the unique solution of (1) is

$$\mathbf{x} = A_1^{-1} P_1 \mathbf{b}. \tag{17}$$

From (a) we note that the range of A is completely determined by the multiplier S and the permutation matrix P. Indeed, the columns of the $m \times r$ matrix

$$P^T \begin{bmatrix} I_r \\ S \end{bmatrix} \tag{18}$$

form a basis for $R(A)$.

Case 3. $A \in \mathbb{C}_r^{m \times n}$, **with $r < m, n$.** This *general case* has some of the characteristics of both Cases 1 and 2, as here we partition both the columns and rows of A.

Since A is of rank r it has at least one nonsingular $r \times r$ submatrix A_{11}, which by a rearrangement of rows and columns can be brought to the top left corner of A, say

$$PAQ = \begin{bmatrix} A_{11} & A_{12} \\ A_{21} & A_{22} \end{bmatrix}, \tag{19}$$

where P and Q are permutation matrices and $A_{11} \in \mathbb{C}_r^{r \times r}$.

By analogy with (2) and (12) we may have to interpret some of these submatrices as absent, e.g., A_{12} and A_{22} are absent if $n = r$.

By analogy with (7) and (13) there are *multipliers* $T \in \mathbb{C}^{r \times (n-r)}$ and $S \in \mathbb{C}^{(m-r) \times r}$ satisfying

$$\begin{bmatrix} A_{12} \\ A_{22} \end{bmatrix} = \begin{bmatrix} A_{11} \\ A_{21} \end{bmatrix} T \quad \text{and} \quad [A_{21} \quad A_{22}] = S[A_{11} \quad A_{12}]. \tag{20}$$

These multipliers are given by

$$T = A_{11}^{-1} A_{12} \quad \text{and} \quad S = A_{21} A_{11}^{-1}. \tag{21}$$

Combining (19) and (20) results in the following *partition* of $A \in \mathbb{C}_r^{m \times n}$,

$$A = P^T \begin{bmatrix} A_{11} & A_{12} \\ A_{21} & A_{22} \end{bmatrix} Q^T$$

$$= P^T \begin{bmatrix} I_r \\ S \end{bmatrix} A_{11}[I_r \quad T]Q^T, \tag{22}$$

where $A_{11} \in \mathbb{C}_r^{r \times r}$, P and Q are permutation matrices, and S and T are given by (21).

As in Cases 1 and 2 we conclude that the multipliers S and T, and the permutation matrices P and Q, carry all the information about the range and null space of A.

LEMMA 1. *Let $A \in \mathbb{C}_r^{m \times n}$ be partitioned as in (22). Then:*

(a) *The columns of the $n \times (n - r)$ matrix*

$$Q \begin{bmatrix} -T \\ I_{n-r} \end{bmatrix} \tag{11}$$

form a basis for $N(A)$.

(b) *The columns of the $m \times r$ matrix*

$$P^T \begin{bmatrix} I_r \\ S \end{bmatrix} \tag{18}$$

form a basis for $R(A)$.   □

Returning to the linear equation (1), it may be partitioned by using (22) and (14), in analogy with (4) and (15), as follows:

$$\begin{bmatrix} I_r \\ S \end{bmatrix} A_{11}[I_r \quad T]Q^T \begin{bmatrix} \mathbf{x}_1 \\ \mathbf{x}_2 \end{bmatrix} = \begin{bmatrix} P_1 \\ P_2 \end{bmatrix} \mathbf{b}. \tag{23}$$

The following theorem summarizes the situation, and includes the results of Cases 1 and 2 as special cases.

THEOREM 1. *Let* $A \in \mathbb{C}_r^{m \times n}$, $\mathbf{b} \in \mathbb{C}^m$ *be given and let the linear equation*

$$A\mathbf{x} = \mathbf{b} \tag{1}$$

be partitioned as in (23). *Then:*

(a) *Equation* (1) *is consistent if and only if*[1]

$$P_2\mathbf{b} = SP_1\mathbf{b} \tag{16}$$

(b) *If* (16) *holds, the general solution of* (1) *is*

$$\begin{bmatrix} \mathbf{x}_1 \\ \mathbf{x}_2 \end{bmatrix} = Q \begin{bmatrix} A_{11}^{-1}P_1\mathbf{b} \\ O \end{bmatrix} + Q \begin{bmatrix} -T \\ I_{n-r} \end{bmatrix} \mathbf{y}, \tag{24}$$

where $\mathbf{y} \in \mathbb{C}^{n-r}$ *is arbitrary.* $\qquad\square$

The partition (22) is also useful for computing generalized inverses. We collect some of these results in the following:

THEOREM 2. *Let* $A \in \mathbb{C}_r^{m \times n}$ *be partitioned as in* (22). *Then:*

(a) *A* $\{1, 2\}$-*inverse of* A *is*

$$A^{(1,2)} = Q \begin{bmatrix} A_{11}^{-1} & O \\ O & O \end{bmatrix} P \quad (\text{Rao } [\mathbf{671}]). \tag{25}$$

(b) *A* $\{1, 2, 3\}$-*inverse of* A *is*

$$A^{(1,2,3)} = Q \begin{bmatrix} A_{11}^{-1} \\ O \end{bmatrix} (I_r + S^*S)^{-1} \begin{bmatrix} I_r & S^* \end{bmatrix} P \tag{26}$$

(Meyer and Painter [**547**]).

(c) *A* $\{1, 2, 4\}$-*inverse of* A *is*

$$A^{(1,2,4)} = Q \begin{bmatrix} I_r \\ T^* \end{bmatrix} (I_r + TT^*)^{-1} \begin{bmatrix} A_{11}^{-1} & O \end{bmatrix} P. \tag{27}$$

(d) *The Moore–Penrose inverse of* A *is*

$$A^\dagger = Q \begin{bmatrix} I_r \\ T^* \end{bmatrix} (I_r + TT^*)^{-1} A_{11}^{-1}(I_r + S^*S)^{-1} \begin{bmatrix} I_r & S^* \end{bmatrix} P \tag{28}$$

(Noble [**614**]).

PROOF. The partition (22) is a full-rank factorization of A (see Lemma 1.4),

$$A = FG, \quad F \in \mathbb{C}_r^{m \times r}, \quad G \in \mathbb{C}_r^{r \times n}, \tag{29}$$

with

$$F = P^T \begin{bmatrix} I_r \\ S \end{bmatrix} A_{11}, \quad G = [I_r \quad T]Q^T, \tag{30}$$

or, alternatively,

$$F = P^T \begin{bmatrix} I_r \\ S \end{bmatrix}, \quad G = A_{11}[I_r \quad T]Q^T. \tag{31}$$

The theorem now follows from Ex. 1.29 and Ex. 1.17 by using (29) with either (30) or (31). $\qquad\square$

---

[1]By convention, (16) is satisfied if $m = r$, in which case P_2 and S are interpreted as absent.

Exercises

Ex. 1. *Schur complements and linear equations* (Cottle [**207**]). Let

$$A = \begin{bmatrix} A_{11} & A_{12} \\ A_{21} & A_{22} \end{bmatrix}, \quad A_{11} \text{ nonsingular}, \tag{32}$$

and let A/A_{11} denote the Schur complement of A_{11} in A, see (0.96).

(a) Let the homogeneous equation $A\mathbf{x} = \mathbf{0}$ be partitioned as

$$\begin{bmatrix} A_{11} & A_{12} \\ A_{21} & A_{22} \end{bmatrix} \begin{bmatrix} \mathbf{x}_1 \\ \mathbf{x}_2 \end{bmatrix} = \begin{bmatrix} \mathbf{0} \\ \mathbf{0} \end{bmatrix}. \tag{33}$$

Eliminating $\mathbf{x}_1$ we get an equation for $\mathbf{x}_2$,

$$(A/A_{11})\mathbf{x}_2 = \mathbf{0}, \tag{34a}$$

and then

$$\mathbf{x}_1 = -A_{11}^{-1}A_{12}\mathbf{x}_2. \tag{34b}$$

(b) Let the equation $A\mathbf{x} = \mathbf{b}$ be partitioned as

$$\begin{bmatrix} A_{11} & A_{12} \\ A_{21} & A_{22} \end{bmatrix} \begin{bmatrix} \mathbf{x}_1 \\ \mathbf{x}_2 \end{bmatrix} = \begin{bmatrix} \mathbf{b}_1 \\ \mathbf{b}_2 \end{bmatrix}. \tag{35}$$

Then (35) is consistent if and only if

$$(A/A_{11})\mathbf{x}_2 = \mathbf{b}_2 - A_{21}A_{11}^{-1}\mathbf{b}_1 \tag{36a}$$

is consistent, in which case a solution is completed by

$$\mathbf{x}_1 = A_{11}^{-1}(\mathbf{b}_1 - A_{12}\mathbf{x}_2). \tag{36b}$$

Ex. 2. Consider the vector $\begin{bmatrix} \mathbf{b}_1 \\ \mathbf{b}_2 \end{bmatrix}$ in RHS(35) as variable. The system (35) gives $\mathbf{b}_1, \mathbf{b}_2$ in terms of $\mathbf{x}_1, \mathbf{x}_2$. We can likewise express $\mathbf{x}_1, \mathbf{b}_2$ in terms of $\mathbf{b}_1, \mathbf{x}_2$,

$$\begin{bmatrix} \mathbf{x}_1 \\ \mathbf{b}_2 \end{bmatrix} = \begin{bmatrix} A_{11}^{-1} & -A_{11}^{-1}A_{12} \\ A_{21}A_{11}^{-1} & (A/A_{11}) \end{bmatrix} \begin{bmatrix} \mathbf{b}_1 \\ \mathbf{x}_2 \end{bmatrix}. \tag{37}$$

The operation that takes

$$\begin{bmatrix} A_{11} & A_{12} \\ A_{21} & A_{22} \end{bmatrix} \quad \text{into} \quad \begin{bmatrix} A_{11}^{-1} & -A_{11}^{-1}A_{12} \\ A_{21}A_{11}^{-1} & (A/A_{11}) \end{bmatrix}$$

is called *pivot operation* or *pivoting*, with the nonsingular submatrix A_{11} as *pivot*.

Ex. 3. Let A, A_{11} be as in (32). Then

$$\operatorname{rank} A = \operatorname{rank} A_{11} \tag{38}$$

if and only if

$$A/A_{11} = O \quad \text{(Brand [135])}. \tag{39}$$

Ex. 4. Let A, A_{11} satisfy (32) and (38).

(a) The general solution of (33) is given by

$$\mathbf{x}_1 = -A_{11}^{-1}A_{12}\mathbf{x}_2, \quad \mathbf{x}_2 \text{ arbitrary}.$$

(b) The linear equation (35) is consistent if and only if

$$A_{21}A_{11}^{-1}\mathbf{b}_1 = \mathbf{b}_2$$

in which case the general solution of (35) is given by

$$\mathbf{x}_1 = A_{11}^{-1}\mathbf{b}_1 - A_{11}^{-1}A_{12}\mathbf{x}_2, \quad \mathbf{x}_2 \text{ arbitrary}.$$

Ex. 5. Let A, A_{11} satisfy (32) and (38). Then

$$A^\dagger = [A_{11} \quad A_{12}]^* T_{11}^* \begin{bmatrix} A_{11} \\ A_{21} \end{bmatrix}^*,$$

where

$$T_{11} = \left([A_{11} \quad A_{12}] A^* \begin{bmatrix} A_{11} \\ A_{21} \end{bmatrix} \right)^{-1} \quad \text{(Zlobec [891]).}$$

Ex. 6. Let $A \in \mathbb{C}_r^{n \times n}$, $r < n$, be partitioned by

$$A = \begin{bmatrix} A_{11} & A_{12} \\ A_{21} & A_{22} \end{bmatrix} = \begin{bmatrix} I_r \\ S \end{bmatrix} A_{11} [I_r \quad T], \quad A_{11} \in \mathbb{C}_r^{r \times r}. \tag{40}$$

Then the group inverse $A^\#$ exists if and only if $I_r + ST$ is nonsingular, in which case

$$A^\# = \begin{bmatrix} I_r \\ S \end{bmatrix} ((I_r + TS) A_{11} (I_r + TS))^{-1} [I_r \quad T] \quad \text{(Robert [695]).} \tag{41}$$

Ex. 7. Let $A \in \mathbb{C}_r^{n \times n}$ be partitioned as in (40). Then A is range-Hermitian if and only if $S = T^*$.

Ex. 8. Let $A \in \mathbb{C}_r^{m \times n}$ be partitioned as in (22). Then the following orthogonal projectors are given in terms of the multipliers S, T and the permutation matrices P, Q as:

(a) $P_{R(A)} = P^T \begin{bmatrix} I_r \\ S \end{bmatrix} (I_r + S^* S)^{-1} [I_r \quad S^*] P;$

(b) $P_{R(A^*)} = Q \begin{bmatrix} I_r \\ T^* \end{bmatrix} (I_r + TT^*)^{-1} [I_r \quad T] Q^T;$

(c) $P_{N(A)} = Q \begin{bmatrix} -T \\ I_{n-r} \end{bmatrix} (I_{n-r} + T^* T)^{-1} [-T^* \quad I_{n-r}] Q^T;$ and

(d) $P_{N(A^*)} = P^T \begin{bmatrix} -S \\ I_{m-r} \end{bmatrix} (I_{m-r} + SS^*)^{-1} [-S \quad I_{m-r}] P.$

REMARK. (a) and (d) are alternative computations since

$$P_{R(A)} + P_{N(A^*)} = I_m.$$

The computation (a) requires inverting the $r \times r$ PD matrix $I_r + S^* S$, while in (d) the dimension of the PD matrix to be inverted is $(m - r) \times (m - r)$. Accordingly, (a) may be preferred if $r < m - r$.

Similarly (b) and (c) are alternative computations since

$$P_{R(A^*)} + P_{N(A)} = I_n$$

with (b) preferred if $r < n - r$.

Ex. 9. (Albert [8]). Recall the notation of Ex. 2.65. Let

$$H = \begin{bmatrix} H_{11} & H_{12} \\ H_{12}^* & H_{22} \end{bmatrix},$$

where H_{11} and H_{22} are Hermitian. Then:

(a) $H \succcurlyeq O$ if and only if

$$H_{11} \succcurlyeq O, \quad H_{11} H_{11}^\dagger H_{12} = H_{12} \quad \text{and} \quad H_{22} - H_{12}^* H_{11}^\dagger H_{12} \succcurlyeq O.$$

(b) $H \succ O$ if and only if

$$H_{11} \succ O, \quad H_{11} - H_{12} H_{22}^\dagger H_{12}^* \succ O \quad \text{and} \quad H_{22} - H_{12}^* H_{11}^{-1} H_{12} \succ O.$$

Ex. 10. (Rohde [705]). Let

$$H = \begin{bmatrix} H_{11} & H_{12} \\ H_{12}^* & H_{22} \end{bmatrix},$$

be Hermitian PSD, and denote

$$H^{(\alpha)} = \begin{bmatrix} H_{11}^{(\alpha)} + H_{11}^{(\alpha)} H_{12} G^{(\alpha)} H_{12}^* H_{11}^{(\alpha)} & -H_{11}^{(\alpha)} H_{12} G^{(\alpha)} \\ -G^{(\alpha)} H_{12}^* H_{11}^{(\alpha)} & G^{(\alpha)} \end{bmatrix}, \tag{42}$$

where

$$G = H_{22} - H_{12}^* H_{11}^{(\alpha)} H_{12}$$

and α is an integer, or a set of integers, to be specified below. Then:

(a) The relation (42) is an identity for $\alpha = 1$ and $\alpha = \{1, 2\}$. This means that RHS(42) is an $\{\alpha\}$-inverse of H if in it one substitutes the $\{\alpha\}$-inverses of H_{11} and G as indicated.

(b) If H_{22} is nonsingular and rank $H =$ rank $H_{11} +$ rank H_{22}, then (42) is an identity with $\alpha = \{1, 2, 3\}$ and $\alpha = \{1, 2, 3, 4\}$.

Ex. 11. (Meyer [544, Lemma 2.1], Campbell and Meyer [159, Theorem 7.7.3]). Let the matrix M be partitioned as

$$M = \begin{bmatrix} A & X \\ O & B \end{bmatrix}$$

where A, B are square. Then M has a group inverse if and only if:

(a) A and B have group inverses; and
(b) $(I - AA^\#)X(I - BB^\#) = O$;

in which case,

$$M^\# = \begin{bmatrix} A^\# & \vdots & A^{\#2}X(I - BB^\#) + (I - AA^\#)XB^{\#2} - A^\#XB^\# \\ \cdots & \cdots & \cdots\cdots\cdots\cdots\cdots\cdots\cdots\cdots\cdots\cdots\cdots\cdots\cdots\cdots\cdots \\ O & \vdots & B^\# \end{bmatrix}.$$

In particular, if A is nonsingular,

$$M^\# = \begin{bmatrix} A^{-1} & \vdots & A^{-2}X(I - BB^\#) - A^{-1}XB^\# \\ \cdots & \cdots & \cdots\cdots\cdots\cdots\cdots\cdots\cdots\cdots\cdots\cdots \\ O & \vdots & B^\# \end{bmatrix}. \tag{43}$$

3. Intersection of Manifolds

For any vector $\mathbf{f} \in \mathbb{C}^n$ and a subspace L of $\mathbb{C}^n$, the set

$$\mathbf{f} + L = \{\mathbf{f} + \boldsymbol{\ell} : \boldsymbol{\ell} \in L\} \tag{44}$$

is called a (linear) manifold (also affine set). The vector $\mathbf{f}$ in (44) is not unique, indeed

$$\mathbf{f} + L = (\mathbf{f} + \boldsymbol{\ell}) + L, \quad \text{for any } \boldsymbol{\ell} \in L.$$

This nonuniqueness suggests singling out the representation

$$(\mathbf{f} - P_L \mathbf{f}) + L = P_{L^\perp} \mathbf{f} + L \tag{45}$$

of the manifold (44) and calling it the orthogonal representation of $\mathbf{f} + L$. We note that $P_{L^\perp} \mathbf{f}$ is the unique vector of least Euclidean norm in the manifold (44).

In this section we study the intersection of two manifolds

$$\{\mathbf{f} + L\} \cap \{\mathbf{g} + M\} \tag{46}$$

for given vectors $\mathbf{f}$ and $\mathbf{g}$ and given subspaces L and M in $\mathbb{C}^n$. The results are needed in Section 4 below where the common solutions of pairs of linear equations are studied. Let such a pair be

$$A\mathbf{x} = \mathbf{a} \tag{47a}$$

and

$$B\mathbf{x} = \mathbf{b}, \tag{47b}$$

where A and B are given matrices with n columns and $\mathbf{a}$ and $\mathbf{b}$ are given vectors. Assuming (47a) and (47b) to be consistent, their solutions are the manifolds

$$A^\dagger \mathbf{a} + N(A) \tag{48a}$$

and

$$B^\dagger \mathbf{b} + N(B), \tag{48b}$$

respectively. If the intersection of these manifolds

$$\{A^\dagger \mathbf{a} + N(A)\} \cap \{B^\dagger \mathbf{b} + N(B)\} \tag{49}$$

is nonempty, then it is the set of common solutions of (47a)–(47b). This is the main reason for our interest in intersections of manifolds, whose study here includes conditions for the intersection (46) to be nonempty, in which case its properties and representations are given.

Since linear subspaces are manifolds, this special case is considered first.

LEMMA 2. *Let L and M be subspaces of $\mathbb{C}^n$, with P_L and P_M the corresponding orthogonal projectors. Then*

$$P_{L+M} = (P_L + P_M)(P_L + P_M)^\dagger$$
$$= (P_L + P_M)^\dagger (P_L + P_M). \tag{50}$$

PROOF. Clearly $L + M = R([P_L \quad P_M])$. Therefore,

$$P_{L+M} = [P_L \quad P_M][P_L \quad P_M]^\dagger$$

$$= [P_L \quad P_M]\begin{bmatrix} P_L \\ P_M \end{bmatrix}[P_L \quad P_M]^\dagger, \quad \text{(by Ex. 12)},$$

$$= (P_L + P_M)(P_L + P_M)^\dagger, \quad \text{since } P_L \text{ and } P_M \text{ are idempotent},$$

$$= (P_L + P_M)^\dagger (P_L + P_M)$$

since a Hermitian matrix commutes with its Moore–Penrose inverse. $\square$

The intersection of any two subspaces L and M in $\mathbb{C}^n$ is a subspace $L \cap M$ in $\mathbb{C}^n$, nonempty since $\mathbf{0} \in L \cap M$. The orthogonal projector $P_{L\cap M}$ is given in terms of P_L and P_M in the following:

THEOREM 3 (Anderson and Duffin [21]). *Let $L, M, P_L,$ and P_M be as in Lemma 2. Then*

$$P_{L\cap M} = 2P_L(P_L + P_M)^\dagger P_M,$$
$$= 2P_M(P_L + P_M)^\dagger P_L. \tag{51}$$

See also Section 8.2.

PROOF. Since $M \subset L + M$, it follows that

$$P_{L+M} P_M = P_M = P_M P_{L+M}, \tag{52}$$

and, by using (50),

$$(P_L + P_M)(P_L + P_M)^\dagger P_M = P_M = P_M (P_L + P_M)^\dagger (P_L + P_M). \tag{53}$$

Subtracting $P_M (P_L + P_M)^\dagger P_M$ from the first and last expressions in (53) gives

$$P_L (P_L + P_M)^\dagger P_M = P_M (P_L + P_M)^\dagger P_L. \tag{54}$$

Now, let

$$H = 2 P_L (P_L + P_M)^\dagger P_M = 2 P_M (P_L + P_M)^\dagger P_L.$$

Evidently, $R(H) \subset L \cap M$ and, therefore,

$$
\begin{aligned}
H &= P_{L \cap M} H = P_{L \cap M} [P_L (P_L + P_M)^\dagger P_M + P_M (P_L + P_M)^\dagger P_L] \\
&= P_{L \cap M} (P_L + P_M)^\dagger (P_L + P_M) \\
&= P_{L \cap M} P_{L+M}, \quad \text{(by Lemma 2)}, \\
&= P_{L \cap M},
\end{aligned}
$$

since $L \cap M \subset L + M$. $\square$

Other expressions for $L \cap M$ are given in the following theorem:

THEOREM 4 (Lent [**507**], Afriat [**3**], Theorem 4.5). *Let L and M be subspaces of $\mathbb{C}^n$. Then:*

(a) $L \cap M = \begin{bmatrix} P_L & O \end{bmatrix} N(\begin{bmatrix} P_L & -P_M \end{bmatrix}) = \begin{bmatrix} O & P_M \end{bmatrix} N(\begin{bmatrix} P_L & -P_M \end{bmatrix})$

(b) $= N(P_{L^\perp} + P_{M^\perp})$

(c) $= N(I - P_L P_M) = N(I - P_M P_L)$.

PROOF. (a) $\mathbf{x} \in L \cap M$ if and only if

$$\mathbf{x} = P_L \mathbf{y} = P_M \mathbf{z}, \quad \text{for some } \mathbf{y}, \mathbf{z} \in \mathbb{C}^n,$$

which is equivalent to

$$\mathbf{x} = \begin{bmatrix} P_L & O \end{bmatrix} \begin{bmatrix} \mathbf{y} \\ \mathbf{z} \end{bmatrix} = \begin{bmatrix} O & P_M \end{bmatrix} \begin{bmatrix} \mathbf{y} \\ \mathbf{z} \end{bmatrix}, \text{ where } \begin{bmatrix} \mathbf{y} \\ \mathbf{z} \end{bmatrix} \in N(\begin{bmatrix} P_L & -P_M \end{bmatrix}).$$

(b) Let $\mathbf{x} \in L \cap M$. Then $P_{L^\perp} \mathbf{x} = P_{M^\perp} \mathbf{x} = \mathbf{0}$, proving that $\mathbf{x} \in N(P_{L^\perp} + P_{M^\perp})$. Conversely, let $\mathbf{x} \in N(P_{L^\perp} + P_{M^\perp})$, i.e.,

$$(I - P_L)\mathbf{x} + (I - P_M)\mathbf{x} = \mathbf{0}$$

or

$$2\mathbf{x} = P_L \mathbf{x} + P_M \mathbf{x}$$

and, therefore,

$$2\|\mathbf{x}\| \le \|P_L \mathbf{x}\| + \|P_M \mathbf{x}\|,$$

by the triangle inequality for norms. But, by Ex. 2.52,

$$\|P_L \mathbf{x}\| \le \|\mathbf{x}\|, \quad \|P_M \mathbf{x}\| \le \|\mathbf{x}\|.$$

Therefore,

$$\|P_L \mathbf{x}\| = \|\mathbf{x}\| = \|P_M \mathbf{x}\|$$

and so, by Ex. 2.52, $P_L \mathbf{x} = \mathbf{x} = P_M \mathbf{x}$, proving $\mathbf{x} \in L \cap M$.

(c) Let $\mathbf{x} \in L \cap M$. Then $\mathbf{x} = P_L\mathbf{x} = P_M\mathbf{x} = P_LP_M\mathbf{x}$ and, therefore, $\mathbf{x} \in N(I - P_LP_M)$. Conversely, let $\mathbf{x} \in N(I - P_LP_M)$ and, therefore,

$$\mathbf{x} = P_LP_M\mathbf{x} \in L. \tag{55}$$

Also,

$$\|P_M\mathbf{x}\|^2 + \|P_{M\perp}\mathbf{x}\|^2 = \|\mathbf{x}\|^2 = \|P_LP_M\mathbf{x}\|^2$$
$$\leq \|P_M\mathbf{x}\|^2, \quad \text{by Ex. 2.52.}$$

Therefore,

$$P_{M\perp}\mathbf{x} = 0, \quad \text{i.e., } \mathbf{x} \in M,$$

and, by (55),

$$\mathbf{x} \in L \cap M.$$

The remaining equality in (c) is proved similarly. $\square$

The intersection of manifolds, which if nonempty is itself a manifold, can now be determined.

THEOREM 5 (Ben-Israel [**65**], Lent [**507**]). *Let* $\mathbf{f}$ *and* $\mathbf{g}$ *be vectors in* $\mathbb{C}^n$ *and let* L *and* M *be subspaces of* $\mathbb{C}^n$. *Then the intersection of manifolds*

$$\{\mathbf{f} + L\} \cap \{\mathbf{g} + M\} \tag{46}$$

is nonempty if and only if

$$\mathbf{g} - \mathbf{f} \in L + M, \tag{56}$$

in which case:

(a) $\{\mathbf{f} + L\} \cap \{\mathbf{g} + M\} = \mathbf{f} + P_L(P_L + P_M)^\dagger(\mathbf{g} - \mathbf{f}) + L \cap M$

(a′) $= \mathbf{g} - P_M(P_L + P_M)^\dagger(\mathbf{g} - \mathbf{f}) + L \cap M$

(b) $= \mathbf{f} + (P_{L\perp} + P_{M\perp})^\dagger P_{M\perp}(\mathbf{g} - \mathbf{f}) + L \cap M$

(b′) $= \mathbf{g} - (P_{L\perp} + P_{M\perp})^\dagger P_{L\perp}(\mathbf{g} - \mathbf{f}) + L \cap M$

(c) $= \mathbf{f} + (I - P_MP_L)^\dagger P_{M\perp}(\mathbf{g} - \mathbf{f}) + L \cap M$

(c′) $= \mathbf{g} - (I - P_LP_M)^\dagger P_{L\perp}(\mathbf{g} - \mathbf{f}) + L \cap M.$

PROOF. $\{\mathbf{f} + L\} \cap \{\mathbf{g} + M\}$ is nonempty if and only if

$$\mathbf{f} + \boldsymbol{\ell} = \mathbf{g} + \mathbf{m}, \quad \text{for some } \boldsymbol{\ell} \in L, \mathbf{m} \in M,$$

which is equivalent to

$$\mathbf{g} - \mathbf{f} = \boldsymbol{\ell} - \mathbf{m} \in L + M.$$

We now prove (a), (b), and (c). The primed statements (a′), (b′), and (c′) are proved similarly to their unprimed counterparts.

(a) The points $\mathbf{x} \in \{\mathbf{f} + L\} \cap \{\mathbf{g} + M\}$ are characterized by

$$\mathbf{x} = \mathbf{f} + P_L\mathbf{u} = \mathbf{g} + P_M\mathbf{v}, \quad \text{for some } \mathbf{u}, \mathbf{v} \in \mathbb{C}^n. \tag{57}$$

Thus

$$[P_L \quad -P_M]\begin{bmatrix}\mathbf{u}\\\mathbf{v}\end{bmatrix} = \mathbf{g} - \mathbf{f}. \tag{58}$$

The linear equation (58) is consistent, since (46) is nonempty and, therefore, the general solution of (58) is

$$\begin{bmatrix} \mathbf{u} \\ \mathbf{v} \end{bmatrix} = [P_L \quad - P_M]^\dagger (\mathbf{g} - \mathbf{f}) + N([P_L \quad - P_M])$$

$$= \begin{bmatrix} P_L \\ -P_M \end{bmatrix} (P_L + P_M)^\dagger (\mathbf{g} - \mathbf{f}) + N([P_L \quad - P_M]), \quad \text{by Ex. 12.} \tag{59}$$

Substituting (59) in (57) gives

$$\mathbf{x} = \mathbf{f} + [P_L \quad O] \begin{bmatrix} \mathbf{u} \\ \mathbf{v} \end{bmatrix}$$

$$= \mathbf{f} + P_L (P_L + P_M)^\dagger (\mathbf{g} - \mathbf{f}) + L \cap M$$

by Theorem 4(a).

(b) Writing (57) as

$$P_L \mathbf{u} - P_M \mathbf{v} = \mathbf{g} - \mathbf{f}$$

and multiplying by $P_{M^\perp}$ gives

$$P_{M^\perp} P_L \mathbf{u} = P_{M^\perp} (\mathbf{g} - \mathbf{f}), \tag{60}$$

which implies

$$(P_{L^\perp} + P_{M^\perp}) P_L \mathbf{u} = P_{M^\perp} (\mathbf{g} - \mathbf{f}). \tag{61}$$

The general solution of (61) is

$$P_L \mathbf{u} = (P_{L^\perp} + P_{M^\perp})^\dagger P_{M^\perp} (\mathbf{g} - \mathbf{f}) + N(P_{L^\perp} + P_{M^\perp})$$

$$= (P_{L^\perp} + P_{M^\perp})^\dagger P_{M^\perp} (\mathbf{g} - \mathbf{f}) + L \cap M,$$

by Theorem 4(b), which when substituted in (57) proves (b).

(c) Equation (60) can be written as

$$(I - P_M P_L) P_L \mathbf{u} = P_{M^\perp} (\mathbf{g} - \mathbf{f})$$

whose general solution is

$$P_L \mathbf{u} = (I - P_M P_L)^\dagger P_{M^\perp} (\mathbf{g} - \mathbf{f}) + N(I - P_M P_L)$$

$$= (I - P_M P_L)^\dagger P_{M^\perp} (\mathbf{g} - \mathbf{f}) + L \cap M,$$

by Theorem 4(c), which when substituted in (57) proves (c). □

Theorem 5 verifies that the intersection (46), if nonempty, is itself a manifold. We note, in passing, that parts (a) and (a′) of Theorem 5 give the same representation of (46); i.e., if (56) holds, then

$$\mathbf{f} + P_L (P_L + P_M)^\dagger (\mathbf{g} - \mathbf{f}) = \mathbf{g} - P_M (P_L + P_M)^\dagger (\mathbf{g} - \mathbf{f}). \tag{62}$$

Indeed, (56) implies that

$$\mathbf{g} - \mathbf{f} = P_{L+M} (\mathbf{g} - \mathbf{f})$$

$$= (P_L + P_M)(P_L + P_M)^\dagger (\mathbf{g} - \mathbf{f}),$$

which gives (62) by rearrangement of terms.

It will now be proved that parts (a), (a′), (b), and (b′) of Theorem 5 give orthogonal representations of

$$\{\mathbf{f} + L\} \cap \{\mathbf{g} + M\} \tag{46}$$

if the representations $\{f + L\}$ and $\{g + M\}$ are orthogonal, i.e., if

$$f \in L^{\perp}, \quad g \in M^{\perp}. \tag{63}$$

COROLLARY 1. *Let L and M be subspaces of $\mathbb{C}^n$ and let*

$$f \in L^{\perp}, \quad g \in M^{\perp}. \tag{63}$$

If (46) is nonempty, then each of the four representations given below is orthogonal:

(a) $\{f + L\} \cap \{g + M\} = f + P_L(P_L + P_M)^{\dagger}(g - f) + L \cap M$

(a') $\qquad\qquad\qquad = g - P_M(P_L + P_M)^{\dagger}(g - f) + L \cap M$

(b) $\qquad\qquad\qquad = f + (P_{L^{\perp}} + P_{M^{\perp}})^{\dagger}P_{M^{\perp}}(g - f) + L \cap M$

(b') $\qquad\qquad\qquad = g - (P_{L^{\perp}} + P_{M^{\perp}})^{\dagger}P_{L^{\perp}}(g - f) + L \cap M.$

PROOF. Each of the above representations is of the form

$$\{f + L\} \cap \{g + M\} = v + L \cap M, \tag{64}$$

which is an orthogonal representation if and only if

$$P_{L \cap M} v = 0. \tag{65}$$

In the proof we use the facts

$$P_{L \cap M} = P_L P_{L \cap M} = P_{L \cap M} P_L = P_M P_{L \cap M} = P_{L \cap M} P_M, \tag{66}$$

which hold since $L \cap M$ is contained in both L and M.

(a) Here $v = f + P_L(P_L + P_M)^{\dagger}(g - f)$. The matrix $P_L + P_M$ is Hermitian and, therefore, $(P_L + P_M)^{\dagger}$ is a polynomial in powers of $P_L + P_M$, by Theorem 4.7. From (66) it follows therefore that

$$P_{L \cap M}(P_L + P_M)^{\dagger} = (P_L + P_M)^{\dagger}P_{L \cap M} \tag{67}$$

and (65) follows from

$$\begin{aligned}
P_{L \cap M} v &= P_{L \cap M} f + P_{L \cap M} P_L(P_L + P_M)^{\dagger}(g - f) \\
&= P_{L \cap M} f + (P_L + P_M)^{\dagger}P_{L \cap M}(g - f), \quad \text{by (66) and (67),} \\
&= 0, \quad \text{by (63).}
\end{aligned}$$

(a') follows from (62) and (a).

(b) Here $v = f + (P_{L^{\perp}} + P_{M^{\perp}})^{\dagger}P_{M^{\perp}}(g - f)$. The matrix $P_{L^{\perp}} + P_{M^{\perp}}$ is Hermitian and, therefore, $(P_{L^{\perp}} + P_{M^{\perp}})^{\dagger}$ is a polynomial in $P_{L^{\perp}} + P_{M^{\perp}}$, which implies that

$$P_{L \cap M}(P_{L^{\perp}} + P_{M^{\perp}})^{\dagger} = O. \tag{68}$$

Finally, (65) follows from

$$\begin{aligned}
P_{L \cap M} v &= P_{L \cap M} f + P_{L \cap M}(P_{L^{\perp}} + P_{M^{\perp}})^{\dagger}P_{M^{\perp}}(g - f) \\
&= 0, \quad \text{by (63) and (68).}
\end{aligned}$$

(b') If (63) holds, then

$$\begin{aligned}
g - f &= P_{L^{\perp} + M^{\perp}}(g - f) \\
&= (P_{L^{\perp}} + P_{M^{\perp}})^{\dagger}(P_{L^{\perp}} + P_{M^{\perp}})(g - f),
\end{aligned}$$

by Lemma 2 and, therefore,

$$f + (P_{L^{\perp}} + P_{M^{\perp}})^{\dagger}P_{M^{\perp}}(g - f) = g - (P_{L^{\perp}} + P_{M^{\perp}})^{\dagger}P_{L^{\perp}}(g - f),$$

which proves (b') identical to (b), if (63) is satisfied.   □

Finally, we characterize subspaces L and M for which the intersection (46) is always nonempty.

COROLLARY 2. *Let L and M be subspaces of $\mathbb{C}^n$. Then the intersection*

$$\{\mathbf{f} + L\} \cap \{\mathbf{g} + M\} \tag{46}$$

is nonempty for all $\mathbf{f}, \mathbf{g} \in \mathbb{C}^n$ if and only if

$$L^\perp \cap M^\perp = \{\mathbf{0}\}. \tag{69}$$

PROOF. The intersection (46) is by Theorem 5 nonempty for all $\mathbf{f}, \mathbf{g} \in \mathbb{C}^n$, if and only if $L + M = \mathbb{C}^n$, which is equivalent to $\{\mathbf{0}\} = (L + M)^\perp = L^\perp \cap M^\perp$, by Ex. 13(b). $\qquad\square$

Exercises

Ex. 12. Let P_L and P_M be $n \times n$ orthogonal projectors. Then

$$[P_L \quad \pm P_M]^\dagger = \begin{bmatrix} P_L \\ \pm P_M \end{bmatrix} (P_L + P_M)^\dagger. \tag{70}$$

PROOF. Use $A^\dagger = A^*(AA^*)^\dagger$ with $A = [P_L \quad \pm P_M]$, and the fact that P_L and P_M are Hermitian idempotents. $\qquad\square$

Ex. 13. Let L and M be subspaces of $\mathbb{C}^n$. Then:

(a) $(L \cap M)^\perp = L^\perp + M^\perp$; and
(b) $(L^\perp \cap M^\perp)^\perp = L + M$.

PROOF. (a) Evidently, $L^\perp \subset (L \cap M)^\perp$ and $M^\perp \subset (L \cap M)^\perp$; hence

$$L^\perp + M^\perp \subset (L \cap M)^\perp.$$

Conversely, from $L^\perp \subset L^\perp + M^\perp$ it follows that

$$(L^\perp + M^\perp)^\perp \subset L^{\perp\perp} = L.$$

Similarly, $(L^\perp + M^\perp)^\perp \subset M$, hence

$$(L^\perp + M^\perp)^\perp \subset L \cap M$$

and, by taking orthogonal complements,

$$(L \cap M)^\perp \subset L^\perp + M^\perp.$$

(b) Replace in (a) L and M by $L^\perp$ and $M^\perp$, respectively. $\qquad\square$

Ex. 14. (von Neumann [840]). Let $L_1, L_2, \ldots, L_k$ be any k linear subspaces of $\mathbb{C}^n$, $k \geq 2$, and let

$$Q = P_{L_k} P_{L_{k-1}} \cdots P_{L_2} P_{L_1} P_{L_2} \cdots P_{L_{k-1}} P_{L_k}. \tag{71}$$

Then the orthogonal projector on $\bigcap_{i=1}^k L_i$ is $\lim_{m \to \infty} Q^m$.

Ex. 15. (Pyle [660]). The matrix Q of (71) is Hermitian, so let its spectral decomposition be given by

$$Q = \sum_{i=1}^q \lambda_i E_i$$

where

$$\lambda_1 \geq \lambda_2 \geq \cdots \geq \lambda_q$$

are the distinct eigenvalues of Q and

$$E_1, E_2, \ldots, E_q$$

are the corresponding orthogonal projectors satisfying

$$E_1 + E_2 + \cdots + E_q = I$$

and

$$E_i E_j = O, \quad \text{if } i \neq j.$$

Then

$$1 \geq \lambda_1 \geq \lambda_2 \geq \cdots \geq \lambda_q \geq 0$$

and

$$\bigcap_{i=1}^{k} L_i \neq \{\mathbf{0}\}, \quad \text{if and only if } \lambda_1 = 1,$$

in which case the orthogonal projector on $\bigcap_{i=1}^{k} L_i$ is E_1.

Ex. 16. *A closed-form expression.* Using the notation of Ex. 15, the orthogonal projector on $\bigcap_{i=1}^{k} L_i$ is

$$Q^\nu + [(Q^{\nu+1} - Q^\nu)^\dagger - (Q^\nu - Q^{\nu-1})^\dagger]^\dagger, \quad \text{for } \nu = 2, 3, \dots . \tag{72}$$

If λ_q, the smallest eigenvalue of Q, is positive, then (72) also holds for $\nu = 1$, in which case Q^0 is taken as I (Pyle [**660**]).

4. Common Solutions of Linear Equations and Generalized Inverses of Partitioned Matrices

Consider the pair of linear equations

$$A\mathbf{x} = \mathbf{a}, \tag{47a}$$

$$B\mathbf{x} = \mathbf{b}, \tag{47b}$$

with given vectors $\mathbf{a}, \mathbf{b}$ and matrices A, B having n columns.

Assuming (47a) and (47b) to be consistent, we study here their common solutions, if any, expressing them in terms of the solutions of (47a) and (47b).

The common solutions of (47a) and (47b) are the solutions of the partitioned linear equation

$$\begin{bmatrix} A \\ B \end{bmatrix} \mathbf{x} = \begin{bmatrix} \mathbf{a} \\ \mathbf{b} \end{bmatrix}, \tag{73}$$

which is often the starting point, the partitioning into (47a) and (47b) being used to reduce the size or difficulty of the problem.

The solutions of (47a) and (47b) constitute the manifolds

$$A^\dagger \mathbf{a} + N(A) \tag{48a}$$

$$\text{and} \quad B^\dagger \mathbf{b} + N(B), \tag{48b}$$

respectively. Thus the intersection

$$\{A^\dagger \mathbf{a} + N(A)\} \cap \{B^\dagger \mathbf{b} + N(B)\} \tag{49}$$

is the set of solutions of (73) and (73) is consistent if and only if (49) is nonempty.

The results of Section 3 are applicable to determining the intersection (49). In particular, Theorem 5 yields the following:

COROLLARY 3. *Let A and B be matrices with n columns and let $\mathbf{a}$ and $\mathbf{b}$ be vectors such that each of the equations (47a) and (47b) is consistent. Then (47a) and (47b) have common solutions if and only if*

$$B^\dagger \mathbf{b} - A^\dagger \mathbf{a} \in N(A) + N(B), \tag{74}$$

in which case the set of common solutions is the manifold:

(a) $A^\dagger \mathbf{a} + P_{N(A)}(P_{N(A)} + P_{N(B)})^\dagger (B^\dagger \mathbf{b} - A^\dagger \mathbf{a}) + N(A) \cap N(B)$

(a') $= B^\dagger \mathbf{b} - P_{N(B)}(P_{N(A)} + P_{N(B)})^\dagger (B^\dagger \mathbf{b} - A^\dagger \mathbf{a}) + N(A) \cap N(B)$

(b) $= (A^\dagger A + B^\dagger B)^\dagger (A^\dagger \mathbf{a} + B^\dagger \mathbf{b}) + N(A) \cap N(B).$

PROOF. Follows from Theorem 5 by substituting

$$\mathbf{f} = A^\dagger \mathbf{a}, \quad L = N(A), \quad \mathbf{g} = B^\dagger \mathbf{b}, \quad M = N(B). \tag{75}$$

Thus (74), (a), and (a') follow directly from (56), (a), and (a') of Theorem 5, respectively, by using (75).

That (b) follows from Theorem 5(b) or (b') is proved as follows. Substituting (75) in Theorem 5(b) gives

$$\{A^\dagger \mathbf{a} + N(A)\} \cap \{B^\dagger \mathbf{b} + N(B)\}$$
$$= A^\dagger \mathbf{a} + (A^\dagger A + B^\dagger B)^\dagger B^\dagger B(B^\dagger \mathbf{b} - A^\dagger \mathbf{a}) + N(A) \cap N(B)$$
$$= (A^\dagger - (A^\dagger A + B^\dagger B)^\dagger B^\dagger B A^\dagger) \mathbf{a} + (A^\dagger A + B^\dagger B)^\dagger B^\dagger \mathbf{b}$$
$$+ N(A) \cap N(B), \tag{76}$$

since $P_{N(X)^\perp} = P_{R(X^*)} = X^\dagger X$ for $X = A, B$.

Now $R(A^\dagger) = R(A^*) \subset R(A^*) + R(B^*)$ and, therefore,

$$A^\dagger = (A^\dagger A + B^\dagger B)^\dagger (A^\dagger A + B^\dagger B) A^\dagger$$

by Lemma 2, from which it follows that

$$A^\dagger - (A^\dagger A + B^\dagger B)^\dagger B^\dagger B A^\dagger = (A^\dagger A + B^\dagger B)^\dagger A^\dagger,$$

which when substituted in (76) gives (b). $\qquad\square$

Since each of the parts (a), (a'), and (b) of Corollary 3 gives the solutions of the partitioned equation (73), these expressions can be used to obtain the generalized inverses of partitioned matrices.

THEOREM 6 (Ben-Israel [65], Katz [472], Mihalyffy [554]). *Let A and B be matrices with n columns. Then each of the following expressions is a $\{1, 2, 4\}$-inverse of the partitioned matrix $\begin{bmatrix} A \\ B \end{bmatrix}$:*

(a) $\quad X = [A^\dagger \ \ O] + P_{N(A)}(P_{N(A)} + P_{N(B)})^\dagger [-A^\dagger \ \ B^\dagger], \tag{77}$

(a') $\quad Y = [O \ \ B^\dagger] - P_{N(B)}(P_{N(A)} + P_{N(B)})^\dagger [-A^\dagger \ \ B^\dagger], \tag{78}$

(b) $\quad Z = (A^\dagger A + B^\dagger B)^\dagger [A^\dagger \ \ B^\dagger]. \tag{79}$

Moreover, if

$$R(A^*) \cap R(B^*) = \{\mathbf{0}\}, \tag{80}$$

then each of the expressions (77), (78), (79) is the Moore–Penrose inverse of $\begin{bmatrix} A \\ B \end{bmatrix}$.

PROOF. From Corollary 3 it follows that whenever

$$\begin{bmatrix} A \\ B \end{bmatrix} \mathbf{x} = \begin{bmatrix} \mathbf{a} \\ \mathbf{b} \end{bmatrix}, \tag{73}$$

is consistent, then $X \begin{bmatrix} \mathbf{a} \\ \mathbf{b} \end{bmatrix}$, $Y \begin{bmatrix} \mathbf{a} \\ \mathbf{b} \end{bmatrix}$, and $Z \begin{bmatrix} \mathbf{a} \\ \mathbf{b} \end{bmatrix}$ are among its solutions. Also the representations (48a) and (48b) are orthogonal and, therefore, by Corollary 1, the representations (a), (a′), and (b) of Corollary 3 are also orthogonal. Thus $X \begin{bmatrix} \mathbf{a} \\ \mathbf{b} \end{bmatrix}$, $Y \begin{bmatrix} \mathbf{a} \\ \mathbf{b} \end{bmatrix}$, and $Z \begin{bmatrix} \mathbf{a} \\ \mathbf{b} \end{bmatrix}$ are all perpendicular to

$$N(A) \cap N(B) = N \begin{bmatrix} A \\ B \end{bmatrix}.$$

By Theorem 3.2, it follows therefore that X, Y, and Z are $\{1, 4\}$-inverses of $\begin{bmatrix} A \\ B \end{bmatrix}$.

We show now that X, Y, and Z are $\{2\}$-inverses of $\begin{bmatrix} A \\ B \end{bmatrix}$.

(a)   From (77) we get

$$X \begin{bmatrix} A \\ B \end{bmatrix} = A^\dagger A + P_{N(A)}(P_{N(A)} + P_{N(B)})^\dagger(-A^\dagger A + B^\dagger B).$$

But

$$(-A^\dagger A + B^\dagger B) = P_{N(A)} - P_{N(B)} = (P_{N(A)} + P_{N(B)}) - 2P_{N(B)}.$$

Therefore, by Lemma 2 and Theorem 3,

$$X \begin{bmatrix} A \\ B \end{bmatrix} = A^\dagger A + P_{N(A)}P_{N(A)+N(B)} - P_{N(A)\cap N(B)}$$

$$= A^\dagger A + P_{N(A)} - P_{N(A)\cap N(B)}, \quad \text{since } N(A) \subset N(A) + N(B),$$

$$= I_n - P_{N(A)\cap N(B)}, \quad \text{since } P_{N(A)} = I - A^\dagger A. \tag{81}$$

Since $R(H^\dagger) = R(H^*) = N(H)^\perp$ for $H = A, B$,

$$P_{N(A)\cap N(B)}A^\dagger = O, \quad P_{N(A)\cap N(B)}B^\dagger = O, \tag{82}$$

and therefore (81) gives

$$X \begin{bmatrix} A \\ B \end{bmatrix} X = X - P_{N(A)\cap N(B)}P_{N(A)}(P_{N(A)} + P_{N(B)})^\dagger[-A^\dagger \quad B^\dagger].$$

Since

$$P_{N(A)\cap N(B)} = P_{N(A)\cap N(B)}P_{N(A)} = P_{N(A)\cap N(B)}P_{N(B)},$$

$$X \begin{bmatrix} A \\ B \end{bmatrix} X = X - \tfrac{1}{2}P_{N(A)\cap N(B)}(P_{N(A)} + P_{N(B)})(P_{N(A)} + P_{N(B)})^\dagger[-A^\dagger \quad B^\dagger]$$

$$= X - \tfrac{1}{2}P_{N(A)\cap N(B)}P_{N(A)+N(B)}[-A^\dagger \quad B^\dagger], \quad \text{by Lemma 2,}$$

$$= X - \tfrac{1}{2}P_{N(A)\cap N(B)}[-A^\dagger \quad B^\dagger],$$

$$\text{since } N(A) \cap N(B) \subset N(A) + N(B),$$

$$= X, \quad \text{(by (82)).}$$

(a′)   That Y, given by (78), is a $\{2\}$-inverse of $\begin{bmatrix} A \\ B \end{bmatrix}$ is similarly proved.

(b)   The proof that Z given by (79) is a $\{2\}$-inverse of $\begin{bmatrix} A \\ B \end{bmatrix}$ is easy since

$$Z \begin{bmatrix} A \\ B \end{bmatrix} = (A^\dagger A + B^\dagger B)^\dagger (A^\dagger A + B^\dagger B)$$

and therefore

$$\begin{aligned}
Z \begin{bmatrix} A \\ B \end{bmatrix} Z &= (A^\dagger A + B^\dagger B)^\dagger (A^\dagger A + B^\dagger B)(A^\dagger A + B^\dagger B)^\dagger [A^\dagger \quad B^\dagger] \\
&= (A^\dagger A + B^\dagger B)^\dagger [A^\dagger \quad B^\dagger] \\
&= Z.
\end{aligned}$$

Finally, we show that (80) implies that $X, Y,$ and Z given by (77), (78), and (79), respectively, are $\{3\}$-inverses of $\begin{bmatrix} A \\ B \end{bmatrix}$. Indeed, (80) is equivalent to

$$N(A) + N(B) = \mathbb{C}^n, \tag{83}$$

since $N(A) + N(B) = \{R(A^*) \cap R(B^*)\}^\perp$ by Ex. 13(b).

(a) From (77) it follows that

$$BX = [BA^\dagger \quad O] + BP_{N(A)}(P_{N(A)} + P_{N(B)})^\dagger [-A^\dagger \quad B^\dagger]. \tag{84}$$

But

$$(P_{N(A)} + P_{N(B)})(P_{N(A)} + P_{N(B)})^\dagger = I_n \tag{85}$$

by (83) and Lemma 2. Therefore,

$$P_{N(A)}(P_{N(A)} + P_{N(B)})^\dagger = B(P_{N(A)} + P_{N(B)} - P_{N(A)})(P_{N(A)} + P_{N(B)})^\dagger = B,$$

and so (84) becomes

$$BX = [O \quad BB^\dagger].$$

Consequently,

$$\begin{bmatrix} A \\ B \end{bmatrix} X = \begin{bmatrix} AA^\dagger & O \\ O & BB^\dagger \end{bmatrix},$$

which proves that X is a $\{3\}$-inverse of $\begin{bmatrix} A \\ B \end{bmatrix}$.

(a′) That Y given by (78) is a $\{3\}$-inverse of $\begin{bmatrix} A \\ B \end{bmatrix}$ whenever (80) holds is similarly proved or, alternatively, (77) and (78) give

$$\begin{aligned}
Y - X &= [-A^\dagger \quad B^\dagger] - (P_{N(A)} + P_{N(B)})(P_{N(A)} + P_{N(B)})^\dagger [-A^\dagger \quad B^\dagger] \\
&= O, \quad \text{by (85)}.
\end{aligned}$$

(b) Finally, we show that Z is the Moore–Penrose inverse of $\begin{bmatrix} A \\ B \end{bmatrix}$ when (80) holds. By Ex. 2.38, the Moore–Penrose inverse of any matrix H is the only $\{1,2\}$-inverse U such that $R(U) = R(H^*)$ and $N(U) = N(H^*)$. Thus, $H^\dagger$ is also the unique matrix $U \in H\{1,2,4\}$ such that $N(H^*) \subset N(U)$. Now, Z has already been shown to be a $\{1,2,4\}$-inverse of $\begin{bmatrix} A \\ B \end{bmatrix}$, and it therefore suffices to prove that

$$N([A^* \quad B^*]) \subset N(Z). \tag{86}$$

Let $\begin{bmatrix} \mathbf{u} \\ \mathbf{v} \end{bmatrix} \in N([A^* \quad B^*])$. Then

$$A^*\mathbf{u} + B^*\mathbf{v} = \mathbf{0},$$

and, therefore,

$$A^*\mathbf{u} = -B^*\mathbf{v} = \mathbf{0}, \tag{87}$$

since, by (80), the only vector common to $R(A^*)$ and $R(B^*)$ is the zero vector. Since $N(H^\dagger) = N(H^*)$ for any H, (87) gives

$$A^\dagger\mathbf{u} = B^\dagger\mathbf{v} = \mathbf{0},$$

and therefore by (79), $Z\begin{bmatrix} \mathbf{u} \\ \mathbf{v} \end{bmatrix} = \mathbf{0}$. Thus (86) is established, and the proof is complete. □

If a matrix is partitioned by columns instead of by rows, then Theorem 6 may still be used. Indeed,

$$[A \quad B] = \begin{bmatrix} A^* \\ B^* \end{bmatrix}^* \tag{88}$$

permits using Theorem 6 to obtain generalized inverses of $\begin{bmatrix} A^* \\ B^* \end{bmatrix}$, which is partitioned by rows and then translating the results to the matrix $[A \quad B]$, partitioned by columns.

In working with the conjugate transposes of a matrix, we note that

$$
\begin{aligned}
X \in A\{i\} &\iff X^* \in A^*\{i\}, \quad (i = 1, 2), \\
X \in A\{3\} &\iff X^* \in A^*\{4\}, \\
X \in A\{4\} &\iff X^* \in A^*\{3\}.
\end{aligned}
\tag{89}
$$

Applying Theorem 6 to $\begin{bmatrix} A^* \\ B^* \end{bmatrix}$ as in (88), and using (89), we obtain the following:

COROLLARY 4. *Let A and B be matrices with n rows. Then each of the following expressions is a $\{1, 2, 3\}$-inverse of the partitioned matrix $[A \quad B]$:*

(a) $$X = \begin{bmatrix} A^\dagger \\ O \end{bmatrix} + \begin{bmatrix} -A^\dagger \\ B^\dagger \end{bmatrix} (P_{N(A^*)} + P_{N(B^*)})^\dagger P_{N(A^*)}, \tag{90}$$

(a') $$Y = \begin{bmatrix} O \\ B^\dagger \end{bmatrix} - \begin{bmatrix} -A^\dagger \\ B^\dagger \end{bmatrix} (P_{N(A^*)} + P_{N(B^*)})^\dagger P_{N(B^*)}, \tag{91}$$

(b) $$Z = \begin{bmatrix} A^\dagger \\ B^\dagger \end{bmatrix} (AA^\dagger + BB^\dagger)^\dagger. \tag{92}$$

Moreover, if

$$R(A) \cap R(B) = \{\mathbf{0}\}, \tag{93}$$

then each of the expressions (90), (91), (92) is the Moore–Penrose inverse of $[A \quad B]$. □

Other and more general results on Moore–Penrose inverses of partitioned matrices were given in Cline [**200**]. However, these results are too formidable for reproduction here.

Exercises

Ex. 17. In the following expressions $^{(1)}$ denotes any $\{1\}$-inverse.

(a) A $\{1\}$-inverse of $\begin{bmatrix} A \\ B \end{bmatrix}$ is $[X \quad Y]$ where

$$Y = (I - A^{(1)}A)(B - BA^{(1)}A)^{(1)}, \quad X = A^{(1)} - YBA^{(1)}. \tag{94}$$

(b) A $\{1\}$-inverse of $[A \quad B]$ is $\begin{bmatrix} X \\ Y \end{bmatrix}$ where

$$Y = (B - AA^{(1)}B)^{(1)}(I - AA^{(1)}), \quad X = A^{(1)} - A^{(1)}BY. \tag{95}$$

Ex. 18. Let the partitioned matrix $\begin{bmatrix} A \\ B \end{bmatrix}$ be nonsingular. Then

(a) $\quad \begin{bmatrix} A \\ B \end{bmatrix}^{-1} = [A^\dagger \quad O] + P_{N(A)}(P_{N(A)} + P_{N(B)})^{-1}[-A^\dagger \quad B^\dagger]$

(a') $\quad\quad\quad\quad = [O \quad B^\dagger] - P_{N(B)}(P_{N(A)} + P_{N(B)})^{-1}[-A^\dagger \quad B^\dagger]$

(b) $\quad\quad\quad\quad = (A^\dagger A + B^\dagger B)^{-1}[A^\dagger \quad B^\dagger].$

PROOF. Follows from Theorem 6. Indeed the nonsingularity of $\begin{bmatrix} A \\ B \end{bmatrix}$ guarantees that (80) is satisfied, and also that the matrices $P_{N(A)} + P_{N(B)}$ and $A^\dagger A + B^\dagger B = P_{R(A^*)} + P_{R(B^*)}$ are nonsingular. $\qquad\square$

Ex. 19. Let $A = [1 \quad 1]$, $B = [1 \quad 2]$. Then $\begin{bmatrix} A \\ B \end{bmatrix} = \begin{bmatrix} 1 & 1 \\ 1 & 2 \end{bmatrix}$ is nonsingular. We calculate now its inverse using Ex. 18(b).

Here

$$A^\dagger = \tfrac{1}{2}\begin{bmatrix} 1 \\ 1 \end{bmatrix}, \quad\quad\quad\quad\quad\quad A^\dagger A = \tfrac{1}{2}\begin{bmatrix} 1 & 1 \\ 1 & 1 \end{bmatrix},$$

$$B^\dagger = \tfrac{1}{5}\begin{bmatrix} 1 \\ 2 \end{bmatrix}, \quad\quad\quad\quad\quad\quad B^\dagger B = \tfrac{1}{5}\begin{bmatrix} 1 & 2 \\ 2 & 4 \end{bmatrix},$$

$$A^\dagger A + B^\dagger B = \tfrac{1}{10}\begin{bmatrix} 7 & 9 \\ 9 & 13 \end{bmatrix}, \quad (A^\dagger A + B^\dagger B)^{-1} = \begin{bmatrix} 13 & -9 \\ -9 & 7 \end{bmatrix},$$

and, finally,

$$\begin{bmatrix} A \\ B \end{bmatrix}^{-1} = (A^\dagger A + B^\dagger B)^{-1}[A^\dagger \quad B^\dagger]$$

$$= \begin{bmatrix} 13 & -9 \\ -9 & 7 \end{bmatrix} \tfrac{1}{10}\begin{bmatrix} 7 & 9 \\ 9 & 13 \end{bmatrix} = \begin{bmatrix} 2 & -1 \\ -1 & 1 \end{bmatrix}.$$

Ex. 20. *Series expansion.* Let the partitioned matrix $\begin{bmatrix} A \\ B \end{bmatrix}$ be nonsingular. Then

$$A^\dagger A + B^\dagger B = I + K, \tag{96}$$

where K is Hermitian and

$$\|K\| < 1. \tag{97}$$

From (96) and (97) it follows that

$$(A^\dagger A + B^\dagger B)^{-1} = \sum_{j=0}^{\infty}(-1)^j K^j, \tag{98}$$

Substituting (98) in Ex. 18(b) gives

$$\begin{bmatrix} A \\ B \end{bmatrix}^{-1} = \sum_{j=0}^{\infty} (-1)^j K^j [A^\dagger \quad B^\dagger]. \tag{99}$$

Similarly,

$$P_{N(A)} + P_{N(B)} = I - A^\dagger A + I - B^\dagger B$$
$$= I - K, \quad \text{with } K \text{ as in (96)},$$

and, therefore,

$$(P_{N(A)} + P_{N(B)})^{-1} = \sum_{j=0}^{\infty} K^j. \tag{100}$$

Substituting (100) in Ex. 18(a) gives

$$\begin{bmatrix} A \\ B \end{bmatrix}^{-1} = [A^\dagger \quad O] + (I - A^\dagger A) \sum_{j=0}^{\infty} K^j [-A^\dagger \quad B^\dagger]. \tag{101}$$

Ex. 21. Let the partitioned matrix $\begin{bmatrix} A \\ B \end{bmatrix}$ be nonsingular. Then the solution of

$$\begin{bmatrix} A \\ B \end{bmatrix} \mathbf{x} = \begin{bmatrix} \mathbf{a} \\ \mathbf{b} \end{bmatrix}, \tag{73}$$

for any given $\mathbf{a}$ and $\mathbf{b}$, is

$$\mathbf{x} = \sum_{j=0}^{\infty} (-1)^j K^j (A^\dagger \mathbf{a} + B^\dagger \mathbf{b}) \tag{102}$$

$$= A^\dagger \mathbf{a} + (I - A^\dagger A) \sum_{j=0}^{\infty} K^j (B^\dagger \mathbf{b} - A^\dagger \mathbf{a}), \tag{103}$$

with K given by (96).

PROOF. Use (99) and (101).   □

REMARK. If the nonsingular matrix $\begin{bmatrix} A \\ B \end{bmatrix}$ is ill-conditioned, then slow convergence may be expected in (98) and (100), and hence in (99) and (101). Even then the convergence of (102) or (103) may be reasonable for certain vectors $\begin{bmatrix} \mathbf{a} \\ \mathbf{b} \end{bmatrix}$. Thus, for example, if $\|B^\dagger \mathbf{b} - A^\dagger \mathbf{a}\|$ is sufficiently small, then (103) may be reasonably approximated by its first few terms.

Ex. 22. *Common solutions for n matrix equations.* For each $i \in \overline{1,n}$ let the matrices $A_i \in \mathbb{C}^{p \times q}$, $B_i \in \mathbb{C}^{p \times r}$ be given, and consider the n matrix equations

$$A_i X = B_i, \quad i \in \overline{1,n}. \tag{104}$$

For $k \in \overline{1,n}$ define recursively

$$C_k = A_k F_{k-1}, \qquad\qquad D_k = B_k - A_k E_{k-1},$$
$$E_k = E_{k-1} + F_{k-1} C_k^\dagger D_k \quad \text{and} \quad F_k = F_{k-1}(I - C_k^\dagger C_k), \tag{105}$$

where

$$E_0 = O_{q \times r}, \quad F_0 = I_q.$$

Then the n matrix equations (104) have a common solution if and only if

$$C_i C_i^\dagger D_i = D_i, \quad i \in \overline{1,n}, \tag{106}$$

in which case the general common solution of (104) is

$$X = E_n + F_n Z, \tag{107}$$

where $Z \in \mathbb{C}^{q \times r}$ is arbitrary (Morris and Odell [583]).

Ex. 23. (Morris and Odell [583]). For $i \in \overline{1,n}$ let $A_i \in \mathbb{C}^{1 \times q}$, and let C_i be defined by (105) for $i \in \overline{1,n}$. Let the vectors $\{A_1, A_2, \ldots, A_k\}$ be linearly independent. Then the vectors $\{A_1, A_2, \ldots, A_{k+1}\}$ are linearly independent if and only if $C_{k+1} = O$.

Ex. 24. (Morris and Odell [583]). For $i \in \overline{1,n}$ let A_i, C_i be as in Ex. 23. For any $k \leq n$ the vectors $\{C_1, C_2, \ldots, C_k\}$ are orthogonal and span the subspace spanned by $\{A_1, A_2, \ldots, A_k\}$.

5. Generalized Inverses of Bordered Matrices

Partitioning was shown above to permit working with submatrices smaller in size and better behaved (e.g., nonsingular) than the original matrix. In this section a nonsingular matrix is obtained from the original matrix by adjoining to it certain matrices. Thus, from a given matrix $A \in \mathbb{C}^{m \times n}$, we obtain the matrix

$$\begin{bmatrix} A & U \\ V^* & O \end{bmatrix}, \tag{108}$$

which, under certain conditions on U and V^*, is nonsingular, and from its inverse $A^\dagger$ can be read off. These ideas find applications in differential equations (Reid [683]) and eigenvalue computation (Blattner [113]).

The following theorem is based on the results of Blattner [113]:

THEOREM 7. *Let* $A \in \mathbb{C}_r^{m \times n}$ *and let the matrices* U *and* V *satisfy:*

(a) $U \in \mathbb{C}_{(m-r)}^{m \times (m-r)}$ *and the columns of* U *are a basis for* $N(A^*)$.

(b) $V \in \mathbb{C}_{(n-r)}^{n \times (n-r)}$ *and the columns of* U *are a basis for* $N(A)$.

Then the matrix

$$\begin{bmatrix} A & U \\ V^* & O \end{bmatrix}, \tag{108}$$

is nonsingular and its inverse is

$$\begin{bmatrix} A^\dagger & V^{*\dagger} \\ U^\dagger & O \end{bmatrix}. \tag{109}$$

PROOF. Premultiplying (109) by (108) gives

$$\begin{bmatrix} AA^\dagger + UU^\dagger & AV^{*\dagger} \\ V^*A^\dagger & V^*V^{*\dagger} \end{bmatrix}. \tag{110}$$

Now, $R(U) = N(A^*) = R(A)^\perp$ by assumption (a) and (0.26) and, therefore,

$$AA^\dagger + UU^\dagger = I_n \tag{111}$$

by Ex. 2.56. Moreover,

$$V^*A^\dagger = V^*A^\dagger AA^\dagger = V^*A^*A^{\dagger *}A^\dagger = (AV)^*A^{\dagger *}A^\dagger = O, \tag{112}$$

by (1.2), (1.4), and assumption (b), while

$$AV^{*\dagger} = AV^{\dagger *} = A(V^{\dagger}VV^{\dagger})^* = A(V^{\dagger}V^{\dagger *}V^*)^*$$
$$= AVV^{\dagger}V^{\dagger *} = O, \tag{113}$$

by (1.2), Ex. 1.18(b), (1.3), and assumption (b). Finally, V^* is of full row rank by assumption (b) and, therefore,

$$V^*V^{*\dagger} = I_{n-r}, \tag{114}$$

by Lemma 1.2(b). By (111)–(114), (110) reduces to I_{m+n-r} and, therefore, (108) is nonsingular and (109) is its inverse.   □

The next two corollaries apply Theorem 7 for the solution of linear equations.

COROLLARY 5. *Let A, U, V be as in Theorem 7, let $\mathbf{b} \in \mathbb{C}^n$, and consider the linear equation*

$$A\mathbf{x} = \mathbf{b}. \tag{1}$$

Then the solution $\mathbf{x}, \mathbf{y}$ of

$$\begin{bmatrix} A & U \\ V^* & O \end{bmatrix} \begin{bmatrix} \mathbf{x} \\ \mathbf{y} \end{bmatrix} = \begin{bmatrix} \mathbf{b} \\ \mathbf{0} \end{bmatrix}. \tag{115}$$

satisfies

$\mathbf{x} = A^{\dagger}\mathbf{b}, \qquad$ *the minimal-norm least squares solution of (1),*

$U\mathbf{y} = P_{N(A^*)}\mathbf{b}, \quad$ *the residual of (1).*

COROLLARY 6 (Cramer's Rule, Ben-Israel [**71**], Verghese [**835**]). *Let $A, U, V, \mathbf{b}$ be as in Corollary 5. Then the minimal-norm least-squares solution $\mathbf{x} = [x_j]$ of (1) is given by*

$$x_j = \frac{\det \begin{bmatrix} A[j \leftarrow \mathbf{b}] & U \\ V^*[j \leftarrow \mathbf{0}] & O \end{bmatrix}}{\det \begin{bmatrix} A & U \\ V^* & O \end{bmatrix}}, \quad j \in \overline{1,n}. \tag{116}$$

PROOF. Apply the proof of Cramer's rule, Ex. 0.59, to (115).   □

Exercises

EX. 25. *A special case of Theorem 7.* Let $A \in \mathbb{C}_r^{m \times n}$ and let the matrices $U \in \mathbb{C}^{m \times (m-r)}$ and $V \in \mathbb{C}^{n \times (n-r)}$ satisfy

$$AV = O, \quad V^*V = I_{n-r}, \quad A^*U = O, \quad \text{and} \quad U^*U = I_{m-r}. \tag{117}$$

Then the matrix

$$\begin{bmatrix} A & U \\ V^* & O \end{bmatrix}, \tag{108}$$

is nonsingular and its inverse is

$$\begin{bmatrix} A^{\dagger} & V \\ U^* & O \end{bmatrix} \quad \text{(Reid [**683**]).} \tag{118}$$

Ex. 26. Let A, U, and V be as in Ex. 25, and let

$$\alpha = \min\{\|A\mathbf{x}\| : \mathbf{x} \in R(A^*), \|\mathbf{x}\| = 1\}, \tag{119}$$

$$\beta = \max\{\|A^\dagger\mathbf{y}\| : \mathbf{y} \in \mathbb{C}^n, \|\mathbf{y}\| = 1\}. \tag{120}$$

Then

$$\alpha\beta = 1 \quad (\text{Reid } [\mathbf{683}]). \tag{121}$$

PROOF. If $\mathbf{y} \in \mathbb{C}^n$, $\|\mathbf{y}\| = 1$, then

$$\mathbf{z} = A^\dagger\mathbf{y}$$

is the solution of

$$A\mathbf{z} = (I_m - UU^*)\mathbf{y}, \quad V^*\mathbf{z} = \mathbf{0}.$$

Therefore,

$$\alpha\|A^\dagger\mathbf{y}\| = \alpha\|\mathbf{z}\| \le \|A\mathbf{z}\|, \quad \text{by (119)},$$
$$= \|(I_m - UU^*)\mathbf{y}\|$$
$$\le \|\mathbf{y}\|, \quad \text{by Ex. 2.52 since } I_m - UU^* \text{ is an orthogonal projector,}$$
$$= 1.$$

Therefore $\alpha\beta \le 1$. On the other hand, let

$$\mathbf{x} \in R(A^*), \quad \|x\| = 1,$$

then

$$\mathbf{x} = A^\dagger A\mathbf{x},$$

so that

$$1 = \|\mathbf{x}\| = \|A^\dagger A\mathbf{x}\| \le \beta\|A\mathbf{x}\|,$$

proving that $\alpha\beta \ge 1$, and completing the proof.   □
See also Exs. 6.4 and 6.7.

Ex. 27. *A generalization of Theorem 7.* Let

$$A = \begin{bmatrix} B & C \\ D & O \end{bmatrix}$$

be nonsingular of order n, where B is $m \times p$, $0 < m < n$, and $0 < p < n$. Then A^{-1} is of the form

$$A^{-1} = \begin{bmatrix} E & F \\ G & O \end{bmatrix}, \tag{122}$$

where E is $p \times m$, if and only if B is of rank $m + p - n$, in which case

$$E = B^{(1,2)}_{N(D),R(C)}, \quad F = D^{(1,2)}_{N(B),\{0\}}, \quad G = C^{(1,2)}_{R(I_{n-p}),R(B)}. \tag{123}$$

PROOF. We first observe that since A is nonsingular, C is of full column rank $n-p$ for, otherwise, the columns of A would not be linearly independent. Similarly, D is of full row rank $n - m$. Since C is $m \times (n - p)$, it follows that $n - p \le m$ or, in other words,

$$m + p \ge n.$$

If: Since A is nonsingular, the $m \times n$ matrix $[B \quad C]$ is of full row rank m, and therefore of column rank m. Therefore, a basis for $\mathbb{C}^m$ can be chosen from among its columns. Moreover, this basis can be chosen so that it includes all $n-p$ columns of C, and the remaining $m + p - n$ basis elements are columns of B. Since B is of rank $m+p-n$, the latter columns span $R(B)$. Therefore $R(B) \cap R(C) = \{\mathbf{0}\}$ and, consequently, $R(B)$ and $R(C)$ are complementary subspaces. Similarly, we can show that $R(B^*)$ and $R(D^*)$ are complementary subspaces of $\mathbb{C}^p$ and, therefore, their orthogonal complements $N(B)$ and $N(D)$ are complementary spaces.

The results of the preceding paragraph guarantee the existence of all $\{1,2\}$-inverses in the right member of (123). If X now denotes RHS(122) with E, F, G given by (123), as easy computation shows that $AX = I_n$.

Only if: It was shown in the "if" part of the proof that rank B is at least $m + p - n$. If A^{-1} is of the form (122) we must have

$$BF = O. \tag{124}$$

Since A^{-1} is nonsingular, it follows from (122) that F is of full column rank $n - m$. Thus, (124) exhibits $n - m$ independent linear relations among the columns of B. Therefore the rank of B is at most $p - (n - m) = m + p - n$. This completes the proof.   □

Ex. 28.

(a) Let $C \in \mathbb{R}_r^{m \times r}$ and let the columns of $U_0 \in \mathbb{R}^{m \times m - r}$ be an o.n. basis of $N(C^T)$. Then

$$\det^2 [C \quad U_0] = \text{vol}^2 (C). \tag{125}$$

(b) Let $R \in \mathbb{R}_r^{r \times n}$ and let the columns of $V_0 \in \mathbb{R}^{n \times (n-r)}$ be an o.n. basis of $N(R)$. Then

$$\det^2 \begin{bmatrix} R \\ V_0^T \end{bmatrix} = \text{vol}^2 (R). \tag{126}$$

PROOF. (a) Follows from $U_0^T U_0 = I$,

$$\det^2 [C \quad U_0] = \det [C \quad U_0]^T \det [C \quad U_0],$$

and

$$[C \quad U_0]^T [C \quad U_0] = \begin{bmatrix} C^T C & O \\ O & U_0^T U_0 \end{bmatrix}.$$

(b) Similarly proved.   □

Ex. 29. Let $A \in \mathbb{R}_r^{m \times n}$, and let $U_0 \in \mathbb{R}^{m \times (m-r)}$ and $V_0 \in \mathbb{R}^{n \times (n-r)}$ be matrices whose columns are o.n. bases of $N(A^T)$ and $N(A)$, respectively. Then:

(a) The m-dimensional volume of $[A \quad U_0]$ equals the r-dimensional volume of A.

(b) The n-dimensional volume of $\begin{bmatrix} A \\ V_0^T \end{bmatrix}$ equals the r-dimensional volume of A.

PROOF. (a) Every $m \times m$ nonsingular submatrix of $[A \quad U_0]$ is of the form

$$[A_{*J} \quad U_0], \quad J \in \mathcal{J}(A),$$

and therefore, $\text{vol}_m^2 [A \quad U_0] = \text{vol}_r^2(A_{*J})$, by Ex. 28(a). The proof is completed by (0.106b).   □

Ex. 30. Let A, U_0, and V_0 be as in Ex. 29.

(a) Consider the bordered matrix

$$\mathbf{B}(A) = \begin{bmatrix} A & U_0 \\ V_0^T & O \end{bmatrix}. \tag{127}$$

Then

$$\text{vol}_r^2(A) = \det^2 \mathbf{B}(A), \tag{128}$$

$$= \det[AA^T + U_0 U_0^T], \tag{129}$$

$$= \det[A^T A + V_0 V_0^T]. \tag{130}$$

(b) If A is square, then

$$\text{vol}^2(A) = \det^2[A + U_0 V_0^T]. \tag{131}$$

PROOF. (a) Since the columns of

$$\begin{bmatrix} V_0 \\ O \end{bmatrix}$$

form an o.n. basis of $N([A \quad U_0]^T)$, we can use Ex. 28(b) and Ex. 29(a) to prove (128),

$$\det^2 \mathbf{B}(A) = \text{vol}_m^2[A \quad U_0] = \text{vol}_r^2(A),$$

and (129),

$$\text{vol}_m^2[A \quad U_0] = \det([A \quad U_0][A \quad U_0]^T) = \det[AA^T + U_0 U_0^T]. \qquad \square$$

Suggested Further Reading

SECTION 2. Ben-Israel [69].

SCHUR COMPLEMENTS. Ando [22], Burns, Carlson, Haynsworth, and Markham [140], Butler and Morley [144], Carlson [165], Carlson, Haynsworth, and Markham [168], Corach, Maestripieri, and Stojanoff [205], Crabtree and Haynsworth [210], Haynsworth [403], Li and Mathias [515], Liu and Wang [520], Neumann [608], Ostrowski [623], Wang, Zhang, and Zhang [845], Zhang [887]. Other references given on p. 39.

SECTION 3. Afriat [3].

SECTION 4. Baksalary and Styan [39], Hartwig [379], Harwood, Lovass–Nagy and Powers [398].

SECTION 5. Further references on bordered matrices are Bapat and Zheng [50], Blattner [113], Germain–Bonne [298], Hearon [406], Reid [683].

CRAMER'S RULE. Further extensions of Cramer's rule are Chen [181], Cimmino [198], Wang [847], [849], Werner [864].

CHAPTER 6

A Spectral Theory for Rectangular Matrices

1. Introduction

We study in this chapter some consequences of the singular value decomposition (SVD), encountered previously in §§ 0.2.14–0.2.15.

The SVD, repeated in Theorem 2, states that for any $A \in \mathbb{C}_r^{m \times n}$ with *singular values* $\sigma(A) = \{\sigma_1, \sigma_2, \ldots, \sigma_r\}$ there exist two unitary matrices $U \in U^{m \times m}$ (the set of $m \times m$ unitary matrices) and $V \in U^{n \times n}$ such that the $m \times n$ matrix

$$\Sigma = U^* A V = \begin{bmatrix} \sigma_1 & & & \vdots & \\ & \ddots & & \vdots & O \\ & & \sigma_r & \vdots & \\ \cdots & \cdots & \cdots & \cdots & \cdots \\ & O & & \vdots & O \end{bmatrix} \tag{1}$$

is diagonal. Thus any $m \times n$ complex matrix is unitarily equivalent to a diagonal matrix

$$A = U \Sigma V^*. \tag{2}$$

The corresponding statement for linear transformations is that for any linear transformations $A : \mathbb{C}^n \to \mathbb{C}^m$ with $\dim R(A) = r$, there exist two orthogonal bases $\mathcal{U} = \{\mathbf{u}_1, \mathbf{u}_2, \ldots, \mathbf{u}_m\}$ and $\mathcal{V} = \{\mathbf{v}_1, \mathbf{v}_2, \ldots, \mathbf{v}_n\}$ of $\mathbb{C}^m$ and $\mathbb{C}^n$, respectively, such that the corresponding matrix representation $A_{\{\mathcal{U}, \mathcal{V}\}}$ is diagonal,

$$A_{\{\mathcal{U}, \mathcal{V}\}} = \mathrm{diag}\,(\sigma_1, \ldots, \sigma_r, 0, \ldots, 0) \in \mathbb{R}^{n \times m}, \tag{0.42}$$

i.e.,

$$\begin{cases} A\mathbf{v}_j = \sigma_j \mathbf{u}_j, & j \in \overline{1, r}, \\ A\mathbf{v}_j = \mathbf{0}, & j \in \overline{r+1, n}. \end{cases} \tag{3}$$

The expression

$$A = U \Sigma V^*, \quad \Sigma = \begin{bmatrix} \sigma_1 & & & \vdots & \\ & \ddots & & \vdots & O \\ & & \sigma_r & \vdots & \\ \cdots & \cdots & \cdots & \cdots & \cdots \\ & O & & \vdots & O \end{bmatrix}, \quad U \in U^{m \times m}, V \in U^{n \times n}, \tag{4}$$

is called the *singular value decomposition* (abbreviated *SVD*) of A.

The SVD is of fundamental importance in the theory and computations of generalized inverses, especially the Moore–Penrose inverse. It is the basis for a generalized spectral theory for rectangular matrices, an extension of the classical spectral theory for normal matrices. This chapter covers the SVD and related topics.

Exercises

Ex. 1. A and A^* have the same singular values.

Ex. 2. Unitarily equivalent matrices have the same singular values.
PROOF. Let $A \in \mathbb{C}^{m \times n}$, and let $U \in U^{m \times m}$ and $V \in U^{n \times n}$ be any two unitary matrices. Then the matrix

$$(UAV)(UAV)^* = UAVV^*A^*U^* = UAA^*U^*$$

is similar to AA^*, and thus has the same eigenvalues. Therefore the matrices UAV and A have the same singular values.   □

Ex. 3. (Lanczos [496]). Let $A \in \mathbb{C}_r^{m \times n}$. Then $\begin{pmatrix} O & A \\ A^* & O \end{pmatrix}$ has $2r$ nonzero eigenvalues given by $\pm \sigma_j(A)$, $j \in \overline{1, r}$.

Ex. 4. *An extremal characterization of singular values.* Let $A \in \mathbb{C}_r^{m \times n}$. Then

$$\sigma_k(A) = \max\{\|A\mathbf{x}\| : \|\mathbf{x}\| = 1, \mathbf{x} \perp \mathbf{x}_1, \dots, \mathbf{x}_{k-1}\}, \quad k \in \overline{1, r}, \tag{5}$$

where

|| || denotes the Euclidean norm,

$\{\mathbf{x}_1, \mathbf{x}_2, \dots, \mathbf{x}_{k-1}\}$ is an o.n. set of vectors in $\mathbb{C}^n$, defined recursively by

$$\|A\mathbf{x}_1\| = \max\{\|A\mathbf{x}\| : \|\mathbf{x}\| = 1\},$$

$$\|A\mathbf{x}_j\| = \max\{\|A\mathbf{x}\| : \|\mathbf{x}\| = 1, \mathbf{x} \perp \mathbf{x}_1, \dots, \mathbf{x}_{j-1}\}, \quad j = 2, \dots, k-1,$$

and RHS(5) is the (attained) supremum of $\|A\mathbf{x}\|$ over all vectors $\mathbf{x} \in \mathbb{C}^n$ with norm one, which are perpendicular to $\mathbf{x}_1, \mathbf{x}_2, \dots, \mathbf{x}_{k-1}$.
PROOF. Follows from the corresponding extremal characterization of the eigenvalues of A^*A, see § 0.2.11(d),

$$\lambda_k(A^*A) = \max\{\langle \mathbf{x}, A^*A\mathbf{x} \rangle : \|\mathbf{x}\| = 1, \mathbf{x} \perp \mathbf{x}_1, \dots, \mathbf{x}_{k-1}\}$$
$$= \langle \mathbf{x}_k, A^*A\mathbf{x}_k \rangle, \quad k = 1, \dots, n,$$

since $\langle \mathbf{x}, A^*A\mathbf{x} \rangle = \langle A\mathbf{x}, A\mathbf{x} \rangle = \|A\mathbf{x}\|^2$. Here the vectors $\{\mathbf{x}_1, \dots, \mathbf{x}_n\}$ are an o.n. set of eigenvectors of A^*A,

$$A^*A\mathbf{x}_k = \lambda_k(A^*A)\mathbf{x}_k, \quad k \in \overline{1, n}. \qquad □$$

The singular values can be characterized equivalently as

$$\sigma_k(A) = \max\{\|A^*\mathbf{y}\| : \|\mathbf{y}\| = 1, \mathbf{y} \perp \mathbf{y}_1, \dots, \mathbf{y}_{k-1}\} = \|A^*\mathbf{y}_k\|,$$

where the vectors $\{\mathbf{y}_1, \dots, \mathbf{y}_r\}$ are an o.n. set of eigenvectors of AA^*, corresponding to its positive eigenvalues

$$AA^*\mathbf{y}_k = \lambda_k(AA^*)\mathbf{y}_k, \quad k \in \overline{1, r}.$$

We can interpret this extremal characterization as follows: let the columns of A be $\mathbf{a}_j$, $j = 1, \dots, n$. Then

$$\|A^*\mathbf{y}_k\|^2 = \sum_{j=1}^{n} |\langle \mathbf{a}_j, \mathbf{y} \rangle|^2.$$

Thus $\mathbf{y}_1$ is a normalized vector maximizing the sum of squares of moduli of its inner products with the columns of A, the maximum value being $\sigma_1^2(A)$, etc.

Ex. 5. The singular values of A are the stationary values of

$$f(\mathbf{x}) = \frac{\|A\mathbf{x}\|}{\|\mathbf{x}\|}. \tag{6}$$

PROOF. From $A^*A\mathbf{x} = \sigma^2\mathbf{x}$ we get $\sigma^2 = \frac{\|A\mathbf{x}\|^2}{\|\mathbf{x}\|^2} = f^2(\mathbf{x})$. Differentiating $f^2(\mathbf{x})$ and equating the gradient to zero, we again get $A^*A\mathbf{x} = f^2(\mathbf{x})\mathbf{x}$. □

Ex. 6. If $A \in \mathbb{C}_r^{n\times n}$ is normal and its eigenvalues are ordered by

$$|\lambda_1(A)| \geq |\lambda_2(A)| \geq \cdots \geq |\lambda_r(A)| > |\lambda_{r+1}(A)| = \cdots = |\lambda_n(A)| = 0,$$

then the singular values of A are $\sigma_j(A) = |\lambda_j(A)|$, $j \in \overline{1,r}$.
Hint. Use Ex. 4 and the spectral theorem for normal matrices, Theorem 2.15.

Ex. 7. Let $A \in \mathbb{C}_r^{m\times n}$ and let the singular values of $A^\dagger$ be ordered by

$$\sigma_1(A^\dagger) \geq \sigma_2(A^\dagger) \geq \cdots \geq \sigma_r(A^\dagger).$$

Then

$$\sigma_j(A^\dagger) = \frac{1}{\sigma_{r-j+1}(A)}, \quad j \in \overline{1,r}. \tag{7}$$

PROOF.

$$\begin{aligned}
\sigma_j^2(A^\dagger) &= \lambda_j(A^{\dagger *}A^\dagger), \quad \text{by definition (0.31b),} \\
&= \lambda_j((AA^*)^\dagger), \quad \text{since } A^{\dagger *}A^\dagger = A^{*\dagger}A^\dagger = (AA^*)^\dagger, \\
&= \frac{1}{\lambda_{r-j+1}(AA^*)} \\
&= \frac{1}{\sigma_{r-j+1}^2(A)}, \quad \text{by definition (0.31a).} \qquad \square
\end{aligned}$$

Ex. 8. Let $\|\ \|_F$ be the Frobenius matrix norm

$$\|A\|_F = (\text{trace } A^*A)^{1/2} = \left(\sum_{i=1}^m \sum_{j=1}^n |a_{ij}|^2\right)^{1/2} \tag{0.50}$$

defined on $\mathbb{C}^{m\times n}$, see, e.g., Ex. 0.34. Then, for any $A \in \mathbb{C}_r^{m\times n}$,

$$\|A\|_F^2 = \sum_{j=1}^r \sigma_j^2(A). \tag{8}$$

PROOF. Follows from trace $A^*A = \sum_{j=1}^r \lambda_j(A^*A)$. □
See also Ex. 62 below.

Ex. 9. Let $\|\ \|_2$ be the *spectral norm*, defined on $\mathbb{C}^{m\times n}$ by

$$\begin{aligned}
\|A\|_2 &= \max\{\sqrt{\lambda} : \lambda \text{ an eigenvalue of } A^*A\} \\
&= \sigma_1(A), \tag{0.14.2}
\end{aligned}$$

see, e.g., Ex. 0.38. Then, for any $A \in \mathbb{C}_r^{m\times n}$, $r \geq 1$,

$$\|A\|_2 \|A^\dagger\|_2 = \frac{\sigma_1(A)}{\sigma_r(A)}. \tag{9}$$

PROOF. Follows from Ex. 7 and definition (0.14.2). □

Ex. 10. *A condition number.* Let A be an $n \times n$ nonsingular matrix, and consider the equation

$$Ax = b \tag{10}$$

for $b \in \mathbb{C}^n$. The sensitivity of the solution of (10) to changes in the right-hand side b is indicated by the *condition number* of A, defined for any multiplicative matrix norm $\| \ \|$ by

$$\text{cond}(A) = \|A\| \|A^{-1}\|. \tag{11}$$

Indeed, changing b to $(b + \delta b)$ results in a change of the solution $x = A^{-1}b$ to $x + \delta x$, with

$$\delta x = A^{-1} \delta b. \tag{12}$$

For any consistent pair of vector and matrix norms (see Exs. 0.35–0.37), it follows from (10) that

$$\|b\| \leq \|A\| \|x\|. \tag{13}$$

Similarly, from (12),

$$\|\delta x\| \leq \|A^{-1}\| \|\delta b\|. \tag{14}$$

From (13) and (14) we get the following bound:

$$\frac{\|\delta x\|}{\|x\|} \leq \|A\| \|A^{-1}\| \frac{\|\delta b\|}{\|b\|} = \text{cond}(A) \frac{\|\delta b\|}{\|b\|} \tag{15}$$

relating the change of the solution to the change in data and the condition number (11).

Ex. 11. The *spectral condition number* corresponding to the spectral norm (0.14.2) is, by (9)

$$\text{cond}(A) = \frac{\sigma_1(A)}{\sigma_n(A)}. \tag{16}$$

Prove that for this condition number

$$\text{cond}(A^*A) = (\text{cond}(A))^2,$$

showing that A^*A is worse conditioned than A, if $\text{cond}(A) > 1$ (Taussky [**798**]).

Ex. 12. *Weyl's inequalities.* Let $A \in \mathbb{C}_r^{n \times n}$ have eigenvalues $\lambda_1, \dots, \lambda_n$ ordered by

$$|\lambda_1| \geq |\lambda_2| \geq \cdots \geq |\lambda_n|$$

and singular values

$$\sigma_1 \geq \sigma_2 \geq \cdots \geq \sigma_r.$$

Then

$$\sum_{j=1}^k |\lambda_j| \leq \sum_{j=1}^k \sigma_j, \tag{17}$$

$$\prod_{j=1}^k |\lambda_j| \leq \prod_{j=1}^k \sigma_j, \tag{18}$$

for $k = 1, \dots, r$ (Weyl [**869**], Marcus and Minc [**534**, pp. 115–116]).

2. The Singular Value Decomposition

There are several ways to approach the SVD, see, e.g., Stewart [783].Our approach follows that of Eckart and Young [248]. First we require the following theorem:

THEOREM 1. *Let $O \neq A \in \mathbb{C}_r^{m \times n}$, let $\sigma(A)$, the singular values of A, be*

$$\sigma_1 \geq \sigma_2 \geq \cdots \geq \sigma_r > 0, \tag{0.32}$$

let $\{\mathbf{u}_1, \mathbf{u}_2, \ldots, \mathbf{u}_r\}$ be an o.n. set of eigenvectors of AA^* corresponding to its nonzero eigenvalues:

$$AA^*\mathbf{u}_i = \sigma_i^2 \mathbf{u}_i, \quad i \in \overline{1, r}, \tag{19a}$$

$$\langle \mathbf{u}_i, \mathbf{u}_j \rangle = \delta_{ij}, \quad i, j \in \overline{1, r}, \tag{19b}$$

and let $\{\mathbf{v}_1, \mathbf{v}_2, \ldots, \mathbf{v}_r\}$ be defined by

$$\mathbf{v}_i = \frac{1}{\sigma_i} A^* \mathbf{u}_i, \quad i \in \overline{1, r}. \tag{20}$$

Then $\{\mathbf{v}_1, \mathbf{v}_2, \ldots, \mathbf{v}_r\}$ is an o.n. set of eigenvectors of A^*A corresponding to its nonzero eigenvalues

$$A^*A\mathbf{v}_i = \sigma_i^2 \mathbf{v}_i, \quad i \in \overline{1, r}, \tag{21a}$$

$$\langle \mathbf{v}_i, \mathbf{v}_j \rangle = \delta_{ij}, \quad i, j \in \overline{1, r}. \tag{21b}$$

Furthermore,

$$\mathbf{u}_i = \frac{1}{\sigma_i} A \mathbf{v}_i, \quad i \in \overline{1, r}. \tag{22}$$

Dually, let the vectors $\{\mathbf{v}_1, \mathbf{v}_2, \ldots, \mathbf{v}_r\}$ satisfy (21) and let the vectors $\{\mathbf{u}_1, \mathbf{u}_2, \ldots, \mathbf{u}_r\}$ be defined by (22). Then $\{\mathbf{u}_1, \mathbf{u}_2, \ldots, \mathbf{u}_r\}$ satisfy (19) and (20).

PROOF. Let $\{\mathbf{v}_i : i \in \overline{1, r}\}$ be given by (20). Then,

$$\begin{aligned} A^*A\mathbf{v}_i &= \frac{1}{\sigma_i} A^* AA^* \mathbf{u}_i \\ &= \sigma_i A^* \mathbf{u}_i, \quad \text{by (19a)}, \\ &= \sigma_i^2 \mathbf{v}_i, \quad \text{by (20)}, \end{aligned}$$

and

$$\begin{aligned} \langle \mathbf{v}_i, \mathbf{v}_i \rangle &= \frac{1}{\sigma_i \sigma_j} \langle A^* \mathbf{u}_i, A^* \mathbf{u}_j \rangle \\ &= \frac{1}{\sigma_i \sigma_j} \langle AA^* \mathbf{u}_i, \mathbf{u}_j \rangle \\ &= \frac{\sigma_i}{\sigma_j} \langle \mathbf{u}_i, \mathbf{u}_j \rangle, \quad \text{by (19a)}, \\ &= \delta_{ij}, \quad \text{by (19b)}. \end{aligned}$$

Equations (22) follow from (20) and (19a). The dual statement follows by interchanging A and A^*. $\square$

An easy consequence of Theorem 1 is the following:

THEOREM 2 (The Singular Value Decomposition). *Let $O \neq A \in \mathbb{C}_r^{m \times n}$ and let*

$$\sigma_1 \geq \sigma_2 \geq \cdots \geq \sigma_r > 0 \qquad (0.32)$$

be the singular values of A.

Then there exist unitary matrices $U \in U^{m \times m}$ and $V \in U^{n \times n}$ such that the matrix

$$\Sigma = U^* A V = \begin{bmatrix} \sigma_1 & & & \vdots & \\ & \ddots & & \vdots & O \\ & & \sigma_r & \vdots & \\ \cdots & \cdots & \cdots & \cdots & \cdots \\ & O & & \vdots & O \end{bmatrix} \qquad (1)$$

is diagonal.

PROOF. For the given $A \in \mathbb{C}_r^{m \times n}$ we construct two such matrices U and V as follows.

Let the vectors $\{\mathbf{u}_1, \ldots, \mathbf{u}_r\}$ in $\mathbb{C}^m$ satisfy (19a) and (19b), and thus form an o.n. basis of $R(AA^*) = R(A)$; see, e.g., Corollary 1.2. Let $\{\mathbf{u}_{r+1}, \ldots, \mathbf{u}_m\}$ be an o.n. basis of $R(A)^\perp = N(A^*)$. Then the set $\{\mathbf{u}_1, \ldots, \mathbf{u}_r, \mathbf{u}_{r+1}, \ldots, \mathbf{u}_m\}$ is an o.n. basis of $\mathbb{C}^m$ satisfying (19a) and

$$A^* \mathbf{u}_i = \mathbf{0}, \quad i \in \overline{r+1, m}. \qquad (23)$$

The matrix U defined by

$$U = [\mathbf{u}_1 \ \ldots \ \mathbf{u}_r \ \mathbf{u}_{r+1} \ \ldots \ \mathbf{u}_m] \qquad (24)$$

is thus an $m \times m$ unitary matrix.

Let now the vectors $\{\mathbf{v}_1, \ldots, \mathbf{v}_r\}$ in $\mathbb{C}^n$ be defined by (20). Then these vectors satisfy (21a) and (21b), and thus form an o.n. basis of $R(A^*A) = R(A^*)$. Let $\{\mathbf{v}_{r+1}, \ldots, \mathbf{v}_n\}$ be an o.n. basis of $R(A^*)^\perp = N(A)$. Then the set $\{\mathbf{v}_1, \ldots, \mathbf{v}_r, \mathbf{v}_{r+1}, \ldots, \mathbf{v}_n\}$ is an o.n. basis of $\mathbb{C}^n$ satisfying (21a) and

$$A\mathbf{v}_i = \mathbf{0}, \quad i \in \overline{r+1, n}. \qquad (25)$$

The matrix V defined by

$$V = [\mathbf{v}_1 \ \ldots \ \mathbf{v}_r \ \mathbf{v}_{r+1} \ \ldots \ \mathbf{v}_n] \qquad (26)$$

is thus an $n \times n$ unitary matrix.

With U and V as given above, the matrix

$$\Sigma = U^* A V = (\Sigma[i, j]), \quad i \in \overline{1, m}, \ j \in \overline{1, n},$$

satisfies

$$\Sigma[i, j] = \mathbf{u}_i^* A \mathbf{v}_j = 0 \quad \text{if } i > r \text{ or } j > r, \quad \text{by (23) and (25)},$$

and, for $i, j = 1, \ldots, r$,

$$\begin{aligned} \Sigma[i, j] &= \mathbf{u}_i^* A \mathbf{v}_j \\ &= \frac{1}{\sigma_j} \mathbf{u}_i^* A A^* \mathbf{u}_j, \quad \text{by (20)}, \\ &= \sigma_j \mathbf{u}_i^* \mathbf{u}_j, \quad \text{by (19a)}, \\ &= \sigma_j \delta_{ij}, \quad \text{by (19b)}, \end{aligned}$$

completing the proof. □

A corresponding decomposition of $A^\dagger$ is given in

COROLLARY 1 (Penrose [635]). *Let A, Σ, U, and V be as in Theorem 2. Then*

$$A^\dagger = V\Sigma^\dagger U^* \qquad (27)$$

where

$$\Sigma^\dagger = \mathrm{diag}\left(\frac{1}{\sigma_1}, \dots, \frac{1}{\sigma_r}, 0, \dots, 0\right) \in \mathbb{R}^{n\times m}. \qquad (28)$$

PROOF. Equation (27) follows from Ex. 1.25. The form (28) for $\Sigma^\dagger$ is obvious. □

Let $A \in \mathbb{C}_r^{m\times n}$ and let the matrices Σ, U, and V be as in Theorem 2. We denote by $U_{(k)}$, $V_{(k)}$, and $\Sigma_{(k)}$ the submatrices

$$U_{(k)} = [\mathbf{u}_1 \ \dots \ \mathbf{u}_k] \in \mathbb{C}^{m\times k}, \quad V_{(k)} = [\mathbf{v}_1 \ \dots \ \mathbf{v}_k] \in \mathbb{C}^{n\times k},$$

$$\Sigma_{(k)} = \begin{bmatrix} \sigma_1 & & \\ & \ddots & \\ & & \sigma_k \end{bmatrix} \in \mathbb{C}^{k\times k}. \qquad (29)$$

Using this notation, the SVD's of A and $A^\dagger$ can be written as

$$A = \sum_{i=1}^r \sigma_i \mathbf{u}_i \mathbf{v}_i^* = U_{(r)}\Sigma_{(r)}V_{(r)}^*, \qquad (30a)$$

$$A^\dagger = \sum_{i=1}^r \frac{1}{\sigma_i} \mathbf{v}_i \mathbf{u}_i^* = V_{(r)}\Sigma_{(r)}^{-1}U_{(r)}^*. \qquad (30b)$$

For $1 \leq k \leq r$ we write, analogously,

$$A_{(k)} = \sum_{i=1}^k \sigma_i \mathbf{u}_i \mathbf{v}_i^* = U_{(k)}\Sigma_{(k)}V_{(k)}^* \in \mathbb{C}_k^{m\times n}. \qquad (30c)$$

In particular, $A = A_{(r)}$.

Exercises

EX. 13. Recall the limit formula for the Moore–Penrose inverse

$$\lim_{\lambda\to 0}(A^*A + \lambda I)^{-1}A^* = A^\dagger. \qquad (3.43)$$

There are cases where stopping at a positive λ is better than going to the limit $\lambda = 0$. Let $A \in \mathbb{C}_r^{m\times n}$ and let $\{\mathbf{u}_1, \mathbf{u}_2, \dots, \mathbf{u}_r\}$ and $\{\mathbf{v}_1, \mathbf{v}_2, \dots, \mathbf{v}_r\}$ be o.n. bases of $R(A^*)$ and $R(A)$, respectively, as in Theorem 1. Consider the equation

$$A\mathbf{x} = \mathbf{b}, \qquad (10)$$

where $\mathbf{b} \in R(A)$ is expressed as

$$\mathbf{b} = \sum_{i=1}^r \beta_i \mathbf{v}_i.$$

The least-norm solution $\mathbf{x} = A^\dagger \mathbf{b}$ is then

$$\mathbf{x} = \sum_{i=1}^r \frac{\beta_i}{\sigma_i} \mathbf{u}_i. \qquad (31)$$

If A is ill-conditioned, this solution is sensitive to errors ε in the smaller singular values, as seen from

$$\frac{1}{\sigma + \varepsilon} \approx \frac{1}{\sigma} - \frac{1}{\sigma^2}\varepsilon + \frac{1}{\sigma^3}\varepsilon^2 + \cdots .$$

Instead of (31), consider the approximate solution

$$\mathbf{x}(\lambda) = (A^*A + \lambda I)^{-1}A^*\mathbf{b}$$

$$= \sum_{i=1}^{r} \frac{\sigma_i \beta_i}{\sigma_i^2 + \lambda}\mathbf{u}_i \tag{32}$$

where λ is positive. It is less sensitive to errors in the singular values, as shown by

$$\frac{(\sigma + \varepsilon)}{(\sigma + \varepsilon)^2 + \lambda} \approx \frac{\sigma}{\sigma^2 + \lambda} - \frac{\sigma^2 - \lambda}{(\sigma^2 + \lambda)^2}\varepsilon + \frac{\sigma(\sigma^2 - 3\lambda)}{(\sigma^2 + \lambda)^3}\varepsilon^2 + \cdots ,$$

where the choice $\lambda = \sigma^2$ gives

$$\frac{(\sigma + \varepsilon)}{(\sigma + \varepsilon)^2 + \lambda} \approx \frac{1}{2\sigma} - \frac{1}{4\sigma^3}\varepsilon^2 + \cdots .$$

See also Section 8.4.

Ex. 14. Let the SVD of $A \in \mathbb{C}_r^{m \times n}$ be written as

$$A = U \begin{bmatrix} \Sigma_{(r)} & O \\ O & O \end{bmatrix} V^*$$

where $\Sigma_{(r)}$ is as in (29). The general $\{1\}$-inverse of A is

$$G = V \begin{bmatrix} \Sigma_{(r)}^{-1} & X \\ Y & Z \end{bmatrix} U^* \tag{33}$$

where X, Y, Z are arbitrary submatrices of appropriate sizes. In particular,

$Z = Y\Sigma_{(r)}X$ gives the general $\{1,2\}$-inverse;
$X = O$ gives the general $\{1,3\}$-inverse; and
$Y = O$ gives the general $\{1,4\}$-inverse;

finally, the Moore–Penrose inverse is (33) with $X = O$, $Y = O$, and $Z = O$.

Ex. 15. *Simultaneous diagonalization.* Let $A_1, A_2 \in \mathbb{C}^{m \times n}$. Then the following are equivalent:

(a) There exist two unitary matrices U, V such that both $\Sigma_1 = U^*A_1V$, $\Sigma_2 = U^*A_2V$ are diagonal real matrices (in which case one of them, say D_1, can be assumed to be nonnegative).

(b) $A_1A_2^*$ and $A_2^*A_1$ are both Hermitian (Eckart and Young [**248**]).

Ex. 16. Let $A_1, A_2 \in \mathbb{C}^{n \times n}$ be Hermitian matrices. Then the following are equivalent:

(a) There is a unitary matrix U such that both $\Sigma_1 = U^*A_1U$, $\Sigma_2 = U^*A_2U$ are diagonal real matrices.

(b) A_1A_2 and A_2A_1 are both Hermitian.

(c) $A_1A_2 = A_2A_1$.

Ex. 17. (Williamson [**876**]). Let $A_1, A_2 \in \mathbb{C}^{m \times n}$. Then the following are equivalent:

(a) There exist two unitary matrices U, V such that both $\Sigma_1 = U^*A_1V$, $\Sigma_2 = U^*A_2V$ are diagonal matrices.

(b) There is a polynomial f such that $A_1A_2^* = f(A_2A_1^*)$, $A_2^*A_1 = f(A_1^*A_2)$.

Ex. 18. *UDV*-decomposition.* In some cases it is convenient to rewrite the SVD (4) as

$$A = UDV^*, \quad D = \begin{bmatrix} d_1 & & & \vdots & \\ & \ddots & & \vdots & O \\ & & d_r & \vdots & \\ \cdots & \cdots & \cdots & \cdots & \cdots \\ & O & & \vdots & O \end{bmatrix}, \quad U \in U^{m \times m}, V \in U^{n \times n}, \quad (34)$$

where the diagonal elements $d(A) = \{d_i : i \in \overline{1,r}\}$ are complex numbers satisfying

$$|d_i| = \sigma_i, \quad i \in \overline{1,r}. \tag{35}$$

Indeed, (34) is obtainable by inserting $1 = e^{i\theta}e^{-i\theta}$, for some $\theta \in \mathbb{R}$, in (4). For example, let $\{\theta_k : k \in \overline{1,r}\}$ be real numbers, let $\Theta = \text{diag}(\theta_k)$, and denote $e^{\pm i\Theta} = \text{diag}(\pm e^{i\theta_k})$. Then:

$$A = U \begin{bmatrix} \Sigma & O \\ O & O \end{bmatrix} V^*, \quad (\Sigma = \text{diag}(\sigma_i), \quad i \in \overline{1,r}),$$

$$= U \begin{bmatrix} \Sigma & O \\ O & O \end{bmatrix} \begin{bmatrix} e^{i\Theta} & O \\ O & O \end{bmatrix} \begin{bmatrix} e^{-i\Theta} & O \\ O & O \end{bmatrix} V^*,$$

$$= U \begin{bmatrix} \Sigma & O \\ O & O \end{bmatrix} \begin{bmatrix} e^{i\Theta} & O \\ O & O \end{bmatrix} \begin{bmatrix} e^{-i\Theta} & O \\ O & I_r \end{bmatrix} V^*,$$

$$= UDW^*,$$

where,

$$D = \begin{bmatrix} \Sigma & O \\ O & O \end{bmatrix} \begin{bmatrix} e^{i\Theta} & O \\ O & O \end{bmatrix} = \begin{bmatrix} \text{diag}(d_k) & O \\ O & O \end{bmatrix}, \quad d_k = \sigma_k e^{i\theta_k}, k \in \overline{1,r}, \quad (36)$$

$$W^* = \begin{bmatrix} e^{-i\Theta} & O \\ O & I_r \end{bmatrix} V^*. \tag{37}$$

The matrix W in (37) is unitary, and will be denoted by V to give (4) and (34) a similar look. We call (34) a *UDV*-decomposition* of A.

Theorems 1–2 can be restated for the *UDV**-decomposition, for example, by replacing (20) and (22) by

$$\mathbf{v}_i = \frac{1}{d_i} A^* \mathbf{u}_i, \quad i \in \overline{1,r}, \tag{20*}$$

and

$$\mathbf{u}_i = \frac{1}{d_i} A \mathbf{v}_i, \quad i \in \overline{1,r}, \text{ respectively.} \tag{22*}$$

Ex. 19. *Normal matrices.* If $O \neq A \in \mathbb{C}_r^{n \times n}$ is normal and its nonzero eigenvalues are ordered by

$$|\lambda_1| \geq |\lambda_2| \geq \cdots \geq |\lambda_r| > 0, \tag{38}$$

then the scalars $d(A) = \{d_1, \ldots, d_r\}$ in (35) can be chosen as the corresponding eigenvalues

$$d_i = \lambda_i, \quad i \in \overline{1,r}. \tag{39}$$

This choice reduces both (20*) and (22*) to

$$\mathbf{u}_i = \mathbf{v}_i, \quad i \in \overline{1,r}. \tag{40}$$

PROOF. The first claim follows from Ex. 6.

Using Exs. 0.22 it can be shown that all four matrices A, A^*, AA^*, and A^*A have common eigenvectors. Therefore, the vectors $\{\mathbf{u}_1, \ldots, \mathbf{u}_r\}$ of (19a) and (19b) are also eigenvectors of A^* and (20*) reduces to (40). □

EX. 20. *Normal matrices.* If $O \neq A \in \mathbb{C}_r^{n \times n}$ is normal, and the scalars $d(A)$ are chosen by (39), then the UDV^*-decomposition (34) of A reduces to the statement that A is unitarily similar to a diagonal matrix

$$A = UDU^*, \quad \text{see Ex. 0.22(a)}.$$

EX. 21. *An alternative definition of the matrix volume.* The volume of a matrix $A \in \mathbb{C}_r^{m \times n}$ with singular values $\{\sigma_i : i \in \overline{1,r}\}$ is the product of these singular values

$$\text{vol } A = \prod_{i=1}^{r} \sigma_i. \tag{41}$$

PROOF. The SVD (30a) is a full-rank factorization of A, $A = CR$, with $C = U_{(r)}\Sigma_{(r)}$, $R = V_{(r)}^*$ (or, alternatively, $A = C_1 R_1$, $C_1 = U_{(r)}$, $R_1 = \Sigma_{(r)} V_{(r)}^*$). Therefore

$$\begin{aligned}
\text{vol } A &= \text{vol } C \text{ vol } R, \quad \text{by Ex. 0.66,} \\
&= \text{vol } U_{(r)} \text{ vol } \Sigma_{(r)} \text{ vol } V_{(r)}^*, \\
&= \text{vol } \Sigma_{(r)}, \quad \text{(by Ex. 0.65 since } U_{(r)}^* U_{(r)} = V_{(r)}^* V_{(r)} = I), \\
&= |\det \Sigma_{(r)}|, \quad \text{since } \Sigma_{(r)} \text{ is nonsingular.}
\end{aligned}$$

□

A geometric interpretation: Equations (3) show that the r-dimensional unit cube $\square(\mathbf{v}_1, \ldots, \mathbf{v}_r)$ is mapped under A into the cube of sides $\sigma_i \mathbf{u}_i$ $(i = 1, \ldots, r)$, whose (r-dimensional) volume is

$$\prod_{i=1}^{r} \sigma_i \,,$$

the volume of the matrix A. Since the singular values are unitarily invariant, it follows that <u>all</u> r-dimensional unit cubes in $R(A^T)$ are mapped under A into parallelepipeds of volume vol A.

See Ex. 8.32 for the special case $m = 3$, $n = r = 2$.

EX. 22. Let $C_k(A)$ be the k-compound of $A \in \mathbb{R}^{m \times n}$. It is an $\binom{m}{k} \times \binom{n}{k}$ matrix of rank $\binom{r}{k}$, and its singular values are all products $\sigma_{i_1} \sigma_{i_2} \cdots \sigma_{i_k}$ of singular values of A. It follows that $C_r(A)$ is of rank 1 and its nonzero singular value equals vol A.

EX. 23. Let o.n. bases of $R(A)$ and $R(A^T)$ be given by the $\{\mathbf{u}_i\}$ and $\{\mathbf{v}_i\}$ of the SVD (3). Then the Plücker coordinates of $R(A)$ are given by

$$\mathbf{u}^\wedge = \mathbf{u}_1 \wedge \cdots \wedge \mathbf{u}_r$$

and those of $R(A^T)$ by

$$\mathbf{v}^\wedge = \mathbf{v}_1 \wedge \cdots \wedge \mathbf{v}_r.$$

Moreover,

$$C_r(A)\, \mathbf{v}^\wedge = \text{vol } A\, \mathbf{u}^\wedge, \quad C_r(A^\dagger)\, \mathbf{u}^\wedge = \frac{1}{\text{vol } A}\, \mathbf{v}^\wedge, \tag{42}$$

correspond to the facts that A is invertible as a mapping: $R(A^T) \rightarrow R(A)$ and $A^\dagger$ is invertible as a mapping: $R(A) \rightarrow R(A^T)$, see also Marcus [**533**]. In particular,

$$C_r(A^\dagger) = (C_r(A))^\dagger, \quad \text{and} \quad \text{vol}(A^\dagger) = \frac{1}{\text{vol } A}.$$

We show next that each singular value of the Moore–Penrose inverse $A^\dagger$ is dominated by a corresponding singular value of any $\{1\}$-inverse of A.

EX. 24. *A minimum property of the Moore–Penrose inverse* (Bapat and Ben-Israel [**47**]). Let G be a $\{1\}$-inverse of $A \in \mathbb{C}_r^{m \times n}$ with singular values

$$\sigma_1(G) \geq \sigma_2(G) \geq \cdots \geq \sigma_s(G), \tag{43}$$

where $s = \text{rank } G \ (\geq \text{rank } A)$. Then

$$\sigma_i(G) \geq \sigma_i(A^\dagger), \quad i = 1, \ldots, r. \tag{44}$$

PROOF. Dropping U, V we write

$$GG^* = \begin{bmatrix} \Sigma^{-1} & X \\ Y & Z \end{bmatrix} \begin{bmatrix} \Sigma^{-1} & Y^* \\ X^* & Z^* \end{bmatrix}$$

$$= \begin{bmatrix} \Sigma^{-2} + XX^* & ? \\ ? & ? \end{bmatrix},$$

where ? denotes a submatrix not needed in this proof. Then, for $i = 1, \ldots, r$,

$$\sigma_i^2(G) := \lambda_i(GG^*)$$

$$\geq \lambda_i(\Sigma^{-2} + XX^*), \quad \text{e.g., [**529**, Chapter 11, Theorem 11]},$$

$$\geq \lambda_i(\Sigma^{-2}), \quad \text{e.g., [**529**, Chapter 11, Theorem 9]},$$

$$= \sigma_i^2(A^\dagger). \qquad \square$$

From (44) and definition (0.114b) we conclude that for each $k = 1, \ldots, r$ the Moore–Penrose inverse $A^\dagger$ is of minimal k-volume among all $\{1\}$-inverses G of A,

$$\text{vol}_k G \geq \text{vol}_k A^\dagger, \quad k = 1, \ldots, r. \tag{45}$$

Moreover, this property is a characterization of $A^\dagger$, as shown next.

EX. 25. Let $A \in \mathbb{R}_r^{m \times n}$ and let k be any integer in $\overline{1, r}$. Then the Moore–Penrose inverse $A^\dagger$ is the unique $\{1\}$-inverse of A with minimal k-volume.

PROOF. We prove this result directly, by solving the k-volume minimization problem, showing it to have the Moore–Penrose inverse as the unique solution.

The easiest case is $k = 1$. The claim is that $A^\dagger$ is the unique solution $X = [x_{ij}]$ of the minimization problem

$$\text{minimize } \tfrac{1}{2} \text{vol}_1^2 X \quad \text{subject to} \quad AXA = A, \tag{P.1}$$

where, by (0.114b),

$$\text{vol}_1^2 [x_{ij}] = \sum_{ij} |x_{ij}|^2 = \text{trace } X^T X.$$

We use the Lagrangian function

$$L(X, \Lambda) := \tfrac{1}{2} \text{trace } X^T X - \text{trace } \Lambda^T (AXA - A) \tag{46}$$

where $\Lambda = [\lambda_{ij}]$ is a matrix Lagrange multiplier. The Lagrangian can be written, using the "vec" notation, as

$$L(X, \Lambda) = \tfrac{1}{2} \langle \text{vec } X, \text{vec } X \rangle - \langle \text{vec } \Lambda, \left(A^T \otimes A \right) \text{vec } X \rangle$$

and its derivative with respect to $\text{vec } X$ is

$$(\nabla_X L(X, \Lambda))^T = (\text{vec } X)^T - (\text{vec } \Lambda)^T (A^T \otimes A),$$

see, e.g., [529]. The necessary condition for optimality is that the derivative vanishes,

$$(\operatorname{vec} X)^T - (\operatorname{vec} \Lambda)^T (A^T \otimes A) = \operatorname{vec} O,$$

or, equivalently,

$$X = A^T \Lambda A^T. \tag{47}$$

This condition is also sufficient, since (P.1) is a problem of minimizing a convex function subject to linear constraints. Indeed, the Moore–Penrose inverse $A^\dagger$ is the unique $\{1\}$-inverse of A satisfying (47) for some Λ (see, e.g., [67]). Therefore $A^\dagger$ is the unique solution of (P.1).

For $1 < k \le r$ the problem analogous to (P.1) is

$$\text{minimize } \tfrac{1}{2} \operatorname{vol}_k^2 X, \quad \text{subject to} \quad AXA = A. \tag{P.k}$$

We note that $AXA = A$ implies

$$C_k(A)C_k(X)C_k(A) = C_k(A). \tag{48}$$

Taking (48) as the constraint in (P.k), we get the Lagrangian

$$L(X,\Lambda) := \tfrac{1}{2} \sum_{\substack{I \in Q_{k,n} \\ J \in Q_{k,m}}} |\det X_{IJ}|^2$$

$$- \operatorname{trace} C_k(\Lambda)^T (C_k(A)C_k(X)C_k(A) - C_k(A)).$$

It follows, in analogy with the case $k = 1$, that a necessary and sufficient condition for optimality of X is

$$C_k(X) = C_k(A^T)C_k(\Lambda)C_k(A^T). \tag{49}$$

Moreover, $A^\dagger$ is the unique $\{1\}$-inverse satisfying (49), and is therefore the unique solution of (P.k). $\qquad\square$

Note: The rank s of a $\{1\}$-inverse G may be greater than r, in which case the volumes

$$\operatorname{vol}_{r+1}(G), \operatorname{vol}_{r+2}(G), \dots, \operatorname{vol}_s(G)$$

are positive. However, the corresponding volumes of $A^\dagger$ are zero, by (0.114c), so the inequalities (45) still hold.

3. The Schmidt Approximation Theorem

The data $A_{(r)} = \{\Sigma_{(r)}, U_{(r)}, V_{(r)}\}$ is of size $r + mr + nr = r(m + n + 1)$. In applications where storage space is restricted, or speed of transmission is important, it would seem desirable to reduce the data size. One such idea is to approximate the original matrix $A = A_{(r)}$ by lower rank matrices $A_{(k)}$, provided the error of approximation is acceptable. This error, using the Frobenius norm (0.50), is

$$\|A - A_{(k)}\|_F = \left\| U\Big(\Sigma - \begin{bmatrix} \Sigma_{(k)} & O \\ O & O \end{bmatrix}\Big)V^* \right\|_F = \left\| \Sigma - \begin{bmatrix} \Sigma_{(k)} & O \\ O & O \end{bmatrix} \right\|_F$$

$$= \| \operatorname{diag}(0, \dots, 0, \sigma_{k+1}, \dots, \sigma_r, 0, \dots, 0) \|_F$$

$$= \Big(\sum_{i=k+1}^{r} \sigma_i^2 \Big)^{1/2}. \tag{50}$$

The question if there is a better approximating matrix of rank k requires the following

DEFINITION 1. Given a matrix $A \in \mathbb{C}_r^{m \times n}$ and an integer k, $1 \leq k \leq r$, a *best rank-k approximation* of A is a matrix $X \in \mathbb{C}_k^{m \times n}$ satisfying

$$\|A - X\|_F = \inf_{Z \in \mathbb{C}_k^{m \times n}} \|A - Z\|_F. \tag{51}$$

The following theorem confirms that $A_{(k)}$ is the best rank-k approximation of A. The theorem has a long history (see Stewart [**783**]) and is often credited to Eckart and Young [**247**] and Mirsky [**558**].

THEOREM 3 (The Schmidt approximation theorem [**727**]). *Let $A \in \mathbb{C}_r^{m \times n}$, let $1 \leq k \leq r$, and let $\Sigma_{(k)}, U_{(k)}, V_{(k)}$ be as above. Then a best rank-k approximation of A is*

$$A_{(k)} = U_{(k)} \Sigma_{(k)} V_{(k)}^*, \tag{30c}$$

which is unique if, and only if, the kth and the $(k+1)^{st}$ singular values of A are distinct:

$$\sigma_k \neq \sigma_{k+1}. \tag{52}$$

The approximation error of $A_{(k)}$ is

$$\|A - A_{(k)}\|_F = \left(\sum_{i=k+1}^{r} \sigma_i^2 \right)^{1/2}. \tag{50}$$

PROOF. For any $X \in \mathbb{C}^{m \times n}$,

$$\|A - X\|_F^2 = \|U^*(A - X)V\|_F^2 = \|\Sigma - Y\|_F^2 = f(Y), \quad \text{say},$$

where

$$Y = U^* X V = [y_{ij}].$$

Let L be any subspace with $\dim L \leq k$ and let P_L denote the orthogonal projector on L. Then the matrix $Y = P_L \Sigma$ minimizes $f(Y)$ among all matrices Y with $R(Y) \subset L$, and the corresponding minimum value is

$$\|\Sigma - P_L \Sigma\|_F^2 = \|Q\Sigma\|_F^2 = \text{trace } \Sigma Q^* Q \Sigma$$

$$= \text{trace } \Sigma Q \Sigma = \sum_{i=1}^{m} \sigma_i^2 q_{ii},$$

where $Q = I - P_L = [q_{ij}]$ is the orthogonal projector on $L^\perp$. Now

$$\inf_{X \in \mathbb{C}_k^{m \times n}} \|A - X\|_F^2 = \inf_{Y \in \mathbb{C}_k^{m \times n}} \|\Sigma - Y\|_F^2$$

$$= \inf \{ \|\Sigma - P_L \Sigma\|_F^2 : \text{over all subspaces } L \text{ with } \dim L \leq k \}$$

$$= \inf \{ \sum_{i=1}^{m} \sigma_i^2 q_{ii} : Q = [q_{ij}] = P_{L\perp}, \dim L \leq k \}$$

and, since $0 \leq q_{ii} \leq 1$ (why?), $\sum_{i=1}^{m} q_{ii} = m - \dim L$, it follows that the minimizing

$$Q = \begin{bmatrix} O & O \\ O & I_{m-k} \end{bmatrix} \quad \text{is unique if and only if} \quad \sigma_k \neq \sigma_{k+1},$$

and the minimizing Y is, accordingly,

$$Y = P_L \Sigma = \begin{bmatrix} I_k & O \\ O & O \end{bmatrix} \Sigma$$

or

$$y_{ij} = \begin{cases} \sigma_i, & \text{if } 1 \leq i = j \leq k, \\ 0, & \text{otherwise.} \end{cases} \qquad \square$$

See Ex. 29 for an alternative proof.

An important application of matrix approximation is the *total least-squares* (TLS) problem, whose development below is based on Nievergelt [612]. Given a linear system

$$A\mathbf{x} = \mathbf{b} \tag{53}$$

the *least-squares problem* is to solve an approximate system

$$A\mathbf{x} = \widetilde{\mathbf{b}} \tag{54}$$

where

$$\widetilde{\mathbf{b}} \in R(A) \quad \text{minimizes} \quad \|\widetilde{\mathbf{b}} - \mathbf{b}\|_2. \tag{55}$$

The TLS problem is to solve an approximate system

$$\widetilde{A}\mathbf{x} = \widetilde{\mathbf{b}} \tag{56}$$

where $\widetilde{\mathbf{b}} \in R(\widetilde{A})$ and the pair $\{\widetilde{A}, \widetilde{\mathbf{b}}\}$ minimizes $\|[\widetilde{A} \vdots \widetilde{\mathbf{b}}] - [A \vdots \mathbf{b}]\|_F$.

Note that in the TLS problem, both the matrix A and the vector $\mathbf{b}$ are modified.

Since (56) is equivalent to

$$\begin{bmatrix} \mathbf{x} \\ -1 \end{bmatrix} \in N([\widetilde{A} \vdots \widetilde{\mathbf{b}}]), \tag{57}$$

the TLS problem is

$$\text{find} \quad [\widetilde{A} \vdots \widetilde{\mathbf{b}}] \in \mathbb{C}^{m \times (n+1)}$$

so as to

$$\text{minimize} \quad \|[\widetilde{A} \vdots \widetilde{\mathbf{b}}] - [A \vdots \mathbf{b}]\|_F \tag{58}$$

subject to (57), for some $\mathbf{x}$.

THEOREM 4. *Let $A \in \mathbb{C}_n^{m \times n}$, let the system (53) be inconsistent, let $[A \vdots \mathbf{b}]$ have the SVD*

$$[A \vdots \mathbf{b}] = U\Sigma V^* = \sum_{i=1}^{n+1} \sigma_i \mathbf{u}_i \mathbf{v}_i^*, \tag{59}$$

and let σ_k be the smallest singular value such that $\mathbf{v}_k$ has nonzero last component $\mathbf{v}_k[n+1]$. Then a solution of the TLS problem is

$$[\widetilde{A} \vdots \widetilde{\mathbf{b}}] = [A \vdots \mathbf{b}] - \sigma_k \mathbf{u}_k \mathbf{v}_k^* \tag{60}$$

and the error of approximation is

$$\|[\widetilde{A} \vdots \widetilde{\mathbf{b}}] - [A \vdots \mathbf{b}]\|_F = \sigma_k. \tag{61}$$

The solution (60) is unique if and only if the smallest singular value σ_k, as above, is unique.

PROOF (Y. Nievergelt[1]). Since (53) is inconsistent, it follows that

$$\text{rank}([A \,\vdots\, \mathbf{b}]) = n + 1 \leq m \quad \text{and} \quad \mathbf{b} \neq \mathbf{0},$$

which in turn guarantees the existence of $\mathbf{v}_j$ with nonzero last component. Denote the index set of such vectors by

$$J = \{j : \mathbf{v}_j[n+1] \neq 0\}, \tag{62}$$

and their number by $|J|$. The unitary matrix $U = [\mathbf{u}_1 \ \cdots \ \mathbf{u}_m]$ in (59) is $m \times m$. Consider its submatrix $\widehat{U} = [\mathbf{u}_1 \ \cdots \ \mathbf{u}_{n+1}]$ whose columns form a basis for $R([A \,\vdots\, \mathbf{b}])$. Let

$$[\widetilde{A} \,\vdots\, \widetilde{\mathbf{b}}] = [A \,\vdots\, \mathbf{b}] - \widehat{U} Y \, \text{diag}\,(\sigma_1, \ldots, \sigma_{n+1}) V^* \tag{63}$$

for some $Y = [y_{ij}] \in \mathbb{C}^{(n+1)\times(n+1)}$, so that

$$\|[\widetilde{A} \,\vdots\, \widetilde{\mathbf{b}}] - [A \,\vdots\, \mathbf{b}]\|_F^2 = \sum_i \Big(\sigma_i^2 \,|y_{ii}|^2 + \sum_{j \neq i} \sigma_j^2 \,|y_{ij}|^2 \Big). \tag{64}$$

Any solution $\begin{bmatrix} \mathbf{x} \\ -1 \end{bmatrix}$ of (57) is a linear combination of the singular vectors $\mathbf{v}_j$, say

$$\begin{bmatrix} \mathbf{x} \\ -1 \end{bmatrix} = \sum_{j \in J} \alpha_j \mathbf{v}_j + \sum_{j \notin J} \beta_j \mathbf{v}_j, \quad \text{with last component} \quad \sum_{j \in J} \alpha_j \,\mathbf{v}_j[n+1] = -1.$$

Then

$$\mathbf{0} = [\widetilde{A} \,\vdots\, \widetilde{\mathbf{b}}] \begin{bmatrix} \mathbf{x} \\ -1 \end{bmatrix} = \Big(\sum_{i=1}^{n+1} \sigma_i \mathbf{u}_i \mathbf{v}_i^* \Big) \Big(\sum_{j \in J} \alpha_j \mathbf{v}_j + \sum_{j \notin J} \beta_j \mathbf{v}_j \Big)$$

$$- \widehat{U} Y \, \text{diag}\,(\sigma_1, \cdots, \sigma_{n+1}) V^* \Big(\sum_{j \in J} \alpha_j \mathbf{v}_j + \sum_{j \notin J} \beta_j \mathbf{v}_j \Big)$$

$$= \sum_{j \in J} \alpha_j \, \sigma_j \, \mathbf{u}_j + \sum_{j \notin J} \beta_j \, \sigma_j \, \mathbf{u}_j - \sum_{j \in J} \alpha_j \, \widehat{U} Y \, \sigma_j \, \mathbf{e}_j - \sum_{j \notin J} \beta_j \, \widehat{U} Y \, \sigma_j \, \mathbf{e}_j.$$

Multiplying on the left by $\{\mathbf{u}_i^* : i \in J\}$ we get $|J|$ equations

$$\alpha_i \, \sigma_i = \sum_{j \in J} \alpha_j \, \mathbf{e}_i^* \, Y \mathbf{e}_j \, \sigma_j + \sum_{j \notin J} \beta_j \, \mathbf{e}_i^* \, Y \mathbf{e}_j \, \sigma_j$$

$$= \sum_{j \in J} y_{ij} \, (\alpha_j \, \sigma_j) + \sum_{j \notin J} y_{ij} \, (\beta_j \, \sigma_j), \quad i \in J. \tag{65}$$

Similarly, premultiplication by $\{\mathbf{u}_i^* : i \notin J\}$ gives the $n + 1 - |J|$ equations

$$\beta_i \, \sigma_i = \sum_{j \in J} y_{ij} \, (\alpha_j \, \sigma_j) + \sum_{j \notin J} y_{ij} \, (\beta_j \, \sigma_j), \quad i \notin J. \tag{66}$$

Equations (65)–(66) show that the matrix $Y = [y_{ij}]$ has an eigenvalue 1, implying

$$\|Y\|_F \geq 1.$$

A matrix $Y = [y_{ij}]$ minimizing (64) has therefore one nonzero, $y_{kk} = 1$ where $\sigma_k = \min\{\sigma_j : j \in J\}$. $\qquad\square$

If $(n+1) \in J$, i.e., if the smallest singular value σ_{n+1} corresponds to a singular vector $\mathbf{v}_{n+1}$ with nonzero last component, then the matrix $[\widetilde{A} \,\vdots\, \widetilde{\mathbf{b}}]$

---

[1]Private communication.

is the best n-rank approximation of $[A : \mathbf{b}]$. In this case Theorem 4 is a corollary of the Schmidt approximation theorem.

See Ex. 36 for the case where A has deficient column rank.

Exercises

Ex. 29 is an alternative proof of the Schmidt approximation theorem, using the inequalities for singular values given in Exs. 26–28. These inequalities were proved by Weyl [868] for integral operators with symmetric kernels and appear here as given in Stewart [783].

Ex. 26. Let $A \in \mathbb{C}_r^{m \times n}$ and let $B \in \mathbb{C}^{m \times n}$ have rank $\leq k < r$. Then

$$\sigma_1(A - B) \geq \sigma_{k+1}(A). \tag{67}$$

PROOF. Let $A = U\Sigma V^*$ be the SVD of A, and let $B = XY^*$ where X, Y have k columns. Denote by $V_{(k+1)}$ the submatrix of the first $k + 1$ columns of V. From Ex. 0.12 with $L = R(V_{(k+1)})$, $M = R(Y)$ it follows that $R(V_{(k+1)}) \cap N(Y^*) \neq \{\mathbf{0}\}$, i.e., there is a linear combination of the first $k + 1$ columns of V,

$$\mathbf{z} = \zeta_1 \mathbf{v}_1 + \zeta_2 \mathbf{v}_2 + \cdots + \zeta_{k+1} \mathbf{v}_{k+1}$$

with $Y^* \mathbf{z} = 0$. Normalizing $\mathbf{z}$ so that $\|\mathbf{z}\|^2 = \zeta_1^2 + \cdots + \zeta_{k+1}^2 = 1$ we get

$$
\begin{aligned}
\sigma_1(A - B)^2 &\geq \mathbf{z}^*(A - B)^*(A - B)\mathbf{z} \\
&= \mathbf{z}^* A^* A \mathbf{z} \\
&= \zeta_1^2 \sigma_1^2 + \zeta_2^2 \sigma_2^2 + \cdots + \zeta_{k+1}^2 \sigma_{k+1}^2 \\
&\geq \sigma_{k+1}^2. \qquad \square
\end{aligned}
$$

Ex. 27. If $A = A' + A''$, then

$$\sigma_{i+j-1} \leq \sigma_i' + \sigma_j'', \tag{68}$$

where σ_i, σ_i', and σ_i'' are the singular values of A, A', and A'', respectively, arranged in descending order.

PROOF. For $i = j = 1$ and $\mathbf{u}_1, \mathbf{v}_1$ as in Theorem 1,

$$\sigma_1 = \langle \mathbf{u}_1, A\mathbf{v}_1 \rangle = \langle \mathbf{u}_1, A'\mathbf{v}_1 \rangle + \langle \mathbf{u}_1, A''\mathbf{v}_1 \rangle \leq \sigma_1' + \sigma_1''.$$

Now, $\sigma_1(A' - A'_{(i-1)}) = \sigma_i(A')$ and $\sigma_1(A'' - A''_{(j-1)}) = \sigma_j(A'')$ where $A'_{(i-1)}$ and $A''_{(j-1)}$ are defined by (30c). Therefore

$$
\begin{aligned}
\sigma_i' + \sigma_j'' &= \sigma_1(A' - A'_{(i-1)}) + \sigma_1(A'' - A''_{(j-1)}) \\
&\geq \sigma_1(A' - A'_{(i-1)} - A''_{(j-1)}) \\
&\geq \sigma_{i+j-1}, \quad \text{by (67), since } \operatorname{rank}(A'_{(i-1)} + A''_{(j-1)}) \leq i + j - 2. \quad \square
\end{aligned}
$$

Ex. 28. If $A \in \mathbb{C}_r^{m \times n}$, $B \in \mathbb{C}_k^{m \times n}$ with $k < r$, then

$$\sigma_i(A - B) \geq \sigma_{k+i}(A), \quad i = 1, 2, \ldots. \tag{69}$$

PROOF. Use Ex. 27 with $A' = A - B$, $A'' = B$, and $j = k + 1$. The conclusion is obtained from (68), since $\sigma_{k+1}(B) = 0$. $\qquad \square$

Ex. 29. If $A \in \mathbb{C}_r^{m \times n}$, $B \in \mathbb{C}_k^{m \times n}$, $k < r$, then it follows from (69) that

$$\|A - B\|_F^2 \geq \sigma_{k+1}^2 + \cdots + \sigma_r^2. \tag{70}$$

This, together with (50), proves the Schmidt approximation theorem.

EX. 30. Let $O \neq A \in \mathbb{C}_r^{m \times n}$ have singular values

$$\sigma_1 \geq \sigma_2 \geq \cdots \geq \sigma_r > 0$$

and let $M_{r-1} = \bigcup_{k=0}^{r-1} \mathbb{C}_k^{m \times n}$ be the set of $m \times n$ matrices of rank $\leq r - 1$. Then the distance, using either the Frobenius norm (0.50) or the spectral norm (0.14.2), of A from M_{r-1} is

$$\inf_{X \in M_{r-1}} \|A - X\| = \sigma_r. \qquad (71)$$

Two easy consequences of (71) are:

(a) Let A be as above and let $B \in \mathbb{C}^{m \times n}$ satisfy

$$\|B\| < \sigma_r,$$

then

$$\mathrm{rank}(A + B) \geq \mathrm{rank}\, A.$$

(b) For any $0 \leq k \leq \min\{m, n\}$, the $m \times n$ matrices of rank $\leq k$ form a closed set in $\mathbb{C}^{m \times n}$.

In particular, the $n \times n$ singular matrices form a closed set in $\mathbb{C}^{n \times n}$. For any nonsingular $A \in \mathbb{C}^{n \times n}$ with singular values

$$\sigma_1 \geq \sigma_2 \geq \cdots \geq \sigma_n > 0,$$

the smallest singular value σ_n is a measure of the nonsingularity of A.

EX. 31. *A minimal rank matrix approximation.* Let $A \in \mathbb{C}^{m \times n}$ and let $\epsilon > 0$. Find a matrix $B \in \mathbb{C}^{m \times n}$ of minimal rank, satisfying

$$\|A - B\|_F \leq \epsilon$$

for the Frobenius norm (0.50).

SOLUTION. Using the notation of (29),

$$B = A_{(k)},$$

where k is determined by

$$\left(\sum_{i=k}^{r} \sigma_i(A)^2 \right)^{1/2} > \epsilon, \quad \left(\sum_{i=k+1}^{r} \sigma_i(A)^2 \right)^{1/2} \leq \epsilon \quad \text{(Golub [304])}. \qquad \square$$

EX. 32. *A unitary matrix approximation.* Let $U^{n \times n}$ denote the set of $n \times n$ unitary matrices. Let $A \in \mathbb{C}_r^{n \times n}$ with an SVD

$$A = U\Sigma V^*, \quad \Sigma = \mathrm{diag}\,(\sigma_1, \ldots, \sigma_r, 0, \ldots, 0) \in \mathbb{R}^{n \times m}.$$

Then

$$\inf_{W \in U^{n \times n}} \|A - W\|_F = \|\Sigma - I\|_F = \sqrt{\sum_{i=1}^{r}(1 - \sigma_i)^2 + n - r}$$

is attained for

$$W = UV^* \quad \text{(Fan and Hoffman [267], Mirsky [558], Golub [304]).}$$

EX. 33. The following generalization of Ex. 32 arises in factor analysis; see, e.g., Green [321] and Schönemann [728].

For given $A, B \in \mathbb{C}^{m \times n}$, find a $W \in U^{n \times n}$ such that

$$\|A - BW\|_F \leq \|A - BX\|_F \quad \text{for any } X \in U^{n \times n}.$$

SOLUTION. $W = UV^*$ where $B^*A = U\Sigma V^*$ is an SVD of B^*A. $\qquad \square$

Ex. 34. Let $A_{(k)}$ be a best rank-k approximation of $A \in \mathbb{C}_r^{m \times n}$ (as given by Theorem 3). Then $A_{(k)}^*$, $A_{(k)}A_{(k)}^*$, and $A_{(k)}^* A_{(k)}$ are best rank-k approximations of A, AA^*, and A^*A, respectively. If A is normal, then $A_{(k)}^j$ is a best rank-k approximation of A^j for all $j = 1, 2 \ldots$ (Householder and Young [**433**]).

Ex. 35. *Real matrices.* If $A \in \mathbb{R}_r^{m \times n}$, then the unitary matrices U and V in the SVD (4) can also be taken to be real, hence orthogonal.

Ex. 36. In Theorem 4, let $A \in \mathbb{C}_r^{m \times n}$ with $r < n$, and let the system (53) be inconsistent. How do the conclusions change?

HINT. Here $\operatorname{rank}([A \vdots \mathbf{b}]) = r + 1$, and the SVD (59) becomes

$$[A \vdots \mathbf{b}] = \sum_{i=1}^{r+1} \sigma_i \, \mathbf{u}_i \mathbf{v}_i^*.$$

Let $\{\mathbf{v}_{r+2}, \ldots, \mathbf{v}_{n+1}\}$ be a basis of $N([A \vdots \mathbf{b}])$. The index set (62) is nonempty, since $\mathbf{b} \neq \mathbf{0}$, but cannot include any $j \in \overline{r+2, n+1}$ for otherwise (53) is consistent. $\square$

4. Partial Isometries and the Polar Decomposition Theorem

A linear transformation $U : \mathbb{C}^n \to \mathbb{C}^m$ is called a *partial isometry* (sometimes also a *subunitary transformation*) if it is norm preserving on the orthogonal complement of its null space, i.e., if

$$\|U\mathbf{x}\| = \|\mathbf{x}\|, \quad \text{for all } \mathbf{x} \in N(U)^\perp = R(U^*), \tag{72}$$

or, equivalently, if it is distance preserving

$$\|U\mathbf{x} - U\mathbf{y}\| = \|\mathbf{x} - \mathbf{y}\|, \quad \text{for all } \mathbf{x}, \mathbf{y} \in N(U)^\perp.$$

Except where otherwise indicated, the norms used here are the Euclidean vector norm and the corresponding spectral norm for matrices, see Ex. 0.38.

Partial isometries in Hilbert spaces were studied extensively by von Neumann [**840**], Halmos [**366**], Halmos and McLaughlin [**367**], Erdelyi [**263**], and others. The results given here are special cases for the finite-dimensional space $\mathbb{C}^n$.

A nonsingular partial isometry is called an *isometry* (or a *unitary transformation*). Thus a linear transformation $U : \mathbb{C}^n \to \mathbb{C}^n$ is an isometry if $\|U\mathbf{x}\| = \|\mathbf{x}\|$ for all $\mathbf{x} \in \mathbb{C}^n$.

We recall that $U \in \mathbb{C}^{n \times n}$ is a unitary matrix if and only if $U^* = U^{-1}$. Analogous characterizations of partial isometries are collected in the following theorem, drawn from Halmos [**366**], Hestenes [**414**], and Erdélyi [**256**].

THEOREM 5. *Let $U \in \mathbb{C}^{m \times n}$. Then the following eight statements are equivalent:*

    (a)   *U is a partial isometry.*
    (a*)   *U^* is a partial isometry.*
    (b)   *U^*U is an orthogonal projector.*
    (b*)   *UU^* is an orthogonal projector.*
    (c)   *$UU^*U = U$.*
    (c*)   *$U^*UU^* = U^*$.*
    (d)   *$U^* = U^\dagger$.*

(e) $U^\dagger$ *is a partial isometry.*

PROOF. We prove (a) $\Longleftrightarrow$ (b), (a) $\Longleftrightarrow$ (e), and (b) $\Longleftrightarrow$ (c) $\Longleftrightarrow$ (d). The obvious equivalence (c) $\Longleftrightarrow$ (c*) then takes care of the dual statements (a*) and (b*).

(a) $\Longrightarrow$ (b)   Since $R(U^*U) = R(U^*)$, (b) can be rewritten as

$$U^*U = P_{R(U^*)}. \tag{73}$$

From Ex. 0.22(b) it follows for any Hermitian $H \in \mathbb{C}^{n \times n}$ that

$$\langle H\mathbf{x}, \mathbf{x} \rangle = 0, \quad \text{for all } \mathbf{x} \in \mathbb{C}^n, \tag{74}$$

implies $H = O$. Consider now the matrix

$$H = P_{R(U^*)} - U^*U.$$

Clearly,

$$\langle H\mathbf{x}, \mathbf{x} \rangle = 0, \quad \text{for all } \mathbf{x} \in R(U^*)^\perp = N(U),$$

while, for $\mathbf{x} \in R(U^*)$,

$$\begin{aligned}
\langle P_{R(U^*)}\mathbf{x}, \mathbf{x} \rangle &= \langle \mathbf{x}, \mathbf{x} \rangle \\
&= \langle U\mathbf{x}, U\mathbf{x} \rangle, \quad \text{by (a)}, \\
&= \langle U^*U\mathbf{x}, \mathbf{x} \rangle.
\end{aligned}$$

Thus (a) implies that the Hermitian matrix $H = P_{R(U^*)} - U^*U$ satisfies (74), which in turn implies (73).

(b) $\Longrightarrow$ (a)   This follows from

$$\begin{aligned}
\langle U\mathbf{x}, U\mathbf{x} \rangle &= \langle U^*U\mathbf{x}, \mathbf{x} \rangle \\
&= \langle P_{R(U^*)}\mathbf{x}, \mathbf{x} \rangle \quad \text{by (73)}, \\
&= \langle \mathbf{x}, \mathbf{x} \rangle, \quad \text{if } \mathbf{x} \in R(U^*).
\end{aligned}$$

(a) $\Longleftrightarrow$ (e)   Since

$$\mathbf{y} = U\mathbf{x}, \quad \mathbf{x} \in R(U^*),$$

is equivalent to

$$\mathbf{x} = U^\dagger\mathbf{y}, \quad \mathbf{y} \in R(U),$$

it follows that

$$\langle U\mathbf{x}, U\mathbf{x} \rangle = \langle \mathbf{x}, \mathbf{x} \rangle, \quad \text{for all } \mathbf{x} \in R(U^*),$$

is equivalent to

$$\langle \mathbf{y}, \mathbf{y} \rangle = \langle U^\dagger\mathbf{y}, U^\dagger\mathbf{y} \rangle, \quad \text{for all } \mathbf{y} \in R(U) = N(U^\dagger)^\perp.$$

(b) $\Longleftrightarrow$ (c) $\Longleftrightarrow$ (d)   The obvious equivalence (c) $\Longleftrightarrow$ (c*) states that $U^* \in U\{1\}$ if, and only if, $U^* \in U\{2\}$. Since U^* is (always) a $\{3,4\}$-inverse of U, it follows that U^* is a $\{1\}$-inverse of U if, and only if, $U^* = U^\dagger$. $\square$

Returning to the SVD of Section 2, we identify some useful partial isometries in the following theorem:

THEOREM 6 (Hestenes [**414**]). *Let $O \neq A \in \mathbb{C}_r^{m \times n}$, and let*

$$A = U\Sigma V^*, \tag{4}$$

the unitary matrices $U \in U^{m \times m}$, $V \in U^{n \times n}$ and the diagonal matrix $\Sigma \in \mathbb{R}_r^{m \times n}$ given as in Theorem 2. Let $U_{(r)}$, $\Sigma_{(r)}$, and $V_{(r)}$ be defined by (29). Then:

(a) *The matrices $U_{(r)}$, $V_{(r)}$ are partial isometries with*

$$U_{(r)}U_{(r)}^* = P_{R(A)}, \qquad U_{(r)}^*U_{(r)} = I_r, \tag{75}$$

$$V_{(r)}V_{(r)}^* = P_{R(A^*)}, \qquad V_{(r)}^*V_{(r)} = I_r. \tag{76}$$

(b) *The matrix*

$$E = U_{(r)}V_{(r)}^* \tag{77}$$

is a partial isometry with

$$EE^* = P_{R(A)}, \qquad E^*E = P_{R(A^*)}. \tag{78}$$

PROOF. (a) That $U_{(r)}$, $V_{(r)}$ are partial isometries is obvious from their definitions and the unitarity of U and V (see, e.g., Ex. 39). Now

$$U_{(r)}^*U_{(r)} = I_r$$

by definition (29), since U is unitary, and

$$\begin{aligned}
P_{R(A^*)} &= A^\dagger A = A_{(r)}^\dagger A_{(r)} \\
&= V_{(r)}\Sigma_{(r)}^{-1}U_{(r)}^*U_{(r)}\Sigma_{(r)}V_{(r)}^*, \quad \text{by (30a) and (30b),} \\
&= V_{(r)}V_{(r)}^*,
\end{aligned}$$

with the remaining statements in (a) similarly proved.

(b) Using (75) and (76), it can be verified that

$$E^\dagger = V_{(r)}U_{(r)}^* = E^*,$$

from which (78) follows easily. □

The partial isometry E thus maps $R(A^*)$ isometrically onto $R(A)$. Since A also maps $R(A^*)$ onto $R(A)$, we should expect A to be a "multiple" of E. This is the essence of the following theorem, proved by Autonne [30] and Williamson [876] for square matrices, by Penrose [635] for rectangular matrices, and by Murray and von Neumann [589] for linear operators in Hilbert spaces.

THEOREM 7 (The Polar Decomposition Theorem). *Let $O \neq A \in \mathbb{C}_r^{m \times n}$. Then A can be written as*

$$A = GE = EH, \tag{79}$$

where $E \in \mathbb{C}^{m \times n}$ is a partial isometry and $G \in \mathbb{C}^{m \times m}$, $H \in \mathbb{C}^{n \times n}$ are Hermitian and PSD.

The matrices E, G, and H are uniquely determined by

$$R(E) = R(G), \tag{80a}$$

$$R(E^*) = R(H), \tag{80b}$$

in which case

$$G^2 = AA^*, \tag{81a}$$

$$H^2 = A^*A, \tag{81b}$$

and E is given by

$$E = U_{(r)}V_{(r)}^*. \tag{77}$$

PROOF. Let

$$A = U\Sigma V^*, \quad \Sigma = \text{diag}\,(\sigma_1, \ldots, \sigma_r, 0, \ldots, 0) \tag{4}$$

be the SVD of A. For any k, $r \leq k \leq \min\{m, n\}$, we use (29) to define the three matrices

$$\Sigma_{(k)} = \text{diag}(\sigma_1, \ldots, \sigma_k) \in \mathbb{R}^{k \times k},$$

$$U_{(k)} = [\mathbf{u}_1 \ \ldots \ \mathbf{u}_k] \in \mathbb{C}^{m \times k}, \quad V_{(k)} = [\mathbf{v}_1 \ \ldots \ \mathbf{v}_k] \in \mathbb{C}^{n \times k}.$$

Then (4) can be rewritten as

$$\begin{aligned}
A &= U_{(k)}\Sigma_{(k)}V_{(k)}^* \\
&= (U_{(k)}\Sigma_{(k)}U_{(k)}^*)(U_{(k)}V_{(k)}^*), \quad \text{since } U_{(k)}^* U_{(k)} = I_k, \\
&= (U_{(k)}V_{(k)}^*)(V_{(k)}\Sigma_{(k)}V_{(k)}^*), \quad \text{since } V_{(k)}^* V_{(k)} = I_k,
\end{aligned}$$

which proves (79) with the partial isometry

$$E = U_{(k)}V_{(k)}^* \tag{82}$$

and the PSD matrices

$$G = U_{(k)}\Sigma_{(k)}U_{(k)}^*, \quad H = V_{(k)}\Sigma_{(k)}V_{(k)}^*. \tag{83}$$

This also shows E to be nonunique if $r < \min\{m, n\}$, in which case G and H are also nonunique, for (83) can then be replaced by

$$G = U_{(k)}\Sigma_{(k)}U_{(k)}^* + \mathbf{u}_{k+1}\mathbf{u}_{k+1}^*,$$

$$H = V_{(k)}\Sigma_{(k)}V_{(k)}^* + \mathbf{v}_{k+1}\mathbf{v}_{k+1}^*,$$

which satisfies (79) for the E given in (82).

Let now E and G satisfy (80a). Then, from (79),

$$AA^* = GEE^*G = GEE^\dagger G = GP_{R(E)}G = G^2,$$

which proves (81a) and the uniqueness of G; see also Ex. 37 below. The uniqueness of E follows from

$$E = EE^\dagger E = GG^\dagger E = G^\dagger GE = G^\dagger A. \tag{84}$$

Similarly (80b) implies (81b) and the uniqueness of H, E.

Finally, from

$$\begin{aligned}
G^2 &= AA^* \\
&= U_{(r)}\Sigma_{(r)}V_{(r)}^*V_{(r)}\Sigma_{(r)}U_{(r)}^*, \quad \text{by (30a),} \\
&= U_{(r)}\Sigma_{(r)}^2 U_{(r)}^*
\end{aligned}$$

we conclude that

$$G = U_{(r)}\Sigma_{(r)}U_{(r)}^*$$

and, consequently,

$$G^\dagger = U_{(r)}\Sigma_{(r)}^{-1}U_{(r)}^*. \tag{85}$$

Therefore,

$$\begin{aligned}
E &= G^\dagger A, \quad \text{by (84)} \\
&= U_{(r)}\Sigma_{(r)}^{-1}U_{(r)}^*U_{(r)}\Sigma_{(r)}V_{(r)}^*, \quad \text{by (85) and (30a),} \\
&= U_{(r)}V_{(r)}^*, \quad \text{proving (77).} \qquad \square
\end{aligned}$$

If one uses a general UDV^*-decomposition of A instead of the SVD, then the matrices G and H defined by (83) are merely normal matrices and need not be Hermitian. Hence, the following corollary:

COROLLARY 2. *Let* $O \neq A \in \mathbb{C}_r^{m \times n}$. *Then, for any choice of the scalars* $d(A)$ *in* (35), *there exist a partial isometry* $E \in \mathbb{C}^{m \times n}$ *and two normal matrices* $G \in \mathbb{C}^{m \times m}$, $H \in \mathbb{C}^{n \times n}$, *satisfying* (79). *The matrices* E, G, *and* H *are uniquely determined by* (80a) *and* (80b), *in which case*

$$GG^* = AA^*, \tag{86}$$

$$H^*H = A^*A, \tag{87}$$

and E *is given by* (77). $\qquad\qquad\square$

Theorem 7 is the matrix analog of the polar decomposition of a complex number

$$z = x + iy, \quad x, y \text{ real},$$

as

$$z = |z|e^{i\theta}, \tag{88}$$

where

$$|z| = (z\bar{z})^{1/2} = (x^2 + y^2)^{1/2}$$

and

$$\theta = \arctan \frac{y}{x}.$$

Indeed, the complex scalar z in (88) corresponds to the matrix A in (79), while $\bar{z}$, $|z|$, and $e^{i\theta}$ correspond to A^*, G (or H), and E, respectively. This analogy is natural since $|z| = (z\bar{z})^{1/2}$ corresponds to the square roots $G = (AA^*)^{1/2}$ or $H = (A^*A)^{1/2}$, while the scalar $e^{i\theta}$ satisfies

$$|ze^{i\theta}| = |z|, \quad \text{for all } z \in \mathbb{C},$$

which justifies its comparison to the partial isometry E; see also Exs. 55 and 59.

Exercises

Ex. 37. *Square roots.* Let $A \in \mathbb{C}_r^{n \times n}$ be Hermitian PSD. Then there exists a unique Hermitian PSD matrix $B \in \mathbb{C}_r^{n \times n}$ satisfying

$$B^2 = A, \tag{89}$$

B is called the *square root* of A, denoted by $A^{1/2}$.
PROOF. Writing A as

$$A = UDU^*, \quad U \text{ unitary}, \quad D = \mathrm{diag}\,(\lambda_1, \dots, \lambda_r, 0, \dots, 0),$$

we see that

$$B = UD^{1/2}U^*, \quad D^{1/2} = \mathrm{diag}\,\left(\lambda_1^{1/2}, \dots, \lambda_r^{1/2}, 0, \dots, 0\right),$$

is a Hermitian PSD matrix satisfying (89). To prove uniqueness, assume that B is a Hermitian matrix satisfying (89). Then, since B and $A = B^2$ commute, it follows from Ex. 16 that

$$B = U\widetilde{D}U^*$$

where $\widetilde{D}$ is diagonal and real, by Ex. 0.22(b), hence

$$\widetilde{D} = D^{1/2}, \quad \text{by (89)}. \qquad \square$$

Partial Isometries

Ex. 38. *Linearity of isometries.* Let X, Y be real normed vector spaces and let $f : X \rightarrow Y$ be isometric, i.e.,

$$\|f(\mathbf{x}_1) - f(\mathbf{x}_2)\|_Y = \|\mathbf{x}_1 - \mathbf{x}_2\|_X, \quad \text{for all } \mathbf{x}_1, \mathbf{x}_2 \in X,$$

where $\| \ \|_X$ and $\| \ \|_Y$ are the norms in X and Y, respectively. If $f(\mathbf{0}) = \mathbf{0}$, then f is a linear transformation (Mazur and Ulam). For extensions and references see Dunford and Schwartz [**246**, p. 91] and Vogt [**837**].

Ex. 39. *Partial isometries.* If the $n \times n$ matrix U is unitary and $U_{(k)}$ is any $n \times k$ submatrix of U, then $U_{(k)}$ is a partial isometry. Conversely, if $W \in \mathbb{C}_k^{n \times k}$ is a partial isometry, then there is an $n \times (n - k)$ partial isometry V such that the matrix $U = [W \quad V]$ is unitary.

Ex. 40. Any matrix unitarily equivalent to a partial isometry is a partial isometry.

PROOF. Let $A = UBV^*$, $U \in U^{m \times m}$, $V \in U^{n \times n}$. Then

$$A^\dagger = VB^\dagger U^*, \quad \text{by Ex. 1.25,}$$
$$= VB^*U^*, \quad \text{if } B \text{ is a partial isometry,}$$
$$= A^*. \qquad \square$$

Ex. 41. Let $A \in \mathbb{C}_r^{m \times n}$ have singular values $\sigma(A) = \{\sigma_i : i \in \overline{1,r}\}$. Then A is a partial isometry if, and only if,

$$\sigma_i = 1, \quad i \in \overline{1,r}.$$

Consequently, in any UDV^*-decomposition (34) of a partial isometry, the diagonal factor D has

$$|d_i| = 1, \quad i \in \overline{1,r}.$$

Ex. 42. A linear transformation $E : \mathbb{C}^n \rightarrow \mathbb{C}^m$ with $\dim R(E) = r$ is a partial isometry if, and only if, there are two o.n. bases $\{\mathbf{v}_1, \ldots, \mathbf{v}_r\}$ and $\{\mathbf{u}_1, \ldots, \mathbf{u}_r\}$ of $R(E^*)$ and $R(E)$, respectively, such that

$$\mathbf{u}_i = E\mathbf{v}_i, \quad i = 1, \ldots, r.$$

Ex. 43. *Contractions.* A matrix $A \in \mathbb{C}^{m \times n}$ is called a *contraction* if

$$\|A\mathbf{x}\| \leq \|\mathbf{x}\|, \quad \text{for all } \mathbf{x} \in \mathbb{C}^n. \tag{90}$$

For any $A \in \mathbb{C}^{\times n}$ the following statements are equivalent:

    (a) A is a contraction.

    (b) A^* is a contraction.

    (c) For any subspace L of $\mathbb{C}^m$ containing $R(A)$, the matrix $P_L - AA^*$ is PSD.

PROOF. (a) $\Longleftrightarrow$ (b)   By Exs. 0.35 and 0.38, (a) is equivalent to

$$\|A\|_2 \leq 1,$$

but

$$\|A\|_2 = \|A^*\|_2, \quad \text{by (0.14.2) and Ex. 1.}$$

    (b) $\Longleftrightarrow$ (c)   By definition (90), the statement (b) is equivalent to

$$0 \leq \langle \mathbf{x}, \mathbf{x} \rangle - \langle A^*\mathbf{x}, A^*\mathbf{x} \rangle$$
$$= \langle (I - AA^*)\mathbf{x}, \mathbf{x} \rangle, \quad \text{for all } \mathbf{x} \in \mathbb{C}^m,$$

which in turn is equivalent to (c).   □

EX. 44. Let $A \in \mathbb{C}^{m \times n}$ be a contraction and let L be any subspace of $\mathbb{C}^m$ containing $R(A)$. Then the $(m + n) \times (m + n)$ matrix $M(A)$ defined by

$$M(A) = \begin{bmatrix} A & \sqrt{P_L - AA^*} \\ O & O \end{bmatrix}$$

is a partial isometry (Halmos and McLaughlin [367], Halmos [366]).

PROOF. The square root $\sqrt{P_L - AA^*}$ exists and is unique by Exs. 43(c) and 37. The proof then follows by verifying that

$$M(A)M(A)^*M(A) = M(A).$$ □

EX. 45. *Eigenvalues of partial isometries.* Let U be an $n \times n$ partial isometry and let λ be an eigenvalue of U corresponding to the eigenvector $\mathbf{x}$. Then

$$|\lambda| = \frac{\|P_{R(U^*)}\mathbf{x}\|}{\|\mathbf{x}\|},$$

hence

$$|\lambda| \leq 1 \quad (\text{Erdélyi } [256]).$$

PROOF. From $U\mathbf{x} = \lambda\mathbf{x}$ we conclude

$$|\lambda|\|\mathbf{x}\| = \|U\mathbf{x}\| = \|UP_{R(U^*)}\mathbf{x}\| = \|P_{R(U^*)}\mathbf{x}\|.$$ □

EX. 46. The partial isometry

$$U = \begin{bmatrix} 1 & 0 & 0 \\ 0 & \frac{\sqrt{3}}{2} & 0 \\ 0 & \frac{1}{2} & 0 \end{bmatrix}$$

has the following eigensystem:

$$\lambda = 0, \quad \mathbf{x} = \begin{bmatrix} 0 \\ 0 \\ 1 \end{bmatrix} \in N(U),$$

$$\lambda = 1, \quad \mathbf{x} = \begin{bmatrix} 1 \\ 0 \\ 0 \end{bmatrix} \in R(U^*),$$

$$\lambda = \tfrac{\sqrt{3}}{2}, \quad \mathbf{x} = \begin{bmatrix} 0 \\ \frac{\sqrt{3}}{2} \\ \frac{1}{2} \end{bmatrix} = \begin{bmatrix} 0 \\ \frac{\sqrt{3}}{2} \\ 0 \end{bmatrix} + \begin{bmatrix} 0 \\ 0 \\ \frac{1}{2} \end{bmatrix}, \quad \begin{bmatrix} 0 \\ \frac{\sqrt{3}}{2} \\ 0 \end{bmatrix} \in R(U^*), \quad \begin{bmatrix} 0 \\ 0 \\ \frac{1}{2} \end{bmatrix} \in N(U).$$

EX. 47. *Normal partial isometries.* Let U be an $n \times n$ partial isometry. Then U is normal if and only if it is range-Hermitian.

PROOF. Since any normal matrix is range-Hermitian, only the "if" part needs proof. Let U be range-Hermitian, i.e., let $R(U) = R(U^*)$. Then $UU^* = U^*U$, by Theorem 5.   □

EX. 48. Let U be an $n \times n$ partial isometry. If U is normal, then its eigenvalues have absolute values 0 or 1.

PROOF. For any nonzero eigenvalue λ of a normal partial isometry U, it follows from $U\mathbf{x} = \lambda\mathbf{x}$ that $\mathbf{x} \in R(U) = R(U^*)$, and therefore

$$|\lambda|\|\mathbf{x}\| = \|U\mathbf{x}\| = \|\mathbf{x}\|.$$ □

EX. 49. The converse of Ex. 48 is false. Consider, for example, the partial isometry $U = \left(\begin{smallmatrix} 0 & 1 \\ 0 & 0 \end{smallmatrix}\right)$.

Ex. 50. Let $E \in \mathbb{C}^{n \times n}$ be a contraction. Then E is a normal partial isometry if, and only if, the eigenvalues of E have absolute values 0 or 1 and rank E = rank E^2 (Erdelyi [260, Lemma 2]).

Ex. 51. A matrix $E \in \mathbb{C}^{n \times n}$ is a normal partial isometry if, and only if,

$$E = U \begin{bmatrix} W & O \\ O & O \end{bmatrix} U^*,$$

where U and W are unitary matrices (Erdelyi [260]).

Polar Decompositions

Ex. 52. Let $A \in \mathbb{C}^{n \times n}$ and let

$$A = GE, \tag{79}$$

where G is PSD and E is a partial isometry satisfying

$$R(E) = R(G). \tag{80a}$$

Then A is normal if, and only if,

$$GE = EG,$$

in which case E is a normal partial isometry (Hearon [408, Theorem 1], Halmos [366, Problem 108]).

Ex. 53. Let $A \in \mathbb{C}^{n \times n}$ have the polar decompositions (79) and (80a). Then A is a partial isometry if and only if G is an orthogonal projector.

PROOF. *If*: Let

$$G = G^* = G^2. \tag{91}$$

Then

$$\begin{aligned} AA^* &= GEE^*G, \quad \text{by (79)}, \\ &= G^2, \quad \text{since } EE^* = P_{R(G)} \text{ by Theorem 5(b}^*) \text{ and (80a)}, \\ &= G, \quad \text{by (91)}, \end{aligned}$$

proving that A is a partial isometry by Theorem 5(b*).

Only if: Let A be a partial isometry and let $A = GE$ be its unique polar decomposition determined by (80a). Then

$$AA^* = G^2$$

is a Hermitian idempotent, by Theorem 5(b*), and hence its square root is also idempotent. $\square$

Ex. 54. (Hestenes [414]). Let $A \in \mathbb{C}^{n \times n}$ have the polar decomposition (79) satisfying (80a) and (80b). Then σ is a singular value of A if, and only if,

$$A\mathbf{x} = \sigma E\mathbf{x}, \quad \text{for some } \mathbf{0} \neq \mathbf{x} \in R(E^*), \tag{92}$$

or, equivalently, if and only if

$$A^*\mathbf{y} = \sigma E^*\mathbf{y}, \quad \text{for some } \mathbf{0} \neq \mathbf{y} \in R(E). \tag{93}$$

PROOF. From (79) it follows that (92) is equivalent to

$$G(E\mathbf{x}) = \sigma(E\mathbf{x}),$$

which, by (81a), is equivalent to

$$AA^*(E\mathbf{x}) = \sigma^2(E\mathbf{x}).$$

The equivalence of (93) is similarly proved. $\square$

Ex. 55. Let z be any complex number with the polar decomposition

$$z = |z|e^{i\theta}. \tag{92}$$

Then, for any real α, the following inequalities are obvious:

$$|z - e^{i\theta}| \leq |z - e^{i\alpha}| \leq |z + e^{i\theta}|.$$

Fan and Hoffman [267] established the following analogous matrix inequalities.
Let $A \in \mathbb{C}^{n \times n}$ be decomposed as

$$A = UH,$$

where U is unitary and H is PSD. Then, for any unitary $W \in U^{n \times n}$, the inequalities

$$\|A - U\| \leq \|A - W\| \leq \|A + U\|$$

hold for every unitarily invariant norm.

Give the analogous inequalities for the polar decomposition of rectangular matrices given in Theorem 7. See also Schönemann [728] and Björck and Bowie [110].

Ex. 56. *Generalized Cayley transforms.* Let L be a subspace of $\mathbb{C}^n$. Then the equations

$$U = (P_L + iH)(P_L - iH)^\dagger, \tag{94}$$

$$H = i(P_L - U)(P_L + U)^\dagger, \tag{95}$$

establish a one-to-one correspondence between all Hermitian matrices H with

$$R(H) \subset L \tag{96}$$

and all normal partial isometries U with

$$R(U) = L \tag{97}$$

whose spectrum excludes -1 (Ben-Israel [62], Pearl [629], [630], Nanda [590]).
PROOF. Note that

$$(P_L \pm iH) \quad \text{and} \quad (P_L + U)$$

map L onto itself for Hermitian H satisfying (96) and normal partial isometries satisfying (97), whose spectrum excludes -1. Since on L, $(P_L \pm iH)$ and $(P_L \pm U)$ reduce to $(I \pm iH)$ and $(I \pm U)$, respectively, the proof follows from the classical theorem; see, e.g., Gantmacher [296, Vol. I, p. 279].   □

Ex. 57. Let H be a given Hermitian matrix. Let L_1 and L_2 be two subspaces containing $R(H)$ and let U_1 and U_2 be the normal partial isometries defined, respectively, by (94). If $L_1 \subset L_2$, then $U_1 = U_2 P_{L_1}$, i.e., U_1 is the restriction of U_2 to L_1. Thus the "minimal" normal partial isometry corresponding to a given Hermitian matrix H is

$$U = (P_{R(H)} + iH)(P_{R(H)} - iH)^\dagger.$$

Ex. 58. A well-known inequality of Fan and Hoffman [267, Theorem 3] is extended to the singular case as follows.
If H_1, H_2 are Hermitian with $R(H_1) = R(H_2)$ and if

$$U_k = (P_{R(H_k)} + iH_k)(P_{R(H_k)} - iH_k)^\dagger, \quad k = 1, 2,$$

then

$$\|U_1 - U_2\| \leq 2\|H_1 - H_2\|$$

for every unitarily invariant norm (Ben-Israel [62]).

Trace Inequalities

EX. 59. Let z be a complex scalar. Then, for any real α, the following inequality is obvious:

$$|z| \geq \Re\{ze^{i\alpha}\}.$$

An analogous matrix inequality can be stated as follows:
Let $H \in \mathbb{C}^{n \times n}$ be Hermitian PSD. Then

$$\text{trace } H \geq \Re\{\text{trace}(HW)\}, \quad \text{for all } W \in U^{n \times n},$$

where $U^{n \times n}$ is the class of $n \times n$ unitary matrices.

PROOF. Suppose there is a $W_0 \in U^{n \times n}$ with

$$\text{trace } H < \Re\{\text{trace}(HW_0)\}. \tag{98}$$

Let

$$H = UDU^* \quad \text{with } U \in U^{n \times n}, \quad D = \text{diag}(\alpha_1, \dots, \alpha_n),$$

where $\{\alpha_1, \dots, \alpha_n\}$ are the eigenvalues of H. Then

$$\sum \alpha_i = \text{trace } H < \Re\{\text{trace}(UDU^*W_0)\}, \quad \text{by (98)},$$
$$= \Re\{\text{trace } A\}, \quad \text{where } A = UDV^*, \ V^* = U^*W_0, \tag{99}$$
$$= \Re\{\sum \lambda_i\}, \quad \text{where } \{\lambda_1, \dots, \lambda_n\} \text{ are the eigenvalues of } A.$$

But $AA^* = UDV^*VDU^* = UD^2U^*$, proving that the nonzero $\{\alpha_i\}$ are the singular values of A. Thus (99) implies that

$$\sum \alpha_i < \sum |\lambda_i|,$$

a contradiction of Weyl's inequality (17).   □

EX. 60. Let $A \in \mathbb{C}_r^{m \times n}$ be given and let $W_r^{m \times n}$ denote the class of all partial isometries in $\mathbb{C}_\ell^{m \times n}$, where $\ell = \min\{m, n\}$. Then

$$\sup_{W \in W_\ell^{m \times n}} \Re\{\text{trace}(AW)\}$$

is attained for some $W_0 \in W_\ell^{m \times n}$. Moreover, AW_0 is Hermitian PSD, and

$$\sup_{W \in W_\ell^{m \times n}} \Re\{\text{trace}(AW)\} = \text{trace}(AW_0) = \sum_{i=1}^r \sigma_i, \tag{100}$$

where $\{\sigma_1, \dots, \sigma_r\}$ are the singular values of A. (For $m = n$, and unitary W, this result is due to von Neumann [839].)

PROOF. Without a loss of generality, assume that $m \leq n$. Let

$$A = GE \tag{79}$$

be a polar decomposition, where the partial isometry E is taken to be of full rank (using (82) with $k = m$), so $E \in W_m^{m \times n}$. Then, for any $W \in W_m^{m \times n}$,

$$\text{trace}(AW) = \text{trace}(GEW)$$

$$= \text{trace}\left(\begin{bmatrix} G & O \\ O & O \end{bmatrix} \begin{bmatrix} E \\ E^\perp \end{bmatrix} [W \quad W^\perp]\right), \tag{101}$$

where the submatrices $E^\perp$ and $W^\perp$ are chosen so as to make

$$\begin{bmatrix} E \\ E^\perp \end{bmatrix} \text{ and } [W \quad W^\perp]$$

unitary matrices; see, e.g., Ex. 39. Since

$$\begin{bmatrix} G & O \\ O & O \end{bmatrix} \text{ is PSD, and } \begin{bmatrix} E \\ E^\perp \end{bmatrix} [W \quad W^\perp]$$

is unitary, it follows from Ex. 59 and (101), that

$$\sup_{W \in W_m^{n \times m}} \Re\{\text{trace}(AW)\}$$

is attained for $W_0 \in W_m^{n \times m}$ satisfying

$$AW_0 = G,$$

and (100) follows from (83).   □

Ex. 61. Let $A \in \mathbb{C}_r^{m \times n}$ and $B \in \mathbb{C}_s^{n \times m}$ have singular values $\alpha_1 \geq \alpha_2 \geq \cdots \geq \alpha_r > 0$ and $\beta_1 \geq \beta_2 \geq \cdots \geq \beta_s > 0$, respectively. Then

$$\sup_{X \in U^{n \times n}, W \in U^{m \times m}} \Re\{\text{trace}(AXBW)\}$$

is attained for some $X_0 \in U^{n \times n}, W_0 \in U^{m \times m}$, and is given by

$$\text{trace}(AX_0 BW_0) = \sum_{i=1}^{\min\{r,s\}} \alpha_i \beta_i.$$

This result was proved by von Neumann [**839**, Theorem 1] for the case $m = n$. The general case is proved by "squaring" the matrices A and B, i.e., adjoining zero rows and columns to make them square.

Gauge Functions and Singular Values

The following two exercises relate gauge functions (Ex. 3.52) to matrix norms and inequalities. The unitarily invariant matrix norms are characterized in Ex. 62 as symmetric gauge functions of the singular values. For square matrices these results were proved by von Neumann [**839**] and Mirsky [**558**].

Ex. 62. *Unitarily invariant matrix norms.* We use here the notation of Ex. 3.52.

Let the functions $\|\ \|_\phi : \mathbb{C}^{m \times n} \to \mathbb{R}$ and $\widehat{\phi} : \mathbb{C}^{mn} \to \mathbb{R}$ be defined, for any function $\phi : \mathbb{R}^\ell \to \mathbb{R}$, $\ell = \min\{m,n\}$, as follows: For any $A = [a_{ij}] \in \mathbb{C}^{m \times n}$ with singular values

$$\sigma_1 \geq \sigma_2 \geq \cdots \geq \sigma_r > 0,$$

$\|A\|_\phi$ and $\widehat{\phi}(a_{11}, \ldots, a_{mn})$ are defined as

$$\|A\|_\phi = \widehat{\phi}(a_{11}, \ldots, a_{mn}) = \phi(\sigma_1, \ldots, \sigma_r, 0, \ldots, 0). \tag{102}$$

Then:

(a) If $\phi : \mathbb{R}^\ell \to \mathbb{R}$ satisfies conditions (G1)–(G3) of Ex. 3.52, so does $\widehat{\phi} : \mathbb{C}^{mn} \to \mathbb{R}$.

(b) $\|UAV\|_\phi = \|A\|_\phi$ for all $A \in \mathbb{C}^{m \times n}$, $U \in U^{m \times m}$, $V \in U^{n \times n}$.

(c) Let $\phi : \mathbb{R}^\ell \to \mathbb{R}$ satisfy conditions (G1)–(G3) of Ex. 3.52 and let $\phi_D : \mathbb{R}^\ell \to \mathbb{R}$ be its dual, defined by (3.130). Then, for any $A \in \mathbb{C}^{m \times n}$, the following supremum is attained and

$$\sup_{X \in \mathbb{C}^{n \times m}, \|X\|_\phi = 1} \Re\{\text{trace}(AX)\} = \|A\|_{\phi_D}. \tag{103}$$

(d) If $\phi : \mathbb{R}^\ell \to \mathbb{R}$ is a symmetric gauge function, then $\widehat{\phi} : \mathbb{C}^{mn} \to \mathbb{R}$ is a gauge function, and $\|\ \|_\phi : \mathbb{C}^{m \times n} \to \mathbb{R}$ is a unitarily invariant norm.

(e) If $\|\ \| : \mathbb{C}^{m \times n} \to \mathbb{R}$ is a unitarily invariant norm, then there is a symmetric gauge function $\phi : \mathbb{R}^\ell \to \mathbb{R}$ such that $\|\ \| = \|\ \|_\phi$.

PROOF. (a) Follows from definition (102).

(b) Obvious by Ex. 2.

(c) For the given $A \in \mathbb{C}^{m \times n}$,

$$\sup_{\substack{X \in \mathbb{C}^{n \times m} \\ \|X\|_\phi = 1}} \Re\{\text{trace}(AX)\}$$

$$= \sup_{\substack{X \in \mathbb{C}^{n \times m} \\ \|X\|_\phi = 1}} \Re\{\text{trace}(AUXV) : U \in U^{n \times n}, V \in U^{m \times m}\}, \quad \text{by (b)},$$

$$= \sup_{\phi(\xi_1, \dots, \xi_\ell) = 1} \sum_i \sigma_i \xi_i, \quad \text{by Ex. 61},$$

$$= \phi_D(\sigma_1, \dots, \sigma_r), \quad \text{by (3.131a) and (3.131c)},$$

$$= \|A\|_{\phi_D}, \quad \text{by (102)},$$

where $\sigma_1 \geq \sigma_2 \geq \cdots \geq \sigma_r > 0$ and $\xi_1 \geq \xi_2 \geq \cdots \geq \xi_\ell > 0$ are the singular values of A and X, respectively.

(d) Let ϕ_D be the dual of ϕ and let $\widehat{(\phi_D)} : \mathbb{C}^{mn} \to \mathbb{R}$ be defined by (102) as

$$\widehat{(\phi_D)}(a_{11}, \dots, a_{mn}) = \|A\|_{\phi_D}, \quad \text{for } A = [a_{ij}].$$

Then

$$\widehat{(\phi_D)}(a_{11}, \dots, a_{mn}) = \|A^*\|_{\phi_D}, \quad \text{by Ex. 1},$$

$$= \sup_{\substack{X = [x_{ij}] \in \mathbb{C}^{m \times n} \\ \widehat{\phi}(x_{11}, \dots, x_{mn}) = 1}} \Re\{\text{trace}(A^* X)\}, \quad \text{by (103)},$$

$$= \sup_{\widehat{\phi}(x_{11}, \dots, x_{mn}) = 1} \sum_{i,j} \overline{a_{ij}}\, x_{ij},$$

proving that $\widehat{(\phi_D)} : \mathbb{C}^{mn} \to \mathbb{R}$ is the dual of $\widehat{\phi} : \mathbb{C}^{mn} \to \mathbb{R}$, by using (3.131a) and (3.131c). Since ϕ is the dual of ϕ_D (by Ex. 3.52(d)), it follows that $\widehat{\phi}$ is the dual of $\widehat{(\phi_D)}$ and, by Ex. 3.52(d), $\widehat{\phi} : \mathbb{C}^{mn} \to \mathbb{R}$ is a gauge function. That $\| \ \|_\phi$ is a unitarily invariant norm then follows from (b) and Ex. 3.56.

(e) Let $\| \ \| : \mathbb{C}^{m \times n} \to \mathbb{R}$ be a unitarily invariant matrix norm and define $\phi : \mathbb{R}^\ell \to \mathbb{R}$ by

$$\phi(\mathbf{x}) = \phi(x_1, x_2, \dots, x_\ell) = \| \text{diag}\, (|x_1|, \cdots, |x_\ell|) \| \in \mathbb{C}^{\ell \times \ell}.$$

Then ϕ is a symmetric gauge function and $\| \ \| = \| \ \|_\phi$. $\qquad \square$

EX. 63. *Inequalities for singular values.* Let $A, B \in \mathbb{C}^{m \times n}$ and let

$$\alpha_1 \geq \cdots \geq \alpha_r > 0$$

and

$$\beta_1 \geq \cdots \geq \beta_s > 0$$

be the singular values of A and B, respectively. Then, for any symmetric gauge function $\phi : \mathbb{R}^\ell \to \mathbb{R}$, $\ell = \min\{m, n\}$, the singular values

$$\gamma_1 \geq \cdots \geq \gamma_t > 0$$

of $A + B$ satisfy

$$\phi(\gamma_1, \dots, \gamma_t, 0, \dots, 0) \leq \phi(\alpha_1, \dots, \alpha_r, 0, \dots, 0) + \phi(\beta_1, \dots, \beta_s, 0, \dots, 0) \quad (104)$$

(von Neumann [839]).

PROOF. The inequality (104) follows from (102) and Ex. 62(d), since

$$\|A + B\|_\phi \leq \|A\|_\phi + \|B\|_\phi. \qquad \square$$

5. Principal Angles Between Subspaces

This section is based on Afriat [**3**], where more details, results, and references can be found.

We use the Euclidean vector norm and the corresponding matrix norm $\|\cdot\|_2$. All results are stated for the real case, analogous results hold for $\mathbb{C}^n$.

DEFINITION 2. (Afriat [**3**]). Let L, M be subspaces in $\mathbb{R}^n$. The subspaces are:

(a) *orthogonal* if $L \subset M^\perp$ and $M \subset L^\perp$, which is equivalent to $P_L P_M = O$;

(b) *inclined* otherwise; and

(c) *orthogonally incident* if $L \cap (L \cap M)^\perp$ and $M \cap (L \cap M)^\perp$ are orthogonal.

(d) L is *completely inclined* to M if $L \cap M^\perp = \{\mathbf{0}\}$; and

(e) L, M are *totally inclined* if they are completely inclined to each other.

(f) The *dimension of inclination* between L, M is

$$r(L, M) = \text{rank}(P_L P_M).$$

(g) The *coefficient of inclination* between L, M is

$$R(L, M) = \text{trace}\,(P_L P_M).$$

(h) A pair of subspaces L_1, M_1 are *reciprocal* in L, M if

$$L_1 = P_L M_1, \quad M_1 = P_M L_1; \quad \text{and}$$

(i) a pair of vectors $\mathbf{x}, \mathbf{y}$ are *reciprocal* if they span reciprocal lines.

In particular, the inclination coefficient between a pair of vectors $\mathbf{x}, \mathbf{y}$,

$$R(\mathbf{x}, \mathbf{y}) = \text{trace}\left(\frac{(\mathbf{x}\mathbf{x}^T)}{(\mathbf{x}^T\mathbf{x})}\frac{(\mathbf{y}\mathbf{y}^T)}{(\mathbf{y}^T\mathbf{y})}\right) = \frac{(\mathbf{x}^T\mathbf{y})^2}{(\mathbf{x}^T\mathbf{x})(\mathbf{y}^T\mathbf{y})} = \cos^2 \angle\{\mathbf{x}, \mathbf{y}\}, \quad (105)$$

giving the angle $0 \leq \angle\{\mathbf{x}, \mathbf{y}\} \leq \pi/2$ between the vectors.

Eigenvalues and eigenvectors of the products $P_L P_M$, $P_M P_L$ are used below. The following properties are stated for complex matrices:

LEMMA 1. *Let L, M be subspaces of $\mathbb{C}^n$ and let $\mathbf{x}_\lambda$ denote an eigenvector of $P_L P_M$ corresponding to the eigenvalue λ. Then:*

(a) *The eigenvalues λ are real and $0 \leq \lambda \leq 1$.*

(b) *If $\lambda \neq \mu$, then $\mathbf{x}_\lambda$, $\mathbf{x}_\mu$ are orthogonal.*

(c) *An eigenvector $\mathbf{x}_1$ (i.e., $P_L P_M \mathbf{x} = \mathbf{x}$) is in $L \cap M$ (the eigenvectors $\mathbf{x}_1$ span $L \cap M$).*

(d) *An eigenvector $\mathbf{x}_0$ (i.e., $P_L P_M \mathbf{x} = \mathbf{0}$) is in $M^\perp \overset{\perp}{\oplus} (M \cap L^\perp)$.*

(e) *If the columns of the matrices $Q_L \in \mathbb{C}^{n \times \ell}$, $Q_M \in \mathbb{C}^{n \times m}$ are o.n. bases for L and M, respectively, then*

$$\lambda(P_L P_M) = \{\sigma^2 : \sigma \in \sigma(Q_L^* Q_M)\},$$

i.e., the eigenvalues of $P_L P_M$ are the squares of singular values of $Q_L^* Q_M$.

PROOF. If $\lambda \neq 0$, then

$$P_L P_M \mathbf{x}_\lambda = \lambda \mathbf{x}_\lambda$$

shows that $\mathbf{x}_\lambda \in L$ and, therefore,

$$P_L P_M P_L \mathbf{x}_\lambda = \lambda \mathbf{x}_\lambda,$$

but $P_L P_M P_L = P_L P_M^2 P_L = P_L P_M (P_L P_M)^*$, showing that $P_L P_M P_L$ is Hermitian and PSD, proving (b) and $0 \leq \lambda$. The inequality $\lambda \leq 1$ is from

$$\lambda = \frac{\langle \mathbf{x}_\lambda, P_L P_M \mathbf{x}_\lambda \rangle}{\langle \mathbf{x}_\lambda, \mathbf{x}_\lambda \rangle}.$$

Part (e) follows from $P_L = Q_L Q_L^*$, $P_M = Q_M Q_M^*$. $\qquad \square$

THEOREM 8 (Afriat [3], Theorem 4.4).

(a) *Any reciprocal vectors* $\mathbf{x}, \mathbf{y}$ *in* L, M *with inclination coefficient* σ^2 *are eigenvectors of* $P_L P_M$, $P_M P_L$, *both with the eigenvalue* σ^2.

(b) *If* $\mathbf{x}$ *is an eigenvector of* $P_L P_M$ *with eigenvalue* σ^2, *then* $\mathbf{y} = P_M \mathbf{x}$ *is an eigenvector of* $P_M P_L$ *with the same eigenvalue, and* $\mathbf{x}, \mathbf{y}$ *are reciprocal vectors in* L, M *with inclination coefficient* σ^2.

PROOF. If $\mathbf{x}, \mathbf{y}$ are reciprocals, then $\lambda \mathbf{x} = P_L \mathbf{y}$, $\mu \mathbf{y} = P_M \mathbf{x}$, for some $\lambda, \mu \in \mathbb{R}$, and

$$\lambda \langle \mathbf{x}, \mathbf{x} \rangle = \langle \mathbf{x}, P_L \mathbf{y} \rangle = \langle \mathbf{x}, \mathbf{y} \rangle. \quad \therefore \ \lambda = \frac{\langle \mathbf{x}, \mathbf{y} \rangle}{\langle \mathbf{x}, \mathbf{x} \rangle}.$$

$$\mu \langle \mathbf{y}, \mathbf{y} \rangle = \langle \mathbf{y}, P_M \mathbf{x} \rangle = \langle \mathbf{y}, \mathbf{x} \rangle. \quad \therefore \ \mu = \frac{\langle \mathbf{x}, \mathbf{y} \rangle}{\langle \mathbf{y}, \mathbf{y} \rangle}.$$

$$\therefore \ \lambda \mu = \frac{\langle \mathbf{x}, \mathbf{y} \rangle^2}{\langle \mathbf{x}, \mathbf{x} \rangle \langle \mathbf{y}, \mathbf{y} \rangle} = \cos^2 \angle \{\mathbf{x}, \mathbf{y}\} = \sigma^2.$$

$$\therefore \ P_L P_M \mathbf{x} = \mu P_L \mathbf{y} = \lambda \mu \mathbf{x} = \sigma^2 \mathbf{x},$$

$$P_M P_L \mathbf{y} = \lambda P_M \mathbf{x} = \lambda \mu \mathbf{y} = \sigma^2 \mathbf{y},$$

proving (a). Conversely, if $\mathbf{x}$ is an eigenvector of $P_L P_M$ for the eigenvalue σ^2 and, if $\mathbf{y} = P_M \mathbf{x}$, then $P_L \mathbf{y} = P_L P_M \mathbf{x} = \sigma^2 \mathbf{x}$ and $P_M P_L \mathbf{y} = \sigma^2 P_M \mathbf{x} = \sigma^2 \mathbf{y}$, so $\mathbf{y}$ is an eigenvector of $P_M P_L$ with eigenvalue σ^2, and $\mathbf{x}, \mathbf{y}$ are reciprocals. $\qquad \square$

THEOREM 9 (Afriat [3], Theorem 5.4). *If* $\mathbf{x}_\lambda, \mathbf{y}_\lambda$ *are reciprocal vectors of* L, M *with inclination* λ, $\lambda = \alpha, \beta \neq 0$, *then the orthogonality conditions*

$$\mathbf{x}_\alpha \perp \mathbf{x}_\beta, \quad \mathbf{x}_\alpha \perp \mathbf{y}_\beta, \quad \mathbf{y}_\alpha \perp \mathbf{x}_\beta, \quad \mathbf{y}_\alpha \perp \mathbf{y}_\beta, \qquad (106)$$

are equivalent, and hold if $\alpha \neq \beta$.

PROOF. Let the vectors be normalized, so that $\langle \mathbf{x}_\lambda, \mathbf{y}_\lambda \rangle = \lambda$ and

$$P_L \mathbf{y}_\lambda = \lambda^{1/2} \mathbf{x}_\lambda, \quad P_M \mathbf{x}_\lambda = \lambda^{1/2} \mathbf{y}_\lambda.$$

Then

$$\langle \mathbf{x}_\mu, \mathbf{y}_\lambda \rangle = \langle \mathbf{x}_\mu, P_L \mathbf{y}_\lambda \rangle = \lambda^{1/2} \langle \mathbf{x}_\mu, \mathbf{x}_\lambda \rangle,$$

showing that $\mathbf{x}_\mu \perp \mathbf{y}_\lambda \iff \mathbf{x}_\mu \perp \mathbf{x}_\lambda$, since $\lambda \neq 0$. The remaining equivalences follow by symmetry. The computation

$$\alpha^{1/2} \beta^{1/2} \langle \mathbf{y}_\alpha, \mathbf{x}_\beta \rangle = \alpha^{1/2} \langle \mathbf{y}_\alpha, P_L \mathbf{y}_\beta \rangle = \alpha^{1/2} \langle P_L \mathbf{y}_\alpha, \mathbf{y}_\beta \rangle = \alpha \langle \mathbf{x}_\alpha, \mathbf{y}_\beta \rangle,$$

$$\alpha^{1/2} \beta^{1/2} \langle \mathbf{y}_\alpha, \mathbf{x}_\beta \rangle = \beta^{1/2} \langle P_M \mathbf{x}_\alpha, \mathbf{x}_\beta \rangle = \beta^{1/2} \langle \mathbf{x}_\alpha, P_M \mathbf{x}_\beta \rangle = \beta \langle \mathbf{x}_\alpha, \mathbf{y}_\beta \rangle,$$

gives

$$\alpha \langle \mathbf{x}_\alpha, \mathbf{y}_\beta \rangle = \beta \langle \mathbf{x}_\alpha, \mathbf{y}_\beta \rangle$$

and (106) holds if $\alpha \neq \beta$. □

Let L, M be subspaces and consider the reciprocal subspaces $P_L M$, $P_M L$. They have the same dimension,

$$\dim P_L M = \dim P_M L = r(L, M), \tag{107}$$

and their coefficient of inclination is

$$R(P_M L, P_L M) = R(L, M). \tag{108}$$

The subspaces $P_L M$, $P_M L$ are spanned by pairs $\mathbf{x}_i, \mathbf{y}_i$ of reciprocal vectors, which are eigenvectors of $P_L P_M$, $P_M P_L$ corresponding to nonzero eigenvalues $\sigma_i^2 = \cos^2 \angle \{\mathbf{x}_i, \mathbf{y}_i\}$.

DEFINITION 3. The angles $\theta_i = \angle \{\mathbf{x}_i, \mathbf{y}_i\}$ between reciprocal vectors,

$$0 \leq \theta_1 \leq \theta_2 \leq \cdots \leq \theta_r \leq \frac{\pi}{2}, \quad r = r(L, M), \tag{109}$$

are called the *principal angles*, or *canonical angles*, between L and M.

Using Lemma 1(a,b) and the extremal characterization of the eigenvalue of Hermitian matrices, §0.2.11(d), we get the following result, an alternative definition of principal angles:

THEOREM 10. *Let L, M be subspaces in $\mathbb{R}^n$, with dimension of inclination $r = r(L, M)$. The principal angles between L and M,*

$$0 \leq \theta_1 \leq \theta_2 \leq \cdots \leq \theta_r \leq \frac{\pi}{2} \tag{109}$$

are given by

$$\cos \theta_i = \frac{\langle \mathbf{x}_i, \mathbf{y}_i \rangle}{\|\mathbf{x}_i\| \|\mathbf{y}_i\|} \tag{110a}$$

$$= \max \left\{ \frac{\langle \mathbf{x}, \mathbf{y} \rangle}{\|\mathbf{x}\| \|\mathbf{y}\|} : \begin{array}{l} \mathbf{x} \in L, \quad \mathbf{x} \perp \mathbf{x}_k, \\ \mathbf{y} \in M, \quad \mathbf{y} \perp \mathbf{y}_k, \end{array} k \in \overline{1, i-1} \right\},$$

where

$$(\mathbf{x}_i, \ \mathbf{y}_i) \in L \times M, \quad i \in \overline{1, r}, \tag{110b}$$

are the corresponding pairs of reciprocal vectors. □

Lemma 1(e) allows using the SVD to compute principal angles as follows:

LEMMA 2 (Björck and Golub [**112**]). *Let the columns of $Q_L \in \mathbb{R}^{n \times \ell}$ and $Q_M \in \mathbb{R}^{n \times m}$ be o.n. bases for L and M, respectively, let $r = r(L, M)$, and let*

$$\sigma_1 \geq \sigma_2 \geq \cdots \geq \sigma_r > 0$$

be the singular values of $Q_M^T Q_L$, then

$$\cos \theta_i = \sigma_i, \quad i \in \overline{1, r}, \tag{111}$$

and

$$\sigma_1 = \cdots = \sigma_k = 1 > \sigma_{k+1} \quad \textit{if and only if} \quad \dim(L \cap M) = k. \quad □$$

The following concept is another measure of the angles between subspaces:

DEFINITION 4. Let L, M be subspaces of $\mathbb{R}^n$. The *distance* between L and M is

$$\text{dist}(L, M) = \|P_L - P_M\|_2 \tag{112a}$$

$$= \sup_{\mathbf{x} \in \mathbb{R}^n} \frac{\|(P_L - P_M)\mathbf{x}\|}{\|\mathbf{x}\|}. \tag{112b}$$

The geometric meaning of $\text{dist}(L, M)$ is given in the next lemma, which follows from the CS-decomposition of Stewart [**781**]. We give a geometric argument.

LEMMA 3. *Let L, M be subspaces of $\mathbb{R}^n$. Then*

$$\text{dist}(L, M) = \sin \theta^* \tag{113}$$

where θ^* is the maximal angle between vectors $\mathbf{x} \in L \cap (L \cap M)^\perp$ and $\mathbf{y} \in M \cap (L \cap M)^\perp$.

PROOF. An elementary geometric argument shows that RHS(112b) ≤ 1, see Ex. 67. If $L \cap M^\perp \neq \{\mathbf{0}\}$ or $M \cap L^\perp \neq \{\mathbf{0}\}$, then $\theta^* = \pi/2$ and RHS(112b) $= 1$ by taking $\mathbf{x}$ in the nonempty intersection. Assume L, M are totally inclined (and therefore dim $L =$ dim M by Ex. 0.12). Then

$$\text{dist}(L, M) = \sup_{\mathbf{x} \in L} \frac{\|\mathbf{x} - P_M \mathbf{x}\|}{\|\mathbf{x}\|}. \tag{114}$$

Indeed,

$$\text{RHS}(112b) \geq \text{RHS}(114), \quad \text{by definition,}$$

and

$$\text{RHS}(112b) \leq \sup_{\mathbf{x} \in \mathbb{R}^n} \frac{\|(P_L - P_M)\mathbf{x}\|}{\|P_L \mathbf{x}\|}, \quad \begin{array}{l}\text{which is infinite in the excluded}\\ \text{case } M \cap L^\perp \neq \{\mathbf{0}\}),\end{array}$$
$$= \text{RHS}(114).$$

Consider the right triangle with sides $\|\mathbf{x}\|, \|\mathbf{x} - P_M \mathbf{x}\|, \|P_M \mathbf{x}\|$. The ratio $\|\mathbf{x} - P_M \mathbf{x}\|/\|\mathbf{x}\|$ is the sine of the angle between $\mathbf{x}$ and $P_M \mathbf{x}$. We conclude that

$$\text{RHS}(114) = \frac{\|\mathbf{x}_r - \mathbf{y}_r\|}{\|\mathbf{x}_r\|},$$

where $\mathbf{x}_r, \mathbf{y}_r = P_M \mathbf{x}_r$ are the reciprocal vectors, corresponding to the largest principal angle between L, M. $\qquad\square$
See also Exs. 68–69 below.

We next discuss inequalities involving principal angles. Let

$$0 \leq \theta_1 \leq \theta_2 \leq \cdots \leq \theta_\ell \leq \frac{\pi}{2} \tag{109}$$

be the principal angles between the subspaces L, M and define

$$\sin\{L, M\} := \prod_{i=1}^{r} \sin \theta_i, \tag{115a}$$

$$\cos\{L, M\} := \prod_{i=1}^{r} \cos \theta_i. \tag{115b}$$

Note that (115a) and (115b) are just notation, and not ordinary trigonometrical functions. In particular,

$$\sin^2\{L, M\} + \cos^2\{L, M\} \leq 1.$$

Let L be a subspace of $\mathbb{R}^n$ of dimension r, spanned by vectors $\{\mathbf{v}_1, \ldots, \mathbf{v}_k\}$. Let $\mathbf{x} \in \mathbb{R}^n$ be decomposed as $\mathbf{x} = \mathbf{x}_L + \mathbf{x}_{L^\perp}$ with $\mathbf{x}_L \in L$, $\mathbf{x}_{L^\perp} \in L^\perp$ and let the one-dimensional subspace M be spanned by $\mathbf{x}$. Then the equation

$$\|\mathbf{x}_{L^\perp}\| = \frac{\mathrm{vol}_{r+1}(\mathbf{v}_1, \ldots, \mathbf{v}_k, \mathbf{x})}{\mathrm{vol}_r(\mathbf{v}_1, \ldots, \mathbf{v}_k)}, \qquad (2.77)$$

can be written as

$$\mathrm{vol}_{r+1}(\mathbf{v}_1, \ldots, \mathbf{v}_k, \mathbf{x}) = \mathrm{vol}_r(\mathbf{v}_1, \ldots, \mathbf{v}_k)\|\mathbf{x}\|_2 \sin\theta, \qquad (116)$$

where θ is the principal angle between $\mathbf{x}$, L. For the general case, Afriat gave the following "equalized" Hadamard inequality.

LEMMA 4 (Afriat [3], Corollary, p. 812). *Let*

$$A = [A_1 \; A_2], \; A_1 \in \mathbb{R}_\ell^{n \times \ell}, \; A_2 \in \mathbb{R}_m^{n \times m}.$$

Then

$$\mathrm{vol}_{\ell+m} A = \mathrm{vol}\, A_1 \,\mathrm{vol}\, A_2 \, \sin\{R(A_1), R(A_2)\}, \qquad (117)$$

where $\sin\{R(A_1), R(A_2)\}$ *is the product of principal sines between* $R(A_1)$ *and* $R(A_2)$. □

In Lemma 4 the matrices A_1, A_2 are of full column rank. A generalization of the Hadamard inequality follows; see also Ex. 70 below.

THEOREM 11 (Miao and Ben-Israel [549], Theorem 4). *Let*

$$A = [A_1 \; A_2], \; A_1 \in \mathbb{R}_\ell^{n \times n_1}, \; A_2 \in \mathbb{R}_m^{n \times n_2}, \; \mathrm{rank}\, A = \ell + m.$$

Then

$$\mathrm{vol}\, A = \mathrm{vol}\, A_1 \,\mathrm{vol}\, A_2 \, \sin\{R(A_1), R(A_2)\}. \qquad (118)$$

PROOF.

$$\mathrm{vol}^2 A = \sum_J \mathrm{vol}^2 A_{*J},$$

where the summation is over all $n \times (\ell + m)$ submatrices of rank $\ell + m$. Since every $n \times (\ell + m)$ submatrix of rank $\ell + m$ has ℓ columns A_{*J_1} from A_1 and m columns A_{*J_2} from A_2, then

$$\mathrm{vol}^2 A = \sum_{J_1} \sum_{J_2} \mathrm{vol}^2[A_{*J_1} \; A_{*J_2}],$$

$$= \sum_{J_1} \sum_{J_2} \mathrm{vol}^2 A_{*J_1} \, \mathrm{vol}^2 A_{*J_2} \, \sin^2\{R(A_1), R(A_2)\}, \quad \text{by Lemma 4,}$$

$$= \mathrm{vol}^2 A_1 \, \mathrm{vol}^2 A_2 \, \sin^2\{R(A_1), R(A_2)\}.$$

□

The Cauchy–Schwarz inequality, Ex. 0.2, is next extended to matrices of full column rank.

THEOREM 12 (Miao and Ben-Israel [**549**], Theorem 5). *Let* $B, C \in \mathbb{R}_r^{n \times r}$. *Then*

$$|\det(B^T C)| = \text{vol}\, B \; \text{vol}\, C \; \cos\{R(B), R(C)\},\qquad(119)$$

where $\cos\{R(B), R(C)\}$ *is the product of principal cosines between* $R(B)$ *and* $R(C)$, *see* (115b).

PROOF. Let Q_B and Q_C be o.n. bases for $R(B)$ and $R(C)$, respectively, so that,

$$B = Q_B R_B, \quad C = Q_C R_C,$$

for some matrices $R_B, R_C \in \mathbb{R}_r^{r \times r}$. Then

$$|\det(B^T C)| = |\det(R_B^T)| \, |\det(R_C)| \, |\det(Q_B^T Q_C)|,$$
$$= \text{vol}\, B \; \text{vol}\, C \; \cos\{R(B), R(C)\}, \quad \text{by Lemma 2.} \qquad \square$$

Exercises

EX. 64. (Afriat [**3**, Theorem 4.6]).

$$L \cap (P_L M)^\perp = L \cap M^\perp = L \cap (P_M L)^\perp.\qquad(120)$$

PROOF. If $\mathbf{u} = P_L \mathbf{v}$, then $P_L \mathbf{u} = \mathbf{u}$ and, therefore,

$$\mathbf{u} = P_L \mathbf{v}, \quad P_M P_L \mathbf{u} = 0 \iff \mathbf{u} = P_L \mathbf{v}, \; P_M \mathbf{u} = 0,$$

proving $L \cap (P_L M)^\perp = L \cap M^\perp$. The other statement is proved similarly. $\square$

EX. 65. (Afriat [**3**, Theorem 4.7]). If the subspaces L, M are inclined, then the reciprocal subspaces $P_L M, P_M L$ are totally inclined.

PROOF. Since $P_L P_M$ and $P_L P_M P_L = P_L P_M (P_L P_M)^T$ have the same range, any $\mathbf{0} \neq \mathbf{x} \in P_L M$ is $\mathbf{x} = P_L P_M P_L \mathbf{v}$ for some $\mathbf{v}$, and therefore $\mathbf{x}$ is not orthogonal to the vector $P_M P_L \mathbf{v} \in P_M L$. $\square$

EX. 66. (Afriat [**3**, Theorem 4.8]). If L, M are inclined subspaces in $\mathbb{R}^n$, then:

(a) $L = (P_L M) \overset{\perp}{\oplus} (L \cap M^\perp)$.

(b) $M = (P_M L) \overset{\perp}{\oplus} (M \cap L^\perp)$.

(c) $P_L M, P_M L$ are totally inclined.

(d) $\dim P_L M = \dim P_M L = r(L, M)$.

(e) $P_M L \perp (L \cap M^\perp)$.

(f) $P_L M \perp (M \cap L^\perp)$.

(g) $(L \cap M^\perp) \perp (M \cap L^\perp)$.

EX. 67. Let L, M be subspaces of $\mathbb{R}^n$. Then

$$\|P_L - P_M\|_2 \leq 1.$$

PROOF. Use the definition (112b), and for any $\mathbf{x} \in \mathbb{R}^n$ consider the right triangle $\{\mathbf{0}, \mathbf{x}, P_L \mathbf{x}\}$. Then the point $P_L \mathbf{x}$ is on a circle of diameter $\|\mathbf{x}\|$ centered at $\mathbf{x}/2$. Similarly, the right triangle $\{\mathbf{0}, \mathbf{x}, P_M \mathbf{x}\}$ shows that the point $P_M \mathbf{x}$ lies on a circle of diameter $\|\mathbf{x}\|$ centered at $\mathbf{x}/2$. Therefore the two points $P_L \mathbf{x}, P_M \mathbf{x}$ are on a sphere with diameter and center as above, and the length of the difference $\|P_L \mathbf{x} - P_M \mathbf{x}\|$ is no greater than $\|\mathbf{x}\|$, the diameter of the sphere. $\square$

EX. 68. Let L, M be inclined subspaces of $\mathbb{R}^n$. Then the following statements are equivalent:

(a) $\text{dist}(L, M) < 1$.

(b) $L \cap M^\perp = \{\mathbf{0}\}$, $M \cap L^\perp = \{\mathbf{0}\}$.

(c) $\dim L = \dim M = \dim (P_L M)$.

PROOF. (a) $\implies$ (b)   If $0 \neq \mathbf{x} \in L \cap M^{\perp}$, then $P_L \mathbf{x} = \mathbf{x}$ and $P_M \mathbf{x} = \mathbf{0}$.
$\therefore$ $(P_L - P_M)\mathbf{x} = \mathbf{x}$, showing that $\|P_L - P_M\|_2 \geq 1$, contradicting (a). The other statement follows by interchanging L and M.

(b) $\implies$ (c)   If $\dim L > \dim M$ then there is $0 \neq \mathbf{x} \in L \cap M^{\perp}$ (see Ex. 0.12). Similarly, $\dim L < \dim M$ is excluded by (b). It remains to show that $\dim L = \dim(P_L M)$. Suppose $\dim(P_L M) < \dim L$. Since $P_L M \subset L$ it follows that there exists $0 \neq \mathbf{x} \in L$ such that $\mathbf{x} \in (P_L M)^{\perp}$. Therefore $\mathbf{x} = P_L \mathbf{x} \in M^{\perp}$, and $\mathbf{x} \in L \cap M^{\perp}$, a contradiction.

(c) $\implies$ (a)   From (c) and Ex. 66(a) we conclude that $L = P_L M$. Similarly, (c) and Ex. 66(b,d) yield $M = P_M L$. Therefore, by Ex. 66(c), L and M are totally inclined, so the maximal angle θ^* between $\mathbf{x} \in L \cap (L \cap M)^{\perp}$ and $\mathbf{y} \in M \cap (L \cap M)^{\perp}$ is acute. Then (a) follows from Lemma 3. $\square$

EX. 69. Let L, M be subspaces of $\mathbb{R}^n$. Then $\operatorname{dist}(L, M) = 0$ if and only if $L = M$.

EX. 70. *A generalized Hadamard inequality.* Let $A = [A_1 \ A_2]$ be a square matrix and let $A_1 \in \mathbb{R}^{n \times \ell}$, $A_2 \in \mathbb{R}^{n \times m}$. Then

$$|\det(A| = \operatorname{vol}_{\ell}(A_1) \, \operatorname{vol}_m(A_2) \, \sin\{R(A_1), R(A_2)\}.$$

PROOF. Follows from Theorem 11. $\square$

In the following exercises we use the notation of §0.5.3, p. 29. In addition: The *basic subspaces* of dimension r of $\mathbb{R}^n$ are the $\binom{n}{r}$ subspaces

$$\mathbb{R}_J^n := \{\mathbf{x} = [x_k] \in \mathbb{R}^n : x_k = 0 \text{ if } k \notin J\}, \quad J \in Q_{r,n}, \tag{121}$$

which, for $r = 1$, reduce to the n coordinate lines

$$\mathbb{R}_{\{j\}}^n := \{\mathbf{x} = [x_k] \in \mathbb{R}^n : x_k = 0 \text{ if } k \neq j\}, \quad j \in \overline{1, n}. \tag{122}$$

EX. 71. (Miao and Ben-Israel [**549**, Corollary 2]). Let $A \in \mathbb{R}_r^{m \times n}$, $I \in Q_{r,m}$. Then

$$\cos\{R(A), \mathbb{R}_I^m\} = \frac{|\det A_{IJ}|}{\operatorname{vol} A_{*J}} \tag{123}$$

for any $J \in \mathcal{J}(A)$.
PROOF. Let $I = \{i_1, i_2, \dots, i_r\}$, $B = [\mathbf{e}_{i_1}, \dots, \mathbf{e}_{i_r}]$, and for any $J \in \mathcal{J}(A)$ let $C = A_{*J}$. Then

$$R(B) = \mathbb{R}_I^m, \quad R(C) = R(A), \quad \text{and} \quad B^T C = A_{IJ},$$

and, by Theorem 12,

$$\cos\{R(A), \mathbb{R}_I^m\} = \cos\{R(B), R(C)\} = \frac{|\det A_{IJ}|}{\operatorname{vol} A_{*J}}. \qquad \square$$

Note that for any $I \in Q_{r,m}$, the ratio $|\det A_{IJ}|/\operatorname{vol} A_{*J}$ is independent of the choice of $J \in \mathcal{J}(A)$.

EX. 72. Let

$$A = \begin{bmatrix} 1 & 2 & 3 \\ 4 & 5 & 6 \\ 7 & 8 & 9 \end{bmatrix},$$

with rank 2 and let $I = \{1, 2\}$, $J = \{1, 2\}$. Then

$$|\det A_{IJ}| = 3, \quad \operatorname{vol} A_{*J} = \sqrt{3^2 + 6^2 + 3^2} = 3\sqrt{6},$$

and, by (123),

$$\cos\{R(A),\ \mathbb{R}^3_{\{1,2\}}\} = \tfrac{1}{\sqrt{6}}.$$

EX. 73. (Miao and Ben-Israel [**549**, Corollary 3]). Let $L \subset \mathbb{R}^m$ be a subspace of dimension r. Then

$$\sum_{I \in Q_{r,m}} \cos^2\{L, \mathbb{R}^m_I\} = 1.$$

PROOF. Follows from (123) since $\text{vol}^2 A_{*J} = \sum_{I \in \mathcal{I}(A)} \det^2 A_{IJ}$. □

EX. 74. (Miao and Ben-Israel [**553**, Theorem 2]). If L and M are subspaces of $\mathbb{R}^n$ of dimension r, then

$$\cos\{L, M\} \le \sum_{J \in Q_{r,n}} \cos\{L, \mathbb{R}^n_J\}\ \cos\{M, \mathbb{R}^n_J\},$$

with equality if and only if the corresponding Plücker coordinates of L and M have the same signs.

EX. 75. Let $A \in \mathbb{R}^{m \times n}_r$. Recall the representation of the Moore–Penrose inverse $A^\dagger$ as a convex combination of basic inverses (see p. 122),

$$A^\dagger = \sum_{(I,J) \in \mathcal{N}(A)} \lambda_{IJ}\ \widehat{A^{-1}_{IJ}}, \tag{3.75}$$

with weights

$$\lambda_{IJ} = \frac{\det^2 A_{IJ}}{\sum_{(K,L) \in \mathcal{N}(A)} \det^2 A_{KL}}, \quad (I, J) \in \mathcal{N}(A). \tag{3.77}$$

Reversing the roles of $A, A^\dagger$, and using (3.104), we get a representation of A as a convex combination

$$A = \sum_{I \in \mathcal{I}(A)} \sum_{J \in \mathcal{J}(A)} \frac{\det^2 A_{IJ}}{\text{vol}^2 A} \widehat{(A^\dagger)^{-1}_{JI}}, \tag{124}$$

where $\widehat{(A^\dagger)^{-1}_{JI}}$ is an $m \times n$ matrix with the inverse of the $(J, I)^{\text{th}}$ submatrix of $A^\dagger$ in position (I, J) and zeros elsewhere. An alternative statement of (124) is given below.

EX. 76. (Miao and Ben-Israel [**553**, Theorem 3]). If $A \in \mathbb{R}^{m \times n}_r$, $r > 0$, then there exist linear operators $\{B_{IJ} : (I, J) \in \mathcal{N}(A)\}$ such that $B_{IJ} : \mathbb{R}^n_J \to \mathbb{R}^m_I$ is one-to-one and onto, $N(B_{IJ}) = (\mathbb{R}^n_J)^\perp$, and

$$A = \sum_{I \in \mathcal{I}(A)} \sum_{J \in \mathcal{J}(A)} \cos^2\{R(A), \mathbb{R}^m_I\}\ \cos^2\{R(A^T), \mathbb{R}^n_J\}\ B_{IJ}. \tag{125}$$

OUTLINE OF PROOF. Let $A = CR$ be a rank factorization and apply (124) to C, R, separately, to get

$$A = \sum_{I \in \mathcal{I}(C)} \frac{\det^2 C_{I*}}{\text{vol}^2 C} \widehat{(C^\dagger)^{-1}_{*I}} \sum_{J \in \mathcal{J}(R)} \frac{\det^2 R_{J*}}{\text{vol}^2 R} \widehat{(R^\dagger)^{-1}_{J*}}. \tag{126}$$

Then use (123) and the facts $\mathcal{I}(A) = \mathcal{I}(C)$, $R(A) = R(C)$, $\mathcal{J}(A) = \mathcal{J}(R)$, $R(A^T) = R(R^T)$, and $A^\dagger = R^\dagger C^\dagger$. □

6. Perturbations

We study the behavior of the Moore–Penrose inverse of a perturbed matrix $A + E$, and its dependence on $A^\dagger$ and on the "error" E. Only a few results are shown, the reader is referred to Stewart [**781**] and Stewart and Sun [**784**, Chapter III] for more details, references, and related results. Again, the results are stated for real matrices, the complex analogs are easy to see.

If $A \in \mathbb{R}^{n \times n}$ is nonsingular, then all matrices

$$\left\{ A + E : \|E\| < 1/\|A^{-1}\| \right\}$$

are nonsingular, with inverse

$$(A + E)^{-1} = A^{-1} - (A^{-1}E)A^{-1} + (A^{-1}E)^2 A^{-1} - \cdots, \qquad (127)$$

see Ex. 77. Therefore the set $\mathbb{R}_n^{n \times n}$ of nonsingular matrices is open (as a subset of $\mathbb{R}^{n \times n}$), and the inverse function $f(A) = A^{-1}$ is continuous in this set,

$$\|E_k\| \to 0 \quad \Longrightarrow \quad (A + E_k)^{-1} \to A^{-1}$$

for every nonsingular A, see also Ex. 78.

The example

$$A = \begin{bmatrix} 1 & 0 \\ 0 & 0 \end{bmatrix}, \quad E_k = \begin{bmatrix} 0 & 0 \\ 0 & \frac{1}{k} \end{bmatrix}, \qquad (128)$$

shows that the Moore–Penrose inverse is not continuous

$$E_k \to O \quad \nRightarrow \quad (A + E_k)^\dagger \to A^\dagger.$$

The main result in this section is

$$(A + E_k)^\dagger \to A^\dagger \quad \Longleftrightarrow \quad \mathrm{rank}(A + E_k) \to \mathrm{rank}(A),$$

a condition violated by (128), where $\mathrm{rank}(A + E_k) = 2$ for all k.

The matrix norms $\| \cdot \|$ used here are unitarily invariant. We denote by $\widetilde{A} = A + E$ a *perturbation* of the matrix $A \in \mathbb{R}_r^{m \times n}$. We can simplify the notation by multiplying with suitable unitary matrices U, V, since $\|U^T A V\| = \|A\|$. If $U = [U_1 \ U_2] \in U^{m \times m}$ and $V = [V_1 \ V_2] \in U^{n \times n}$ are unitary matrices with $R(A) = R(U_1)$, $R(A^T) = R(V_1)$, then

$$U^T A V = \begin{bmatrix} U_1^T A V_1 & U_1^T A V_2 \\ U_2^T A V_1 & U_2^T A V_2 \end{bmatrix} = \begin{bmatrix} A_{11} & O \\ O & O \end{bmatrix} \qquad (129a)$$

where A_{11} is $r \times r$ and nonsingular. Applying the same matrices U, V to E and $\widetilde{A}$ we get

$$U^T E V = \begin{bmatrix} U_1^T E V_1 & U_1^T E V_2 \\ U_2^T E V_1 & U_2^T E V_2 \end{bmatrix} = \begin{bmatrix} E_{11} & E_{12} \\ E_{21} & E_{22} \end{bmatrix}, \qquad (129b)$$

$$U^T \widetilde{A} V = \begin{bmatrix} A_{11} + E_{11} & E_{12} \\ E_{21} & E_{22} \end{bmatrix} = \begin{bmatrix} \widetilde{A}_{11} & E_{12} \\ E_{21} & E_{22} \end{bmatrix}, \qquad (129c)$$

where $[A_{11} + E_{11} \ \ E_{12}] \in \mathbb{R}^{r \times n}$.

DEFINITION 5. (Wedin [**855**]). The matrix $\widetilde{A} = A + E$ is an *acute perturbation* if

$$\|P_{R(A)} - P_{R(\widetilde{A})}\|_2 < 1 \quad \text{and} \quad \|P_{R(A^T)} - P_{R(\widetilde{A}^T)}\|_2 < 1, \tag{130}$$

in which case we say that A and $\widetilde{A}$ are *acute*.

THEOREM 13 (Stewart and Sun [**784**], Theorem III.3.1). *Let* $A, E \in \mathbb{R}^{m \times n}$, $\widetilde{A} = A + E$. *The following statements are equivalent:*

(a) $\|P_{R(A)} - P_{R(\widetilde{A})}\|_2 < 1$;

(b) $R(A) \cap R(\widetilde{A})^\perp = \{\mathbf{0}\}$ *and* $R(\widetilde{A}) \cap R(A)^\perp = \{\mathbf{0}\}$; *and*

(c) $\operatorname{rank}(A) = \operatorname{rank}(\widetilde{A}) = \operatorname{rank}(P_{R(A)}\widetilde{A})$.

Corresponding statements hold for the subspaces $R(A^T)$, $R(\widetilde{A}^T)$.

PROOF. See Ex. 68 above.   □

For $\widetilde{A}^\dagger$ to remain well behaved, the perturbation $\widetilde{A}$ cannot stray too far from the original matrix A. This is the case for acute perturbations.

THEOREM 14 (Stewart and Sun [**784**], Theorem III.3.3). *Let* $A \in \mathbb{R}_r^{m \times n}$. *Then* $\widetilde{A}$ *is an acute perturbation if and only if the* $r \times r$ *submatrix* $\widetilde{A}_{11}$ *is nonsingular and*

$$E_{22} = E_{21}\widetilde{A}_{11}^{-1}E_{12} \tag{131}$$

in (129c), *in which case,*

$$\widetilde{A} = \begin{bmatrix} I_r \\ S \end{bmatrix} \widetilde{A}_{11}[I_r \ T] \tag{132}$$

with

$$T = E_{21}\widetilde{A}_{11}^{-1}, \quad S = \widetilde{A}_{11}^{-1}E_{12}, \tag{133}$$

and

$$\widetilde{A}^\dagger = [I_r \ T]^\dagger \, \widetilde{A}_{11}^{-1} \begin{bmatrix} I_r \\ S \end{bmatrix}^\dagger. \tag{134}$$

PROOF. If $\widetilde{A}_{11}$ is singular, then Theorem 13(b) is violated. Condition (131) is then equivalent to Theorem 13(c). The rest follows as in (5.22) and Theorem 5.2(d).   □

It follows that $\lim_{\widetilde{A} \to A} \operatorname{rank}(\widetilde{A}) = \operatorname{rank}(A)$ if and only if $\widetilde{A}$ is eventually an acute perturbation and the Moore–Penrose inverse is continuous on the set of acute perturbations. The following theorem shows what to expect in the nonacute case:

THEOREM 15 (Wedin [**855**]). *If* $A, \widetilde{A}$ *are not acute, then*

$$\|\widetilde{A}^\dagger - A^\dagger\|_2 \geq \frac{1}{\|E\|_2}, \tag{135a}$$

$$\|\widetilde{A}^\dagger\|_2 \geq \frac{1}{\|E\|_2}. \tag{135b}$$

PROOF. Let $\tilde{A}$ be nonacute. Then one of the two equations in Theorem 13(b), or one of the analogous statements for the subspaces $R(A^T)$, $R(\tilde{A}^T)$, is violated. Suppose there is a $\mathbf{0} \neq \mathbf{x} \in R(\tilde{A}) \cap R(A)^\perp$ (the other three cases are similarly treated) and assume $\|\mathbf{x}\|_2 = 1$. Then

$$1 = \mathbf{x}^T\mathbf{x} = \mathbf{x}^T P_{R(\tilde{A})}\mathbf{x} = \mathbf{x}^T \tilde{A}\tilde{A}^\dagger\mathbf{x} = \mathbf{x}^T (A + E)\tilde{A}^\dagger\mathbf{x} = \mathbf{x}^T E\tilde{A}^\dagger\mathbf{x}$$

$$\leq \|E\|_2\|\tilde{A}^\dagger\mathbf{x}\|_2.$$

$$\therefore \ \|\tilde{A}^\dagger\|_2 \geq \|\tilde{A}^\dagger\mathbf{x}\|_2 \geq \frac{1}{\|E\|_2},$$

proving (135b). Since $A^\dagger\mathbf{x} = \mathbf{0}$ (from $\mathbf{x} \in R(A)^\perp = N(A^\dagger)$) we have

$$\|\tilde{A}^\dagger - A^\dagger\|_2 \geq \|(\tilde{A}^\dagger - A^\dagger)\mathbf{x}\|_2 = \|\tilde{A}^\dagger\mathbf{x}\|_2 \geq \frac{1}{\|E\|_2},$$

completing the proof. □

Explicit expressions for $\tilde{A}^\dagger$ are available:

THEOREM 16.

$$\tilde{A}^\dagger - A^\dagger = -\tilde{A}^\dagger EA^\dagger + \tilde{A}^\dagger P_{N(A^T)} - P_{N(\tilde{A})}A^\dagger \tag{136a}$$

$$= -\tilde{A}^\dagger P_{R(\tilde{A})}EP_{R(A^T)}A^\dagger + \tilde{A}^\dagger P_{R(\tilde{A})}P_{N(A^T)}$$
$$- P_{N(\tilde{A})}P_{R(A^T)}A^\dagger \tag{136b}$$

$$= -\tilde{A}^\dagger P_{R(\tilde{A})}EP_{R(A^T)}A^\dagger + (\tilde{A}^T\tilde{A})^\dagger P_{R(\tilde{A}^T)}E^T P_{N(A^T)}$$
$$- P_{N(\tilde{A})}E^T P_{R(A)}(AA^T)^\dagger. \tag{136c}$$

If $\tilde{A}$ is acute,

$$\tilde{A}^\dagger - A^\dagger = -\tilde{A}^\dagger EA^\dagger. \tag{136d}$$

PROOF. That $A^\dagger + \text{RHS}(136a)$ is the Moore–Penrose inverse of $A + E$ can be checked directly. The last two terms in RHS(136a) drop if $\tilde{A}$ is acute, giving (136d). Finally, (136b)–(136c) are obtained from (136a) by inserting harmless factors, such as $P_{R(A^T)}$ in front of $A^\dagger$. □

Expression (136c), with E, $A^\dagger$, and $\tilde{A}^\dagger$ appearing in all terms, allows writing an error bound for LHS(136a) in the form:

THEOREM 17 (Wedin [855]). If $\|\cdot\|$ is a unitarily invariant norm, then

$$\|\tilde{A}^\dagger - A^\dagger\| \leq \mu \max\{\|A^\dagger\|^2, \|\tilde{A}^\dagger\|^2\}\|E\|, \tag{137}$$

where $\mu = 3$ (sharper values of μ are given for specific norms). □

In the acute case, error bounds for $\|\tilde{A}^\dagger - A^\dagger\|$ are obtained from (134).

Exercises

EX. 77. Let $A \in \mathbb{R}^{n \times n}$ be nonsingular and let $\|\ \|$ be any multiplicative matrix norm (see p. 13). Then $A + E$ is nonsingular for any matrix E satisfying

$$\|E\| < \frac{1}{\|A^{-1}\|}, \tag{138}$$

and its inverse is (127).
PROOF. From

$$A + E = A(I + A^{-1}E)$$

and Ex. 0.47, it follows that $A + E$ is nonsingular if $A^{-1}E$ is convergent which, by Ex. 0.41, is implied by $\|A^{-1}E\| < 1$, and therefore by $\|A^{-1}\|\|E\| < 1$. The expansion (127) is obtained by taking the inverse

$$(A + E)^{-1} = (I + A^{-1}E)^{-1}A^{-1}$$

and expanding $(I + A^{-1}E)^{-1}$, as in (0.66).   □
See also Ex. 30.

EX. 78. The inverse function $f(A) = A^{-1}$ is differentiable in $\mathbb{R}_n^{n \times n}$,

$$f'(A)\,dX = -A^{-1}\,dX\,A^{-1}. \tag{139}$$

PROOF.

$$\lim_{\|dX\| \to 0} \frac{\|(A + dX)^{-1} - A^{-1} - (A^{-1}\,dX\,A^{-1})\|}{\|dX\|} = 0$$

by (127).   □

EX. 79. If $A \in \mathbb{C}_r^{m \times n}$, $E \in \mathbb{C}^{m \times n}$, then the last $n-r$ singular values of $\widetilde{A} = A + E$ satisfy

$$\widetilde{\sigma}_{r+1}^2 + \cdots + \widetilde{\sigma}_n^2 \leq \|E\|_F^2.$$

PROOF. Use (70).   □

EX. 80. If the matrices A, E in $\mathbb{R}^{m \times n}$ satisfy

$$R(E) \subset R(A), \tag{140}$$

$$R(E^T) \subset R(A^T), \tag{141}$$

and

$$\|A^\dagger E\| < 1 \tag{142}$$

for any multiplicative matrix norm, then

$$(A + E)^\dagger = (I + A^\dagger E)^{-1}A^\dagger. \tag{143}$$

PROOF. The matrix $B = I + A^\dagger E$ is nonsingular by (142) and Exs. 0.41 and 0.47. Since

$$A + E = A + AA^\dagger E, \quad \text{by (140)},$$
$$= A(I + A^\dagger E),$$

it suffices to show that the matrices A and $B = I + A^\dagger$ have the "reverse order" property (4.31),

$$(A(I + A^\dagger E))^\dagger = (I + A^\dagger E)^{-1}A^\dagger,$$

which by Ex. 4.22 is equivalent to

$$R(A^T AB) \subset R(B) \tag{144}$$

and

$$R(BB^T A^T) \subset R(A^T). \tag{145}$$

Now (144) holds since B is nonsingular and (145) follows from

$$R(BB^T A^T) = R((I + A^\dagger E)(I + A^\dagger E)^T A^T)$$
$$= R(A^T + E^T A^{\dagger T} A^T + A^\dagger E(I + A^\dagger E)^T A^T)$$
$$\subset R(A^T), \quad \text{by (141)}.$$ □

Ex. 81. *Error bounds for generalized inverses* (Ben-Israel [**63**]). Let A, E satisfy
(140)–(142) or, equivalently, let $A, \tilde{A}$ be acute with zero blocks E_{12}, E_{21}, E_{22} in
(129b). Then,

$$\|(A + E)^\dagger - A^\dagger\| \leq \frac{\|A^\dagger E\|\|A^\dagger\|}{1 - \|A^\dagger E\|}. \tag{146}$$

If (140) and (141) hold, but (142) is replaced by

$$\|A^\dagger\|\|E\| < 1, \tag{147}$$

then

$$\|(A + E)^\dagger - A^\dagger\| \leq \frac{\|A^\dagger\|^2\|E\|}{1 - \|A^\dagger E\|}. \tag{148}$$

PROOF. From Ex. 80 it follows that

$$(A + E)^\dagger - A^\dagger = (I + A^\dagger E)^{-1} A^\dagger - A^\dagger$$

$$= \sum_{k=0}^{\infty} (-1)^k (A^\dagger E)^k A^\dagger - A^\dagger, \quad \text{by (142) and Ex. 0.47,}$$

$$= \sum_{k=1}^{\infty} (-1)^k (A^\dagger E)^k A^\dagger$$

and, hence,

$$\|(A + E)^\dagger - A^\dagger\| \leq \sum_{k=1}^{\infty} \|(A^\dagger E)\|^k \|A^\dagger\|$$

$$= \frac{\|A^\dagger E\|\|A^\dagger\|}{1 - \|A^\dagger E\|}, \quad \text{by (142).}$$

The condition (147), which is stronger than (142), then implies (148).   □

7. A Spectral Theory for Rectangular Matrices

The following theorem, due to Penrose [**635**], is a generalization to rectangular matrices of the classical spectral theorem for normal matrices (Theorem 2.15).

THEOREM 18 (Spectral Theorem for Rectangular Matrices). *Let $O \neq A \in \mathbb{C}_r^{m \times n}$ and let $d(A) = \{d_1, \ldots, d_r\}$ be complex scalars satisfying*

$$|d_i| = \sigma_i, \quad i \in \overline{1, r}, \tag{35}$$

where

$$\sigma_1 \geq \sigma_2 \geq \cdots \geq \sigma_r > 0 \tag{0.32}$$

are the singular values, $\sigma(A)$, of A.
    Then there exist r partial isometries $\{E_i : i \in \overline{1, r}\}$ in $\mathbb{C}_1^{m \times n}$ satisfying

$$E_i E_j^* = O, \quad E_i^* E_j = O, \quad 1 \leq i \neq j \leq r, \tag{149a}$$

$$E_i E^* A = A E^* E_i, \quad i \in \overline{1, r}, \tag{149b}$$

where

$$E = \sum_{i=1}^{r} E_i \tag{150}$$

is the partial isometry given by (77) *and*

$$A = \sum_{i=1}^{r} d_i E_i. \tag{151}$$

Furthermore, for each $i = 1, \ldots, r$, *the partial isometry* $(\overline{d_i}/|d_i|)E_i$ *is unique if the corresponding singular value is simple, i.e., if* $\alpha_i < \alpha_{i-1}$ *and* $\alpha_i > \alpha_{i+1}$ *for* $2 \leq i \leq r$ *and* $1 \leq i \leq r-1$, *respectively.*

PROOF. Let the vectors $\{u_1, u_2, \ldots, u_r\}$ satisfy (19a) and (19b), let vectors $\{v_1, v_2, \ldots, v_r\}$ be defined by (20), and let

$$E_i = u_i v_i^*, \quad i = 1, \ldots, r. \tag{152}$$

E_i is a partial isometry by Theorem 5(c), since $E_i E_i^* E_i = E_i$ by (19b) and (21b), from which (149a) also follows. The statement on uniqueness follows from (152), (35), (19a), (19b), and (20). The result (151) follows from (30a), which also shows the matrix E of (77) to be given by (150). Finally, (149b) follows from (150), (151), and (149a). ☐

As shown by the proof of Theorem 18, the spectral representation (151) of A is just a way of rewriting its SVD. The following spectral representation of $A^\dagger$ similarly follows from Corollary 1.

COROLLARY 3. *Let* A, d_i, *and* E_i, $i = 1, \ldots, r$, *be as in Theorem* 18. *Then*

$$A^\dagger = \sum_{i=1}^{r} \frac{1}{d_i} E_i. \tag{153}$$

☐

If $A \in \mathbb{C}_r^{n \times n}$ is normal with nonzero eigenvalues $\{\lambda_i : i \in \overline{1,r}\}$ ordered by

$$|\lambda_1| \geq |\lambda_2| \geq \cdots \geq |\lambda_r|, \tag{38}$$

then, by Ex. 19, the choice

$$d_i = \lambda_i, \quad i \in \overline{1,r}, \tag{39}$$

guarantees that

$$u_i = v_i, \quad i \in \overline{1,r}, \tag{40}$$

and, consequently, the partial isometries E_i of (152) are orthogonal projectors

$$P_i = u_i u_i^*, \quad i \in \overline{1,r}, \tag{154}$$

and (151) reduces to

$$A = \sum_{i=1}^{r} \lambda_i P_i, \tag{155}$$

giving the spectral theorem for normal matrices as a special case of Theorem 18.

The classical spectral theory for square matrices (see, e.g., Dunford and Schwartz [**246**, pp. 556–565]) makes extensive use of matrix functions $f : \mathbb{C}^{n \times n} \to \mathbb{C}^{n \times n}$, induced by scalar functions $f : \mathbb{C} \to \mathbb{C}$, according to

Definition 2.1. Similarly, the spectral theory for rectangular matrices given here uses matrix functions $f : \mathbb{C}^{m \times n} \to \mathbb{C}^{m \times n}$ which correspond to scalar functions $f : \mathbb{C} \to \mathbb{C}$, according to the following:

DEFINITION 6. Let $f : \mathbb{C} \to \mathbb{C}$ be any scalar function. Let $A \in \mathbb{C}_r^{m \times n}$ have a spectral representation

$$A = \sum_{i=1}^{r} d_i E_i \tag{151}$$

as in Theorem 18. Then the *matrix function* $f : \mathbb{C}^{m \times n} \to \mathbb{C}^{m \times n}$ *corresponding to* $f : \mathbb{C} \to \mathbb{C}$ is defined at A by

$$f(A) = \sum_{i=1}^{r} f(d_i) E_i. \tag{156}$$

Note that the value of $f(A)$ defined by (156) depends on the particular choice of the scalars $d(A)$ in (35). In particular, for a normal matrix $A \in \mathbb{C}^{n \times n}$, the choice of $d(A)$ by (39) reduces (156) to the classical definition – see (170) below – in the case that $f(0) = 0$ or that A is nonsingular.

Let

$$A = U_{(r)} D_{(r)} V_{(r)}^*, \quad D_{(r)} = \begin{bmatrix} d_1 & & \\ & \ddots & \\ & & d_r \end{bmatrix}, \tag{30a}$$

be a UDV^*-decomposition of a given $A \in \mathbb{C}_r^{m \times n}$. Then Definition 6 gives $f(A)$ as

$$f(A) = U_{(r)} f(D_{(r)}) V_{(r)}^*, \quad f(D_{(r)}) = \begin{bmatrix} f(d_1) & & \\ & \ddots & \\ & & f(d_r) \end{bmatrix}. \tag{157}$$

An easy consequence of Theorem 18 and Definition 6 is the following:

THEOREM 19. *Let* $f, g, h : \mathbb{C} \to \mathbb{C}$ *be scalar functions and let* $f, g, h : \mathbb{C}^{m \times n} \to \mathbb{C}^{m \times n}$ *be the corresponding matrix functions defined by Definition 6.*

Let $A \in \mathbb{C}_r^{m \times n}$ *have a* UDV^*-decomposition

$$A = U_{(r)} D_{(r)} V_{(r)}^* \tag{30a}$$

and let the partial isometry E *be given by*

$$E = U_{(r)} V_{(r)}^*. \tag{77}$$

Then:

  (a) *If* $f(z) = g(z) + h(z)$, *then* $f(A) = g(A) + h(A)$.
  (b) *If* $f(z) = g(z)h(z)$, *then* $f(A) = g(A)E^*h(A)$.
  (c) *If* $f(z) = g(h(z))$, *then* $f(A) = g(h(A))$.

PROOF. Parts (a) and (c) are obvious by Definition (156).

(b) If $f(z) = g(z)h(z)$, then

$$g(A)E^*h(A) = \left(\sum_{i=1}^{r} g(d_i)E_i\right)\left(\sum_{j=1}^{r} E_j^*\right)\left(\sum_{k=1}^{r} h(d_k)E_i\right),$$

by (156) and (150),

$$= \sum_{i=1}^{r} g(d_i)h(d_i)E_i, \quad \text{by (149a) and Theorem 5(c),}$$

$$= \sum_{i=1}^{r} f(d_i)E_i = f(A). \qquad \square$$

For matrix functions defined as above, an analog of Cauchy's integral theorem is given in Corollary 4 below. First we require

LEMMA 5. *Let $A \in \mathbb{C}_r^{m \times n}$ be represented by*

$$A = \sum_{i=1}^{r} d_i E_i. \tag{151}$$

Let $\{\widehat{d}_j : j \in \overline{1,q}\}$ be the set of distinct $\{d_i : i \in \overline{1,r}\}$ and let

$$\widehat{E}_j = \sum_{i} \{E_i : d_i = \widehat{d}_j\}, \quad j = 1, \dots, q. \tag{158}$$

For each $j \in \overline{1,q}$ let Γ_j be a contour (i.e., a closed rectifiable Jordan curve, positively oriented in the customary way) surrounding $\widehat{d}_j$ but no other $\widehat{d}_k$. Then:

(a) *For each $j \in \overline{1,q}$, $\widehat{E}_j$ is a partial isometry and*

$$\widehat{E}_j^* = \frac{1}{2\pi i} \int_{\Gamma_i} (zE - A)^\dagger \, dz. \tag{159}$$

(b) *If $f : \mathbb{C} \to \mathbb{C}$ is analytic in a domain containing the set surrounded by*

$$\Gamma = \bigcup_{j=1}^{q} \Gamma_j,$$

then

$$\sum_{j=1}^{r} f(d_j)E_j^* = \frac{1}{2\pi i} \int_{\Gamma} f(z)(zE - A)^\dagger \, dz, \tag{160}$$

in particular,

$$A^\dagger = \frac{1}{2\pi i} \int_{\Gamma} \frac{1}{z}(zE - A)^\dagger \, dz. \tag{161}$$

PROOF. (a) From (149a) and Theorem 5 it follows that $\widehat{E}_j$ and $\widehat{E}_j^*$ are partial isometries for each $j \in \overline{1,q}$. Also, from (150), (151), and Corollary 3,

$$(zE - A)^\dagger = \sum_{k=1}^{r} \frac{1}{z - d_k} E_k^*, \tag{162}$$

hence

$$\frac{1}{2\pi i} \int_{\Gamma_j} (zE - A)^\dagger \, dz = \sum_{k=1}^{r} \Big(\frac{1}{2\pi i} \int_{\Gamma_j} \frac{dz}{z - d_k} \Big) E_k^* = \sum_{\{d_k = \widehat{d}_j\}} E_k^*,$$

by the assumptions on Γ_j and Cauchy's integral theorem,

$$= \widehat{E}_j^{\,*}, \quad \text{by (158)}.$$

(b) Similarly, we calculate

$$\frac{1}{2\pi i} \int_{\Gamma} f(z)(zE - A)^\dagger \, dz = \sum_{j=1}^{q} \sum_{k=1}^{r} \Big(\frac{1}{2\pi i} \int_{\Gamma_j} \frac{f(z)}{z - d_k} \Big) E_k^* = \sum_{j=1}^{q} f(\widehat{d}_j) \widehat{E}_j^{\,*}$$

$$= \sum_{j=1}^{r} f(d_j) E_j^*, \quad \text{proving (160)}.$$

Finally, (161) follows from (160) and Corollary 3.   □

Cartan's formula for matrix functions

$$f(A) = \frac{1}{2\pi i} \int_{\Gamma} f(z)(zI - A)^{-1} \, dz \tag{2.53}$$

is extended to rectangular matrices as follows:

COROLLARY 4. *Let A, E, Γ, and f be as in Lemma 5. Then*

$$f(A) = E \Big(\frac{1}{2\pi i} \int_{\Gamma} f(z)(zE - A)^\dagger \, dz \Big) E. \tag{163}$$

PROOF. Using (150) and (160) we calculate

$$E \Big(\frac{1}{2\pi i} \int_{\Gamma} f(z)(zE - A)^\dagger \, dz \Big) E = \Big(\sum_{i=1}^{r} E_i \Big) \Big(\sum_{j=1}^{r} f(d_j) E_j^* \Big) \Big(\sum_{k=1}^{r} E_k \Big)$$

$$= \sum_{j=1}^{r} f(d_j) E_j, \quad \text{by (149a) and Theorem 5(c)},$$

$$= f(A).   □$$

The *generalized resolvent* of a matrix $A \in \mathbb{C}^{m \times n}$ is the function $\widehat{R}(z, A) : \mathbb{C} \to \mathbb{C}^{n \times m}$ given by

$$\widehat{R}(z, A) = (zE - A)^\dagger, \tag{164}$$

where the partial isometry E is given as in Theorem 18. This definition is suggested by the classical definition of the *resolvent* of a square matrix as

$$R(z, A) = (zI - A)^{-1}, \quad \text{for all } z \notin \lambda(A). \tag{2.55}$$

In analogy to the classical case – see (2.56) – we state the following identity, known as the *(first) resolvent equation*.

LEMMA 6. *Let $A \in \mathbb{C}_r^{m \times n}$ and let $d(A)$ and E be as in Theorem 18. Then*

$$\widehat{R}(\lambda, A) - \widehat{R}(\mu, A) = (\mu - \lambda) \widehat{R}(\lambda, A) \widehat{R}(\mu, A) \tag{165}$$

for any scalars $\lambda, \mu \notin d(A)$.

PROOF.

$$\widehat{R}(\lambda, A) - \widehat{R}(\mu, A) = (\lambda E - A)^\dagger - (\mu E - A)^\dagger, \quad \text{by (164)},$$

$$= \sum_{k=1}^{r} \Big(\frac{1}{\lambda - d_k} - \frac{1}{\mu - d_k}\Big) E_k^*, \quad \text{by (162)},$$

$$= \sum_{k=1}^{r} \Big(\frac{\mu - \lambda}{(\lambda - d_k)(\mu - d_k)}\Big) E_k^*$$

$$= (\mu - \lambda) \Big(\sum_{k=1}^{r} \frac{1}{\lambda - d_k} E_k^*\Big) E \Big(\sum_{\ell=1}^{r} \frac{1}{\mu - d_\ell} E_\ell^*\Big),$$

by (149a), (150) and Theorem 5(c),

$$= (\mu - \lambda) \widehat{R}(\lambda, A) \widehat{R}(\mu, A), \quad \text{by (162)}. \qquad \square$$

The resolvent equation, (165), is used in the following lemma, based on Lancaster [**494**, p. 552].

LEMMA 7. *Let $A \in \mathbb{C}^{m \times n}$, let $d(A)$ and E be given as in Theorem 18, and let the scalar functions $f, g : \mathbb{C} \to \mathbb{C}$ be analytic in a domain $\mathcal{D}$ containing $d(A)$. If Γ is a contour surrounding $d(A)$ and lying in the interior of $\mathcal{D}$, then*

$$\Big(\frac{1}{2\pi i} \int_\Gamma f(\lambda) \widehat{R}(\lambda, A) \, d\lambda\Big) E \Big(\frac{1}{2\pi i} \int_\Gamma g(\lambda) \widehat{R}(\lambda, A) \, d\lambda\Big)$$

$$= \frac{1}{2\pi i} \int_\Gamma f(\lambda) g(\lambda) \widehat{R}(\lambda, A) \, d\lambda. \quad (166)$$

PROOF. Let Γ_1 be a contour surrounding Γ and still lying in the interior of $\mathcal{D}$. Then

$$\frac{1}{2\pi i} \int_\Gamma g(\lambda) \widehat{R}(\lambda, A) \, d\lambda = \frac{1}{2\pi i} \int_{\Gamma_1} g(\mu) \widehat{R}(\mu, A) \, d\mu,$$

which when substituted in LHS(166) gives

$$\Big(\frac{1}{2\pi i} \int_\Gamma f(\lambda) \widehat{R}(\lambda, A) \, d\lambda\Big) E \Big(\frac{1}{2\pi i} \int_{\Gamma_1} g(\mu) \widehat{R}(\mu, A) \, d\mu\Big)$$

$$= -\frac{1}{4\pi^2} \int_{\Gamma_1} \int_\Gamma f(\lambda) g(\mu) \widehat{R}(\lambda, A) E \widehat{R}(\mu, A) \, d\lambda \, d\mu$$

$$= \frac{1}{4\pi^2} \int_{\Gamma_1} \int_\Gamma f(\lambda) g(\mu) \frac{\widehat{R}(\lambda, A) - \widehat{R}(\mu, A)}{\lambda - \mu} \, d\lambda \, d\mu, \quad \text{by (165)},$$

$$= \frac{1}{4\pi^2} \int_\Gamma f(\lambda) \widehat{R}(\lambda, A) \Big(\int_{\Gamma_1} \frac{g(\mu)}{\lambda - \mu} d\mu\Big) d\lambda$$

$$\qquad - \frac{1}{4\pi^2} \int_{\Gamma_1} \Big(\int_\Gamma \frac{f(\lambda)}{\lambda - \mu} d\lambda\Big) g(\mu) \widehat{R}(\mu, A) \, d\mu$$

$$= \frac{1}{2\pi i} \int_\Gamma f(\lambda) g(\lambda) \widehat{R}(\lambda, A) \, d\lambda,$$

since $\int_{\Gamma_1} [g(\mu)/(\lambda - \mu)] \, d\mu = -2\pi i \, g(\lambda)$ and $\int_\Gamma [f(\lambda)/(\lambda - \mu)] \, d\lambda = 0$, by our assumptions on Γ, Γ_1. $\qquad \square$

We illustrate now the application of the above concepts to the solution of the matrix equation

$$AXB = D \qquad (167)$$

studied in Theorem 2.1. Here the matrices $A \in \mathbb{C}^{m \times n}$, $B \in \mathbb{C}^{k \times \ell}$, and $D \in \mathbb{C}^{m \times \ell}$ are given and, in addition, the matrices A and B have spectral representations, given by Theorem 18 as follows:

$$A = \sum_{i=1}^{p} d_i^A E_i^A, \qquad E^A = \sum_{i=1}^{p} E_i^A, \quad p = \operatorname{rank} A, \qquad (168a)$$

and

$$B = \sum_{i=1}^{q} d_i^B E_i^B, \qquad E^B = \sum_{i=1}^{q} E_i^B, \quad q = \operatorname{rank} B. \qquad (168b)$$

THEOREM 20. *Let A, B, D be as above, and let Γ_1 and Γ_2 be contours surrounding $d(A) = \{d_1^A, \dots, d_p^A\}$ and $d(B) = \{d_1^B, \dots, d_q^B\}$, respectively. If (167) is consistent, then it has the following solution:*

$$X = -\frac{1}{4\pi^2} \int_{\Gamma_1} \int_{\Gamma_2} \frac{\widehat{R}(\lambda, A) D \widehat{R}(\mu, B)}{\lambda \mu} \, d\mu \, d\lambda. \qquad (169)$$

PROOF. From (163) it follows that

$$A = E^A \Big(\frac{1}{2\pi i} \int_{\Gamma_1} \lambda \widehat{R}(\lambda, A) \, d\lambda \Big) E^A$$

and

$$B = E^B \Big(\frac{1}{2\pi i} \int_{\Gamma_2} \mu \widehat{R}(\mu, B) d\mu \Big) E^B.$$

Therefore,

$$AXB = A \Big[\frac{1}{2\pi i} \int_{\Gamma_1} \frac{\widehat{R}(\lambda, A)}{\lambda} D \Big(\frac{1}{2\pi i} \int_{\Gamma_2} \frac{\widehat{R}(\mu, B)}{\mu} \, d\mu \Big) d\lambda \Big] B$$

$$= E^A \Big[\frac{1}{2\pi i} \int_{\Gamma_1} \widehat{R}(\lambda, A) d\lambda \Big] D \Big[\frac{1}{2\pi i} \int_{\Gamma_2} \widehat{R}(\mu, B) \, d\mu \Big] E^B,$$

$$\text{by a double application of Lemma 7,}$$

$$= E^A (E^A)^* D (E^B)^* E^B, \quad \text{by (160) with } f \equiv 1,$$

$$= P_{R(A)} D P_{R(B^*)}, \quad \text{by (78)},$$

$$= A A^\dagger D B^\dagger B$$

$$= D, \quad \text{if and only if (167) is consistent, by Theorem 2.1.}$$

Alternatively, it follows from (161) and (169) that $X = A^\dagger D B^\dagger$, a solution of (167) if consistent. $\qquad \square$

For additional results along these lines see Lancaster [494] and Wimmer and Ziebur [879].

Exercises

Ex. 82. If $A \in \mathbb{C}^{n \times n}$ is normal with a spectral representation

$$A = \sum_{i=1}^{r} \lambda_i P_i \qquad (155)$$

then, for any $f \in \mathcal{F}(A)$, definition (2.42) gives

$$f(A) = \sum_{i=1}^{r} f(\lambda_i) P_i + f(0) P_{N(A)}, \qquad (170)$$

since the eigenvalues of a normal matrix have index one.

Ex. 83. *Generalized powers.* The matrix function $f : \mathbb{C}^{m \times n} \to \mathbb{C}^{m \times n}$ corresponding to the scalar function

$$f(z) = z^k, \quad k \text{ any integer,}$$

is denoted by

$$f(A) = A^{\langle k \rangle}$$

and called the *generalized* k^{th} *power* of $A \in \mathbb{C}^{m \times n}$. Definition 6 shows that

$$A^{\langle k \rangle} = \sum_{i=1}^{r} d_i^k E_i, \quad \text{by (156),} \qquad (171a)$$

or, equivalently,

$$A^{\langle k \rangle} = U_{(r)} D_{(r)}^k V_{(r)}^*, \quad \text{by (157).} \qquad (171b)$$

The generalized powers of A satisfy

$$A^{\langle k \rangle} = \begin{cases} E, & k = 0, \\ A^{\langle k-1 \rangle} E^* A, & k \geq 1, \quad \text{in particular, } A^{\langle 1 \rangle} = A, \\ A^{\langle k+1 \rangle} E^* A^{\langle -1 \rangle}, & k \leq -1. \end{cases} \qquad (172)$$

Ex. 84. If in Theorem 18 the scalars $d(A)$ are chosen as the singular values of A, i.e., if $d(A) = \sigma(A)$, then for any integer k,

$$A^{*\langle k \rangle} = A^{\langle k \rangle *}, \qquad (173a)$$

$$A^{\langle 2k+1 \rangle} = A(A^*A)^k = (AA^*)^k A, \qquad (173b)$$

in particular,

$$A^{\langle -1 \rangle} = A^{*\dagger}. \qquad (173c)$$

Ex. 85. If $A \in \mathbb{C}_r^{n \times n}$ is normal and if the scalars $d(A)$ are chosen as the eigenvalues of A, i.e., if $d(A) = \lambda(A)$, then

$$A^{\langle k \rangle} = \begin{cases} A^k, & k \geq 1, \\ P_{R(A)}, & k = 0, \\ (A^\dagger)^k, & k \leq -1. \end{cases} \qquad (174)$$

Ex. 86. *Ternary powers.* From (173b) follows the definition of a polynomial in ternary powers of $A \in \mathbb{C}^{m \times n}$, as a polynomial

$$\sum_k p_k A^{\langle 2k+1 \rangle} = \sum_k p_k (AA^*)^k A.$$

Such polynomials were studied by Hestenes [**417**] in the more general context of ternary algebras.

In (177) below, we express $A^\dagger$ as a polynomial in ternary powers of A^*. First we require the following:

Ex. 87. (Albert [9, p. 75]). Let $A \in \mathbb{C}^{n \times n}$ be Hermitian and let a vanishing polynomial of A, i.e., a polynomial $m(\lambda)$ satisfying $m(A) = O$, be given in the form

$$m(\lambda) = c\lambda^{\ell}(1 - \lambda q(\lambda)) \qquad (4.13)$$

where $c \neq 0$, $\ell \geq 0$, and the leading coefficient of q is 1.

Then

$$A^{\dagger} = q(A) + q(O)[Aq(A) - I], \qquad (175)$$

and, in particular,

$$A^{-1} = q(A), \quad \text{if } A \text{ is nonsingular.} \qquad (0.131)$$

PROOF. From (4.13) it follows that

$$A^{\ell} = A^{\ell+1}q(A)$$

and since A is Hermitian

$$\begin{aligned}
A^{\dagger} &= (A^{\dagger})^{\ell+1}A^{\ell} = AA^{\dagger}q(A) \\
&= AA^{\dagger}[q(A) - q(O)] + AA^{\dagger}q(O) \\
&= q(A) - q(O) + AA^{\dagger}q(O) \qquad (176)
\end{aligned}$$

since $q(A) - q(O)$ contains only positive powers of A. Postmultiplying (176) by A gives

$$\begin{aligned}
A^{\dagger}A &= [q(A) - q(O)]A + Aq(O) \\
&= q(A)A = Aq(A),
\end{aligned}$$

which, when substituted in (176), gives (175). □

Alternatively, (175) can be shown to follow from the results of Section 4.6, since here $A^{D} = A^{\dagger}$.

Ex. 88. Let $A \in \mathbb{C}^{m \times n}$ and let

$$m(\lambda) = c\lambda^{\ell}(1 - \lambda q(\lambda)) \qquad (4.13)$$

be a vanishing polynomial of $A^{*}A$, as in Ex. 87. Then

$$A^{\dagger} = q(A^{*}A)A^{*} \qquad (177)$$

(Penrose [635], Hestenes [417], Ben-Israel and Charnes [77]).

PROOF. From (175) it follows that

$$(A^{*}A)^{\dagger} = q(A^{*}A) + q(O)[A^{*}Aq(A^{*}A) - I],$$

so, by Ex. 1.18(d),

$$A^{\dagger} = (A^{*}A)^{\dagger}A^{*} = q(A^{*}A)A^{*}. \qquad □$$

A computational method based on (177) is given in Decell [223] and in Albert [9].

Ex. 89. *Partial isometries.* Let $W \in \mathbb{C}^{m \times n}$. Then W is a partial isometry if and only if

$$W = e^{iA}$$

for some $A \in \mathbb{C}^{m \times n}$.

PROOF. Follows from (157) and Exs. 40–41. □

Ex. 90. Let $U \in \mathbb{C}^{n \times n}$. Then U is a unitary matrix if and only if

$$U = e^{iH} + P_{N(H)} \tag{178}$$

for some Hermitian matrix $H \in \mathbb{C}^{n \times n}$. Note that the exponential in (178) is defined according to Definition 6. For the classical definition 2.1 (page 68), Eq. (178) should be replaced by

$$U = e^{iH}. \tag{178'}$$

Ex. 91. *Polar decompositions.* Let $A \in \mathbb{C}_r^{m \times n}$ and let

$$A = GE = EH \tag{79}$$

be a polar decomposition of A, given as in Corollary 2. Then, for any function f, Definition 6 gives

$$f(A) = f(G)E = Ef(H), \tag{179}$$

in particular,

$$A^{\langle k \rangle} = G^k E = E H^k, \quad \text{for any integer } k. \tag{180}$$

8. Generalized Singular Value Decompositions

This section is based on Van Loan ([**831**]. Two generalizations of the SVD are described. For more details, and other generalizations, see the suggested reading list on p. 256.

8.1. The B-SVD. The SVD concerns the diagonalization of an $m \times n$ matrix using unitary transformations. The generalized SVD described here is about a simultaneous diagonalization of two n-columned matrices.

The singular values of a matrix A are defined as the elements of the set

$$\sigma(A) = \{\sigma : \sigma \geq 0, \quad \det(A^*A - \sigma^2 I) = 0\}.$$

A natural generalization is the following:

DEFINITION 7. Let $A \in \mathbb{C}^{a \times n}$, $B \in \mathbb{C}^{b \times n}$. The B-*singular values* of A are the elements of the set

$$\mu(A, B) = \{\mu : \mu \geq 0, \quad \det(A^*A - \mu^2 B^*B) = 0\}. \tag{181}$$

A corresponding generalization of the SVD, see Theorem 2, is the following theorem.

THEOREM 21 (The B-Singular Value Decomposition, Van Loan [**831**], Theorem 2). *Let $A \in \mathbb{C}^{a \times n}$, $B \in \mathbb{C}^{b \times n}$ and $a \geq n$. Then there exist unitary matrices $U \in U^{a \times a}$ and $V \in U^{b \times b}$ and a nonsingular matrix $X \in \mathbb{C}^{n \times n}$ such that*

$$U^*AX = \Sigma_A = \text{diag}(\alpha_1, \dots, \alpha_n), \quad \alpha_i \geq 0, \tag{182a}$$
$$V^*BX = \Sigma_B = \text{diag}(\beta_1, \dots, \beta_q), \quad \beta_i \geq 0, \tag{182b}$$

where $q = \min\{b, n\}$, $r = rank(B)$, and

$$\beta_1 \geq \cdots \geq \beta_r > \beta_{r+1} = \cdots = \beta_q = 0. \tag{182c}$$

Moreover,

$$\mu(A, B) = \begin{cases} \{\mu : \mu \geq 0\}, & \text{if } \alpha_j = 0 \text{ for any } j \in \overline{r+1, n}, \\ \left\{ \dfrac{\alpha_i}{\beta_i} : i \in \overline{1, r} \right\}, & \text{otherwise.} \end{cases} \tag{183}$$

PROOF. Let $k = \text{rank} \begin{bmatrix} A \\ B \end{bmatrix}$ and consider the SVD,

$$Y^* \begin{bmatrix} A \\ B \end{bmatrix} Z = \text{diag}(\gamma_1, \dots, \gamma_n) = \begin{bmatrix} D & O \\ O & O \end{bmatrix}, \tag{184}$$

where Y, Z are unitary, $\gamma_1 \geq \cdots \geq \gamma_k > \gamma_{k+1} = \cdots = \gamma_n = 0$, and $D = \text{diag}(\gamma_1, \dots, \gamma_k)$ is nonsingular. Let $Z_{(k)}$ denote the submatrix formed by the first k columns of Z, and define A_1, B_1 by

$$\begin{bmatrix} A_1 \\ B_1 \end{bmatrix} = \begin{bmatrix} A \\ B \end{bmatrix} Z_{(k)} D^{-1}.$$

It follows from (184) that

$$A_1^* A_1 + B_1^* B_1 = I_k. \tag{185}$$

Let the SVD of B_1 be

$$V^* B_1 W = \text{diag}(\beta_1, \dots, \beta_p), \quad \beta_1 \geq \cdots \geq \beta_p, \tag{186}$$

where V, W are unitary and $p = \min\{b, k\}$. Define the nonsingular matrix

$$X = Z \begin{bmatrix} D^{-1} W & O \\ O & I_{n-k} \end{bmatrix}. \tag{187}$$

It follows then from (184) and (186) that, with $q = \min\{b, n\}$,

$$V^* BX = \text{diag}(\beta_1, \dots, \beta_q), \quad \beta_{p+1} = \cdots = \beta_q = 0, \tag{188}$$

A comparison of (186) and (188) shows that rank $B = \text{rank } B_1$ and (188) reduces to (182b)–(182c).

The columns of $A_1 W$ form an orthogonal set

$$(A_1 W)^* (A_1 W) = W^* (I_k - B_1^* B_1) W, \quad \text{by (185)},$$
$$= \text{diag}(1 - \beta_1^2, \dots, 1 - \beta_k^2), \quad \text{by (186)}.$$

Therefore there exists a unitary matrix $U \in U^{a \times a}$ and real α_i such that

$$A_1 W = U \, \text{diag}(\alpha_1, \dots, \alpha_k) \in \mathbb{C}^{a \times k},$$

where the α_i can be assumed nonnegative and ordered: $\alpha_1 \geq \cdots \geq \alpha_k$. Defining $\alpha_i = 0$ for $i \in \overline{k+1, n}$, we write

$$U^* AX = U^* AZ \begin{bmatrix} D^{-1} W & O \\ O & I_{n-k} \end{bmatrix} = \text{diag}(\alpha_1, \dots, \alpha_n)$$

which is (182a). Finally, it follows from (182a)–(182b) that

$$\det(A^* A - \mu^2 B^* B) = \det(X)^{-2} \prod_{i=1}^{r} (\alpha_i^2 - \mu^2 \beta_i^2) \prod_{i=r+1}^{n} \alpha_i^2.$$

which implies (183).   □

REMARK 1. (a) If $B \in \mathbb{C}^{n \times n}$ is nonsingular, with singular values $\beta_1 \geq \cdots \geq \beta_n > 0$, there is a unitary matrix Q such that

$$Q^* (A^* A - \mu^2 B^* B) Q = (AQ)^* (AQ) - \mu^2 \, \text{diag}(\beta_1^2, \dots, \beta_n^2).$$

The B-singular values of A are then, in agreement with (183),

$$\mu_i = \frac{\alpha_i}{\beta_i},$$

where the α_i are the singular values of A.

(b) If both $a < n$ and $b < n$, then the conclusions of Theorem 21 need not hold, as shown by the example $A = [1 \quad 0]$, $B = [0 \quad 1]$.

(c) The last $n - k$ columns of X, which by (187) are the last $n - k$ columns of Z, form an o.n. basis of a subspace that is orthogonal to the subspace spanned by the first k columns of X.

(d) If $\alpha_j = 0$ for any $j \in \overline{r+1, n}$, then the jth column of X satisfies $A\mathbf{x} = \mathbf{0}$, $B\mathbf{x} = \mathbf{0}$. It follows that the case $\{\mu : \mu \geq 0\}$ in (183) is equivalent to $N(A) \cap N(B) \neq \{\mathbf{0}\}$.

The next example illustrates the usefulness of the B-SVD. See also Ex. 92 below.

EXAMPLE 1. (Lawson and Hanson [**504**]). Consider the quadratic function

$$f(\mathbf{x}) = \|A\mathbf{x} - \mathbf{b}\|^2 + \lambda^2 \|B\mathbf{x} - \mathbf{c}\|^2, \tag{189}$$

where $A \in \mathbb{R}^{a \times n}$, $B \in \mathbb{R}^{b \times n}$, $a \geq n$, and $\lambda > 0$. It is required to find the (unique) minimum-norm minimizer of f, i.e., the vector $\widehat{\mathbf{x}} = \widehat{x}(\lambda)$ such that

$$f(\widehat{\mathbf{x}}) = \min_{\mathbf{x} \in \mathbb{R}^n} f(x) \quad \text{and} \quad \|\widehat{\mathbf{x}}\| = \min\{\|x\| : f(\mathbf{x}) = f(\widehat{\mathbf{x}})\}.$$

Applying Theorem 21 and using

$$\mathbf{y} = X^{-1}\mathbf{x}, \quad \widetilde{b} = U^*\mathbf{b}, \quad \widetilde{c} = V^*\mathbf{c}, \tag{190}$$

we get

$$f(\mathbf{x}) = \|\Sigma_A \, \mathbf{y} - \widetilde{\mathbf{b}}\|^2 + \lambda^2 \|\Sigma_B \, \mathbf{y} - \widetilde{\mathbf{c}}\|^2$$

$$= \sum_{i=1}^{n} (\alpha_i y_i - \widetilde{b}_i)^2 + \lambda^2 \sum_{j=1}^{q} (\beta_j y_j - \widetilde{c}_j)^2 := \widetilde{f}(\mathbf{y}). \tag{191}$$

The minimum-norm minimizer $\widehat{\mathbf{y}} = [\widehat{y}_j]$ of $\widetilde{f}(\mathbf{y})$ can be read from (191)

$$\widehat{y}_j = \begin{cases} \dfrac{\alpha_j \widetilde{b}_j + \lambda^2 \beta_j \widetilde{c}_j}{\alpha_j^2 + \lambda^2 \beta_j^2}, & j = 1, \cdots, r = \operatorname{rank} B, \\[2mm] \dfrac{\widetilde{b}_j}{\alpha_j}, & j \in \overline{r+1, n}, \; \alpha_j \neq 0, \\[2mm] 0, & j \in \overline{r+1, n}, \; \alpha_j = 0, \end{cases} \tag{192}$$

and $\widehat{\mathbf{x}} = X\widehat{\mathbf{y}}$ is a minimizer of $f(\mathbf{x})$. We now show that $\widehat{\mathbf{x}}$ is the minimum-norm minimizer. Denote by $X_{(k)}$ the submatrix formed by the first k columns of X. Similarly, $\widehat{\mathbf{x}}_{(k)}$ and $\widehat{\mathbf{y}}_{(k)}$ denote the first k-component subvectors of $\widehat{\mathbf{x}}$ and $\widehat{\mathbf{y}}$, respectively. It follows from $r = \operatorname{rank} B \leq k = \operatorname{rank} \begin{bmatrix} A \\ B \end{bmatrix}$ and Remark 1(d) that

$$\widehat{\mathbf{x}}_{(k)} = X_{(k)}\widehat{\mathbf{y}}_{(k)} \in R(A^*) + R(B^*),$$

while, by Remark 1(c),

$$\sum_{j=k+1}^{n} \widehat{x}_j^2 = \sum_{j=k+1}^{n} \widehat{y}_j^2,$$

completing the proof.

8.2. The $\{W, Q\}$-SVD. The singular values of $A \in \mathbb{C}^{m \times n}$ are the stationary values of $\|A\mathbf{x}\|/\|\mathbf{x}\|$, see Ex. 5. A natural generalization is the following:

DEFINITION 8. Let $A \in \mathbb{C}^{m \times n}$, let $W \in \mathbb{C}^{m \times m}$, and let $Q \in \mathbb{C}^{n \times n}$ be positive definite. The $\{W, Q\}$-*singular values* of A are the elements of the set

$$\mu_{W,Q}(A) = \left\{ \mu : \mu \text{ is a stationary value of } \frac{\|A\mathbf{x}\|_W}{\|\mathbf{x}\|_Q} \right\}, \qquad (193)$$

where, necessarily, $\mu \geq 0$.

One more definition is needed:

DEFINITION 9. Let $Q \in \mathbb{C}^{n \times n}$ be positive definite. A matrix $V \in \mathbb{C}^{n \times n}$ is Q-*orthogonal* if $V^* Q V = I$.

The analog of the SVD is then

THEOREM 22 (The $\{W, Q\}$-singular value decomposition, Van Loan [**831**], Theorem 3). *Let $A \in \mathbb{C}^{m \times n}$, and let $W \in \mathbb{C}^{m \times m}$ and $Q \in \mathbb{C}^{n \times n}$ be positive definite. Then there exist a W-orthogonal matrix $U \in \mathbb{C}^{m \times m}$ and a Q-orthogonal matrix $V \in \mathbb{C}^{n \times n}$ such that*

$$U^{-1} A V = D_A = \text{diag}(\mu_1, \dots, \mu_n) \qquad (194)$$

where

$$\mu_{W,Q}(A) = \{\mu_1, \dots, \mu_n\}. \qquad (195)$$

PROOF. Let $B = W^{1/2} A Q^{-1/2}$ and let

$$\widetilde{U}^* B \widetilde{V} = D$$

be the SVD of B. Then

$$U = W^{-1/2} \widetilde{U}, \text{ and } V = Q^{-1/2} \widetilde{V} \qquad (196)$$

are W-orthogonal and Q-orthogonal, respectively, and satisfy (194). Using Lagrange multipliers we can see that the stationary values of $\|A\mathbf{x}\|_W/\|\mathbf{x}\|_Q$ are the zeros of $\det(A^* W A - \mu^2 Q)$. The calculation

$$\det(A^* W A - \mu^2 Q) = \det(Q) \det(B^* B - \mu^2 I)$$
$$= \det(Q) \det(D^* D - \mu^2 I)$$
$$= \det(Q) \prod_{i=1}^{n} (\mu_i^2 - \mu^2)$$

then proves (195). $\qquad \square$

Exercises

Ex. 92. *Constrained least-squares.* Let $A \in \mathbb{R}^{a \times n}$, $B \in \mathbb{R}^{b \times n}$, $a \geq n$. It is required to find a minimum-norm solution of the constrained least-squares problem

$$\begin{array}{cc} \text{minimize} & \|A\mathbf{x} - \mathbf{b}\|^2, \\ \text{subject to} & B\mathbf{x} = \mathbf{c}. \end{array} \tag{197}$$

Using Theorem 21 and the transformation (190), the problem becomes

$$\begin{array}{cc} \text{minimize} & \|\Sigma_A \, \mathbf{y} - \widetilde{\mathbf{b}}\|^2, \\ \text{subject to} & \Sigma_B \, \mathbf{y} = \widetilde{\mathbf{c}}. \end{array} \tag{198}$$

A minimum-norm solution of the transformed problem (198) is

$$\widehat{y}_j = \begin{cases} \dfrac{\widetilde{c}_j}{\beta_j}, & j = 1, \cdots, r = \text{rank}\, B, \\[2mm] \dfrac{\widetilde{b}_j}{\alpha_j}, & j \in \overline{r+1, n}, \ \alpha_j \neq 0, \\[2mm] 0, & j \in \overline{r+1, n}, \ \alpha_j = 0. \end{cases} \tag{199}$$

Then, reasoning as in Example 1, $\widehat{\mathbf{x}} = X\widehat{\mathbf{y}}$ is the minimum-norm solution of the original problem (197).

Ex. 93. *Weighted least-squares.* Let $A \in \mathbb{R}^{m \times n}$, $\mathbf{b} \in \mathbb{R}^m$, and let $W \in \mathbb{C}^{m \times m}$ and $Q \in \mathbb{C}^{n \times n}$ be PD. It is required to find a vector $\widehat{\mathbf{x}}$ minimizing $\|A\mathbf{x} - \mathbf{b}\|_W$ and of minimal $\|\mathbf{x}\|_Q$-norm among all such minimizers. This problem was solved in Corollary 3.4 using the $\{W, Q\}$-weighted $\{1, 2\}$-inverse,

$$\widehat{\mathbf{x}} = A^{(1,2)}_{(W,Q)} \, \mathbf{b}. \tag{200}$$

An alternative expression of this solution is enabled by the $\{W, Q\}$-SVD of A,

$$U^{-1}AV = D_A = \text{diag}\,(\mu_1, \ldots, \mu_n) \tag{194}$$

giving

$$\widehat{\mathbf{x}} = V D_A^\dagger U^{-1} \mathbf{b}. \tag{201}$$

Indeed, using (196), we can show that

$$V D_A^\dagger U^{-1} = Q^{-1/2} (W^{1/2} A Q^{-1/2})^\dagger W^{1/2}$$

in agreement with (3.68a).

Suggested Further Reading

HISTORICAL NOTES. The singular value decomposition (SVD) was proved by Beltrami [**57**], Jordan [**452**], [**453**], and Sylvester [**790**], [**791**] for square real matrices, by Autonne [**30**] for square complex matrices, and by Eckart and Young [**248**] for rectangular matrices. In parallel, singular values of integral operators were studied by Schmidt [**727**] and Weyl [**868**]. For the history see Horn and Johnson [**429**, Chapter 3] and Stewart [**783**].

SECTION 2. Businger and Golub [**142**], Golub and Kahan [**306**], Golub and Reinsch [**309**], Good [**314**], Hartwig [**381**], Hestenes [**414**], Lanczos [**496**], Pan and Sigmon [**626**], Roch and Silbermann [**701**], Wedin [**854**].

APPLICATIONS OF SVD. Hanson [**375**], Hanson and Norris [**376**], Höskuldsson [**430**].

SECTION 3. Antoulas [**23**], Chipman [**188**], Householder and Young [**433**], Golub and Kahan [**306**]; Gaches, Rigal, and Rousset de Pina [**294**], Franck [**289**], Nievergelt [**611**].

TOTAL LEAST SQUARES. Golub and Van Loan [**311**], De Moor [**219**], Jiang and Berry [**446**], Nievergelt [**610**], [**612**], Van Huffel and Vanderwalle [**830**].

SECTION 4. Björck and Bowie [110], Erdélyi [256], [262], [260], [263], [261], Erdélyi and Miller [266], Halmos and Wallen [368], Hearon [408], [407], Hestenes [414], [415], [416], [417], Poole and Boullion [644].

SECTION 5. Afriat [2], [3], Davis and Kahan [214], Deutsch [229], Hotelling [431], Ipsen and Meyer [438], Scharnhorst [725], Seidel [743], Sun [786], Wedin [856].

PRINCIPAL ANGLES. A brief history of principal angles appears in Stewart and Sun [784, p. 45]. An excellent reference is Afriat [3], a study of the geometry of subspaces in $\mathbb{R}^n$ in terms of their orthogonal and oblique projectors. An important application of principal angles in Statistics is the canonical correlation theory of Hotelling [431], see also Baksalary, Puntanen, and Yanai [38].

GEOMETRY OF SUBSPACES. Groß and Trenkler [352], Jiang [447], Robinson [699].

SECTION 6. Golub and Pereyra [308], Moore and Nashed [578], Nashed [594], Pereyra [638], Stewart [779], [781], Wedin [854], [855], Wei and Wang [860].

SECTION 7. Hawkins and Ben-Israel [402], Hestenes [414], [415], [416], [417]), Lanczos [496] and Penrose [635], Rose [709], Rothblum [715].

SECTION 8. Chu, Funderlic, and Golub [196], De Lathauwer, De Moor, and Vanderwalle [216], [217], Eldén [250], Levy [513], Paige and Saunders [625], Van Loan [831], [832], Wimmer [878].

CHAPTER 7

Computational Aspects of Generalized Inverses

1. Introduction

There are three principal situations in which it is required to obtain numerically a generalized inverse of a given matrix:

(i) the case in which any $\{1\}$-inverse will suffice;
(ii) the cases in which any $\{1,3\}$-inverse (or sometimes any $\{1,4\}$-inverse) will do; and
(iii) the case in which a $\{2\}$-inverse having a specified range and null space is required.

The inverse desired in case (iii) is, in the majority of cases, the Moore–Penrose inverse, which is the unique $\{2\}$-inverse of the given matrix A having the same range and null space as A^*. The Drazin inverse can also be fitted into this pattern, being the unique $\{2\}$-inverse of A having the same range and null space as A^ℓ, where ℓ is any integer not less than the index of A. When $\ell = 1$, this is the group inverse.

Generalized inverses are closely associated with linear equations, orthonormalization, least-squares solutions, singular values, and various matrix factorizations. In particular, the *QR-factorization* and the *Singular Value Decomposition* (SVD) figure prominently in the computation of the Moore–Penrose inverse.

The two principal ways of computing the QR-factorization are

(1) Using a Gram–Schmidt type of orthogonalization; see, e.g., Rice [**689**] and Björck [**106**] where a detailed error analysis is given for least-squares solutions.
(2) Using Householder transformations or other rotations; see, e.g., Wilkinson [**872**], Parlett [**628**], and Golub [**305**].

QR factorization is implicit in the main algorithm for SVD, of Golub and Kahan [**306**], where the matrix in question is first transformed, by rotations, to an upper bidiagonal form.

These topics have been studied extensively and many excellent references are available in the numerical analysis literature, see, e.g., Björck [**109**], Golub and Van Loan [**311**], and Lawson and Hanson [**504**]. For this reason we can keep this chapter brief, restricting our efforts to listing some computational methods for generalized inversion, and discussing the mathematics behind these methods. No error analysis is attempted.

2. Computation of Unrestricted {1}- and {1,2}-Inverses

Let A be a given matrix for which a {1}-inverse is desired, when any {1}-inverse will suffice. If it should happen that A is of such a structure, or has risen in such a manner, that a nonsingular submatrix of maximal order is known, we can write

$$PAQ = \begin{bmatrix} A_{11} & A_{12} \\ A_{21} & A_{22} \end{bmatrix}, \tag{5.19}$$

where A_{11} is nonsingular and P and Q are permutation matrices used to bring the nonsingular submatrix into the upper left position. (If A is of full (column or row) rank, some of the submatrices in (5.19) will be absent.) Since rank A is the order of A_{11}, this implies that

$$A_{22} = A_{21}A_{11}^{-1}A_{12} \quad \text{(Brand [135])} \tag{5.39}$$

and a {1,2}-inverse of A is

$$A^{(1,2)} = Q \begin{bmatrix} A_{11}^{-1} & O \\ O & O \end{bmatrix} P \quad \text{(C.R. Rao [671]).} \tag{5.25}$$

In the more usual case in which a nonsingular submatrix of maximal order is not known and, likewise, rank A is not known, perhaps the simplest method is that of Section 1.2, using Gaussian elimination to bring A to Hermite normal form,

$$EAP = \begin{bmatrix} I_r & K \\ O & O \end{bmatrix} \tag{0.72}$$

(with modifications in the case that A is of full rank), where E is nonsingular and P is a permutation matrix, then

$$A^{(1)} = P \begin{bmatrix} I_r & O \\ O & L \end{bmatrix} E \tag{1.5}$$

is a {1}-inverse of A for arbitrary L. Of course, the simplest choice is $L = O$, which gives the {1,2}-inverse

$$A^{(1,2)} = P \begin{bmatrix} I_r & O \\ O & O \end{bmatrix} E. \tag{1.11}$$

On the other hand, when A is square, a nonsingular {1}-inverse may sometimes be desired. This is obtained by taking L in (1.5) to be nonsingular. The simplest choice for L is a unit matrix, which gives

$$A^{(1)} = PE.$$

In applications involving linear equations, it is often the case that a particular solution suffices. The above results can be easily adapted to obtain such a solution, whenever it exists.

THEOREM 1. *Given $A \in \mathbb{C}_r^{m \times n}$ with $r < m$, and $\mathbf{b} \in \mathbb{C}^m$, consider the linear equation*

$$A\mathbf{x} = \mathbf{b}. \tag{1}$$

Let E be a nonsingular matrix, P a permutation matrix, such that

$$EAP = \begin{bmatrix} I_r & K \\ O & O \end{bmatrix} \tag{0.72}$$

and let $\widehat{\mathbf{b}}$ and $\mathbf{z}$ be defined by

$$E[A \quad \mathbf{b}]\begin{bmatrix} P & 0 \\ 0^T & 1 \end{bmatrix} = \begin{bmatrix} I_r & K & \widehat{\mathbf{b}} \\ O & O & \mathbf{z} \end{bmatrix}. \tag{2}$$

Then (1) has a solution if and only if $\mathbf{z} = \mathbf{0}$, in which case a particular solution is[1]

$$\mathbf{x} = A^{(1,2)}\mathbf{b} = P\widehat{\mathbf{b}}. \tag{3}$$

PROOF. The first statement is obvious, the second follows from (1.11). □

As in the nonsingular case, the accuracy may depend on the choice of pivots used in the Gaussian elimination. (For a discussion of pivoting see, e.g., Golub and Van Loan [**311**, § 3.4]; for a simple illustration, see Ex. 1 below.)

Exercises

Ex. 1. Consider the two nonsingular matrices

$$A = \begin{bmatrix} \epsilon & 1 \\ 0 & 1 \end{bmatrix}, \quad B = \begin{bmatrix} \epsilon & 1 \\ 1 & 1 \end{bmatrix},$$

where ϵ is a small, positive number. Compare the various ways (i.e., choices of pivots) of transforming A and B to their Hermite normal forms. The objective is a numerically stable process, which here means to avoid, or to postpone, division by ϵ.

Ex. 2. *An iterative method for computing a {1}-inverse.* Exercise 5.17(b) suggests a finite iterative method for computing a {1}-inverse of a given $A \in \mathbb{C}^{m \times n}$. The method requires n iterations (an analogous method based on Ex. 5.17(a) would use m iterations). At the kth iteration ($k = 1, 2, \ldots, n$) it computes $A_k^{(1)}$, where A_k is the submatrix of A consisting of its first k columns.

First we need some notation. For $k = 2, \ldots, n$ the matrix A_k is partitioned as

$$A_k = [A_{k-1} \quad \mathbf{a}_k] \tag{4}$$

where $\mathbf{a}_k$ is the kth column of A. For $k = 2, \ldots, n$ let the vectors $\mathbf{d}_k$, $\mathbf{c}_k$, and $\mathbf{b}_k$ be defined by

$$\mathbf{d}_k = A_{k-1}^{(1)}\mathbf{a}_k, \tag{5}$$

$$\mathbf{c}_k = \mathbf{a}_k - A_{k-1}\mathbf{d}_k, \tag{6}$$

$$\mathbf{b}_k^* = \begin{cases} \mathbf{0}^T, & \text{if } \mathbf{c}_k = \mathbf{0}, \\ (\mathbf{c}_k)^{(1)}(I - A_{k-1}A_{k-1}^{(1)}), & \text{otherwise}, \end{cases} \tag{7}$$

where $A_{k-1}^{(1)}$ is any {1}-inverse of A_{k-1}. Then a {1}-inverse of A_k is

$$\begin{bmatrix} A_{k-1}^{(1)} - \mathbf{d}_k\mathbf{b}_k^* \\ \mathbf{b}_k^* \end{bmatrix} \tag{8}$$

Note that at each iteration we only need a {1}-inverse of a vector (the vector $\mathbf{c}_k$ if nonzero), a trivial task.

Ex. 3. Modify the iterative method in Ex. 2 so as to get a {1, 2}-inverse.

---

[1]NOTE: If $r = m$, then for any $\mathbf{b}$ the vector $\mathbf{z}$ is absent, and (3) is a solution.

3. Computation of Unrestricted $\{1,3\}$-Inverses

Let $A \in \mathbb{C}_r^{m \times n}$ and let

$$A = FG \tag{9}$$

be a full-rank factorization. Then, by Ex. 1.29(b),

$$X = G^{(1)} F^\dagger, \tag{10}$$

where $G^{(1)}$ is an arbitrary element of $G\{1\}$, is a $\{1,2,3\}$-inverse of A.

Recall the $\widetilde{Q}\widetilde{R}$ factorization of A,

$$A = \widetilde{Q}\widetilde{R} \tag{0.39}$$

where the columns of $\widetilde{Q} = [\mathbf{q}_1 \ \cdots \ \mathbf{q}_r]$ are o.n., and $\widetilde{R}$ is upper triangular. Then (10) gives

$$A^{(1,2,3)} = \widetilde{R}^{(1)} \widetilde{Q}^\dagger. \tag{11}$$

This is useful, since the Moore–Penrose inverse of $\widetilde{Q}$ is simply

$$\widetilde{Q}^\dagger = \begin{bmatrix} \mathbf{q}_1^* \\ \mathbf{q}_2^* \\ \cdots \\ \mathbf{q}_r^* \end{bmatrix}, \tag{12}$$

and a $\{1\}$-inverse of $\widetilde{R}$ is also easily obtained: $\widetilde{R}$ is of full row rank and it can be written (by permuting columns as necessary) as

$$\widetilde{R}P = [T \ \ K]$$

where T is nonsingular and upper triangular and P is a permutation matrix. Then, see Ex. 1.13,

$$\widetilde{R}^{(1)} = P \begin{bmatrix} T^{-1} \\ O \end{bmatrix}, \tag{13}$$

here T^{-1} is upper triangular and obtained from T by back substitution.

Exercises

EX. 4. If the factorization (9) has been obtained from the Hermite normal form of A by the procedure described in § 0.4.4, then

$$F = AP_1, \tag{14}$$

where P_1 denotes the first r columns of the permutation matrix P. Moreover, we may take $G^{(1)} = P_1$, and (10) gives

$$X = P_1 F^\dagger. \tag{15}$$

Since F is of full column rank,

$$F^\dagger = (F^* F)^{-1} F^*, \tag{16}$$

by (1.19). Thus (14), (16), and (15), in that order, give a $\{1,2,3\}$-inverse of A.

Observe that (15) shows that each of the r rows of $F^\dagger$ is a row of X (in general, not the corresponding row), while the remaining $n - r$ rows of X are rows of zeros. Thus, in the language of linear programming, X is a "basic" $\{1,2,3\}$-inverse of A.

Ex. 5. Obtain a $\{1, 2, 3\}$-inverse of

$$A = \begin{bmatrix} 1 & 0 & 0 & 1 \\ 1 & 1 & 0 & 0 \\ 0 & 1 & 1 & 0 \\ 0 & 0 & 1 & 1 \end{bmatrix}$$

using the method outlined in equations:
  (a) (11), (12), and (13).
  (b) (14), (16), and (15).

4. Computation of {2}-Inverses with Prescribed Range and Null Space

Let $A \in \mathbb{C}_r^{m \times n}$, let $A\{2\}_{S,T}$ contain a nonzero matrix X, and let U and V be such that $R(U) = R(X)$, $N(V) = N(X)$, and the product VAU is defined. Then, by Theorems 2.13 and 2.14, rank $U = $ rank $V = $ rank VAU, and

$$X = U(VAU)^{(1)}V, \tag{17}$$

where $(VAU)^{(1)}$ is an arbitrary element of $(VAU)\{1\}$. This is the basic formula for the case considered in this section. Zlobec's formula

$$A^\dagger = A^*(A^*AA^*)^{(1)}A^* \tag{18}$$

(see Ex. 2.39) and Greville's formula

$$A^D = A^\ell(A^{2\ell+1})^{(1)}A^\ell, \tag{4.48}$$

where ℓ is a positive integer not less than the index of A, are particular cases. Formula (17) has the advantage that it does not require inversion of any nonsingular matrix. Aside from matrix multiplication, only the determination of a $\{1\}$-inverse of VAU is needed, and this can be obtained by the method of Section 1.2.

It should be noted, however, that when ill-conditioning of A is a problem, this is accentuated by forming products like A^*AA^* or $A^{2\ell+1}$ and, in such cases, other methods are preferable.

In the case of the Moore–Penrose inverse, Noble's formula

$$A^\dagger = Q \begin{bmatrix} I_r \\ T^* \end{bmatrix} (I_r + TT^*)^{-1} A_{11}^{-1} (I_r + S^*S)^{-1} [I_r \quad S^*] P \tag{5.28}$$

is available, if a maximal nonsingular submatrix A_{11} is known, where the permutation matrices P and Q and the "multipliers" S and T are defined by

$$A = P^T \begin{bmatrix} A_{11} & A_{12} \\ A_{21} & A_{22} \end{bmatrix} Q^T$$

$$= P^T \begin{bmatrix} I_r \\ S \end{bmatrix} A_{11}[I_r \quad T]Q^T, \quad \text{see Ex. 6.} \tag{0.78}$$

Otherwise, it is probably best to use the method of §0.4.4 to obtain a full-rank factorization

$$A = FG. \tag{0.78}$$

Then, the Moore–Penrose inverse is

$$A^{\dagger} = G^*(F^*AG^*)^{-1}F^*, \tag{1.17}$$

while the group inverse is

$$A^{\#} = F(GF)^{-2}G, \tag{4.22}$$

whenever GF is nonsingular.

Full-rank factorization can also be used for the computation of the Drazin inverse. If A is a square matrix of index k, Cline's method (Cline [201]) employs full-rank factorization of matrices of successively smaller order,

$$A = B_1G_1, \tag{4.46a}$$

$$G_iB_i = B_{i+1}G_{i+1} \quad (i = 1, 2, \ldots, k-1), \tag{4.46b}$$

until the nonsingular matrix G_kB_k is obtained. Then

$$A^D = B_1B_2 \cdots B_k(G_kB_k)^{-k-1}G_kG_{k-1} \cdots G_1. \tag{4.46c}$$

Exercises

Ex. 6. *Noble's method.* Let the nonzero matrix $A \in \mathbb{C}_r^{m \times n}$ be transformed to a column-permuted Hermite normal form

$$PEAQ = \begin{bmatrix} I_r & T \\ O & O \end{bmatrix} = (PEP^T)(PAQ), \tag{19}$$

where P and Q are permutation matrices and E is a product of elementary row matrices of types (i) and (ii) (see Section 1.2),

$$E = E_kE_{k-1} \cdots E_2E_1,$$

which does not involve permutation of rows.

Then E can be chosen so that

$$PE[A \quad I_m] \begin{bmatrix} Q & O \\ O & P^T \end{bmatrix} = \begin{bmatrix} I_r & T & A_{11}^{-1} & O \\ O & O & -S & I_{m-r} \end{bmatrix} \tag{20}$$

giving all the matrices P, Q, T, S, and A_{11}^{-1} which appear in (5.28). Note that after the left-hand portion of RHS(20) has been brought to the form (19), still further row operations may be needed to bring the right-hand portion to the required form (Noble [614]).

Ex. 7. *Singular value decomposition.* Let

$$A = U_{(r)}D_{(r)}V_{(r)}^* \tag{6.30a}$$

be an SVD of $A \in \mathbb{C}^{m \times n}$. Then

$$A^{\dagger} = V_{(r)}D_{(r)}^{-1}U_{(r)}^*$$
$$= V_{(r)}(U_{(r)}^*AV_{(r)})^{-1}U_{(r)}^* \tag{6.30b}$$

is shown to be a special case of (17) by taking

$$U = V_{(r)}, \quad V = U_{(r)}^*.$$

A method for computing the Moore–Penrose inverse, based on (6.30b), has been developed by Golub and Kahan [306]. See also Businger and Golub [141], [142] and Golub and Reinsch [309].

Ex. 8. *Gram–Schmidt orthonormalization.* The GSO of Ex. 0.7 can be modified to compute the Moore–Penrose inverse. This method is due to Rust and Burrus and Schneeberger [**716**]; see also Albert [**9**, Chapter V].

Ex. 9. For the matrix A of Ex. 5, calculate $A^\dagger$ by:
  (a) Zlobec's formula (18).
  (b) Noble's formula (5.28).
  (c) MacDuffee's formula (1.17).

Ex. 10. For the matrix A of Ex. 5, calculate $A^\#$ by:
  (a) Cline's formula (4.22).
  (b) Formula (4.29).

5. Greville's Method and Related Results

Greville's method for computing the Moore–Penrose inverse $A^\dagger$ of a matrix $A \in \mathbb{C}^{m \times n}$ is a finite iterative method. At the kth iteration ($k = 1, 2, \ldots, n$) it computes $A_k^\dagger$, where A_k is the submatrix of A consisting of its first k columns. An analogous method for computing a $\{1\}$-inverse was encountered in Ex. 2. As there, we partition

$$A_k = [A_{k-1} \quad \mathbf{a}_k] \tag{4}$$

where $\mathbf{a}_k$ is the kth column of A. For $k = 2, \ldots, n$ let the vectors $\mathbf{d}_k$ and $\mathbf{c}_k$ be defined by

$$\mathbf{d}_k = A_{k-1}^\dagger \mathbf{a}_k, \tag{21}$$

$$\mathbf{c}_k = \mathbf{a}_k - A_{k-1}\mathbf{d}_k \tag{22}$$

$$= \mathbf{a}_k - A_{k-1}A_{k-1}^\dagger \mathbf{a}_k$$

$$= \mathbf{a}_k - P_{R(A_{k-1})}\mathbf{a}_k$$

$$= P_{N(A_{k-1}^*)}\mathbf{a}_k.$$

THEOREM 2 (Greville [**325**]). *Let $A \in \mathbb{C}^{m \times n}$. Using the above notation, the Moore–Penrose inverse of A_k ($k = 2, \ldots, n$) is*

$$[A_{k-1} \quad \mathbf{a}_k]^\dagger = \begin{bmatrix} A_{k-1}^\dagger - \mathbf{d}_k\mathbf{b}_k^* \\ \mathbf{b}_k^* \end{bmatrix}, \tag{23}$$

where

$$\mathbf{b}_k^* = \mathbf{c}_k^\dagger, \quad \text{if } \mathbf{c}_k \neq \mathbf{0}, \tag{24a}$$

$$\mathbf{b}_k^* = (1 + \mathbf{d}_k^*\mathbf{d}_k)^{-1}\mathbf{d}_k^* A_{k-1}^\dagger, \quad \text{if } \mathbf{c}_k = \mathbf{0}. \tag{24b}$$

PROOF. Let $A_k^\dagger = [A_{k-1} \quad \mathbf{a}_k]^\dagger$ be partitioned as

$$A_k^\dagger = \begin{bmatrix} B_k \\ \mathbf{b}_k^* \end{bmatrix} \tag{25}$$

where $\mathbf{b}_k^*$ is the kth row of $A_k^\dagger$. Multiplying (4) and (25) gives

$$A_k A_k^\dagger = A_{k-1}B_k + \mathbf{a}_k\mathbf{b}_k^*. \tag{26}$$

Now by (4), Ex. 2.38, and Corollary 2.7,

$$N(A_{k-1}^\dagger) = N(A_{k-1}^*) \supset N(A_k^*) = N(A_k^\dagger) = N(A_k A_k^\dagger),$$

and it follows from Ex. 2.20 that

$$A_{k-1}^\dagger A_k A_k^\dagger = A_{k-1}^\dagger. \tag{27}$$

Moreover, since

$$R(A_k^\dagger) = R(A_k^*)$$

by Ex. 2.38, it follows from (4), (25), and Corollary 2.7 that

$$R(B_k) \subset R(A_{k-1}^*) = R(A_{k-1}^\dagger) = R(A_{k-1}^\dagger A_{k-1}),$$

and, therefore,

$$A_{k-1}^\dagger A_{k-1} B_k = B_k \tag{28}$$

by Ex. 2.20. It follows from (27) and (28) that premultiplication of (26) by $A_{k-1}^\dagger$ gives

$$A_{k-1}^\dagger = B_k + A_{k-1}^\dagger \mathbf{a}_k \mathbf{b}_k^*$$
$$= B_k + \mathbf{d}_k \mathbf{b}_k^*, \tag{29}$$

by (21). Thus we may write

$$[A_{k-1} \quad \mathbf{a}_k]^\dagger = \begin{bmatrix} A_{k-1}^\dagger - \mathbf{d}_k \mathbf{b}_k^* \\ \mathbf{b}_k^* \end{bmatrix}, \tag{23}$$

with $\mathbf{b}_k^*$ still to be determined. We distinguish two cases according as $\mathbf{a}_k$ is or is not in $R(A_{k-1})$, i.e., according as $\mathbf{c}_k$ is or is not $\mathbf{0}$.

Case I ($\mathbf{c}_k \neq \mathbf{0}$)

By using (29), (26) becomes

$$A_k A_k^\dagger = A_{k-1} A_{k-1}^\dagger + (\mathbf{a}_k - A_{k-1} \mathbf{d}_k) \mathbf{b}_k^*$$
$$= A_{k-1} A_{k-1}^\dagger + \mathbf{c}_k \mathbf{b}_k^* \tag{30}$$

by (22). Since $A_k A_k^\dagger$ is Hermitian, it follows from (30) that $\mathbf{c}_k \mathbf{b}_k^*$ is Hermitian and, therefore,

$$\mathbf{b}_k^* = \delta \mathbf{c}_k^*, \tag{31}$$

where δ is some real number. From (4) and (22) we obtain

$$A_k = A_k A_k^\dagger A_k = [A_{k-1} + \mathbf{c}_k \mathbf{b}_k^* A_{k-1} \quad \mathbf{a}_k - \mathbf{c}_k + (\mathbf{b}_k^* \mathbf{a}_k) \mathbf{c}_k],$$

and comparison with (4) shows that

$$\mathbf{b}_k^* \mathbf{a}_k = 1, \tag{32}$$

since $\mathbf{c}_k \neq \mathbf{0}$. Now, by (22),

$$\mathbf{c}_k = P \mathbf{a}_k,$$

where P denotes the orthogonal projector on $N(A_{k-1}^*)$. Therefore, (31) and (32) give

$$1 = \mathbf{b}_k^* \mathbf{a}_k = \delta \mathbf{c}_k^* \mathbf{a}_k = \delta \mathbf{a}_k^* P \mathbf{a}_k$$
$$= \delta \mathbf{a}_k^* P^2 \mathbf{a}_k = \delta \mathbf{c}_k^* \mathbf{c}_k, \tag{33}$$

since P is idempotent. By (31), (33), and Ex. 1.19(a),

$$\mathbf{b}_k^* = \delta \mathbf{c}_k^* = \mathbf{c}_k^\dagger.$$

Case II ($\mathbf{c}_k = \mathbf{0}$)

Here $R(A_k) = R(A_{k-1})$, and so, by (25) and (0.27),

$$N(\mathbf{b}_k^*) \supset N(A_k^\dagger) = N(A_k^*) = N(A_{k-1}^*) = N(A_{k-1}^\dagger)$$
$$= N(A_{k-1}A_{k-1}^\dagger).$$

Therefore, by Ex. 2.20,

$$\mathbf{b}_k^* A_{k-1} A_{k-1}^\dagger = \mathbf{b}_k^*. \tag{34}$$

Now, (4) and (23) give

$$A_k^\dagger A_k = \begin{bmatrix} A_{k-1}^\dagger - \mathbf{d}_k \mathbf{b}_k^* A_{k-1} & (1-\alpha)\mathbf{d}_k \\ \mathbf{b}_k^* A_{k-1} & \alpha \end{bmatrix}, \tag{35}$$

where

$$\alpha = \mathbf{b}_k^* \mathbf{a}_k \tag{36}$$

is a scalar (real, in fact, since it is a diagonal element of a Hermitian matrix). Since (35) is Hermitian we have

$$\mathbf{b}_k^* A_{k-1} = (1-\alpha)\mathbf{d}_k^*.$$

Thus, by (34),

$$\mathbf{b}_k^* = \mathbf{b}_k^* A_{k-1} A_{k-1}^\dagger = (1-\alpha)\mathbf{d}_k^* A_{k-1}^\dagger. \tag{37}$$

Substitution of (37) in (36) gives

$$\alpha = (1-\alpha)\mathbf{d}_k^* \mathbf{d}_k, \tag{38}$$

by (21). Adding $1 - \alpha$ to both sides of (38) gives

$$(1-\alpha)(1 + \mathbf{d}_k^* \mathbf{d}_k) = 1$$

and substitution for $1 - \alpha$ in (37) gives (24b). □

Greville's method as described above, computes $A^\dagger$ recursively in terms of $A_k^\dagger$ ($k = 1, 2, \ldots, n$). This method was adapted by Greville [**325**] for the computation of $A^\dagger \mathbf{y}$, for any $\mathbf{y} \in \mathbb{C}^m$, without computing $A^\dagger$. This is done as follows:

Let

$$\widetilde{A} = [A \quad \mathbf{y}]. \tag{39}$$

Then (23) gives

$$A_k^\dagger \widetilde{A} = \begin{bmatrix} A_{k-1}^\dagger \widetilde{A} - \mathbf{d}_k \mathbf{b}_k^* \widetilde{A} \\ \mathbf{b}_k^* \widetilde{A} \end{bmatrix}. \tag{40}$$

By (21) it follows that $\mathbf{d}_k$ is the kth column of $A_{k-1}^\dagger \widetilde{A}$ for $k = 2, \ldots, n$. Therefore only the vector $\mathbf{b}_k^* \widetilde{A}$ is needed to get $A_k^\dagger \widetilde{A}$ from $A_{k-1}^\dagger \widetilde{A}$ by (40).

If $\mathbf{c}_k = \mathbf{0}$, then (24b) gives $\mathbf{b}_k^* \widetilde{A}$ as

$$\mathbf{b}_k^* \widetilde{A} = (1 + \mathbf{d}_k^* \mathbf{d}_k)^{-1} \mathbf{d}_k^* A_{k-1}^\dagger \widetilde{A}. \tag{41}$$

If $\mathbf{c}_k \neq \mathbf{0}$, then from (24a),

$$\mathbf{b}_k^* \widetilde{A} = (\mathbf{c}_k^* \mathbf{c}_k)^{-1} \mathbf{c}_k^* \widetilde{A}. \tag{42}$$

The computation of (42) is simplified by noting that the kth element of the vector $\mathbf{c}_k^* \widetilde{A}$ is $\mathbf{c}_k^* \mathbf{a}_k$ $(k = 1, 2, \ldots, n)$. Premultiplying (22) by $\mathbf{c}_k^*$ we obtain

$$\mathbf{c}_k^* \mathbf{c}_k = \mathbf{c}_k^* \mathbf{a}_k, \tag{43}$$

since $\mathbf{c}_k^* A_{k-1} = \mathbf{0}$ by (22). Thus the vector (42) may be computed by computing $\mathbf{c}_k^* \widetilde{A}$ and normalizing it by dividing by its kth element.

In the Greville method as described above, the matrix to be inverted is modified at each iteration by adjoining an additional column. This is the natural approach to some applications. Consider, for example, the *least-squares polynomial approximation problem* where a real function $y(t)$ is to be approximated by polynomials $\sum_{j=0}^{k} x_j t^j$. In the discrete version of this problem, the function $y(t)$ is represented by the m-dimensional vector

$$\mathbf{y} = [y_i] = [y(t_i)] \quad (i = 1, \ldots, m), \tag{44}$$

whose i^{th} component is the function y evaluated at $t = t_i$, where the points $t_1, t_2, \ldots, t_m$ are given. Similarly, the polynomial t^j $(j = 0, 1, \ldots)$ is represented by the m-dimensional vector

$$\mathbf{a}_{j+1} = [a_{i,j+1}] = [(t_i)^j] \quad (i = 1, \ldots, m). \tag{45}$$

The problem is, therefore, for a given approximation error $\epsilon > 0$ to find an integer $k = k(\epsilon)$ and a vector $\mathbf{x} \in \mathbb{R}^{k-1}$ such that

$$\|A_{k-1}\mathbf{x} - \mathbf{y}\| \le \epsilon, \tag{46}$$

where $\mathbf{y}$ is given by (44) and $A_{k-1} \in \mathbb{R}^{m \times (k-1)}$ is the matrix

$$A_{k-1} = [\mathbf{a}_1 \ \mathbf{a}_2 \ \cdots \ \mathbf{a}_{k-1}] \tag{47}$$

for $\mathbf{a}_j$ given by (45). For any k, the Euclidean norm $\|A_{k-1}\mathbf{x} - \mathbf{y}\|$ is minimized by

$$\mathbf{x} = A_{k-1}^\dagger \mathbf{y}. \tag{48}$$

If, for a given k, the vector (48) does not satisfy (46), i.e., if

$$\|A_{k-1} A_{k-1}^\dagger \mathbf{y} - \mathbf{y}\| > \epsilon, \tag{49}$$

then we try achieving (46) with the matrix

$$A_k = [A_{k-1} \quad \mathbf{a}_k], \tag{4}$$

where, in effect, the degree of the approximating polynomial has been increased from $k - 2$ to $k - 1$. Greville's method described above computes $A_k^\dagger \mathbf{y}$ in terms of $A_{k-1}^\dagger \mathbf{y}$, and is thus the natural method for solving the above polynomial approximation problem and similar problems in approximation and regression.

There are applications on the other hand which call for modifying the matrix to be inverted by adjoining additional rows. Consider, for example, the problem of solving (or approximating the solution of) the following linear equation:

$$\sum_{j=1}^{n} A_{ij} x_j = y_i \quad (i = 1, \ldots, k-1), \tag{50}$$

where n is fixed and the data $\{a_{ij}, y_i : i = 1, \ldots, k - 1, j = 1, \ldots, n\}$ are the result of some experiment or observation repeated $k - 1$ times, with the row

$$[a_{i1}\ a_{i2}\ \cdots\ a_{in}\ y_i]\quad (i = 1, \ldots, k - 1)$$

the result of the ith experiment.

Let $\widehat{x}_{k-1}$ be the least-squares solution of (50), i.e.,

$$\widehat{x}_{k-1} = A^{\dagger}_{(k-1)}\mathbf{y}_{(k-1)}, \tag{51}$$

where $A_{(k-1)} = [a_{ij}]$ and $\mathbf{y}_{(k-1)} = [y_i]$, $i = 1, \ldots, k - 1$, $j = 1, \ldots, n$. If the results of an additional experiment or observation become available after (50) is solved, then it is necessary to update the solution (51) in light of the additional information. This explains the need for the variant of Greville's method described in Corollaries 1 and 2 below, for which some notation is needed.

Let n be fixed and let $A_{(k)} \in \mathbb{C}^{k \times n}$ be partitioned as

$$A_{(k)} = \begin{bmatrix} A_{(k-1)} \\ \mathbf{a}^*_k \end{bmatrix}, \quad \mathbf{a}^*_k \in \mathbb{C}^{1 \times n}. \tag{52}$$

Also, in analogy with (21) and (22), let

$$\mathbf{d}^*_k = \mathbf{a}^*_k A^{\dagger}_{(k-1)}, \tag{53}$$

$$\mathbf{c}^*_k = \mathbf{a}^*_k - \mathbf{d}^*_k A_{(k-1)}. \tag{54}$$

COROLLARY 1 (Kishi [476]). *Using the above notation*

$$A^{\dagger}_{(k)} = [A^{\dagger}_{(k-1)} - \mathbf{b}_k\mathbf{d}^*_k \quad \mathbf{b}_k], \tag{55}$$

where

$$\mathbf{b}_k = \mathbf{c}^{*\dagger}_k, \quad \textit{if}\ \mathbf{c}^*_k \neq \mathbf{0}, \tag{56}$$

$$\mathbf{b}_k = (1 + \mathbf{d}^*_k\mathbf{d}_k)^{-1}A^{\dagger}_{(k-1)}\mathbf{d}_k, \quad \textit{if}\ \mathbf{c}^*_k = \mathbf{0}. \tag{57}$$

PROOF. Follows by applying Theorem 2 to the conjugate transpose of the matrix (52). □

In some applications it is necessary to compute

$$\widehat{x}_k = A^{\dagger}_{(k)}\mathbf{y}_{(k)}, \quad \text{for given}\ \mathbf{y}_{(k)} \in \mathbb{C}^k,$$

but $A^{\dagger}_{(k)}$ is not needed. Then $\widehat{x}_k$ may be obtained from $\widehat{x}_{k-1}$, in analogy with Corollary 1, as follows:

COROLLARY 2 (Albert and Sittler [12]). *Let the vector* $\mathbf{y}_{(k)} \in \mathbb{C}^k$ *be partitioned as*

$$\mathbf{y}_{(k)} = \begin{bmatrix} \mathbf{y}_{(k-1)} \\ y_k \end{bmatrix}, \quad y_k \in \mathbb{C}, \tag{58}$$

and let

$$\widehat{x}_k = A^{\dagger}_{(k)}\mathbf{y}_{(k)}, \quad \widehat{x}_{k-1} = A^{\dagger}_{(k-1)}\mathbf{y}_{(k-1)}, \tag{59}$$

using the notation (52). Then

$$\widehat{x}_k = \widehat{x}_{k-1} + (y_k - \mathbf{a}^*_k\widehat{x}_{k-1})\mathbf{b}_k, \tag{60}$$

with $\mathbf{b}_k$ *given by* (56) *or* (57).

PROOF. Follows directly from Corollary 1.   □

Exercises

EX. 11. Use Greville's method to calculate $A^\dagger$ for the matrix A of Ex. 5.

EX. 12. *A converse of Theorem* 2. Let the matrix $A_{k-1} \in \mathbb{C}^{m \times (k-1)}$ be obtained from $A_k \in \mathbb{C}^{m \times k}$ by deleting its kth column $\mathbf{a}_k$. If A_k is of full column rank,

$$\begin{bmatrix} A^\dagger_{(k-1)} \\ \mathbf{0}^T \end{bmatrix} = A^\dagger_k - \frac{A^\dagger_k \mathbf{b}_k \mathbf{b}^*_k}{\mathbf{b}^*_k \mathbf{b}_k}, \tag{61}$$

where $\mathbf{b}^*_k$ is the last row of $A^\dagger_k$ (Fletcher [**277**]).

EX. 13. Let

$$A_2 = \begin{bmatrix} \mathbf{a}_1 & \mathbf{a}_2 \end{bmatrix} = \begin{bmatrix} 1 & 0 \\ 2 & 1 \\ 0 & -1 \end{bmatrix}.$$

Then

$$A^\dagger_1 = \mathbf{a}^\dagger_1 = \begin{bmatrix} 1 \\ 2 \\ 0 \end{bmatrix}^\dagger = \tfrac{1}{5} \begin{bmatrix} 1 & 2 & 0 \end{bmatrix},$$

$$\mathbf{d}_2 = A^\dagger_1 \mathbf{a}_2 = \tfrac{1}{5} \begin{bmatrix} 1 & 2 & 0 \end{bmatrix} \begin{bmatrix} 0 \\ 1 \\ -1 \end{bmatrix} = \tfrac{2}{5},$$

$$\mathbf{c}_2 = \mathbf{a}_2 - A_1 \mathbf{d}_2 = \begin{bmatrix} 0 \\ 1 \\ -1 \end{bmatrix} - \begin{bmatrix} 1 \\ 2 \\ 0 \end{bmatrix} \tfrac{2}{5} = \begin{bmatrix} -\tfrac{2}{5} \\ \tfrac{1}{5} \\ -1 \end{bmatrix},$$

and, by (24a),

$$\mathbf{b}^*_2 = \mathbf{c}^\dagger_2 = \tfrac{1}{6} \begin{bmatrix} -2 & 1 & -5 \end{bmatrix}.$$

$A^\dagger_2$ is now computed by (23) as

$$A^\dagger_2 = \tfrac{1}{5} \begin{bmatrix} 1 & 2 & 0 \\ 0 & 0 & 0 \end{bmatrix} + \begin{bmatrix} -\tfrac{2}{5} \\ 1 \end{bmatrix} \begin{bmatrix} -\tfrac{2}{6} & \tfrac{1}{6} & -\tfrac{5}{6} \end{bmatrix}$$

$$= \begin{bmatrix} \tfrac{1}{3} & \tfrac{1}{3} & \tfrac{1}{3} \\ -\tfrac{1}{3} & \tfrac{1}{6} & -\tfrac{5}{6} \end{bmatrix}.$$

Let now $\mathbf{a}^\dagger_2$ be computed by (61), i.e., by deleting $\mathbf{a}_1$ from A_2. Interchanging columns of A_2 and rows of $A^\dagger_2$ we obtain

$$A^\dagger_2 \mathbf{b}_2 = \begin{bmatrix} -\tfrac{1}{3} & \tfrac{1}{6} & -\tfrac{5}{6} \\ \tfrac{1}{3} & \tfrac{1}{3} & \tfrac{1}{3} \end{bmatrix} \begin{bmatrix} \tfrac{1}{3} \\ \tfrac{1}{3} \\ \tfrac{1}{3} \end{bmatrix} = \begin{bmatrix} -\tfrac{1}{3} \\ \tfrac{1}{3} \end{bmatrix}$$

and

$$\mathbf{b}^*_2 \mathbf{b}_2 = \begin{bmatrix} \tfrac{1}{3} & \tfrac{1}{3} & \tfrac{1}{3} \end{bmatrix} \begin{bmatrix} \tfrac{1}{3} \\ \tfrac{1}{3} \\ \tfrac{1}{3} \end{bmatrix} = \tfrac{1}{3},$$

and, finally, from (61),

$$\begin{bmatrix} A_1^\dagger \\ \mathbf{0}^T \end{bmatrix} = \begin{bmatrix} -\frac{1}{3} & \frac{1}{6} & -\frac{5}{6} \\ \frac{1}{3} & \frac{1}{3} & \frac{1}{3} \end{bmatrix} - \begin{bmatrix} -1 \\ 1 \end{bmatrix} \begin{bmatrix} \frac{1}{3} & \frac{1}{3} & \frac{1}{3} \end{bmatrix}$$

$$= \begin{bmatrix} 0 & \frac{1}{2} & -\frac{1}{2} \\ 0 & 0 & 0 \end{bmatrix},$$

or

$$\mathbf{a}_2^\dagger = \begin{bmatrix} 0 \\ 1 \\ -1 \end{bmatrix}^\dagger = \frac{1}{2} \begin{bmatrix} 0 \\ 1 \\ -1 \end{bmatrix} \quad \text{(Fletcher [277]).}$$

6. Computation of Least-Squares Solutions

The Euclidean norm is used throughout. Given the equation

$$A\mathbf{x} = \mathbf{b} \tag{1}$$

and a vector $\mathbf{x}$, the *residual* $\mathbf{r} = \mathbf{r}(\mathbf{x})$ is

$$\mathbf{r} = \mathbf{b} - A\mathbf{x}, \tag{62}$$

and $\mathbf{x}$ is a *least-squares solution* of (1) if $\|\mathbf{r}(\mathbf{x})\|$ is minimum. We recall that a vector $\mathbf{x}$ is a least-squares solution if and only if it is a solution of the *normal equation*

$$A^*A\mathbf{x} = A^*\mathbf{b}. \tag{3.9}$$

NOTE. The use of the normal equation in finding least-squares solutions is limited by the fact that the matrix A^*A is ill-conditioned and very sensitive to roundoff errors, see, e.g., Taussky [798] and Ex. 6.7. Methods for computing least-squares solutions which take account of this difficulty have been studied by several authors. We mention in particular Björck [106], [105], [107], Björck and Golub [111], Businger and Golub [141], [142], Golub and Wilkinson [312], and Noble [615]. These methods can be used, with slight modifications, to compute the generalized inverse.

To avoid the normal equation, let A be factorized as

$$A = FG \tag{63}$$

where G is of full row rank. Then the normal equation is equivalent to

$$F^*A\mathbf{x} = F^*\mathbf{b}, \tag{64}$$

a useful fact, if the system (64) is not ill-conditioned, or at least not worse-conditioned than (1).

In particular, recall the QR factorization of the matrix $A \in \mathbb{C}_n^{m \times n}$ (full column rank is assumed here for convenience; the modifications required for the general case are the subject of Ex. 15),

$$A = \widetilde{Q}\widetilde{R} \tag{0.39}$$

$$= QR, \tag{0.40}$$

here $Q \in \mathbb{C}^{m \times m}$ is unitary (i.e., $Q^*Q = I$), $R = \begin{bmatrix} \widetilde{R} \\ O \end{bmatrix}$ where $\widetilde{R}$ is an $n \times n$ upper triangular matrix and $\widetilde{Q}$ consists of the first n columns of Q, forming an o.n. basis of $R(A)$.

It follows that the normal equation (3.9) is equivalent to

$$Q^* Ax = Q^* \mathbf{b} \tag{65}$$

or to

$$\widetilde{R}x = \widetilde{Q}^* \mathbf{b}, \tag{66}$$

since $\widetilde{Q}^* \widetilde{Q} = I_n$. Now $\widetilde{R}$ is upper triangular and thus (66) is solved by backward substitution. The system (65) is not worse-conditioned than the original system $Ax = \mathbf{b}$; indeed the coefficient matrix $Q^* A$ is obtained by "rotating" the columns of A.

Exercises

Ex. 14. Using the above notation, let

$$Q^* \mathbf{b} = \mathbf{c} = [c_i], \quad i \in \overline{1, m}.$$

Show that the minimum value of $\|Ax - \mathbf{b}\|^2$ is $\sum_{i=n+1}^{m} |c_i|^2$.
Hint: $\|Ax - \mathbf{b}\|^2 = \|Q^*(Ax - \mathbf{b})\|^2$ since Q is unitary.

Ex. 15. Modify the above results for the case $A \in \mathbb{C}_r^{m \times n}$, $r < n$.

Ex. 16. Show that the $\widetilde{Q}\widetilde{R}$-factorization for the matrix of Ex. 3.4, is

$$A = \begin{bmatrix} 1 & 1 \\ \epsilon & 0 \\ 0 & \epsilon \end{bmatrix} \approx \mathrm{fl}(\widetilde{Q})\,\mathrm{fl}(\widetilde{R}) = \begin{bmatrix} 1 & \frac{\epsilon}{\sqrt{2}} \\ \epsilon & -\frac{1}{\sqrt{2}} \\ 0 & \frac{1}{\sqrt{2}} \end{bmatrix} \begin{bmatrix} 1 & 1 \\ 0 & \epsilon\sqrt{2} \end{bmatrix}.$$

Use this to compute the least-squares solution of

$$\begin{bmatrix} 1 & 1 \\ \epsilon & 0 \\ 0 & \epsilon \end{bmatrix} \begin{bmatrix} x_1 \\ x_2 \end{bmatrix} = \begin{bmatrix} 1 \\ \epsilon \\ 2\epsilon \end{bmatrix}.$$

SOLUTION. The (rounded) least-squares solution obtained by using (66) with the rounded matrices $\mathrm{fl}(\widetilde{Q})$ and $\mathrm{fl}(\widetilde{R})$ is

$$x_1 = 0, \quad x_2 = 1.$$

The exact least-squares solution is

$$x_1 = \frac{\epsilon^2}{2 + \epsilon^2}, \quad x_2 = \frac{2(1 + \epsilon^2)}{2 + \epsilon^2}.$$

7. Iterative Methods for Computing $A^\dagger$

An iterative method for computing $A^\dagger$ is a set of instructions for generating a sequence $\{X_k : k = 1, 2, \dots\}$ converging to $A^\dagger$. The instructions specify how to select the initial approximation X_0, how to proceed from X_k to X_{k+1} for each k, and when to stop, having obtained a reasonable approximation of $A^\dagger$.

The rate of convergence of such an iterative method is determined in terms of the corresponding sequence of *residuals*

$$R_k = P_{R(A)} - AX_k, \quad k = 0, 1, \dots \tag{67}$$

which converges to O as $X_k \to A^\dagger$. An iterative method is said to be a pth-*order method*, for some $p > 1$, if there is a positive constant c such that

$$\|R_{k+1}\| \le c\|R_k\|^p, \quad k = 0, 1, \dots \tag{68}$$

for any multiplicative matrix norm; see, e.g., Ex. 0.34.

In analogy with the nonsingular case – see, e.g., Householder [**432**, pp. 94–95] – we consider iterative methods of the type

$$X_{k+1} = X_k + C_k R_k, \quad k = 0, 1, \dots , \tag{69}$$

where $\{C_k : k = 0, 1, \dots \}$ is a suitable sequence and X_0 is the initial approximation (to be specified).

One objection to (69) as an iterative method for computing $A^\dagger$ is that (69) requires at each iteration the residual R_k, for which one needs the projection $P_{R(A)}$, whose computation is a task comparable to computing $A^\dagger$. This difficulty will be overcome here by choosing the sequence $\{C_k\}$ in (69) to satisfy

$$C_k = C_k P_{R(A)}, \quad k = 0, 1, \dots . \tag{70}$$

For such a choice we have

$$\begin{aligned} C_k R_k &= C_k \left(P_{R(A)} - AX_k\right), \quad \text{by (67)}, \\ &= C_k \left(I - AX_k\right), \quad \text{by (70)}, \end{aligned} \tag{71}$$

and (69) can therefore be rewritten as

$$X_{k+1} = X_k + C_k T_k, \quad k = 0, 1, \dots , \tag{72}$$

where

$$T_k = I - AX_k, \quad k = 0, 1, \dots . \tag{73}$$

The iterative method (69), or (72), is suitable for the case where A is an $m \times n$ matrix with $m \le n$, for then R_k and T_k are $m \times m$ matrices. However, if $m > n$, the following dual version of (69) is preferable to it

$$X'_{k+1} = X'_k + R'_k C'_k, \quad k = 0, 1, \dots , \tag{69'}$$

where

$$R'_k = P_{R(A^*)} - X'_k A \tag{67'}$$

and $\{C'_k : k = 0, 1, \dots \}$ is a suitable sequence satisfying

$$C'_k = P_{R(A^*)} C'_k, \quad k = 0, 1, \dots , \tag{70'}$$

a condition which allows rewriting (69') as

$$X'_{k+1} = X'_k + T'_k C'_k, \quad k = 0, 1, \dots , \tag{72'}$$

where

$$T'_k = I - X'_k A, \quad k = 0, 1, \dots . \tag{73'}$$

Indeed, if $m > n$, then (72') is preferable to (72), for the former method uses the $n \times n$ matrix T'_k while the latter uses T_k, which is an $m \times m$ matrix.

Since all the results and proofs pertaining to the iterative method (69) or (72) hold true, with obvious modifications, for the dual method (69') or (72'), we will, for the sake of convenience, restrict the discussion to the case

$$m \leq n, \tag{74}$$

leaving to the reader the details of the complementary case.

A *first-order iterative method for computing* $A^\dagger$, of type (72), is presented in the following:

THEOREM 3. *Let* $O \neq A \in \mathbb{C}^{m \times n}$ *and let the initial approximation* X_0 *and its residual* R_0 *satisfy*

$$X_0 \in R(A^*, A^*) \tag{75}$$

(i.e., $X_0 = A^* B A^*$ *for some* $B \in \mathbb{C}^{m \times n}$, *see Ex. 3.15, p. 110), and*

$$\rho(R_0) < 1, \tag{76}$$

respectively. Then the sequence

$$\begin{aligned} X_{k+1} &= X_k + X_0 T_k \\ &= X_k + X_0(I - AX_k), \quad k = 0, 1, \ldots, \end{aligned} \tag{77}$$

converges to $A^\dagger$ *as* $k \to \infty$ *and the corresponding sequence of residuals satisfies*

$$\|R_{k+1}\| \leq \|R_0\| \|R_k\|, \quad k = 0, 1, \ldots, \tag{78}$$

for any multiplicative matrix norm.

PROOF. The sequence (77) is obtained from (72) by choosing

$$C_k = X_0, \quad k = 0, 1, \ldots, \tag{79}$$

a choice which, by (75), satisfies (70), and allows rewriting (77) as

$$\begin{aligned} X_{k+1} &= X_k + X_0 R_k \\ &= X_k + X_0(P_{R(A)} - AX_k), \quad k = 0, 1, \ldots. \end{aligned} \tag{80}$$

From (80) we compute the residual

$$\begin{aligned} R_{k+1} &= P_{R(A)} - AX_{k+1} \\ &= P_{R(A)} - AX_k - AX_0 R_k \\ &= R_k - AX_0 R_k \\ &= P_{R(A)} R_k - AX_0 R_k, \quad \text{by (67),} \\ &= R_0 R_k, \quad k = 0, 1, \ldots, \\ &= R_0^{k+2}, \quad \text{by repeating the argument.} \end{aligned} \tag{81}$$

For any multiplicative matrix norm, it follows from (81) that

$$\|R_{k+1}\| \leq \|R_0\| \|R_k\|. \tag{78}$$

From

$$R_{k+1} = R_0^{k+2}, \quad k = 0, 1, \ldots, \tag{81}$$

it also follows, by using (76) and Ex. 0.44, that the sequence of residuals converges to the zero matrix:

$$P_{R(A)} - AX_k \to O, \quad \text{as } k \to \infty. \tag{82}$$

We will now prove that the sequence (77) converges. Rewriting the sequence (77) as

$$X_{k+1} = X_k + X_0 R_k, \tag{80}$$

it follows from (81) that

$$
\begin{aligned}
X_{k+1} &= X_k + X_0 R_0^{k+1} \\
&= X_{k_1} + X_0 R_0^k + X_0 R_0^{k+1} \\
&= X_0(I + R_0 + R_0^2 + \cdots + R_0^{k+1}), \quad k = 0, 1, \dots,
\end{aligned} \tag{83}
$$

which, by (76) and Exs. 0.44–0.45, converges to a limit X_∞.

Finally, we will show that $X_\infty = A^\dagger$. From (82) it follows that

$$AX_\infty = P_{R(A)},$$

and in particular, that X_∞ is a $\{1\}$-inverse of A. From (75) and (77) it is obvious that all X_k lie in $R(A^*, A^*)$ and, therefore,

$$X_\infty \in R(A^*, A^*),$$

proving that $X_\infty = A^\dagger$, since $A^\dagger$ is the unique $\{1\}$-inverse which lies in $R(A^*, A^*)$; see Ex. 3.19.   □

For any integer $p \geq 2$, a pth-order iterative method for computing $A^\dagger$, of type (72), is described in the following:

THEOREM 4. Let $O \neq A \in \mathbb{C}^{m \times n}$ and let the initial approximation X_0 and its residual R_0 satisfy (75) and (76), respectively. Then, for any integer $p \geq 2$, the sequence

$$
\begin{aligned}
X_{k+1} &= X_k(I + T_k + T_k^2 + \cdots + T_k^{p-1}) \\
&= X_k(I + (I - AX_k) + (I - AX_k)^2 + \cdots + (I - AX_k)^{p-1}), \\
&\qquad k = 0, 1, \dots,
\end{aligned} \tag{84}
$$

converges to $A^\dagger$ as $k \to \infty$ and the corresponding sequence of residuals satisfies

$$\|R_{k+1}\| \leq \|R_k\|^p, \quad k = 0, 1, \dots. \tag{85}$$

PROOF. The sequence (84) is obtained from (72) by choosing

$$C_k = X_k(I + T_k + T_k^2 + \cdots + T_k^{p-2}). \tag{86}$$

From (75) and (84) it is obvious that all the X_k lie in $R(A^*, A^*)$, and therefore the sequence $\{C_k\}$, given by (86), satisfies (70), proving that the sequence (84) can be rewritten in the form (69),

$$X_{k+1} = X_k(I + R_k + R_k^2 + \cdots + R_k^{p-1}), \quad k = 0, 1, \dots. \tag{87}$$

From (87) we compute

$$
\begin{aligned}
R_{k+1} &= P_{R(A)} - AX_{k+1} \\
&= P_{R(A)} - AX_k(I + R_k + R_k^2 + \cdots + R_k^{p-1}) \\
&= R_k - AX_k(R_k + R_k^2 + \cdots + R_k^{p-1}).
\end{aligned} \tag{88}
$$

Now, for any $j = 1, \ldots, p - 1$,

$$
\begin{aligned}
R_k^j - AX_k R_k^j &= P_{R(A)} R_k^j - AX_k R_k^j \\
&= R_k R_k^j = R_k^{j+1},
\end{aligned}
$$

and, therefore, the last line in (88) collapses to

$$
R_{k+1} = R_k^p, \tag{89}
$$

which implies (85). The remainder of the proof, namely, that the sequence (87) converges to $A^\dagger$, is analogous to the proof of Theorem 3. □

The iterative methods (77) and (84) are related by the following:

THEOREM 5. *Let $O \neq A \in \mathbb{C}^{m \times n}$ and let the sequence $\{X_k : k = 0, 1, \ldots\}$ be constructed as in Theorem 3. Let p be any integer ≥ 2 and let a sequence $\{\widetilde{X}_j : j = 0, 1, \ldots\}$ be constructed as in Theorem 4 with the same initial approximation X_0 as the first sequence*

$$
\begin{aligned}
\widetilde{X}_0 &= X_0, \\
\widetilde{X}_{j+1} &= \widetilde{X}_j (I + \widetilde{T}_j + \widetilde{T}_j^2 + \cdots + \widetilde{T}_j^{p-1}), \quad j = 0, 1, \ldots,
\end{aligned} \tag{84}
$$

where

$$
\widetilde{T}_j = I - A\widetilde{X}_k, \quad j = 0, 1, \ldots. \tag{73}
$$

Then

$$
\widetilde{X}_j = X_{p^j - 1}, \quad j = 0, 1, \ldots. \tag{90}
$$

PROOF. We use induction on j to prove (90), which obviously holds for $j = 0$. Assuming

$$
\widetilde{X}_j = X_{p^j - 1}, \tag{90}
$$

we will show that

$$
\widetilde{X}_{j+1} = X_{p^{j+1} - 1}.
$$

From

$$
X_k = X_0(I + R_0 + R_0^2 + \cdots + R_0^k) \tag{83}
$$

and (90), it follows that

$$
\widetilde{X}_j = X_0(I + R_0 + R_0^2 + \cdots + R_0^{p^j - 1}). \tag{91}
$$

Rewriting (84) as

$$
\widetilde{X}_{j+1} = \widetilde{X}_j(I + \widetilde{R}_j + \widetilde{R}_j^2 + \cdots + \widetilde{R}_j^{p-1}), \tag{87}
$$

it follows from

$$
\begin{aligned}
\widetilde{R}_j &= P_{R(A)} - A\widetilde{X}_j \\
&= P_{R(A)} - AX_{p^j - 1}, \quad \text{by (90)}, \\
&= R_{p^j - 1} \\
&= R_0^{p^j}, \quad \text{by (81)},
\end{aligned}
$$

that

$$\widetilde{X}_{j+1} = \widetilde{X}_j(I + R_0^{p^j} + R_0^{2p^j} + \cdots + R_0^{(p-1)p^j})$$

$$= X_0(I + R_0 + R_0^2 + \cdots + R_0^{p^j-1})(I + R_0^{p^j} + R_0^{2p^j} + \cdots + R_0^{(p-1)p^j}),$$

by (91),

$$= X_0(I + R_0 + R_0^2 + \cdots + R_0^{p^{j+1}-1})$$

$$= X_{p^{j+1}-1}, \quad \text{by (83).} \qquad \square$$

Theorem 5 shows that an approximation $\widetilde{X}_j$ obtained by the pth-order method (84) in j iterations will require $p^j - 1$ iterations of the first-order method (77), both methods using the same initial approximation. For any two iterative methods of different orders, the higher order method will, in general, require fewer iterations but more computations per iteration. A discussion of the optimal order p for methods of type (84) is given in Ex. 23.

Exercises

Ex. 17. The condition

$$X_0 \in R(A^*, A^*) \tag{75}$$

is necessary for the convergence of the iterative methods (77) and (84): let

$$A = \tfrac{1}{2}\begin{bmatrix} 1 & 1 \\ 1 & 1 \end{bmatrix}, \quad B = \epsilon \begin{bmatrix} 1 & -1 \\ -1 & 1 \end{bmatrix}, \quad \epsilon \neq 0,$$

and let

$$X_0 = A + B.$$

Then

$$R_0 = P_{R(A)} - AX_0 = O$$

and, in particular, (76) holds, but

$$X_0 \notin R(A^*, A^*)$$

and both sequences (77) and (84) reduce to

$$X_k = X_0, \quad k = 0, 1, \ldots,$$

without converging to $A^\dagger$.

Ex. 18. Let $O \neq A \in \mathbb{C}^{m \times n}$ and let X_0 and $R_0 = P_{R(A)} - AX_0$ satisfy

$$X_0 \in R(A^*, A^*), \tag{75}$$

$$\rho(R_0) < 1. \tag{76}$$

Then

$$A^\dagger = X_0(I - R_0)^{-1}. \tag{92}$$

PROOF. The proof of Theorem 3 shows $A^\dagger$ to be the limit of

$$X_k = X_0(I + R_0 + R_o^2 + \cdots + R_0^k) \tag{83}$$

as $k \to \infty$. But (83) converges, by Ex. 0.45–46, to RHS(92). $\square$

The Special Case $X_0 = \beta A^*$

A frequent choice of the initial approximation X_0, in the iterative methods (77) and (84), is

$$X_0 = \beta A^* \qquad (93)$$

for a suitable real scalar β. This special case is treated in the following three exercises:

Ex. 19. Let $O \neq A \in \mathbb{C}_r^{m \times n}$, let β be a real scalar, and let

$$R_0 = P_{R(A)} - \beta AA^*,$$
$$T_0 = I - \beta AA^*.$$

Then the following are equivalent:

(a) The scalar β satisfies

$$0 < \beta < \frac{2}{\lambda_1(AA^*)}, \qquad (94)$$

where

$$\lambda_1(AA^*) \geq \lambda_2(AA^*) \geq \cdots \geq \lambda_r(AA^*) > 0$$

are the nonzero eigenvalues of AA^*.

(b) $\rho(R_0) < 1$.

(c) $\rho(T_0) \leq 1$ and $\lambda = -1$ is not an eigenvalue of T_0.

PROOF. The nonzero eigenvalues of R_0 and T_0 are among

$$\{1 - \beta\lambda_i(AA^*) : i = 1, \ldots, r\}$$

and

$$\{1 - \beta\lambda_i(AA^*) : i = 1, \ldots, m\},$$

respectively. The equivalence of (a), (b), and (c) then follows from the observation that (94) is equivalent to

$$|1 - \beta\lambda_i(AA^*)| < 1, \quad i = 1, \ldots, r. \qquad \square$$

Ex. 20. Let $O \neq A \in \mathbb{C}_r^{m \times n}$ and let the real scalar β satisfy

$$0 < \beta < \frac{2}{\lambda_1(AA^*)}. \qquad (94)$$

Then:

(a) The sequence

$$X_0 = \beta A^*, \quad X_{k+1} = X_k(I - \beta AA^*) + \beta A^*, \quad k = 0, 1, \ldots, \qquad (95)$$

or, equivalently,

$$X_k = \beta \sum_{j=0}^{k} A^*(I - \beta AA^*)^j, \quad k = 0, 1, \ldots, \qquad (96)$$

is a first-order method for computing $A^\dagger$.

(b) The corresponding residuals $R_k = P_{R(A)} - AX_k$ are given by

$$R_k = (P_{R(A)} - \beta AA^*)^{k+1}, \quad k = 0, 1, \ldots. \qquad (97)$$

(c) For any k, the spectral norm of R_k, $\|R_k\|_2$, is minimized by choosing

$$\beta = \frac{2}{\lambda_1(AA^*) + \lambda_r(AA^*)}, \tag{98}$$

in which case the minimal $\|R_k\|_2$ is

$$\|R_k\|_2 = \left(\frac{\lambda_1(AA^*) - \lambda_r(AA^*)}{\lambda_1(AA^*) + \lambda_r(AA^*)}\right)^{k+1}, \quad k = 0, 1, \dots. \tag{99}$$

PROOF. (a) Substituting (93) in (77) results in (95) or, equivalently, in (96).

(b) Follows from (96).

(c) R_k is Hermitian and therefore

$$\|R_k\|_2 = \rho(R_k), \quad \text{by Ex. 0.44,}$$
$$= \rho(R_0^{k+1}), \quad \text{by (81),}$$
$$= \rho^{k+1}(R_0), \quad \text{by Ex. 0.43.}$$

Thus, $\|R_k\|_2$ is minimized by the same β that minimizes $\rho(R_0)$. Since the nonzero eigenvalues of $R_0 = P_{R(A)} - \beta AA^*$ are

$$1 - \beta\lambda_i(AA^*), \quad i = 1, \dots, r,$$

it is clear that β minimizes

$$\rho(R_0) = \max\{|1 - \beta\lambda_i(AA^*)| : i = 1, \dots, r\}$$

if and only if

$$-(1 - \beta\lambda_1(AA^*)) = 1 - \beta\lambda_r(AA^*), \tag{100}$$

which is (98). Finally (99) is obtained by substituting (98) in

$$\rho(R_k) = \max\{|1 - \beta\lambda_i(AA^*)|^{k+1} : i = 1, \dots, r\}, \quad \text{by (97),}$$
$$= |1 - \beta\lambda_r(AA^*)|^{k+1}, \quad \text{for } \beta \text{ satisfying (100).} \qquad \square$$

EX. 21. Let A, β be as in Ex. 20. Then, for any integer $p \geq 2$, the sequence

$$X_{k+1} = X_k(I + T_k + T_k^2 + \cdots + T_k^{p-1}) \tag{84}$$

with

$$X_0 = \beta A^* \tag{93}$$

is a pth-order method for computing $A^\dagger$. The corresponding residuals are

$$R_k = (P_{R(A)} - \beta AA^*)^{p^{k+1}}$$

and their spectral norms are minimized by β of (98). The iterative methods of Exs. 20 and 21 were studied by Ben-Israel and Cohen [81], Petryshyn [641], and Zlobec [889].

EX. 22. *A second order iterative method.* An important special case of Theorem 4 is the case $p = 2$, resulting in the following second-order iterative method for computing $A^\dagger$. Let $O \neq A \in \mathbb{C}^{m \times n}$ and let the initial approximation X_0 and its residual R_0 satisfy (75) and (76), respectively. Then the sequence

$$X_{k+1} = X_k(2I - AX_k), \quad k = 0, 1, \dots \tag{101}$$

converges to $A^\dagger$ as $k \to \infty$ and the corresponding sequence of residuals satisfies

$$\|R_{k+1}\| \leq \|R_k\|^2, \quad k = 0, 1, \dots. \tag{102}$$

The iterative method (101) is a generalization of the well-known method of Schulz [731] for the iterative inversion of a nonsingular matrix, see, e.g., Householder

[**432**, p. 95]. The method (101) was studied by Ben-Israel [**61**], Ben-Israel and Cohen [**81**], Petryshyn [**641**], and Zlobec [**889**]. A detailed error analysis of (101) is given in Söderstörm and Stewart [**766**].

EX. 23. *Discussion of the optimum order p.* As in Theorem 5 we denote by $\{X_k\}$ and $\{\widetilde{X}_k\}$ the sequences generated by the first-order method (77) and by the pth-order method (84), respectively, using the same initial approximation $X_0 = \widetilde{X}_0$. Taking the sequence $\{X_k\}$ as the standard for comparing different orders p in (84), we use (90) to conclude that, for each $k = 0, 1, \ldots$, the smallest integer $\widetilde{k}$ such that the iterate $\widetilde{X}_{\widetilde{k}}$ is beyond X_k is the smallest integer $\widetilde{k}$ satisfying

$$p^{\widetilde{k}} - 1 \geq k$$

and, therefore,

$$\widetilde{k} = \langle \ln(k+1)/\ln p \rangle, \tag{103}$$

where, for any real α, $\langle \alpha \rangle$ is the smallest integer $\geq \alpha$.

In assessing the work per iteration, we assume that the computational effort required to add or subtract an identity matrix is negligible compared to the effort to perform a matrix multiplication. Assuming (74) and hence the usage of the methods (77) and (84), rather than their duals based on (72'), we define a unit of computational effort as the effort required to multiply two $m \times m$ matrices. Accordingly, premultiplying an $n \times m$ matrix by an $m \times n$ matrix requires n/m units, as does the premultiplication of an $m \times m$ matrix by an $n \times m$ matrix. The iteration

$$X_{k+1} = X_k(I + T_k + T_k^2 + \cdots + T_k^{p-1}) \tag{84}$$
$$= X_k(I + T_k(I + \cdots + T_k(I + T_k) \cdots))$$

thus requires:

   n/m units of effort to compute T_k;
   $p - 2$ units of effort to compute $T_k(I + \cdots + T_k(I + T_k) \cdots))$; and
   n/m units of effort to multiply $X_k(I + \cdots + T_k^{p-1})$;

adding to

$$p - 2 + 2\frac{n}{m} \tag{104}$$

units of effort.

The figure (104) can be improved for certain p. For example, the iteration (84) can be written for $p = 2^q$, $q = 1, 2, \ldots$, as

$$X_{k+1} = X_k \prod_{j=1}^{2^{q-1}} (I + T_k^j)$$
$$= X_k(I + T_k)(I + T_k^2)(I + T_k^4) \cdots (I + T_k^{2^{q-1}}) \tag{105}$$

requiring only

$$2(q - 1) + 2\frac{n}{m} \tag{106}$$

units of effort, improving on (104) for all $q \geq 3$; see also Lonseth [**523**].

In comparing the first-order iterative method (77) and the second-order method (101) (obtained from (84) for $p = 2$) one sees that both methods require $2(n/m)$ units of effort per iteration. Therefore, by Theorem 5, the second-order method (101) is superior to the first order method (77).

For a given integer $k = 1, 2, \ldots$ we define the *optimal order p* as the order of the iterative method (84) which, starting with an initial approximation

X_0, minimizes the computational effort required to obtain, or go beyond, the approximation X_k, obtained by the first-order method (77) in k iterations.

Combining (103), (104), and (105) it follows that, for a given k, the optimal p is the integer p minimizing

$$\left(p - 2 + 2\frac{n}{m}\right)\left\langle\frac{\ln(k+1)}{\ln p}\right\rangle, \ p = 2, 3, \ldots, \ p \neq 2^q, \ q = 1, 2, \ldots \tag{107a}$$

or

$$\left(2q - 2 + 2\frac{n}{m}\right)\left\langle\frac{\ln(k+1)}{q\ln 2}\right\rangle, \ p = 2^q, \ q = 1, 2, \ldots. \tag{107b}$$

Lower bounds for (107a) and (107b) are

$$\ln(k+1)\frac{p - 2 + 2(n/m)}{\ln p}, \ p = 2, 3, \ldots, \ p \neq 2^q, \ q = 1, 2, \ldots, \tag{107a$'$}$$

and

$$\ln(k+1)\frac{2q - 2 + 2(n/m)}{q\ln 2}, \ p = 2^q, \ q = 1, 2, \ldots, \tag{107b$'$}$$

respectively, suggesting the following definition which is independent of k. The *approximate optimum order* p is the integer $p \geq 2$ minimizing

$$f(p) = \begin{cases} \dfrac{p - 2 + 2(n/m)}{\ln p}, & p \neq 2^q, \ p = 1, 2, \ldots, \\ \dfrac{2(q - 1 + (n/m))}{q\ln 2}, & p = 2^q, \ q = 1, 2, \ldots. \end{cases} \tag{108}$$

The approximate optimum order p depends on the ratio n/m.

Ex. 24. *Iterative methods for computing projections.* Since $AA^\dagger = P_{R(A)}$, it follows that for any sequence $\{X_k\}$, the sequence $\{Y_k = AX_k\}$ satisfies

$$Y_k \rightarrow P_{R(A)}, \quad \text{if } X_k \rightarrow A^\dagger.$$

Thus, for any iterative method for computing $A^\dagger$, defined by a sequence of successive approximations $\{X_k\}$, there is an associated iterative method for computing $P_{R(A)}$ defined by the sequence $\{Y_k = AX_k\}$. Similarly, an iterative method for computing $P_{R(A^*)}$ is given by the sequence $\{Y_k' = X_k A\}$ since $A^\dagger A = P_{R(A^*)}$.

The residuals R_k, $k = 0, 1, \ldots$, of the sequence $\{Y_k\}$ will still be defined by (67) or, equivalently,

$$R_k = P_{R(A)} - Y_k, \quad k = 0, 1, \ldots. \tag{109}$$

Therefore, the iterative method $\{Y_k = AX_k\}$ for computing $P_{R(A)}$ is of the same order as the iterative method $\{X_k\}$ for computing $A^\dagger$.

In particular, a pth-order iterative method for computing $P_{R(A)}$, based on Theorem 4, is given as follows:

Let $O \neq A \in \mathbb{C}^{m \times n}$ and let the initial approximation Y_0 and its residual R_0 satisfy

$$Y_0 \in R(A, A^*), \tag{110}$$

$$\rho(R_0) < 1, \tag{76}$$

respectively. Then, for any integer $p \geq 2$, the sequence

$$Y_{k+1} = Y_k(I + T_k + T_k^2 + \cdots + T_k^{p-1}) \tag{111}$$

with

$$T_k = I - Y_k, \quad k = 0, 1, \ldots,$$

converges to $P_{R(A)}$ as $k \to \infty$ and the corresponding sequence of residuals (109) satisfies

$$\|R_{k+1}\| \leq \|R_k\|^p, \quad k = 0, 1, \ldots, \tag{85}$$

for any multiplicative matrix norm.

EX. 25. *A monotone property of* (111). Let $O \neq A \in \mathbb{C}^{m \times n}$, let p be an even positive integer, and let the sequence $\{Y_k\}$ be given by (111), (110), (76), and the additional condition that Y_0 be Hermitian. Then the sequence $\{$trace $Y_k : k = 1, 2, \ldots \}$ is monotone increasing and converges to rank A.

PROOF. From (109), (81), and Theorem 5 it follows that

$$Y_k = P_{R(A)} - R_0^{p^k}. \tag{112}$$

From the fact that the trace of a matrix equals the sum of its eigenvalues, it follows that

$$\text{trace } P_{R(A)} = \dim R(A) = \text{rank } A$$

and

$$\text{trace } R_0^{p^k} = \sum_{i=1}^{m} \lambda_i(R_0^{p^k})$$
$$= \sum_{i=1}^{m} \lambda_i^{p^k}(R_0),$$

which is a monotone decreasing sequence converging to zero, since p is even, R_0 is Hermitian (by (109) and the assumption that Y_0 is Hermitian), and therefore its eigenvalues $\lambda_i(R_0)$, which by (76) have moduli less than 1, are real. The proof is completed by noting that, by (112),

$$\text{trace } Y_k = \text{trace } P_{R(A)} - \text{trace } R_0^{p^k}$$
$$= \text{rank } A - \sum_{i=1}^{m} \lambda_i^{p^k}(R_0). \qquad \square$$

EX. 26. *A lower bound on rank A.* Let $O \neq A \in \mathbb{C}^{m \times n}$ and let the sequence $\{Y_k : k = 0, 1, \ldots \}$ be as in Ex. 25. Then

$$\text{rank } A \geq \langle \text{trace } Y_k \rangle, \quad k = 1, 2, \ldots, \tag{113}$$

where $\langle \alpha \rangle$ is the smallest integer $\geq \alpha$.

EX. 27. *Iterative methods for computing matrix products involving generalized inverses.* In some applications one has to compute a matrix product $A^\dagger B$ or $BA^\dagger$, where $A \in \mathbb{C}^{m \times n}$ is given and B is a given matrix or vector. The iterative methods for computing $A^\dagger$ given above can be adapted for computing such products.

Consider, for example, the iterative method

$$X_{k+1} = X_k(I + T_k + T_k^2 + \cdots + T_k^{p-1}), \quad k = 0, 1, \ldots \tag{84}$$

where p is an integer ≥ 2,

$$T_k = I - AX_k, \quad k = 0, 1, \ldots, \tag{73}$$

and the initial approximation X_0 satisfies (75) and (76). A corresponding iterative method for computing $BA^\dagger$, for a given $B \in \mathbb{C}^{q \times n}$, is given as follows:

Let $X_0 \in \mathbb{C}^{n \times m}$ satisfy (75) and (76) and let the sequence $\{Z_k\}$ be given by

$$Z_0 = BX_0, \tag{114}$$

$$Z_{k+1} = Z_k M_k, \quad k = 0, 1, \ldots \tag{115}$$

where

$$M_k = I + T_k + T_k^2 + \cdots + T_k^{p-1}, \quad k = 0, 1, \ldots, \tag{116}$$

$$T_{k+1} = I + M_k(T_k - I), \quad k = 0, 1, \ldots, \tag{117}$$

and

$$T_0 = I - AX_0.$$

Then the sequence $\{Z_k\}$ converges to $BA^\dagger$ as $k \to \infty$ (Garnett, Ben-Israel, and Yau [**297**]).

Suggested Further Reading

SECTION 4. Albert [**9**], Ben-Israel and Wersan [**83**], Businger and Golub [**142**], Decell [**222**], Germain–Bonne [**298**], Golub and Reinsch [**309**], Graybill, Meyer, and Painter [**320**], Ijiri [**436**], Kublanovskaya [**488**], Noble [**614**], [**616**], Pereyra and Rosen [**639**], Peters and Wilkinson [**640**], Pyle [**659**], Shinozaki, Sibuya, and Tanabe [**753**], Stallings and Boullion [**774**], Tewarson [**801**], [**802**], [**803**], Urquhart [**824**],Willner [**877**].

COMPUTATION OF THE DRAZIN INVERSE. Hartwig [**386**], Kuang [**487**], Rose [**708**], Rothblum [**714**], Sidi [**759**], [**760**], Sidi and Kluzner [**761**].

SECTION 5. Meyer [**542**], Udwadia and Kalaba [**823**], Wang [**846**].

SECTION 6. Abdelmalek [**1**], Businger and Golub [**141**], James [**444**], Lawson and Hanson [**504**], Peters and Wilkinson [**640**], Sreedharan [**771**], Stewart [**781**], Žukovskiĭ [**896**], Žukovskiĭ and Lipcer [**897**].

NONLINEAR LEAST SQUARES PROBLEMS. Golub and Pereyra [**307**], [**308**], Pereyra [**637**].

SECTION 7. Kammerer and Nashed [**464**], [**463**], [**465**], [**466**], Nashed [**591**], [**592**], Shinozaki, Sibuya, and Tanabe [**754**], Showalter [**755**], Showalter and Ben-Israel [**756**], Whitney and Meany [**870**], Zlobec [**889**], [**892**].

CHAPTER 8

Miscellaneous Applications

1. Introduction

The selection of a few applications to represent the scope and diversity of generalized inverses is by no means obvious, except for one item, the linear statistical model, that would be on everyone's list.

We hope the applications selected here illustrate the richness and potential of generalized inverses.

Section 2 is an introduction to the important operation of *parallel sums* with applications in electrical networks, orthogonal projections, and infimal convolutions.

The *linear statistical model*, the main application of generalized inverses, is the subject of Section 3.

Section 5 describes an application to *Newton methods* for solving systems of nonlinear equations, without requiring nonsingularity of the Jacobian matrix. Interestingly, this application calls for a {2}-inverse, the "most trivial" of generalized inverses.

An application to *linear system theory* is briefly outlined in Section 6.
Section 7 applies the group inverse to *finite Markov chains*.

The Drazin inverse is a natural tool for solving *singular linear difference equations*. This application is described in Section 8.

The last two sections deal with the matrix volume, a concept closely related to the Moore–Penrose inverse. It is applied in Section 9 to surface integrals and in Section 10 to probability distributions.

2. Parallel Sums

"What can you do with generalized inverses that you could not do without?" My best answer, unchanged in 30 years, is the explicit formula for the orthogonal projection $P_{L \cap M}$ on the intersection of subspaces L, M, in terms of their respective projections

$$P_{L \cap M} = 2P_L(P_L + P_M)^\dagger P_M \tag{5.51}$$
$$= 2P_M(P_L + P_M)^\dagger P_L, \quad \text{(Anderson and Duffin [21]).}$$

This formula, extended to infinite-dimensional spaces in Filmore and Williams [272], answers the following question in Halmos's *Hilbert Space Problem Book*, [366, p. 58]

> Problem 96. *If E and F are projections with ranges M
> and N, find the projection $E \wedge F$ with range $M \cap N$.* [1]

---

[1]Halmos goes on, apparently not believing that an *explicit* answer is possible:

Halmos refers the reader to von Neumann [**840**, p. 55] where the projection $E \wedge F$ is obtained in the limit as an infinite product of the respective projections E and F, each projection appearing infinitely often, see Ex. 5.14,

$$E \wedge F = \lim_{n \to \infty} (EFE)^n, \quad [\mathbf{366}, \text{ p. 257}], \tag{1}$$

an idea used in the Kaczmarz iterative method [**454**] for solving linear equations, and elsewhere.

The Anderson–Duffin result is a special case of their *parallel sum*, defined for $A, B \in \mathbb{C}^{m \times n}$ by

$$A : B = A(A + B)^\dagger B, \quad [\mathbf{21}]. \tag{2}$$

Other definitions of the *parallel sum* are

$$A \boxplus B = (A^\dagger + B^\dagger)^\dagger, \quad \text{Rao and Mitra } [\mathbf{678}, \text{ p. 187}], \tag{3}$$

and if A, B are PSD, $A : B$ is defined by an extremal principle due to Morley [**580**],

$$\langle x, A : B\mathbf{x} \rangle = \inf \left\{ \langle \mathbf{y}, A\mathbf{y} \rangle + \langle \mathbf{z}, B\mathbf{z} \rangle : \mathbf{y} + \mathbf{z} = \mathbf{x} \right\}, \tag{4}$$

shown by Morley to be equivalent to the Anderson–Duffin definition.

Definitions (2)–(4) have a common physical motivation: if two resistances R_1 and R_2 are connected in parallel, the resulting resistance is the parallel sum

$$R_1 : R_2 = \left(\frac{1}{R_1} + \frac{1}{R_2} \right)^{-1}, \quad \text{compare with (3)},$$

$$= \frac{R_1 R_2}{R_1 + R_2}, \quad \text{the scalar version of (2)}.$$

A current $\mathbf{x}$ through $R_1 : R_2$ splits into currents $\mathbf{y}$ and $\mathbf{z}$, through R_1 and R_2 respectively, so as to minimize the dissipated power

$$\langle \mathbf{y}, R_1\mathbf{y} \rangle + \langle \mathbf{z}, R_2\mathbf{z} \rangle, \quad \text{the motivation for (4)}.$$

In physical and engineering applications the matrices A, B are PSD, and so they are here. The projection result (5.51) is then a special case

$$P_{L \cap M} = 2P_L : P_M, \tag{5}$$

and Morley's result gives

$$\tfrac{1}{2} \langle x, P_{L \cap M}\mathbf{x} \rangle = \inf \left\{ \langle \mathbf{y}, P_L\mathbf{y} \rangle + \langle \mathbf{z}, P_M\mathbf{z} \rangle : \mathbf{y} + \mathbf{z} = \mathbf{x} \right\}.$$

Here is a convex-analytic proof of the equivalence of definitions (3)–(4) if A, B are PSD. The proof is stated for real matrices.

"The problem is to find an "expression" for the projection described. Although most mathematicians would read the statement of such a problem with sympathetic understanding, it must be admitted that rigorously speaking it does not really mean anything. The most obvious way to make it precise is to describe certain classes of operators by the requirement that they be closed under some familiar algebraic and topological operations, and then try to prove that whenever E and F belong to such a class, then so does $E \wedge F$, etc." (ibid, pp. 58–59).

First some preliminaries (the word *convex* is used below to mean *proper convex*, see, e.g., [**703**]). The *conjugate* of a convex function $f : \mathbb{R}^n \to \mathbb{R}$ is

$$f^*(\boldsymbol{\xi}) := \sup_{\mathbf{x} \in \mathbb{R}^n} \{\langle \boldsymbol{\xi}, \mathbf{x} \rangle - f(\mathbf{x})\}, \quad [\textbf{703}, \text{ p. 104}]. \tag{6}$$

f^* is convex (even if f in (6) is not), and $f = f^{**}$ if and only if f is a closed convex function [**703**, Theorem 12.2]. Two cases where the conjugate is readily available:

(a) If A is PSD, then the conjugate of $f(\mathbf{x}) = \frac{1}{2}\langle \mathbf{x}, A\mathbf{x} \rangle$ is

$$f^*(\boldsymbol{\xi}) = \begin{cases} \frac{1}{2}\langle \boldsymbol{\xi}, A^\dagger \boldsymbol{\xi} \rangle, & \text{if } \boldsymbol{\xi} \in R(A), \\ +\infty, & \text{otherwise.} \end{cases} \tag{7}$$

PROOF. $\nabla\{\langle \boldsymbol{\xi}, \mathbf{x} \rangle - f(\mathbf{x})\} = \boldsymbol{\xi} - A\mathbf{x}$, etc. □

(b) The *infimal convolution* of two convex functions $f, g : \mathbb{R}^n \to \mathbb{R}$,

$$(f \,\square\, g)(\mathbf{x}) := \inf_{\mathbf{y}}\{f(\mathbf{y}) + g(\mathbf{x} - \mathbf{y})\} \tag{8}$$

is convex and

$$(f \,\square\, g)^*(\boldsymbol{\xi}) = f^*(\boldsymbol{\xi}) + g^*(\boldsymbol{\xi}), \quad [\textbf{703}, \text{ p. 145}]. \tag{9}$$

PROOF OF (3) $\Longleftrightarrow$ (4):

RHS(4) $= 2(f \,\square\, g)(\mathbf{x})$ where $f(\mathbf{x}) = \frac{1}{2}\langle \mathbf{x}, A\mathbf{x} \rangle$ and $g(\mathbf{x}) = \frac{1}{2}\langle \mathbf{x}, B\mathbf{x} \rangle$. The proof follows then from (9), (7), and $(f \,\square\, g) = (f \,\square\, g)^{**}$. □

3. The Linear Statistical Model

This section is based on Albert [**10**]. Given a random vector $\mathbf{y} = [y_i]$ with *expected value* $\mathrm{E}\,\mathbf{y} = \boldsymbol{\mu} = [\mu_i]$, its *covariance matrix* is

$$\mathrm{Cov}\,\mathbf{y} = \mathrm{E}\left\{(\mathbf{y} - \boldsymbol{\mu})(\mathbf{y} - \boldsymbol{\mu})^T\right\} = [\mathrm{E}\,(y_i - \mu_i)(y_j - \mu_j)].$$

A common situation is statistics is where a random vector $\mathbf{y}$ depends linearly on a vector of parameters,

$$\mathbf{y} = X\boldsymbol{\beta} + \boldsymbol{\varepsilon} \tag{10}$$

where:

- $\mathbf{y} \in \mathbb{R}^n$ is observed or measured in some experimental set-up;
- the *parameters* $\boldsymbol{\beta} \in \mathbb{R}^p$ are unknown;
- the matrix $X \in \mathbb{R}^{n \times p}$ (the *design matrix*) is given; and
- $\boldsymbol{\varepsilon} \in \mathbb{R}^n$ is a random vector representing the *errors* of observing $\mathbf{y}$, which are not systematic, i.e.,

$$\mathrm{E}\,\boldsymbol{\varepsilon} = \mathbf{0}, \tag{11}$$

and have covariance,

$$\mathrm{Cov}\,\boldsymbol{\varepsilon} = V^2, \tag{12}$$

where the matrix V, assumed known, is symmetric PSD.

Note that, as a consequence of (10)–(12), we have

$$\mathrm{E}\,\mathbf{y} = X\beta, \quad \mathrm{Cov}\,\mathbf{y} = V^2. \tag{13}$$

This situation falls under several names, including: *linear statistical model* (abbreviated *linear model*), *linear regression*, and the *Gauss–Markov model*. We denote this model by $(\mathbf{y}, X\beta, V^2)$.

The problem is to estimate a linear function of the parameters, say

$$B\beta, \quad \text{for a given matrix } B \in \mathbb{R}^{m \times p}, \tag{14}$$

from the observed $\mathbf{y}$ (the problem of estimating the variance V^2, if unknown, is not treated here.)

A *linear estimator* (abbreviated LE) of $B\beta$ is

$$A\mathbf{y}, \quad \text{for some } A \in \mathbb{R}^{m \times n}. \tag{15}$$

It is a *linear unbiased estimator* (abbreviated LUE) if

$$\mathrm{E}\,\{A\mathbf{y}\} = B\beta, \quad \text{for all } \beta \in \mathbb{R}^p, \tag{16}$$

and it is the *best linear unbiased estimator* (abbreviated BLUE) if its variance is minimal, in some sense, among all LUEs. In general, not all linear functions have LUEs, see Ex. 2.

The function $B\beta$ is called *estimable* if it has an LUE, i.e., if there is a matrix $A \in \mathbb{R}^{m \times n}$ such that (16) holds (Bose [**119**], Scheffé [**726**, p. 13]). See also Ex. 4 .

The unbiasedness condition (16) reduces to an identity

$$AX\beta = B\beta, \quad \text{for all } \beta,$$

and, consequently, a linear equation

$$AX = B, \tag{17}$$

giving a necessary and sufficient condition[2] for the estimability of $B\beta$. The estimability of $B\beta$ is thus equivalent to the statement that the rows of B are linear combinations of the rows of X.

There are two cases for the design matrix $X \in \mathbb{R}_r^{n \times p}$: it is either of full column rank $(r = p)$ or not $(r < p)$. For the matrix V^2 (PSD because it is a covariance matrix) there are two cases: it is either nonsingular (i.e., PD) or it is singular. There are therefore four main cases for the linear model $(\mathbf{y}, X\beta, V^2)$, corresponding to the dichotomies on X and V^2, and many special cases with a huge literature, see selected references in p. 328. The simplest case is studied next.

3.1. X of Full Column Rank, V Nonsingular. Consider the special case where the $n \times p$ matrix X is of full column rank, i.e., $R(X^T) = \mathbb{R}^p$. Then any linear function $B\beta$ is estimable. In particular, for $B = I$, the linear equation (17) reduces to $AX = I$, and we conclude that $A\mathbf{y}$ is an LUE of β whenever A is a left inverse of X. The set of LUEs of β is therefore

$$\mathrm{LUE}(\beta) = \{X^{(1)}\mathbf{y} : X^{(1)} \in X\{1\}\}.$$

---

[2]For other conditions see Alalouf and Styan [**6**], [**7**] and their references.

and the mimimum-norm LUE of β is

$$\widehat{\beta} = \left(X^T X\right)^{-1} X^T \mathbf{y} = X^\dagger \mathbf{y}. \tag{18}$$

Without loss of generality (see Ex. 6) we can assume that the variance $V^2 = \sigma^2 I$ (i.e., the errors have equal variances and are uncorrelated). We now state a basic version of the main result on the linear statistical model.

THEOREM 1 (The Gauss–Markov Theorem). *Consider the linear model* $(\mathbf{y}, X\beta, \sigma^2 I)$ *with X of full column rank. Then for any $B \in \mathbb{R}^{m \times p}$:*

(a) *The linear function $B\beta$ is estimable.*

(b) *The estimator $B\widehat{\beta} = BX^\dagger \mathbf{y}$ is BLUE in the sense that*

$$\operatorname{Cov} A\mathbf{y} \;\succcurlyeq\; \operatorname{Cov} B\widehat{\beta} \tag{19}$$

for any other LUE $A\mathbf{y}$ of $B\beta$.

(c) *The BLUE $B\widehat{\beta} = BX^\dagger \mathbf{y}$ belongs to the class of estimators*

$$\mathcal{E}(X) := \{A\mathbf{y} : A = KX^T, \text{ for some matrix } K\}. \tag{20}$$

If $A\mathbf{y}$ is any LUE in $\mathcal{E}(X)$ (i.e., the rows of A are in $R(X)$), then

$$A\mathbf{y} = B\widehat{\beta}, \quad \text{with probability } 1. \tag{21}$$

PROOF. (a) was shown above. To prove (b) let $A\mathbf{y}$ be any LUE of $B\beta$. The covariance of $A\mathbf{y}$ is

$$\operatorname{Cov} A\mathbf{y} = \sigma^2 A A^T, \quad \text{see Ex. 1,}$$

while the covariance of $B\widehat{\beta}$ is

$$\operatorname{Cov} B\left(X^T X\right)^{-1} X^T \mathbf{y} = \sigma^2 B\left(X^T X\right)^{-1} B^T$$

$$= \sigma^2 A X\left(X^T X\right)^{-1} X^T A^T, \quad (B = AX \text{ by (17))}.$$

$$\therefore \; \operatorname{Cov} A\mathbf{y} - \operatorname{Cov} B\widehat{\beta} = \sigma^2 A\left(I - X\left(X^T X\right)^{-1} X^T\right) A^T, \tag{22}$$

which is PSD.

(c) The estimate $BX^\dagger \mathbf{y}$ is in $\mathcal{E}(X)$ since $X^\dagger = (X^T X)^\dagger X^T$. Then (21) follows from

$$\operatorname{RHS}(22) = \sigma^2 A P_{N(X^T)} A^T = O,$$

if $A = KX^T$ for some K. □

Note that all LUEs in $\mathcal{E}(X)$ are indistinguishable, by (21), while for any LUE $A\mathbf{y}$ outside $\mathcal{E}(X)$ the difference $\operatorname{Cov} A\mathbf{y} - \operatorname{Cov} B\widehat{\beta}$ in (22) is a nonzero PSD matrix. In this sense, $B\widehat{\beta}$ can be said to be the *unique BLUE*.

Consider now the problem of estimating linear functionals $\langle \mathbf{b}, \beta \rangle$. A linear estimate $\langle \mathbf{a}, \mathbf{y} \rangle$ is in the class $\mathcal{E}(X)$ if and only if $\mathbf{a} \in R(X^T)$. Theorem 1 then reduces to

COROLLARY 1. *Let $(\mathbf{y}, X\beta, \sigma^2 I)$ and X be as in Theorem 1. Then for any $\mathbf{b} \in \mathbb{R}^p$:*

(a) *The linear functional $\langle \mathbf{b}, \beta \rangle$ is estimable.*

(b) *The estimator* $\langle \mathbf{b}, \widehat{\beta} \rangle = \langle \mathbf{b}, BX^\dagger\mathbf{y} \rangle$ *is* BLUE *in the sense that*

$$\mathrm{Var}\,\langle \mathbf{a}, \mathbf{y} \rangle \;\geq\; \mathrm{Var}\,\langle \mathbf{b}, \widehat{\beta} \rangle$$

for any other LUE $\langle \mathbf{a}, \mathbf{y} \rangle$ *of* $\langle \mathbf{b}, \widehat{\beta} \rangle$.

(c) *If* $\langle \mathbf{a}, \mathbf{y} \rangle$ *is any* LUE *of* $\langle \mathbf{b}, \widehat{\beta} \rangle$ *with* $\mathbf{a} \in R(X^T)$, *then*
$\langle \mathbf{a}, \mathbf{y} \rangle = \langle \mathbf{b}, \widehat{\beta} \rangle$ *with probability 1.* $\qquad\square$

3.2. The General Case. An analog of Theorem 1, for the general linear model, is the following:

THEOREM 2 (The Generalized Gauss–Markov Theorem, Albert [**10**], Zyskind and Martin [**901**]). *Let* $(\mathbf{y}, X\beta, V^2)$ *be a linear model, and let* $\langle \mathbf{b}, \beta \rangle$ *be any estimable functional. Then:*

(a) $\langle \mathbf{b}, \beta \rangle$ *has a unique* BLUE $\langle \mathbf{b}, \widetilde{\beta} \rangle$ *where*

$$\widetilde{\beta} = X^\dagger \left(I - (VP_{N(X^T)})^\dagger V \right)^T \mathbf{y}. \tag{23}$$

(b) $\widetilde{\beta} \in R(X^T)$, *and if* β^* *is any other* LUE *in* $R(X^T)$,

$$\mathrm{Cov}\,\beta^* \;\succcurlyeq\; \mathrm{Cov}\,\widetilde{\beta}. \tag{24}$$

PROOF. The projection $P_{N(X^T)}$ appears repeatedly and will be denoted by Q.

(a) The unbiasedness condition (17) here becomes

$$X^T\mathbf{a} = \mathbf{b} \tag{25}$$

with general solution

$$\mathbf{a}(\mathbf{w}) = X^{T\dagger}\mathbf{b} - Q\mathbf{w}, \quad \mathbf{w}\ \text{arbitrary.} \tag{26}$$

The general LUE of $\langle \mathbf{b}, \beta \rangle$ is therefore

$$\langle \mathbf{a}(\mathbf{w}), \mathbf{y} \rangle, \quad \text{with } \mathbf{a}(\mathbf{w}) \text{ given by (26),}$$

and its variance

$$\mathrm{Var}\,\langle \mathbf{a}(\mathbf{w}), \mathbf{y} \rangle = \mathbf{a}(\mathbf{w})^T V^2 \mathbf{a}(\mathbf{w})$$
$$= \|VX^{T\dagger}\mathbf{b} - VQ\mathbf{w}\|^2$$

is minimized if and only if

$$\mathbf{w}(\mathbf{u}) = (VQ)^\dagger VX^{T\dagger}\mathbf{b} + P_{N(VQ)}\mathbf{u}, \quad \mathbf{u}\ \text{arbitrary.}$$

Substituting in (26) we get the general BLUE

$$\mathbf{a}(\mathbf{w}(\mathbf{u})) = (I - Q(VQ)^\dagger V)X^{T\dagger}\mathbf{b} - QP_{N(VQ)}\mathbf{u}. \tag{27}$$

The special case with $\mathbf{u} = 0$ is denoted

$$\mathbf{a}^* = (I - Q(VQ)^\dagger V)X^{T\dagger}\mathbf{b}. \tag{28}$$

We now prove that

$$\langle \mathbf{a}(\mathbf{w}(\mathbf{u})), \mathbf{y} \rangle = \langle \mathbf{a}^*, \mathbf{y} \rangle, \quad \text{with probability 1.} \tag{29}$$

Indeed,

$$\mathrm{E}\,\langle \mathbf{a}^* - \mathbf{a}(\mathbf{w}(\mathbf{u})), \mathbf{y} \rangle = \mathbf{u}^T P_{N(VQ)}QX\beta = 0, \tag{30}$$

since $QX = O$. The variance

$$\text{Var} \langle \mathbf{a}^* - \mathbf{a}(\mathbf{w}(\mathbf{u})), \mathbf{y} \rangle = \mathbf{u}^T P_{N(VQ)} QVQ P_{N(VQ)} \mathbf{u}$$

$$= \| VQ P_{N(VQ)} \mathbf{u} \|^2 = 0 \qquad (31)$$

and (29) follows from (30)–(31). Therefore the unique BLUE for $\langle \mathbf{b}, \boldsymbol{\beta} \rangle$ is $\langle \mathbf{a}^*, \mathbf{y} \rangle$ with $\mathbf{a}^*$ given by (28) or, equivalently,

$$\mathbf{a}^* = (I - (VQ)^\dagger V) X^{T\dagger} \mathbf{b}, \qquad (32)$$

using $Q(VQ)^\dagger = (VQ)^\dagger$, which follows since $R((VQ)^\dagger) = R((VQ)^T) \subset R(Q)$. Substituting (32) in $\langle \mathbf{a}^*, \mathbf{y} \rangle$ we get

$$\langle \mathbf{a}^*, \mathbf{y} \rangle = \langle \mathbf{b}, X^\dagger (I - (VQ)^\dagger V)^T \mathbf{y} \rangle$$

$$= \langle \mathbf{b}, \widetilde{\boldsymbol{\beta}} \rangle, \quad \text{by (23)}.$$

(b) $\widetilde{\boldsymbol{\beta}} \in R(X^T)$ since $R(X^T) = R(X^\dagger)$. Let $\boldsymbol{\beta}^*$ be any LUE in $R(X^T)$ and let $\mathbf{c}$ be an arbitrary vector. Then $\langle X^\dagger X \mathbf{c}, \boldsymbol{\beta} \rangle$ is estimable and $\langle X^\dagger X \mathbf{c}, \boldsymbol{\beta}^* \rangle$ is an LUE with variance

$$\text{Var} \langle X^\dagger X \mathbf{c}, \boldsymbol{\beta}^* \rangle = \mathbf{c}^T (\text{Cov } X^\dagger X \boldsymbol{\beta}^*) \mathbf{c}$$

$$= \mathbf{c}^T (\text{Cov } \boldsymbol{\beta}^*) \mathbf{c}, \text{ since } \boldsymbol{\beta}^* \in R(X^T).$$

By its construction, $\widetilde{\boldsymbol{\beta}}$ is an LUE with smaller variance. Therefore

$$\mathbf{c}^T (\text{Cov } \boldsymbol{\beta}^*) \mathbf{c} \geq \mathbf{c}^T (\text{Cov } \widetilde{\boldsymbol{\beta}}) \mathbf{c},$$

for all $\mathbf{c}$, proving (24). $\qquad \square$

The following corollary gives alternative expressions for $\widetilde{\boldsymbol{\beta}}$, using the covariance V^2 rather than its square root V.

COROLLARY 2 (Alternative forms of the BLUE). *The BLUE $\widetilde{\boldsymbol{\beta}}$ of Theorem 2 is*

$$\widetilde{\boldsymbol{\beta}} = X^\dagger (I - (V P_{N(X^T)})^\dagger V)^T \mathbf{y} \qquad (23)$$

$$= X^\dagger (I - (P_{N(X^T)} V^2 P_{N(X^T)})^\dagger P_{N(X^T)} V^2)^T \mathbf{y}, \qquad (33)$$

and, if V^2 is nonsingular,

$$\widetilde{\boldsymbol{\beta}} = (X^T V^{-2} X)^\dagger X^T V^{-2} \mathbf{y}. \qquad (34)$$

PROOF. Denote $Q := P_{N(X^T)}$. The equivalence of (23) and (33) follows from $(VQ)^\dagger = (QV^2Q)^\dagger QV$, which is Ex. 1.18(d) with $A = VQ$, using the symmetry of V, Q. To prove (34) we simplify

$$(VQ)^{T\dagger} V = (VQ(VQ)^T)^\dagger VQV$$

$$= (VQ(VQ)^T)^\dagger VQ(VQ)^T, \quad \text{since } Q = Q^2 = Q^T,$$

$$= VQ(VQ)^\dagger,$$

and rewrite (23) as

$$\widetilde{\boldsymbol{\beta}} = X^\dagger V (I - (VQ)^{T\dagger} V) V^{-1} \mathbf{y}.$$

$$\therefore \ \widetilde{\boldsymbol{\beta}} = X^\dagger V (I - VQ(VQ)^\dagger) V^{-1} \mathbf{y}$$

$$= (V^{-1} X)^\dagger V^{-1} \mathbf{y}, \quad \text{(by Ex. 9)},$$

$$= (X^T V^{-2} X)^\dagger X^T V^{-2} \mathbf{y}$$

using Ex. 1.18(d) with $A = V^{-1} X$. $\qquad \square$

In particular, for the model $(\mathbf{y}, X\boldsymbol{\beta}, \sigma^2 I)$, the BLUE (34) reduces to

$$\widehat{\boldsymbol{\beta}} = X^\dagger \mathbf{y}, \tag{35}$$

called here the *naive least-square estimator*, see Ex. 8.

Exercises

Ex. 1. Given a linear model $(\mathbf{y}, X\boldsymbol{\beta}, V^2)$ with $X \in \mathbb{R}^{n\times p}$, a matrix $A \in \mathbb{R}^{m\times n}$, and a vector $\mathbf{a} \in \mathbb{R}^m$:

$$\mathrm{E}\,(A\mathbf{y} + \mathbf{a}) = AX\boldsymbol{\beta} + \mathbf{a},$$
$$\mathrm{Cov}\,(A\mathbf{y} + \mathbf{a}) = AV^2 A^T.$$

Ex. 2. Let $(\mathbf{y}, X\boldsymbol{\beta}, V^2)$ be a linear model with $X \in \mathbb{R}^{n\times p}_r$, $r < p$. Then:

(a) $\boldsymbol{\beta}$ does not have an LUE, for otherwise,

$$\mathrm{E}\,A\mathbf{y} = AX\boldsymbol{\beta} = \boldsymbol{\beta}, \quad \text{for all } \boldsymbol{\beta}.$$
$$\therefore \ AX = I. \quad \therefore \ \mathrm{rank}\,X = p, \quad \text{a contradiction.}$$

(b) Let $\mathbf{0} \neq \mathbf{b} \in N(X)$. Then $\langle \mathbf{b}, \boldsymbol{\beta} \rangle$ is not estimable for, otherwise, there is $\mathbf{a} \in \mathbb{R}^n$ such that

$$\mathrm{E}\,\langle \mathbf{a}, \mathbf{y} \rangle = \langle \mathbf{a}, X\boldsymbol{\beta} \rangle = \langle \mathbf{b}, \boldsymbol{\beta} \rangle, \quad \text{for all } \boldsymbol{\beta},$$

and $\boldsymbol{\beta} = \mathbf{b}$ gives

$$0 = \|\mathbf{b}\|^2.$$

Ex. 3. *Constrained linear models.* Let $(\mathbf{y}, X\boldsymbol{\beta}, V^2)$ be a linear model with $X \in \mathbb{R}^{n\times p}$. The parameters $\boldsymbol{\beta}$ can be "forced" to lie in $R(X^T)$ by appropriate linear constraints,

$$C\boldsymbol{\beta} = \mathbf{0} \tag{36}$$

where the rows of C span $N(X)$. The linear equation

$$\mathbf{a}^T X = \mathbf{b}^T \tag{25}$$

must hold if $\boldsymbol{\beta}$ satisfies (36). By Ex. 2.45 there is a vector $\boldsymbol{\lambda}$ such that

$$(\mathbf{a}^T X - \mathbf{b}^T) = \boldsymbol{\lambda}^T C$$

or

$$\mathbf{b} = X^T \mathbf{a} - C^T \boldsymbol{\lambda}$$

proving that $\langle \mathbf{b}, \boldsymbol{\beta} \rangle$ is estimable for all $\mathbf{b}$.

Ex. 4. *Constrained linear models* (continued). Let $(\mathbf{y}, X\boldsymbol{\beta}, V^2)$ be a linear model with nonhomogeneous linear constraints on the parameters

$$C\boldsymbol{\beta} = \mathbf{c}, \tag{37}$$

where the matrix C is arbitrary and $\mathbf{c} \in R(C)$. It is required to estimate $\langle \mathbf{a}, \boldsymbol{\beta} \rangle$ subject to (37). To accommodate such constraints we replace the linear estimates $\langle \mathbf{a}, \mathbf{y} \rangle$ by affine functions

$$\langle \mathbf{a}, \mathbf{y} \rangle + \alpha, \quad \alpha \text{ scalar.}$$

The functional $\langle \mathbf{b}, \boldsymbol{\beta} \rangle$ is *estimable* if there exist a vector $\mathbf{a}$ and a scalar α such that

$$\mathrm{E}\,\{\langle \mathbf{a}, \mathbf{y} \rangle + \alpha\} = \langle \mathbf{b}, \boldsymbol{\beta} \rangle \tag{38}$$

whenever (37) holds. This is equivalent to

$$C\beta = \mathbf{c} \quad \Longrightarrow \quad (\mathbf{a}^T X - \mathbf{b}^T)\beta = -\alpha.$$

It follows, as in Ex. 3, that there exists a vector λ such that

$$(\mathbf{a}^T X - \mathbf{b}^T) = \lambda^T C, \quad -\alpha = \langle \lambda, \mathbf{c} \rangle.$$

Therefore the linear function $\langle \mathbf{b}, \beta \rangle$ is estimable (subject to (37)) if and only if there exists a λ such that

$$\mathbf{b} = X^T \mathbf{a} - C^T \lambda$$

in which case any affine unbiased estimate is of the form $\langle \mathbf{a}, \mathbf{y} \rangle - \langle \lambda, \mathbf{c} \rangle$.

Ex. 5. (Aitken [4]). Let $(\mathbf{y}, X\beta, \sigma^2 V^2)$ be a linear model with X of full column rank. Then the BLUE of β is the solution $\hat{\beta}$ of

$$\min_{\beta} (X\beta - \mathbf{y})^T V^{-2} (X\beta - \mathbf{y}). \tag{39}$$

If $V^2 = CC^T$ is a factorization of the covariance V^2, then (39) becomes

$$\min_{\beta, \mathbf{u}} \mathbf{u}^T \mathbf{u}, \quad \text{subject to } X\beta + C\mathbf{u} = \mathbf{y}.$$

See also Kourouklis and Paige [482].

Ex. 6. Consider a linear model $(\mathbf{y}, X\beta, V^2)$, V^2 $n \times n$, PSD. Then there is an orthogonal matrix U such that

$$U^T V^2 U = \begin{bmatrix} \lambda_1^2 & & & 0 & \cdots & 0 \\ & \ddots & & \vdots & & \vdots \\ & & \lambda_r^2 & 0 & \cdots & 0 \\ 0 & \cdots & 0 & 0 & \cdots & 0 \\ \vdots & & \vdots & \vdots & \ddots & \vdots \\ 0 & \cdots & 0 & 0 & \cdots & 0 \end{bmatrix} = \begin{bmatrix} \Lambda^2 & O \\ O & O \end{bmatrix},$$

where $r = \operatorname{rank} V^2$, the columns of U are eigenvectors of V^2, and λ_i^2 are its nonzero eigenvalues. Let $T = \begin{bmatrix} \Lambda^{-1} & O \\ O & I \end{bmatrix} U^T$. Then T is nonsingular and

$$\operatorname{Cov} T\mathbf{y} = \begin{bmatrix} \Lambda^{-1} & O \\ O & I \end{bmatrix} U^T V^2 U \begin{bmatrix} \Lambda^{-1} & O \\ O & I \end{bmatrix} = \begin{bmatrix} I & O \\ O & O \end{bmatrix},$$

showing that the model $(\mathbf{y}, X\beta, V^2)$ is equivalent to

$$\left(T\mathbf{y}, TX\beta, \begin{bmatrix} I & O \\ O & O \end{bmatrix} \right). \tag{40}$$

Ex. 7. Let $(\mathbf{y}, X\beta, V^2)$, U, and T be as in Ex. 6 and let the covariance matrix V^2 be singular. Partition the matrix $U = [U_1 \quad U_2]$, where the columns of $U_2 \in \mathbb{R}^{n \times (n-r)}$ are a basis of the null space of V^2. The model (40) is partitioned accordingly as

$$\Lambda^{-1} U_1^T \mathbf{y} = \Lambda^{-1} U_1^T X\beta + \Lambda^{-1} U_1^T \varepsilon, \tag{40.a}$$

$$U_2^T \mathbf{y} = U_2^T X\beta + U_2^T \varepsilon. \tag{40.b}$$

Since $\operatorname{E} U_2^T \varepsilon = \mathbf{0}$, $\operatorname{Cov} U_2^T \varepsilon = O$, (40.b) implies

$$U_2^T \mathbf{y} = U_2^T X\beta, \quad \text{with probability 1,}$$

the "deterministic" part of the model, a consequence of the singularity of V.

EX. 8. The BLUE $\widetilde{\beta}$ of (34), and the naive least-squares estimator $\widehat{\beta}$ of (35), coincide for the model $(\mathbf{y}, X\beta, \sigma^2 I)$. Are there other cases where these two estimators coincide for all estimable functionals? This would mean

$$X^{\dagger T} X^T \mathbf{y} = (I - (V P_{N(X^T)})^{\dagger} V) X^{\dagger T} X^T \mathbf{y}, \quad \text{for all } \mathbf{y},$$

or

$$(V P_{N(X^T)})^{\dagger} V X X^{\dagger} = O, \tag{41}$$

a necessary and sufficient condition for $\widetilde{\beta} = \widehat{\beta}$. Show that (41) is equivalent to

$$R(V^2 X) \subset R(X).$$

Other necessary and sufficient conditions for $\widetilde{\beta} = \widehat{\beta}$ are given in Albert [**10**, p. 186].

EX. 9. (Albert [**10**, Lemma, p. 185]). Let $X, Q,$ and V be as in Corollary 2. If V is nonsingular, then

$$(V^{-1}X)^{\dagger} = X^{\dagger} V (I - VQ(VQ)^{\dagger}). \tag{42}$$

PROOF. For any vector $\mathbf{z}$, the minimum-norm least-squares solution of

$$(V^{-1}X)^T \mathbf{x} = \mathbf{z} \tag{a}$$

is

$$\widehat{\mathbf{x}} = (V^{-1}X)^{T\dagger} \mathbf{z}.$$

If $\mathbf{u}^*$ is a least squares solution of

$$X^T \mathbf{u} = \mathbf{z}, \tag{b}$$

then $\mathbf{x}^* = V\mathbf{u}^*$ is a least-squares solution of (a) and therefore

$$\|\mathbf{x}^*\| = \|V\mathbf{u}^*\| > \|\widehat{\mathbf{x}}\|, \quad \text{unless } \widehat{\mathbf{x}} = V\mathbf{u}^*. \tag{c}$$

Since $Q = P_{N(X^T)}$, the general least-squares solution of (b) is

$$\mathbf{u}(\mathbf{w}) = X^{T\dagger} \mathbf{z} - Q\mathbf{w}, \tag{d}$$

with arbitrary $\mathbf{w}$. Therefore

$$\|V\mathbf{u}(\mathbf{w})\|^2 = \|V X^{T\dagger} \mathbf{z} - VQ\mathbf{w}\|^2 \tag{e}$$

is minimized when $\mathbf{u} = \mathbf{u}(\widehat{\mathbf{w}})$, where

$$\widehat{\mathbf{w}} = (VQ)^{\dagger} V X^{T\dagger} \mathbf{z}. \tag{f}$$

Suppose

$$\|V\mathbf{u}(\widehat{\mathbf{w}})\| > \|\widehat{\mathbf{x}}\|.$$

Since $\widehat{\mathbf{u}} = V^{-1}\widehat{\mathbf{x}}$ is also a least-squares solution of (b), we conclude

$$\|V\widehat{\mathbf{u}}\| = \|\widehat{\mathbf{x}}\| < \|V\mathbf{u}(\widehat{\mathbf{w}})\|,$$

a contradiction to $\mathbf{u}(\widehat{\mathbf{w}})$ being a mimimizer of $\|V\mathbf{u}\|$.

$$\therefore \ \|V\mathbf{u}(\widehat{\mathbf{w}})\| = \|\widehat{\mathbf{x}}\|,$$

and, by (c),

$$V\mathbf{u}(\widehat{\mathbf{w}}) = \widehat{\mathbf{x}}. \tag{g}$$

From (d), (f), and (g) we get

$$(V^{-1}X)^\dagger \mathbf{z} = X^\dagger V(I - VQ(VQ)^\dagger)\mathbf{z},$$

and (42) follows since $\mathbf{z}$ is arbitrary.   □

4. Ridge Regression

Consider the linear model

$$\mathbf{y} = X\beta + \varepsilon \tag{10}$$

with $X \in \mathbb{R}_p^{n\times p}$ (X is of full column rank), and the error ε is normally distributed with mean $\mathbf{0}$ and covariance matrix $\sigma^2 I$, a fact denoted by $\varepsilon \sim N(\mathbf{0}, \sigma^2 I)$.

If $X^T X$ is ill-conditioned, then the BLUE of β,

$$\widehat{\beta} = (X^T X)^{-1} X^T \mathbf{y}, \tag{18}$$

may be unsatisfactory. To see this, consider the SVD of X,

$$U^T X V = \Lambda = \begin{bmatrix} \lambda_1 & & & & \\ & \lambda_2 & & & \\ & & \ddots & & \\ & & & \lambda_p & \\ 0 & \cdots & & \cdots & 0 \\ \vdots & & & & \vdots \\ 0 & \cdots & & \cdots & 0 \end{bmatrix}, \tag{43}$$

where the singular values are denoted by λ_i. The transformation

$$\mathbf{z} = U^T \mathbf{y}, \quad \gamma = V^T \beta, \quad \nu = U^T \varepsilon, \tag{44}$$

then takes (10) into

$$\mathbf{z} = \Lambda \gamma + \nu, \tag{45}$$

where the parameters to be estimated are $\gamma = [\gamma_i]$. Since the matrix V is orthogonal, it follows that $\nu \sim N(\mathbf{0}, \sigma^2 I)$ and the components z_i of $\mathbf{z}$ are also normally distributed

$$z_i \sim N(\lambda_i \gamma_i, \sigma^2), \quad i \in \overline{1, p}, \tag{46a}$$

$$z_i \sim N(0, \sigma^2), \quad i \in \overline{p+1, n}. \tag{46b}$$

For $i \in \overline{1, p}$, the BLUE (18) of γ_i is

$$\widehat{\gamma_i} = \frac{z_i}{\lambda_i}, \tag{47a}$$

with variance

$$\operatorname{Var} \widehat{\gamma_i} = \operatorname{E}\left(\frac{z_i}{\lambda_i} - \gamma_i\right)^2 = \frac{\sigma^2}{\lambda_i^2}, \tag{47b}$$

which may be unacceptably large if the singular value λ_i is small, as is typically the case when $X^T X$ is ill-conditioned. See also Ex. 6.13.

Hoerl and Kennard [**423**] suggested a way of reducing the variance by dropping the unbiasedness of the estimator. Their *ridge regression estimator* (abbreviated RRE) of $\boldsymbol{\beta}$ is

$$\widehat{\boldsymbol{\beta}}(k) = (X^T X + kI)^{-1} X^T \mathbf{y}, \tag{48}$$

where k is a positive parameter. The RRE is actually a family of estimators $\{\widehat{\boldsymbol{\beta}}(k) : k > 0\}$ parametrized by k. The value $k = 0$ gives the BLUE (18) $= \widehat{\boldsymbol{\beta}}(0)$.

For the transformed model (45), the RRE of $\boldsymbol{\gamma}$ is

$$\widehat{\boldsymbol{\gamma}}(k) = (\Lambda^T \Lambda + kI)^{-1} \Lambda^T \mathbf{z}, \tag{49}$$

and for the ith component,

$$\widehat{\gamma}_i(k) = \frac{\lambda_i z_i}{\lambda_i^2 + k}, \quad i \in \overline{1, p}. \tag{49.i}$$

The RRE thus shrinks every component of the observation vector $\mathbf{z}$. The shrinkage factor is

$$c(\lambda_i, k) = \frac{\lambda_i}{\lambda_i^2 + k}, \tag{50}$$

see Ex. 12.

If $\boldsymbol{\beta}^*$ is an estimator of a parameter $\boldsymbol{\beta}$, the *bias* of $\boldsymbol{\beta}^*$ is

$$\text{bias}(\boldsymbol{\beta}^*) = \text{E}\,\boldsymbol{\beta}^* - \boldsymbol{\beta},$$

and its *mean square error*, abbreviated MSE, is

$$\text{MSE}(\boldsymbol{\beta}^*) = \text{E}\left(\boldsymbol{\beta}^* - \boldsymbol{\beta}\right)^2$$

which is equal to the variance of $\boldsymbol{\beta}^*$ if $\boldsymbol{\beta}^*$ is unbiased, see Ex. 10.

The RRE (49) is biased, with[3]

$$\text{bias}(\widehat{\boldsymbol{\gamma}}(k)) = -k(\Lambda^T \Lambda + kI)^{-1} \boldsymbol{\gamma}, \tag{51}$$

and for the ith component,

$$\text{bias}(\widehat{\gamma}_i(k)) = -k\frac{\gamma_i}{\lambda_i^2 + k}, \quad i \in \overline{1, p}. \tag{51.i}$$

The variance of $\widehat{\gamma}_i(k)$ is

$$\text{Var}(\widehat{\gamma}_i(k)) = \frac{\lambda_i^2 \sigma^2}{(\lambda_i^2 + k)^2}, \tag{52}$$

---

[3]A weakness of the RRE is that the bias depends on the unknown parameter to be estimated.

and the MSE of $\widehat{\gamma}(k)$ is, by Ex. 10,

$$
\begin{aligned}
\mathrm{MSE}(\widehat{\gamma}(k)) &= \sum_{i=1}^{p} \frac{\lambda_i^2 \sigma^2}{(\lambda_i^2 + k)^2} + \sum_{i=1}^{p} \frac{k^2 \gamma_i^2}{(\lambda_i^2 + k)^2} \\
&= \sum_{i=1}^{p} \frac{\lambda_i^2 \sigma^2 + k^2 \gamma_i^2}{(\lambda_i^2 + k)^2}.
\end{aligned}
\tag{53}
$$

An advantage of the RRE is indicated in the following theorem:

THEOREM 3. *There is a $k > 0$ for which the MSE of the RRE is smaller than that of the BLUE,*

$$
\mathrm{MSE}(\widehat{\boldsymbol{\beta}}(k)) < \mathrm{MSE}(\widehat{\boldsymbol{\beta}}(0)).
$$

PROOF. Let $f(k) = \mathrm{RHS}(53)$. We have to show that f is decreasing at zero, i.e., $f'(0) < 0$. This follows since

$$
f'(k) = 2 \sum_{i=1}^{p} \frac{\lambda_i^2 (k \gamma_i^2 - \sigma^2)}{(\lambda_i^2 + k)^3}.
\tag{54}
$$

$\square$

An optimal RRE $\widehat{\boldsymbol{\beta}}(k^*)$ may be defined as corresponding to a value k^* where $f(k)$ is minimum. There are two difficulties with this concept: it requires solving the nonlinear equation $f'(k) = 0$, and k^* depends on the unknown parameter γ (since $f'(k)$ does). For a discussion of the good choices of RREs, see Goldstein and Smith [303], Obenchain [618], Vinod [836], and their references.

Exercises

Ex. 10. If $\boldsymbol{\beta}^*$ is an estimator of a parameter $\boldsymbol{\beta}$, then

$$
\mathrm{MSE}(\boldsymbol{\beta}^*) = \mathrm{trace}\ \mathrm{Cov}\ \boldsymbol{\beta}^* + \|\mathrm{bias}(\boldsymbol{\beta}^*)\|^2,
$$

where $\| \cdot \|$ is the Euclidean norm.

Ex. 11. A generalization of the RRE (48) is defined, for a vector $\mathbf{k} = [k_i]$ of positive parameters, by

$$
\widehat{\boldsymbol{\beta}}(\mathbf{k}) = (X^T X + \mathrm{diag}(\mathbf{k}))^{-1} X^T \mathbf{y},
\tag{55}
$$

where $\mathrm{diag}(\mathbf{k})$ is the diagonal matrix with diagonal elements $\{k_i : i \in \overline{1,p}\}$. Replacing k by k_i in (51.i)–(52) we get

$$
\mathrm{MSE}(\widehat{\gamma}(\mathbf{k})) = \sum_{i=1}^{p} \frac{\lambda_i^2 \sigma^2 + k_i^2 \gamma_i^2}{(\lambda_i^2 + k_i)^2},
$$

and the analog of (54) is

$$
\nabla f(\mathbf{k}) = 2 \Big(\frac{\lambda_i^2 (k_i \gamma_i^2 - \sigma^2)}{(\lambda_i^2 + k_i)^3} \Big),
$$

showing that $\mathrm{MSE}(\widehat{\gamma}(\mathbf{k}))$ is minimized for $\mathbf{k} = [\sigma^2 / \gamma_i^2]$.

Ex. 12. (Goldstein and Smith [303]). A shrinkage estimator of γ in the model (45) is defined by

$$
\widehat{\gamma_i}^* = c(\lambda_i, k)\, z_i,
\tag{56}
$$

where $c(\lambda_i, k)$ is a suitable shrinkage function. A comparison with (49.i) shows that the RRE is a shrinkage estimator with

$$c(\lambda_i, k) = \frac{\lambda_i}{\lambda_i^2 + k}. \qquad (50)$$

The shrinkage function is assumed to have the properties:

- (a) $c(\lambda_i, 0) = 1/\lambda_i$;
- (b) $|c(\lambda_i, k)|$ is, for any fixed λ_i, a continuous, monotone decreasing function of k; and
- (c) $c(\lambda_i, k)$ has the same sign as λ_i.

Then, for each γ, there is a $k > 0$ such that the shrinkage estimator $\widehat{\gamma}_i^* = c(\lambda_i, k) z_i$ has smaller MSE than the BLUE (47a), for all $i \in \overline{1, p}$.

5. An Application of {2}-Inverses in Iterative Methods for Solving Nonlinear Equations

One of the best-known methods for solving a single equation in a single variable, say

$$f(x) = 0, \qquad (57)$$

is the *Newton* (also *Newton–Raphson*) *method*

$$x^{k+1} = x^k - \frac{f(x^k)}{f'(x^k)}, \quad k = 0, 1, \ldots . \qquad (58)$$

Under suitable conditions on the function f and the initial approximation x^0, the sequence (58) converges to a solution of (57); see, e.g., Ortega and Rheinboldt [620]. The *modified Newton method* uses the iteration

$$x^{k+1} = x^k - \frac{f(x^k)}{f'(x^0)}, \quad k = 0, 1, \ldots , \qquad (59)$$

instead of (58).

Newton's method for solving a system of m equations in n variables

$$\left.\begin{cases} f_1(x_1, \ldots, x_n) & = 0 \\ \quad \cdots \\ f_m(x_1, \ldots, x_n) & = 0 \end{cases}\right\}, \quad \text{or} \quad \mathbf{f}(\mathbf{x}) = \mathbf{0}, \qquad (60)$$

is similarly given, for the case $m = n$, by

$$\mathbf{x}^{k+1} = \mathbf{x}^k - \mathbf{f}'(\mathbf{x}^k)^{-1}\mathbf{f}(\mathbf{x}^k), \quad k = 0, 1, \ldots , \qquad (61)$$

where $\mathbf{f}'(\mathbf{x}^k)$ is the *derivative* of $\mathbf{f}$ at $\mathbf{x}^k$, represented by the matrix of partial derivatives (the *Jacobian* matrix)

$$\mathbf{f}'(\mathbf{x}^k) = \left(\frac{\partial f_i}{\partial x_j}(\mathbf{x}^k)\right), \qquad (62)$$

and denoted below by $J_{\mathbf{f}}(\mathbf{x})$ or by $J_{\mathbf{x}}$.

The reader is referred to the excellent texts by Ortega and Rheinboldt [620] and Rall [667] for iterative methods in nonlinear analysis and, in particular, for the many variations and extensions of Newton's method (61).

If the nonsingularity of $\mathbf{f}'(\mathbf{x}^k)$ cannot be assumed for every $\mathbf{x}^k$ and, in particular, if the number of equations (60) is different from the number of unknowns, then it is natural to inquire whether a generalized inverse of $\mathbf{f}'(\mathbf{x}^k)$ can be used in (61), still resulting in a sequence converging to a solution of (60).

Interestingly, the $\{2\}$-inverse is the natural tool for use here. This may seem surprising because until now we used the definition

$$XAX = X \tag{1.2}$$

in conjunction with other, more useful, properties. Indeed, the $\{2\}$-inverse is trivial in the sense that $X = O$ satisfies (1.2), and is a $\{2\}$-inverse.

In this section we illustrate the use of $\{2\}$-inverses in Newton's method (Theorems 4–5 and Ex. 14).

Except in Corollary 3, where we use the Euclidean norm, we denote by $\|\ \|$ both a given (but arbitrary) vector norm in $\mathbb{R}^n$ and a matrix norm in $\mathbb{R}^{m \times n}$ consistent with it; see, e.g., Ex. 0.35. For a given point $\mathbf{x}^0 \in \mathbb{R}^n$ and a positive scalar r we denote by

$$B(\mathbf{x}^0, r) = \{\mathbf{x} \in \mathbb{R}^n : \|\mathbf{x} - \mathbf{x}^0\| < r\}$$

the *open ball* with *center* $\mathbf{x}^0$ and *radius* r. The corresponding *closed ball* is

$$\overline{B(\mathbf{x}^0, r)} = \{\mathbf{x} \in \mathbb{R}^n : \|\mathbf{x} - \mathbf{x}^0\| \leq r\}.$$

The following theorem establishes the quadratic convergence for a Newton method, using suitable $\{2\}$-inverses of Jacobian matrices. However, the iterates converge not to a solution of (60) but to an "approximate solution," the degree of approximation depending on the $\{2\}$-inverse used.

THEOREM 4 (Levin and Ben-Israel [**509**], Theorem 1). *Let* $\mathbf{x}^0 \in \mathbb{R}^n$, $r > 0$ *and let* $\mathbf{f} : \mathbb{R}^n \to \mathbb{R}^m$ *be differentiable in the open ball* $B(\mathbf{x}^0, r)$. *Let* $M > 0$ *be such that*

$$\|J_{\mathbf{u}} - J_{\mathbf{v}}\| \leq M \|\mathbf{u} - \mathbf{v}\| \tag{63}$$

for all $\mathbf{u}, \mathbf{v} \in B(\mathbf{x}^0, r)$, *where* $J_{\mathbf{u}}$ *is the Jacobian of* $\mathbf{f}$ *at* $\mathbf{u}$. *Further, assume that, for all* $\mathbf{x} \in \overline{B(\mathbf{x}^0, r)}$, *the Jacobian* $J_{\mathbf{x}}$ *has a* $\{2\}$-*inverse* $T_{\mathbf{x}} \in \mathbb{R}^{n \times m}$,

$$T_{\mathbf{x}} J_{\mathbf{x}} T_{\mathbf{x}} = T_{\mathbf{x}} \tag{64a}$$

such that

$$\|T_{\mathbf{x}^0}\| \left\|\mathbf{f}(\mathbf{x}^0)\right\| < \alpha, \tag{64b}$$

and, for all $\mathbf{u}, \mathbf{v} \in B(\mathbf{x}^0, r)$,

$$\|(T_{\mathbf{u}} - T_{\mathbf{v}})\mathbf{f}(\mathbf{v})\| \leq N \|\mathbf{u} - \mathbf{v}\|^2, \tag{64c}$$

and

$$\frac{M}{2} \|T_{\mathbf{u}}\| + N \leq K < 1, \tag{64d}$$

for some positive scalars N, K, *and* α, *and*

$$h := \alpha K < 1, \quad r > \frac{\alpha}{1 - h}. \tag{64e}$$

Then:

(a) *Starting at* $\mathbf{x}^0$, *all iterates*

$$\mathbf{x}^{k+1} = \mathbf{x}^k - T_{\mathbf{x}^k}\mathbf{f}(\mathbf{x}^k), \quad k = 0, 1, \ldots, \tag{65}$$

lie in $B(\mathbf{x}^0, r)$.

(b) *The sequence* $\{\mathbf{x}^k\}$ *converges, as* $k \to \infty$, *to a point* $\mathbf{x}^\infty \in \overline{B(\mathbf{x}^0, r)}$, *that is, a solution of*

$$T_{\mathbf{x}^\infty}\mathbf{f}(\mathbf{x}) = 0. \tag{66}$$

(c) *For all* $k \geq 0$,

$$\|\mathbf{x}^k - \mathbf{x}^\infty\| \leq \alpha \frac{h^{2^k - 1}}{1 - h^{2^k}}.$$

Since $0 < h < 1$, *the above method is at least quadratically convergent.*

PROOF. PART 1. Using induction on k we prove that the sequence (65) satisfies, for $k = 0, 1, \ldots$,

$$\mathbf{x}^k \in B(\mathbf{x}^0, r), \tag{67a}$$

$$\left\|\mathbf{x}^{k+1} - \mathbf{x}^k\right\| \leq \alpha h^{2^k - 1}. \tag{67b}$$

For $k = 0$, (67b), and for $k = 1$, (67a), follow from (64b). Assuming (67b) holds for $0 \leq j \leq k - 1$ we get

$$\|\mathbf{x}^{k+1} - \mathbf{x}^0\| \leq \sum_{j=1}^{k+1} \|\mathbf{x}^j - \mathbf{x}^{j-1}\| \leq \alpha \sum_{j=0}^{k} h^{2^j - 1} < \frac{\alpha}{1 - h} < r$$

which proves (67a). To prove (67b) we write

$$\mathbf{x}^{k+1} - \mathbf{x}^k = -T_{\mathbf{x}^k}\mathbf{f}(\mathbf{x}^k) = \mathbf{x}^k - \mathbf{x}^{k-1} - T_{\mathbf{x}^k}\mathbf{f}(\mathbf{x}^k) + T_{\mathbf{x}^{k-1}}\mathbf{f}(\mathbf{x}^{k-1}), \quad \text{by (65)},$$

$$= T_{\mathbf{x}^{k-1}}J_{\mathbf{x}^{k-1}}(\mathbf{x}^k - \mathbf{x}^{k-1}) - T_{\mathbf{x}^k}\mathbf{f}(\mathbf{x}^k) + T_{\mathbf{x}^{k-1}}\mathbf{f}(\mathbf{x}^{k-1}),$$

since $TJT = T$ implies $TJ\mathbf{x} = \mathbf{x}$ for every $\mathbf{x} \in R(T)$,

$$= -T_{\mathbf{x}^{k-1}}(\mathbf{f}(\mathbf{x}^k) - \mathbf{f}(\mathbf{x}^{k-1}) - J_{\mathbf{x}^{k-1}}(\mathbf{x}^k - \mathbf{x}^{k-1})) + (T_{\mathbf{x}^{k-1}} - T_{\mathbf{x}^k})\mathbf{f}(\mathbf{x}^k).$$

$$\therefore \ \|\mathbf{x}^{k+1} - \mathbf{x}^k\| \leq \| - T_{\mathbf{x}^{k-1}}(\mathbf{f}(\mathbf{x}^k) - \mathbf{f}(\mathbf{x}^{k-1}) - J_{\mathbf{x}^{k-1}}(\mathbf{x}^k - \mathbf{x}^{k-1}))\|$$

$$+ \|(T_{\mathbf{x}^{k-1}} - T_{\mathbf{x}^k})\mathbf{f}(\mathbf{x}^k)\|$$

$$\leq \left(\frac{M}{2}\|T_{\mathbf{x}^{k-1}}\| + N\right)\|\mathbf{x}^k - \mathbf{x}^{k-1}\|^2,$$

by (63), Ex. 13 below and (64c),

$$\leq K\left\|\mathbf{x}^k - \mathbf{x}^{k-1}\right\|^2, \quad \text{by (64d)}. \tag{68}$$

Consequently, $\|\mathbf{x}^{k+1} - \mathbf{x}^k\| \leq \alpha h^{2^k - 1}$. This inequality is valid for $k = 0$ because of (64b). Assuming it holds for any $k \geq 0$, its validity for $k + 1$ follows, since (68) implies

$$\|\mathbf{x}^{k+1} - \mathbf{x}^k\| \leq K\|\mathbf{x}^k - \mathbf{x}^{k-1}\|^2 \leq K\alpha^2 h^{2^k - 2} \leq \alpha h^{2^k - 1}, \text{ proving (67b)}.$$

PART 2. From (67b) it follows for $m \geq n$

$$\|\mathbf{x}^{m+1} - \mathbf{x}^n\| \ \leq \ \|\mathbf{x}^{m+1} - \mathbf{x}^m\| + \|\mathbf{x}^m - \mathbf{x}^{m-1}\| + \cdots + \|\mathbf{x}^{n+1} - \mathbf{x}^n\|$$

$$\leq \ \alpha h^{2^n - 1}(1 + h^{2^n} + (h^{2^n})^2 + \cdots) < \frac{\alpha h^{2^n - 1}}{1 - h^{2^n}} < \varepsilon \tag{69}$$

for sufficiently large $n \geq N(\varepsilon)$, because $0 < h < 1$. Therefore $\{\mathbf{x}^k\}$ is a Cauchy sequence and its limit $\lim_{k \to \infty}\mathbf{x}^k = \mathbf{x}^\infty$ lies in the closure $\overline{B(\mathbf{x}^0, r)}$ since $\mathbf{x}^k \in$

$B(\mathbf{x}^0, r)$, for all $k \geq 0$. Now, let us show that $\mathbf{x}^\infty$ is a zero of $T_{\mathbf{x}} \mathbf{f}(\mathbf{x}) = 0$ in $\overline{B(\mathbf{x}^0, r)}$. From $\left\| T_{\mathbf{x}^k} \mathbf{f}(\mathbf{x}^k) \right\| = \left\| \mathbf{x}^{k+1} - \mathbf{x}^k \right\|$ it follows that $\lim_{k \to \infty} \left\| T_{\mathbf{x}^k} \mathbf{f}(\mathbf{x}^k) \right\| = 0$. Note,

$$
\begin{aligned}
\| T_{\mathbf{u}} \mathbf{f}(\mathbf{u}) - T_{\mathbf{v}} \mathbf{f}(\mathbf{v}) \| &= \| (T_{\mathbf{u}} - T_{\mathbf{v}}) \mathbf{f}(\mathbf{v}) + T_{\mathbf{u}} (\mathbf{f}(\mathbf{u}) - \mathbf{f}(\mathbf{v})) \| \\
&\leq \| (T_{\mathbf{u}} - T_{\mathbf{v}}) \mathbf{f}(\mathbf{v}) \| + \| T_{\mathbf{u}} (\mathbf{f}(\mathbf{u}) - \mathbf{f}(\mathbf{v})) \| \\
&\leq N \| \mathbf{u} - \mathbf{v} \|^2 + \| T_{\mathbf{u}} \| C \| \mathbf{u} - \mathbf{v} \|, \quad \text{where } C \text{ is a constant,} \\
&\qquad\qquad \text{by (64c) and the fact that } \mathbf{f}(\mathbf{x}) \text{ is differentiable,} \\
&\leq N \| \mathbf{u} - \mathbf{v} \|^2 + C' \| \mathbf{u} - \mathbf{v} \|, \\
&\qquad\qquad \text{by (64d), where } C' \text{ is a constant.}
\end{aligned}
$$

Therefore, since $\mathbf{f}$ is continuous at $\mathbf{x}^\infty$, $T_{\mathbf{x}^k} \mathbf{f}(\mathbf{x}^k) \to T_{\mathbf{x}^\infty} \mathbf{f}(\mathbf{x}^\infty)$ as $\mathbf{x}^k \to \mathbf{x}^\infty$, and

$$
\lim_{k \to \infty} T_{\mathbf{x}^k} \mathbf{f}(\mathbf{x}^k) = T_{\mathbf{x}^\infty} \mathbf{f}(\mathbf{x}^\infty) = \mathbf{0},
$$

i.e., $\mathbf{x}^\infty$ is a zero of $T_{\mathbf{x}} \mathbf{f}(\mathbf{x})$ in $\overline{B(\mathbf{x}^0, r)}$.

PART 3. Taking the $\lim_{m \to \infty}$ in (69) we get

$$
\lim_{m \to \infty} \| \mathbf{x}^{m+1} - \mathbf{x}^n \| = \| \mathbf{x}^\infty - \mathbf{x}^n \| \leq \frac{\alpha h^{2^n - 1}}{1 - h^{2^n}}.
$$

Since $0 < h < 1$, the above method is at least quadratically convergent. $\qquad\square$

The limit $\mathbf{x}^\infty$ of the sequence (65) is a solution of (66), but in general not of (60), unless $T_{\mathbf{x}^\infty}$ is of full column rank (in which case (66) and (60) are equivalent.) Thus, the choice of the $\{2\}$-inverses $T_{\mathbf{x}^k}$ in Theorem 4, which by Section 1.7 can have any rank between 0 and rank $J_{\mathbf{x}^k}$, will determine the extent to which $\mathbf{x}^\infty$ resembles a solution of (60). In general, the greater the rank of the $\{2\}$-inverses $T_{\mathbf{x}^k}$, the more faithful is (66) to the original equation. The "worst" choice is the trivial $T = O$, in which case any $\mathbf{x}$ is a solution of (66) and the iterations (65) stop at $\mathbf{x}^0$. The best choice is therefore a $\{1, 2\}$-inverse. In particular, the Moore–Penrose inverse is useful in solving least-squares problems.

COROLLARY 3. *Under conditions analogous to Theorem 4, the limit* $\mathbf{x}^\infty$ *of the iterates*

$$
\mathbf{x}^{k+1} = \mathbf{x}^k - J_{\mathbf{x}^k}^\dagger \mathbf{f}(\mathbf{x}^k), \quad k = 0, 1, \dots, \tag{70}
$$

is a stationary point of the sum-of-squares $\| \mathbf{f}(\mathbf{x}) \|^2 = \sum_{i=1}^m |f_i(\mathbf{x})|^2$, *i.e.,*

$$
\nabla \| \mathbf{f} \|^2 (\mathbf{x}^\infty) = \mathbf{0}, \tag{71}
$$

where $\| \cdot \|$ *denotes the Euclidean norm.*

PROOF. Condition (64c) expresses a continuity of the $\{2\}$-inverse used. If this holds for the Moore–Penrose inverse, then $(J_{\mathbf{x}^k})^\dagger \to (J_{\mathbf{x}^\infty})^\dagger$ and (66) becomes

$$
(J_{\mathbf{x}^\infty})^\dagger \mathbf{f}(\mathbf{x}^\infty) = \mathbf{0},
$$

and, since $N(A^\dagger) = N(A^*)$,

$$
J_{\mathbf{x}^\infty}^* \mathbf{f}(\mathbf{x}^\infty) = \mathbf{0},
$$

and the proof is completed by noting that

$$
\nabla \| \mathbf{f}(\mathbf{x}) \|^2 = 2 J_{\mathbf{x}}^* \mathbf{f}(\mathbf{x}). \qquad\square
$$

REMARK 1. The standard setup for a Newton method (58) is that the initial value $f(x^0)$ is not too large, and the moduli of derivatives $|f'(x)|$ and $|f''(x)|$ stay bounded below and above, respectively, throughout the computations. These standard assumptions are present in Theorem 4.

(a) For any nontrivial T_{x^0}, the inequality (64b) bounds the value of $\mathbf{f}$ at $\mathbf{x}^0$ as follows:

$$\| \mathbf{f}(\mathbf{x}^0) \| < \frac{\alpha}{\| T_{\mathbf{x}^0} \|},$$

showing that $f(\mathbf{x}^0)$ cannot be too large.

(b) The inequality (64d) guarantees that $\|J_\mathbf{u}\|$ is bounded below for all $\mathbf{u} \in B(\mathbf{x}^0, r)$.

(c) Condition (63) is Lipschitz continuity of the derivative in $B(\mathbf{x}^0, r)$. It stands here for the boundedness above of the second derivative (even though $\mathbf{f}$ is not assumed twice differentiable).

The following theorem establishes the convergence of a modified Newton method, using the same {2}-inverse throughout.

THEOREM 5. *Let the following be given:*

$$\mathbf{x}^0 \in \mathbb{R}^n, \quad r > 0,$$
$$\mathbf{f} : B(\mathbf{x}^0, r) \to \mathbb{R}^m \quad \textit{a function,}$$
$$A \in \mathbb{R}^{m \times n}, \quad T \in \mathbb{R}^{n \times m} \ \textit{matrices,}$$
$$\epsilon > 0, \quad \delta > 0 \ \textit{positive scalars,}$$

such that

$$TAT = T, \tag{72a}$$
$$\epsilon \|T\| = \delta < 1, \tag{72b}$$
$$\|T\| \|\mathbf{f}(\mathbf{x}^0)\| < (1 - \delta)r, \tag{72c}$$

and, for all $\mathbf{u}, \mathbf{v} \in B(\mathbf{x}^0, r)$,

$$\|\mathbf{f}(\mathbf{u}) - \mathbf{f}(\mathbf{v}) - A(\mathbf{u} - \mathbf{v})\| \leq \epsilon \|\mathbf{u} - \mathbf{v}\|. \tag{72d}$$

Then the sequence

$$\mathbf{x}^{k+1} = \mathbf{x}^k - T\mathbf{f}(\mathbf{x}^k) \tag{73}$$

converges to a point

$$\mathbf{x}^\infty \in \overline{B(\mathbf{x}^0, r)} \tag{74}$$

satisfying

$$T\mathbf{f}(\mathbf{x}) = \mathbf{0}. \tag{75}$$

PROOF. Using induction on k we prove that the sequence (73) satisfies, for $k = 0, 1, \ldots$,

$$\mathbf{x}^k \in B(\mathbf{x}^0, r), \tag{76}$$
$$\|\mathbf{x}^{k+1} - \mathbf{x}^k\| \leq \delta^k (1 - \delta)r. \tag{77}$$

We denote by (76.k) and (77.k) the validity of (76) and (77), respectively, for the given value of k. Now (77.0) and (76.1) follow from (72c). Assuming (77.j) for $0 \leq j \leq k - 1$ we get

$$\|\mathbf{x}^k - \mathbf{x}^0\| \leq \sum_{j=0}^{k-1} \|\mathbf{x}^{j+1} - \mathbf{x}^j\| \leq (1 - \delta)r \sum_{j=0}^{k-1} \delta^j = (1 - \delta^k)r,$$

which proves (76.k). To prove (77.k) we write

$$\begin{aligned}
\mathbf{x}^{k+1} - \mathbf{x}^k &= -T\mathbf{f}(\mathbf{x}^k) \\
&= -T\mathbf{f}(\mathbf{x}^{k-1}) - T\left[\mathbf{f}(\mathbf{x}^k) - \mathbf{f}(\mathbf{x}^{k-1})\right] \\
&= T\left[A(\mathbf{x}^k - \mathbf{x}^{k-1}) - \mathbf{f}(\mathbf{x}^k) + \mathbf{f}(\mathbf{x}^{k-1})\right], \quad \text{by (72a) and (73).}
\end{aligned}$$

From (72d) and (72b) it therefore follows that

$$\begin{aligned}
\|\mathbf{x}^{k+1} - \mathbf{x}^k\| &\leq \|T\|\|\mathbf{f}(\mathbf{x}^k) - \mathbf{f}(\mathbf{x}^{k-1}) - A(\mathbf{x}^k - \mathbf{x}^{k-1})\| \\
&\leq \delta\|\mathbf{x}^k - \mathbf{x}^{k-1}\|,
\end{aligned}$$

proving (77.k). $\square$

REMARK 2. $\mathbf{f}$ is *differentiable* at $\mathbf{x}^0$ and the linear transformation A is its *derivative* at $\mathbf{x}^0$, if

$$\lim_{\mathbf{x} \to \mathbf{x}^0} \frac{\|\mathbf{f}(\mathbf{x}) - \mathbf{f}(\mathbf{x}^0) - A(\mathbf{x} - \mathbf{x}^0)\|}{\|\mathbf{x} - \mathbf{x}^0\|} = 0.$$

Comparing this with (72d) we conclude that the linear transformation A in Theorem 5 is an "approximate derivative," and can be chosen as the derivative of $\mathbf{f}$ at $\mathbf{x}^0$ if $\mathbf{f}$ is continuously differentiable in $B(\mathbf{x}^0, r)$. See also Ex. 14 below.

REMARK 3. Note that (72d) needs to hold only for $\mathbf{u}, \mathbf{v} \in B(\mathbf{x}^0, r)$ such that $\mathbf{u} - \mathbf{v} \in R(T)$, and the limit $\mathbf{x}^\infty$ of (73) lies in

$$\overline{B(\mathbf{x}^0, r)} \cap \{\mathbf{x}^0 + R(T)\}.$$

Exercises

Ex. 13. Let C be a convex subset of $\mathbb{R}^n$, let $\mathbf{f} : C \to \mathbb{R}^m$ be differentiable, and let $M > 0$ satisfy

$$\|J_{\mathbf{f}}(\mathbf{x}) - J_{\mathbf{f}}(\mathbf{y})\| \leq M\|\mathbf{x} - \mathbf{y}\|, \quad \text{for all } \mathbf{x}, \mathbf{y} \in C.$$

Then

$$\|\mathbf{f}(\mathbf{x}) - \mathbf{f}(\mathbf{y}) - J_{\mathbf{f}}(\mathbf{y})(\mathbf{x} - \mathbf{y})\| \leq \frac{M}{2}\|\mathbf{x} - \mathbf{y}\|^2, \quad \text{for all } \mathbf{x}, \mathbf{y} \in C.$$

PROOF. For any $\mathbf{x}, \mathbf{y} \in C$, the function $g : [0, 1] \to \mathbb{R}^m$, defined by $g(t) := \mathbf{f}(\mathbf{y} + t(\mathbf{x} - \mathbf{y}))$, is differentiable for all $0 \leq t \leq 1$, and its derivative is

$$g'(t) = J_{\mathbf{f}}(\mathbf{y} + t(\mathbf{x} - \mathbf{y}))(\mathbf{x} - \mathbf{y}).$$

So, for all $0 \leq t \leq 1$,

$$\|g'(t) - g'(0)\| = \|J_{\mathbf{f}}(\mathbf{y} + t(\mathbf{x} - \mathbf{y}))(\mathbf{x} - \mathbf{y}) - J_{\mathbf{f}}(\mathbf{y})(\mathbf{x} - \mathbf{y})\|$$
$$\leq \|J_{\mathbf{f}}(\mathbf{y} + t(\mathbf{x} - \mathbf{y})) - J_{\mathbf{f}}(\mathbf{y})\|\|\mathbf{x} - \mathbf{y}\|$$
$$\leq Mt\|\mathbf{x} - \mathbf{y}\|^2.$$

$$\therefore \ \|\mathbf{f}(\mathbf{x}) - \mathbf{f}(\mathbf{y}) - J_{\mathbf{f}}(\mathbf{y})(\mathbf{x} - \mathbf{y})\| = \|g(1) - g(0) - g'(0)\|$$
$$= \left\| \int_0^1 (g'(t) - g'(0))\, dt \right\|$$
$$\leq \int_0^1 \|g'(t) - g'(0)\|\, dt \leq M \int_0^1 t\, dt \|\mathbf{x} - \mathbf{y}\|^2$$
$$= \frac{M}{2}\|\mathbf{x} - \mathbf{y}\|^2. \qquad \square$$

Ex. 14. *A Newton method using {2}-inverses of approximate derivatives.* Let the following be given:

$$\mathbf{x}^0 \in \mathbb{R}^n, \quad r > 0,$$
$$\mathbf{f} : B(\mathbf{x}^0, r) \to \mathbb{R}^m \quad \text{a function,}$$
$$\epsilon > 0, \quad \delta > 0, \quad \eta > 0 \quad \text{positive scalars,}$$

and, for any $\mathbf{x} \in \overline{B(\mathbf{x}^0, r)}$, let

$$A_{\mathbf{x}} \in \mathbb{R}^{m \times n}, \quad T_{\mathbf{x}} \in \mathbb{R}^{n \times m},$$

be matrices satisfying, for all $\mathbf{u}, \mathbf{v} \in B(\mathbf{x}^0, r)$,

$$\|\mathbf{f}(\mathbf{u}) - \mathbf{f}(\mathbf{v}) - A_{\mathbf{v}}(\mathbf{u} - \mathbf{v})\| \leq \epsilon \|\mathbf{u} - \mathbf{v}\|, \tag{78a}$$
$$T_{\mathbf{u}} A_{\mathbf{u}} T_{\mathbf{u}} = T_{\mathbf{u}}, \tag{78b}$$
$$\|(T_{\mathbf{u}} - T_{\mathbf{v}})\mathbf{f}(\mathbf{v})\| \leq \eta \|\mathbf{u} - \mathbf{v}\|, \tag{78c}$$
$$\epsilon \|T_{\mathbf{u}}\| + \eta \leq \delta < 1, \tag{78d}$$
$$\|T_{\mathbf{x}^0}\|\|\mathbf{f}(\mathbf{x}^0)\| < (1 - \delta)r. \tag{78e}$$

Then the sequence

$$\mathbf{x}^{k+1} = \mathbf{x}^k - T_{\mathbf{x}^k}\mathbf{f}(\mathbf{x}^k), \quad k = 0, 1, \ldots, \tag{65}$$

converges to a point

$$\mathbf{x}^\infty \in \overline{B(\mathbf{x}^0, r)} \tag{74}$$

which is a solution of

$$T_{\mathbf{x}^\infty}\mathbf{f}(\mathbf{x}) = \mathbf{0}. \tag{79}$$

PROOF. As in the proof of Theorem 5 we use induction on k to prove that the sequence (65) satisfies

$$\mathbf{x}^k \in B(\mathbf{x}^0, r), \tag{76.k}$$
$$\|\mathbf{x}^{k+1} - \mathbf{x}^k\| \leq \delta^k(1 - \delta)r. \tag{77.k}$$

Again (77.0) and (76.1) follow from (78e) and, assuming (77.j) for $0 \leq j \leq k-1$, we get (76.k). To prove (77.k) we write

$$
\begin{aligned}
\mathbf{x}^{k+1} - \mathbf{x}^k &= -T_{\mathbf{x}^k}\mathbf{f}(\mathbf{x}^k) \\
&= \mathbf{x}^k - \mathbf{x}^{k-1} - T_{\mathbf{x}^k}\mathbf{f}(\mathbf{x}^k) + T_{\mathbf{x}^{k-1}}\mathbf{f}(\mathbf{x}^{k-1}), \quad \text{by (65),} \\
&= T_{\mathbf{x}^{k-1}}A_{\mathbf{x}^{k-1}}(\mathbf{x}^k - \mathbf{x}^{k-1}) - T_{\mathbf{x}^k}\mathbf{f}(\mathbf{x}^k) + T_{\mathbf{x}^{k-1}}\mathbf{f}(\mathbf{x}^{k-1}), \\
&\qquad \text{since } TAT = T \text{ implies } TA\mathbf{x} = \mathbf{x} \text{ for every } \mathbf{x} \in R(T), \\
&= T_{\mathbf{x}^{k-1}}[A_{\mathbf{x}^{k-1}}(\mathbf{x}^k - \mathbf{x}^{k-1}) - \mathbf{f}(\mathbf{x}^k) + \mathbf{f}(\mathbf{x}^{k-1})] + (T_{\mathbf{x}^{k-1}} - T_{\mathbf{x}^k})\mathbf{f}(\mathbf{x}^k).
\end{aligned}
$$

Therefore

$$
\|\mathbf{x}^{k+1} - \mathbf{x}^k\| \leq (\epsilon\|T_{\mathbf{x}^{k-1}}\| + \eta)\|\mathbf{x}^k - \mathbf{x}^{k-1}\|, \quad \text{by (78a) and (78c),}
$$

$$
\leq \delta\|\mathbf{x}^k - \mathbf{x}^{k-1}\|, \quad \text{by (78d),}
$$

which proves (77.k). $\qquad\square$

6. Linear Systems Theory

Systems modeled by linear differential equations call for symbolic computation of generalized inverses for matrices whose elements are rational functions. Such algorithms were given by Karampetakis [**469**], [**470**] and Jones, Karampetakis, and Pugh [**451**].

As an example, consider the homogeneous system

$$
A(D)\mathbf{x}(t) = \mathbf{0} \tag{80}
$$

where $\mathbf{x}(t) : [0-, \infty) \to \mathbb{R}^n$, $D := \frac{d}{dt}$,

$$
A(D) = A_q D^q + \cdots + A_1 D + A_0, \tag{81}
$$

and $A_i \in \mathbb{R}^{m \times n}$, $i = 0, 1, \ldots, q$. Let $\mathcal{L}$ denote the *Laplace transform* and let $\widehat{\mathbf{x}}(s) = \mathcal{L}(\mathbf{x}(t))$. The system (80) transforms to

$$
A(s)\widehat{x}(s) =
$$

$$
= [s^{q-1}I_m \quad s^{q-2}I_m \quad \cdots \quad I_m]
\begin{bmatrix} A_q & & O \\ \vdots & \ddots & \\ A_1 & \cdots & A_q \end{bmatrix}
\begin{bmatrix} \mathbf{x}(0-) \\ \vdots \\ \mathbf{x}^{(q-1)}(0-) \end{bmatrix}
:= \widehat{\mathbf{b}}(s), \tag{82}
$$

that allows casting the solution in a familiar form.

THEOREM 6 (Jones, Karampetakis, and Pugh [**451**], Theorem 4.2). *The system (80) has a solution if and only if*

$$
A(s)A(s)^\dagger\widehat{\mathbf{b}}(s) = \widehat{\mathbf{b}}(s) \tag{83}
$$

in which case the general solution is

$$
\mathbf{x}(t) = \mathcal{L}^{-1}(\widehat{\mathbf{x}}(s)) = \mathcal{L}^{-1}\left\{ A(s)^\dagger\widehat{\mathbf{b}}(s) + (I_n - A(s)^\dagger A(s))\mathbf{y}(s) \right\} \tag{84}
$$

where $\mathbf{y}(s) \in \mathbb{R}^n(s)$ *is arbitrary.* $\qquad\square$

For more details and other applications to linear systems theory, see the papers cited above, and the references on p. 329.

7. Application of the Group Inverse in Finite Markov Chains

This section is based on Meyer [**544**] and Campbell and Meyer [**159**, Chapter 8]. A system capable of being in one of N states, denoted $1, 2, \ldots, N$, is observed at discrete times $t = 0, 1, 2, \ldots$. The state of the system at time t is denoted by $\mathbf{X}_t$. The system is a *finite Markov chain* (or just *chain*) if there is a matrix $P = (p_{ij}) \in \mathbb{R}^{N \times N}$ such that

$$\text{Prob}\{\mathbf{X}_{t+1} = j \,|\, \mathbf{X}_t = i\} = p_{ij}, \quad \forall\, i, j \in \overline{1, N}, \ \forall\, t = 0, 1, \ldots. \quad (85)$$

In words: the probability that the system at time $t + 1$ is in state j, given it is in state i at time t, is equal to p_{ij} (independently of t). The numbers p_{ij} are called the *transition probabilities* and the matrix P is called the *transition matrix*. It follows that the transition probabilities must satisfy

$$\sum_{j=1}^{N} p_{ij} = 1, \quad i \in \overline{1, N}, \quad (86a)$$

in addition to

$$p_{ij} \geq 0, \quad i, j \in \overline{1, N}. \quad (86b)$$

A square matrix $P = [p_{ij}]$ satisfying (86) is called *stochastic*. Condition (86a) can be written as

$$P\mathbf{e} = \mathbf{e} \quad (87)$$

where $\mathbf{e}$ is the vector of ones. This shows that 1 is an eigenvalue of P.

Let $p_{ij}^{(n)}$ denote the n-step transition probability

$$p_{ij}^{(n)} = \text{Prob}\{\mathbf{X}_{t+n} = j \,|\, \mathbf{X}_t = i\},$$

where $p_{ij}^{(1)}$ are the previous p_{ij} and, by convention, $p_{ij}^{(0)} := \delta_{ij}$. The n-step transition probabilities satisfy

$$p_{ij}^{(2)} = \sum_{k=1}^{N} p_{ik}\, p_{kj},$$

and, inductively,

$$p_{ij}^{(n)} = \sum_{k=1}^{N} p_{ik}\, p_{kj}^{(n-1)} = \sum_{k=1}^{N} p_{ik}^{(n-1)}\, p_{kj}, \quad (88)$$

giving the *Chapman–Kolmogorov equations*,

$$p_{ij}^{(m+n)} = \sum_{k=1}^{N} p_{ik}^{(m)}\, p_{kj}^{(n)}, \quad \text{for all positive integers } m, n. \quad (89)$$

We conclude that $p_{ij}^{(n)}$ is the $(i, j)^{\text{th}}$ element of P^n.

Some terminology: a state i *leads* to state j, denoted by $i \to j$, if $p_{ij}^{(n)} > 0$ for some $n \geq 1$, i.e., there is a positive probability that j can be reached from i in n steps. Two states i, j are said to *communicate*, a fact denoted by $i \rightleftarrows j$, if each state leads to the other. Communication is not reflexive (a state i need not communicate with any state, including itself), but is symmetric and transitive, see Ex. 17.

A set S of states is *closed* if $i, j \in S \implies i \rightleftarrows j$ ($i = j$ is possible). A Markov chain is *irreducible* if its only closed set is the set of all states and is *reducible* otherwise. A chain is reducible if and only if its transition matrix is reducible, see Ex. 18.

A single state forming a closed set is called an *absorbing state*: a state i is absorbing if and only if $p_{ii} = 1$. A reducible chain is *absorbing* if each of its closed sets consists of a single (necessarily absorbing) state.

A chain with a transition matrix P is *regular* if, for some k, P^k is a positive matrix.

A state i has *period* τ if $p_{ii}^{(n)} = 0$ except when $n = \tau, 2\tau, 3\tau, \ldots$. The period of i is denoted $\tau(i)$. If $\tau(i) = 1$, the state i is called *aperiodic*.

Let $f_{ij}^{(n)}$ denote the probability that starting from state i, the system reaches state j for the first time at the nth step, and let

$$f_{ij} = \sum_{n=1}^{\infty} f_{ij}^{(n)}, \quad \text{the probability that } j \text{ is eventually reached from } i,$$

$$\mu_{ij} = \sum_{n=1}^{\infty} n f_{ij}^{(n)}, \quad \text{the mean first passage time from } i \text{ to } j,$$

in particular,

$$\mu_{kk} = \sum_{n=1}^{\infty} n f_{kk}^{(n)}, \quad \text{the mean return time of state } k, \text{ see Ex. 20.}$$

A state i is *recurrent* if $f_{ii} = 1$ and is *transient* otherwise. A chain is *recurrent* if all its states are recurrent.

If i is recurrent and $\mu_{ii} = \infty$, i is called a *null state*, see Exs. 21–22. All states in a closed set are of the same type, see Ex. 23.

A state is *ergodic* if it is recurrent and aperiodic, but not a null state. If a chain is irreducible, all its states have the same period. An irreducible aperiodic chain is called *ergodic*.

The n-step transition probabilities and the mean return times are related as follows:

THEOREM 7 (Chung [**197**], Theorem 9, p. 275). *For any states i, j of an irreducible recurrent chain*

$$\lim_{n \to \infty} \frac{1}{n+1} \sum_{k=0}^{n} p_{ij}^{(k)} = \frac{1}{\mu_{jj}}. \tag{90}$$

PROBABILISTIC REASON: Both sides of (90) give the expected number of returns to state j in unit time.   □

If a chain is irreducible, the matrix P^T is also irreducible, and has spectral radius 1 by Ex. 15 and Ex. 0.73. The Perron–Frobenius Theorem 0.5 then guarantees a positive eigenvector $\mathbf{x}$,

$$P^T \mathbf{x} = \mathbf{x}, \tag{91}$$

corresponding to the eigenvalue 1, with algebraic multiplicity 1. The vector $\mathbf{x}$ can be normalized to give a probability vector $\widehat{\boldsymbol{\pi}} = [\widehat{\pi}_i]$,

$$\widehat{\boldsymbol{\pi}} := \frac{\mathbf{x}}{\mathbf{e}^T \mathbf{x}}, \tag{92}$$

satisfying (91), written in detail as,

$$\widehat{\pi}_k = \sum_{i=1}^{N} p_{ik}\,\widehat{\pi}_i, \quad k \in \overline{1,N}. \tag{93}$$

$\widehat{\pi}$ is called a *stationary distribution* or *steady state*. In an irreducible recurrent chain, the stationary distribution is, simply,

$$\widehat{\pi}_k = \frac{1}{\mu_{kk}}, \quad k \in \overline{1,N}, \tag{94}$$

see Ex. 24.

Let $\pi_i^{(t)}$ denote the probability that the system is in state i at time t and collect these probabilities, called *state probabilities* or *distribution*, in the vector $\boldsymbol{\pi}^{(t)} = [\pi_i^{(t)}]$. These satisfy

$$\pi_k^{(t)} = \sum_{i=1}^{N} p_{ik}\,\pi_i^{(t-1)}, \quad k \in \overline{1,N},\ t = 0,1,\dots, \tag{95}$$

or

$$\boldsymbol{\pi}^{(t)} = P^T \boldsymbol{\pi}^{(t-1)}, \quad t = 0,1,\dots, \tag{96}$$

and, inductively,

$$\boldsymbol{\pi}^{(t)} = (P^T)^t \boldsymbol{\pi}^{(0)}, \quad t = 0,1,\dots, \tag{97}$$

where $\boldsymbol{\pi}^{(0)} = [\pi_i^{(0)}]$ is the *initial distribution*, assumed known. In particular, if the initial state $\mathbf{X}_0$ is deterministic, say $\mathbf{X}_0 = i$, then $\boldsymbol{\pi}^{(0)} = \mathbf{e}_i$, the ith unit vector.

The transition matrix P and the initial distribution $\boldsymbol{\pi}^{(0)}$ suffice for studying a Markov chain.

If the chain starts with a stationary distribution, i.e., if $\boldsymbol{\pi}^{(0)} = \widehat{\boldsymbol{\pi}}$, then $\boldsymbol{\pi}^{(t)} = \widehat{\boldsymbol{\pi}}$ for all t. This justifies the name *stationary distribution*. On the other hand, if the probabilities $\pi_k^{(t)} \to \pi_k^{(\infty)}$, as $t \to \infty$, then the limiting probabilities $\boldsymbol{\pi}^{(\infty)}$ satisfy (93) and are the stationary distribution $\widehat{\boldsymbol{\pi}}$.

The existence of a stationary distribution $\widehat{\boldsymbol{\pi}}$ does not mean that the system converges to that distribution, see Ex. 25. However, if a chain is ergodic the system converges to its stationary distribution from any initial distribution.

THEOREM 8 (Feller [**271**], Theorem 2, p. 325). *Let $P \in \mathbb{R}^{N \times N}$ be the transition matrix of an ergodic chain. Then:*

(a) *For every pair $j,k \in \overline{1,N}$, the limit*

$$\lim_{n \to \infty} p_{jk}^{(n)} = \pi_k > 0 \tag{98}$$

exists and is independent of j.

(b) *π_k is the reciprocal of the mean return time μ_{kk} of the state k,*

$$\pi_k = \frac{1}{\mu_{kk}}. \tag{99}$$

(c) *The numbers* $\{\pi_k : k \in \overline{1, N}\}$ *are probabilities,*

$$\pi_k > 0, \quad \sum \pi_k = 1. \tag{100}$$

(d) *The probabilities* π_k *are a stationary distribution of the chain*

$$\pi_k = \sum_{i=1}^{N} p_{ik} \pi_i, \quad k \in \overline{1, N}. \tag{93}$$

The distribution $\{\pi_k\}$ *is uniquely determined by* (93) *and* (100).

In an ergodic chain, the stationary probabilities are given by

$$\widehat{\pi}_j = \lim_{n \to \infty} \frac{1}{n+1} \sum_{k=0}^{n} p_{ij}^{(k)}, \quad j \in \overline{1, N}, \tag{101}$$

which follows by comparing (90) and (99).

For an ergodic chain, Theorem 8(a) states that all rows of P^n converge, as $n \to \infty$, to the stationary distribution $\widehat{\pi}$,

$$\lim_{n \to \infty} P^n = \widehat{\Pi}^T = \begin{bmatrix} \widehat{\pi}^T \\ \widehat{\pi}^T \\ \cdots \\ \widehat{\pi}^T \end{bmatrix}. \tag{102}$$

This implies that the *Cesaro means* of P^n also converge to the same limit

$$\lim_{n \to \infty} \frac{1}{n+1} \sum_{k=0}^{n} P^k = \widehat{\Pi}^T, \tag{103}$$

which is statement (101). The converse, (103) $\implies$ (102) is in general false, see Ex. 26.

The stationary probabilities and other objects of interest can be computed by applying the group inverse. First we require:

THEOREM 9 (Meyer [**544**], Theorem 2.1). *If P is a stochastic matrix, then $Q = I - P$ has group inverse.*

PROOF. If P is irreducible it has 1 as eigenvalue with algebraic multiplicity 1, by the Perron–Frobenius Theorem. Therefore Q has the eigenvalue 0 with algebraic multiplicity 1, has index 1 by Theorem 4.1(c), and has a group inverse by Theorem 4.2.

If P is reducible, it can written in the form (108). Each of the blocks $\{R_{ii} : i \in \overline{1, q}\}$ has spectral radius $\rho(R_{ii}) < 1$ by Ex. 16. Therefore, each $I - R_{ii}$ is nonsingular and Q has a group inverse by a repeated use of (5.43). $\square$

THEOREM 10 (Meyer [**544**], Theorem 2.2). *Let P be the transition matrix of a finite Markov chain and let $Q = I - P$. Then:*

$$I - QQ^{\#} = \begin{cases} \lim\limits_{n \to \infty} \frac{1}{n+1} \sum\limits_{k=0}^{n} P^k, \\ \lim\limits_{n \to \infty} (\alpha I + (1 - \alpha) P)^n, & \text{for any } 0 < \alpha < 1, \\ \lim\limits_{n \to \infty} P^n, & \text{if the chain is regular or absorbing.} \end{cases} \tag{104}$$

PROOF. By Ex. 27(a), there exists a nonsingular matrix S such that

$$P = S^{-1} \begin{bmatrix} I_k & O \\ O & K \end{bmatrix} S, \quad \text{with } 1 \notin \lambda(K). \tag{105}$$

$$\therefore \; Q = S^{-1} \begin{bmatrix} O & O \\ O & I-K \end{bmatrix} S, \quad Q^{\#} = S^{-1} \begin{bmatrix} O & O \\ O & (I-K)^{-1} \end{bmatrix} S.$$

$$\therefore \; I - QQ^{\#} = S^{-1} \begin{bmatrix} I_k & O \\ O & O \end{bmatrix} S. \tag{106}$$

Since $(I - K)$ is nonsingular,

$$\frac{I + K + K^2 + \cdots + K^n}{n+1} = \frac{(I - K^{n+1})(I - K)^{-1}}{n+1}. \tag{107}$$

From (105)–(107) it follows that

$$\frac{I + P + P^2 + \cdots + P^{n+1}}{n+1} = \frac{(I - P^{n+1})Q^{\#}}{n+1} + I - QQ^{\#},$$

and, since $\|(I - P^{n+1})Q^{\#}\|_{\infty}$ is bounded,

$$\lim_{n \to \infty} \frac{I + P + P^2 + \cdots + P^{n+1}}{n+1} = I - QQ^{\#},$$

proving the first line in (104). The second line is proved similarly, noting that $(\alpha I + (1-\alpha)P)$ is a stochastic matrix and its spectral radius $\rho(\alpha I + (1-\alpha)P) < 1$, so $(\alpha I + (1 - \alpha)P)^n \to O$ as $n \to \infty$.

If the chain is regular then, by Ex. 27(b,c), $k = 1$ and $\lim_{n \to \infty} K^n = O$.

$$\therefore \; \lim_{n \to \infty} P^n = S^{-1} \begin{bmatrix} 1 & O \\ O & O \end{bmatrix} S = I - QQ^{\#}.$$

If the chain is absorbing, with exactly r absorbing states, then by rearranging states as in Ex. 27(d),

$$P^n = \begin{bmatrix} I_r & O \\ (\sum_{k=0}^{n-1} K^k)R & K^n \end{bmatrix}.$$

$$\therefore \; \lim_{n \to \infty} P^n = \begin{bmatrix} I_r & O \\ (I - K)^{-1}R & O \end{bmatrix},$$

and, since Q is of the form

$$Q = \begin{bmatrix} O & O \\ -R & (I - K) \end{bmatrix},$$

we have, by Ex. 5.11,

$$Q^{\#} = \begin{bmatrix} O & O \\ -(I - K)^{-2}R & (I - K)^{-1} \end{bmatrix}.$$

$$\therefore \; I - QQ^{\#} = \begin{bmatrix} I_r & O \\ (I - K)^{-1}R & O \end{bmatrix} = \lim_{n \to \infty} P^n,$$

completing the proof. $\square$

From (102) and Theorem 10 it follows, for an ergodic chain, that

$$I - QQ^{\#} = \widehat{\Pi}^T = \begin{bmatrix} \widehat{\pi}^T \\ \widehat{\pi}^T \\ \cdots \\ \widehat{\pi}^T \end{bmatrix},$$

giving the stationary distribution $\widehat{\pi}$ in terms of the group inverse of $(I - P)$.

Exercises

EX. 15. If P is a stochastic matrix, its spectral radius $\rho(P) = 1$.

PROOF. From (0.59) and the fact that the matrix norm (0.56.1) is multiplicative it follows that $\rho(P) \leq 1$. On the other hand, $\rho(P) \geq 1$ by (87). $\square$

EX. 16. (Campbell and Meyer [159, Lemma 8.2.1]). Let $P = A + B$ be a stochastic matrix, $A \geq O$ and irreducible, $B \geq O$ and nonzero. Then $\rho(A) < 1$.

PROOF. As in Ex. 15, $\rho(A) \leq 1$. If $\rho(A) \geq 1$, then $\rho(A) = 1$, and the Perron–Frobenius theorem guarantees a positive eigenvector $\mathbf{x}$,

$$A^T \mathbf{x} = \mathbf{x}. \tag{a}$$

From (87)

$$A\mathbf{e} + B\mathbf{e} = \mathbf{e}. \tag{b}$$

Multiplying (a) by $\mathbf{e}$ and (b) by $\mathbf{x}$ we get

$$\mathbf{e}^T A^T \mathbf{x} = \mathbf{e}^T \mathbf{x} = \mathbf{x}^T A \mathbf{e} + \mathbf{x}^T B \mathbf{e}.$$

$$\therefore \ \mathbf{x}^T B \mathbf{e} = 0,$$

a contradiction since $O \neq B \geq O$, $\mathbf{x} > \mathbf{0}$, $\mathbf{e} > \mathbf{0}$. $\square$

EX. 17. Let i, j, k denote states of a finite Markov chain. Then:

(a) $i \rightleftarrows j \implies j \rightleftarrows i$.

(b) $i \rightleftarrows j$ and $j \rightleftarrows k \implies i \rightleftarrows k$.

(c)   Consider $P = \begin{bmatrix} 1 & 0 & 0 \\ \frac{1}{2} & \frac{1}{2} & 0 \\ \frac{1}{3} & \frac{2}{3} & 0 \end{bmatrix}$. Here state 1 is absorbing, states 2 and 3 are transient, and state 3 does not communicate with any state.

PROOF. (b) follows from $p_{ik}^{(n+m)} \geq p_{ij}^{(n)} p_{jk}^{(m)}$. $\square$

EX. 18. A chain is reducible if and only if the states can be rearranged so that the transition matrix is

$$P = \begin{bmatrix} P_{11} & P_{12} \\ O & P_{22} \end{bmatrix}.$$

Here P_{22} gives the transition probabilities for states in a closed set. Compare with (0.135).

EX. 19. If a chain has m minimal closed sets, its transition matrix $P = [p_{ij}]$ can be written, by rearranging states, as

$$P = \begin{bmatrix} P_{00} & P_{01} & P_{02} & \cdots & P_{0m} \\ O & P_{11} & O & \cdots & O \\ O & O & P_{22} & \cdots & O \\ \vdots & & & \ddots & \vdots \\ O & O & O & \cdots & P_{mm} \end{bmatrix}, \text{ where } P_{00} = \begin{bmatrix} R_{11} & R_{12} & \cdots & R_{1q} \\ O & R_{22} & \cdots & R_{2q} \\ O & O & \ddots & \vdots \\ O & O & \cdots & R_{qq} \end{bmatrix}, \tag{108}$$

here $\{P_{ii} : i \in \overline{1,m}\}$ are irreducible transition matrices of appropriate sizes, for each of the minimal closed sets, and the matrices $\{R_{ii} : i \in \overline{1,q}\}$ are irreducible.

EX. 20. For any i, j and $1 \leq n < \infty$,

$$p_{ij}^{(n)} = \sum_{t=1}^{n} f_{ij}^{(t)} p_{jj}^{(n-t)}.$$

EX. 21. (Feller [**271**, Corollary, p. 324]). A finite Markov chain can have no null states and it is impossible for all states to be transient.

EX. 22. (Feller [**271**, p. 321]). Let i be a state in a finite Markov chain and consider the series $\sum_{n=1}^{\infty} p_{ii}^{(n)}$. Then:

(a) The state i is transient if and only if the series converges.
(b) If i is a null state the series diverges and $\lim_{n \to \infty} p_{ii}^{(n)} = 0$.
(c) If i is ergodic, then $\lim_{n \to \infty} p_{ii}^{(n)} = 1/\mu_{ii}$.
(d) If i is recurrent nonnull with period τ, then $\lim_{n \to \infty} p_{ii}^{(n\tau)} = \tau/\mu_{ii}$.

EX. 23. (Feller [**271**, Theorem, p. 322]). If two states communicate, they are of the same type (transient, recurrent, periodic with same period).

EX. 24. (Chung [**197**, Theorem 12, p. 278]). Let P be the transition matrix of an irreducible chain and let $\boldsymbol{\pi} = [\pi_k]$ be given by (94). Then:

(a) $\boldsymbol{\pi}$ is a solution of (93);
(b) $\sum_k \pi_k = 1$;
(c) $\pi_k > 0$ for all k; and
(d) any solution of (93) is a multiple of $\boldsymbol{\pi}$.

PROOF. For every $t \geq 0$,

$$p_{ik}^{(t+1)} = \sum_j p_{ij}^{(t)} p_{jk}.$$

Averaging over t we get

$$\frac{1}{n+1} \sum_{t=0}^{n} p_{ik}^{(t+1)} = \sum_j \left(\frac{1}{n+1} \sum_{t=0}^{n} p_{ij}^{(t)} \right) p_{jk},$$

whose limit, as $n \to \infty$, is, by (90) and (94),

$$\pi_k = \sum_{j=1}^{N} \pi_j p_{jk}, \tag{93}$$

proving (a). □

EX. 25. Consider the stochastic matrix

$$P = \begin{bmatrix} 0 & 1 \\ 1 & 0 \end{bmatrix} \quad \text{and its stationary distribution} \quad \widehat{\boldsymbol{\pi}} = \tfrac{1}{2} \begin{bmatrix} 1 \\ 1 \end{bmatrix}.$$

Then $(P^T)^t \boldsymbol{\pi}^{(0)}$ does not converge, as $t \to \infty$, for any $\boldsymbol{\pi}^{(0)} \neq \widehat{\boldsymbol{\pi}}$. In particular,

$$(P^T)^t \mathbf{e}_1 = \begin{cases} \mathbf{e}_1, & t \text{ even,} \\ \mathbf{e}_2, & t \text{ odd,} \end{cases} \quad \text{where } \mathbf{e}_1 = \begin{bmatrix} 1 \\ 0 \end{bmatrix}, \mathbf{e}_2 = \begin{bmatrix} 0 \\ 1 \end{bmatrix}.$$

EX. 26. For P of Ex. 25, (103) holds, and

$$\lim_{n \to \infty} \frac{1}{n+1} \sum_{k=0}^{n} P^k = \begin{bmatrix} \frac{1}{2} & \frac{1}{2} \\ \frac{1}{2} & \frac{1}{2} \end{bmatrix}.$$

EX. 27. Let P be a stochastic matrix. Then:

(a) P is similar to a matrix of the form

$$\begin{bmatrix} I_k & O \\ O & K \end{bmatrix}, \quad \text{where } 1 \notin \lambda(K).$$

(b) If the Markov chain (of which P is the transition matrix) is ergodic, then $k = 1$, i.e., P is similar to
$$\begin{bmatrix} 1 & O \\ O & K \end{bmatrix}, \quad \text{where } 1 \notin \lambda(K).$$

(c) If the chain is regular, $\lim_{n \to \infty} K^n = O$.

(d) If the chain is absorbing, with r absorbing states, then P can be written, by rearranging states, as follows:
$$P = \begin{bmatrix} I_r & O \\ R & K \end{bmatrix}, \quad \text{where } \rho(K) < 1.$$

8. An Application of the Drazin Inverse to Difference Equations

In this section, based on Campbell and Meyer [159, §9.3], we study an application of the Drazin inverse to singular linear difference equations. Applications to singular linear differential equations use similar ideas, but require more details than is perhaps justified here. For these, the reader is referred to Boyarintsev [130] and Campbell and Meyer [159, Chapter 9], and their references, as well as references at the end of this chapter.

The following simple example illustrates how the Drazin inverse arises in such applications.

EXAMPLE 1. Consider the difference equation
$$A\mathbf{x}_{t+1} = \mathbf{x}_t, \quad t = 0, 1, \dots, \tag{109}$$
where A is singular, of index k. Since $\mathbf{x}_{t+1} = A\mathbf{x}_{t+2} = \cdots = A^k\mathbf{x}_{t+k+1}$ it follows that a solution $\mathbf{x}_{t+1}$ of (109), for given $\mathbf{x}_t$, must belong to $R(A^k)$ and, therefore, by Ex. 4.28,
$$\mathbf{x}_{t+1} = A^D x_t, \quad t = 0, 1, \dots . \tag{110}$$
A sequence $\{\mathbf{x}_t\}$ described by (109) is therefore restricted to $R(A^k)$, where A^D is an inverse of A, see (4.41). The representations (109) and (110) are equivalent. □

The difference equations studied here are of the form
$$A\mathbf{x}_{t+1} = B\mathbf{x}_t + \mathbf{f}_t, \tag{111}$$
$$\text{s.t.} \quad \mathbf{x}_0 = \mathbf{c} \quad \text{(the \textit{initial condition})},$$
where $A, B \in \mathbb{C}^{n \times n}$, A is singular, $\mathbf{c} \in \mathbb{C}^n$, and $\{\mathbf{f}_t : t = 0, 1, \dots\} \subset \mathbb{C}^n$. An equation is *homogeneous* if $\mathbf{f}_t = \mathbf{0}$ for all t.

DEFINITION 1.

(a) The initial vector $\mathbf{c}$ is called *consistent* if, given $\mathbf{x}_t$, (111) has a solution.

(b) The difference equation (111) is called *tractable* if there is a unique solution for each consistent $\mathbf{c}$.

In the homogeneous case, tractability has a simple characterization:

THEOREM 11 (Campbell and Meyer [159], Theorem 9.3.1). *The homogeneous difference equation*
$$A\mathbf{x}_{t+1} = B\mathbf{x}_t, \quad t = 0, 1, \dots, \tag{112}$$
is tractable if and only if there is a scalar $\lambda \in \mathbb{C}$ such that $(\lambda A - B)$ is nonsingular.

PROOF. *If:* Let $\lambda, \widehat{A}_\lambda$, and $\widehat{B}_\lambda$ be as in Ex. 28. The tractability of (112) is equivalent to that of

$$\widehat{A}_\lambda \mathbf{x}_{t+1} = \widehat{B}_\lambda \mathbf{x}_t, \quad t = 0, 1, \dots . \tag{113}$$

Using Ex. 4.42, we write

$$\widehat{A}_\lambda = X \begin{bmatrix} J_1 & O \\ O & J_0 \end{bmatrix} X^{-1},$$

where J_1 is nonsingular and J_0 is nilpotent. From (121) it follows that

$$\widehat{B}_\lambda = X \begin{bmatrix} \lambda J_1 - I & O \\ O & \lambda J_0 - I \end{bmatrix} X^{-1},$$

and partition $\mathbf{x}$ accordingly,

$$\mathbf{x} = \begin{bmatrix} \mathbf{x}^{(1)} \\ \mathbf{x}^{(0)} \end{bmatrix} .$$

The equation $J_1 \mathbf{x}_{t+1}^{(1)} = (\lambda J_1 - I) \mathbf{x}_t^{(1)}$ is tractable, since J_1 is nonsingular. It therefore remains to show that

$$J_0 \mathbf{x}_{t+1}^{(0)} = (\lambda J_0 - I) \mathbf{x}_t^{(0)}, \quad t = 0, 1, \dots , \tag{114}$$

is tractable. Let J_0 have index k and multiply (114) by J_0^{k-1} to get $(\lambda J_0 - I)J_0^{k-1} \mathbf{x}_t^{(0)} = \mathbf{0}$. Therefore $J_0^{k-1} \mathbf{x}_t^{(0)} = \mathbf{0}$ for all t. Similarly, multiplying (114) by J_0^{k-2}, we show $J_0^{k-2} \mathbf{x}_t^{(0)} = \mathbf{0}$ for all t. Repeating this argument we show that $\mathbf{x}_t^{(0)} = \mathbf{0}$ for all t, and (114) is trivially tractable.

Only if: Suppose (112) is tractable, but $(\lambda A - B)$ is singular for all $\lambda \in \mathbb{C}$, i.e., for each λ there is a nonzero $\mathbf{v}^\lambda \in \mathbb{C}^n$ such that $(\lambda A - B) \mathbf{v}^\lambda = \mathbf{0}$. Let $\{\mathbf{v}^{\lambda_1}, \mathbf{v}^{\lambda_2}, \dots, \mathbf{v}^{\lambda_s}\}$ be a linearly dependent set of such vectors and let $\{\alpha_1, \dots, \alpha_s\}$ be scalars, not all zero, such that $\sum_{i=1}^s \alpha_i \mathbf{v}^{\lambda_i} = \mathbf{0}$. Define $\mathbf{x}_t^{\lambda_i} = \lambda_i^t \mathbf{v}^{\lambda_i}$. Then the sequence $\{\mathbf{x}_t := \sum_{i=1}^s \alpha_i \mathbf{x}_t^{\lambda_i} : t = 0, 1, \dots\}$ is not identically zero, and is a solution of (112), satisfying the initial condition $\mathbf{x}_0 = \sum_{i=1}^s \alpha_i \mathbf{v}^{\lambda_i} = \mathbf{0}$. However, the sequence $\{\mathbf{x}_t = \mathbf{0} : t = 0, 1, \dots\}$ is another such solution, a contradiction to the assumed tractability. $\square$

In what follows, let $\lambda \in \mathbb{C}$ be such that $(\lambda A - B)$ is nonsingular, and denote $\widehat{A}_\lambda = (\lambda A - B)^{-1} A$ by $\widehat{A}$ and $\widehat{B}_\lambda = (\lambda A - B)^{-1} B$ by $\widehat{B}$. Dropping the subscript λ is justified since the results below do not depend on λ, as shown by Ex. 29.

THEOREM 12 (Campbell and Meyer [**159**], Theorem 9.3.2). *If the homogeneous equation*

$$A\mathbf{x}_{t+1} = B\mathbf{x}_t, \quad t = 0, 1, \dots , \tag{112}$$

is tractable, then its general solution is given by

$$\mathbf{x}_t = \begin{cases} \widehat{A}\widehat{A}^D \mathbf{y}, & \text{if } t = 0, \\ (\widehat{A}^D \widehat{B})^t \mathbf{y}, & \text{if } t = 1, 2, \dots , \end{cases} \tag{115}$$

where $\mathbf{y} \in \mathbb{C}^n$ is arbitrary. Moreover, the initial vector $\mathbf{c}$ is consistent for (112) if and only if $\mathbf{c} \in R(\widehat{A}^k)$, where $k = \text{index } \widehat{A}$. In this case, the solution of (112), subject to $\mathbf{x}_0 = \mathbf{c}$, is

$$\mathbf{x}_t = (\widehat{A}^D \widehat{B})^t \mathbf{c}, \quad t = 0, 1, \dots . \tag{116}$$

PROOF. As in the proof of Theorem 11, we transform (112) to an equivalent system

$$\begin{bmatrix} J_1 & O \\ O & J_0 \end{bmatrix} \begin{bmatrix} \mathbf{x}_{t+1}^{(1)} \\ \mathbf{x}_{t+1}^{(0)} \end{bmatrix} = \begin{bmatrix} \lambda J_1 - I & O \\ O & \lambda J_0 - I \end{bmatrix} \begin{bmatrix} \mathbf{x}_t^{(1)} \\ \mathbf{x}_t^{(0)} \end{bmatrix}, \tag{117}$$

where J_1 is nonsingular and J_0 is nilpotent, of index k. It follows that

$$\mathbf{x}_t^{(0)} = (\lambda J_0 - I)^{-k} J_0^k \mathbf{x}_{t+k}^{(0)} = \mathbf{0}, \quad \mathbf{x}_t^{(1)} = J_1^{-t}(\lambda J_1 - I)^t \mathbf{x}_0^{(0)}, \tag{118}$$

which proves (115), using Theorem 4.8. The remaining statements also follow from (118). $\qquad\square$

We turn to the nonhomogeneous case.

THEOREM 13 (Campbell and Meyer [**159**], Theorem 9.3.2). *If the equation*

$$A\mathbf{x}_{t+1} = B\mathbf{x}_t + \mathbf{f}_t, \tag{111}$$

$$s.t. \quad \mathbf{x}_0 = \mathbf{c},$$

is tractable, its general solution is, for $t \geq 1$,

$$\mathbf{x}_t = (\widehat{A}^D \widehat{B})^t \widehat{A}\widehat{A}^D \mathbf{y} + \widehat{A}^D \sum_{i=0}^{t-1} (\widehat{A}^D \widehat{B})^{t-i-1} \widehat{\mathbf{f}}_i$$

$$- (I - \widehat{A}\widehat{A}^D) \sum_{i=0}^{k-1} (\widehat{A}\widehat{B}^D)^i \widehat{B}^D \widehat{\mathbf{f}}_{t+i}, \tag{119}$$

where $\widehat{A} = (\lambda A - B)^{-1} A$, $\widehat{B} = (\lambda A - B)^{-1} B$, $\widehat{\mathbf{f}}_i = (\lambda A - B)^{-1}\mathbf{f}_i$, $k = \mathrm{Ind}(\widehat{A})$ and $\mathbf{y} \in \mathbb{C}^n$ is arbitrary. The solution (119) is independent of λ.

The initial vector $\mathbf{c}$ is consistent if and only if

$$\mathbf{c} + (I - \widehat{A}\widehat{A}^D) \sum_{i=0}^{k-1} (\widehat{A}\widehat{B}^D)^i \widehat{B}^D \widehat{\mathbf{f}}_{t+i} \in R(\widehat{A}^k). \tag{120}$$

PROOF. A comparison with (115) shows that the first term of (119) is the general solution of the homogeneous equation (112). The proof that the remaining two terms are a particular solution of (111), and the remaining statements, are left as exercise. $\qquad\square$

Exercises

Ex. 28. (Campbell and Meyer [**159**, Lemma 9.2.1]). Let $A, B \in \mathbb{C}^{n \times n}$, let $\lambda \in \mathbb{C}$ be such that $(\lambda A - B)$ is nonsingular, and define $\widehat{A}_\lambda = (\lambda A - B)^{-1} A$, $\widehat{B}_\lambda = (\lambda A - B)^{-1} B$. Then $\widehat{A}_\lambda \widehat{B}_\lambda = \widehat{B}_\lambda \widehat{A}_\lambda$.
PROOF. $\widehat{A}_\lambda$ and $\widehat{B}_\lambda$ commute since

$$\lambda \widehat{A}_\lambda - \widehat{B}_\lambda = I. \tag{121}$$

$\qquad\square$

Ex. 29. (Campbell and Meyer [**159**, Theorem 9.2.2]). Let $A, B \in \mathbb{C}^{n \times n}$, let λ, $\widehat{A}_\lambda$, $\widehat{B}_\lambda$ be as in Ex. 28, and let $\widehat{\mathbf{f}}_\lambda = (\lambda A - B)^{-1}\mathbf{f}$. For all α, β, for which $(\alpha A - B)$ and $(\beta A - B)$ are nonsingular, the following statements hold:
   (a) $\widehat{A}_\alpha \widehat{A}_\alpha^D = \widehat{A}_\beta \widehat{A}_\beta^D$.
   (b) $\widehat{A}_\alpha^D \widehat{B}_\alpha = \widehat{A}_\beta^D \widehat{B}_\beta$ and $\widehat{A}_\alpha \widehat{B}_\alpha^D = \widehat{A}_\beta \widehat{B}_\beta^D$.
   (c) $\mathrm{Ind}(\widehat{A}_\alpha) = \mathrm{Ind}(\widehat{A}_\beta)$ and $R(\widehat{A}_\alpha) = R(\widehat{A}_\beta)$.
   (d) $\widehat{A}_\alpha^D \widehat{\mathbf{f}}_\alpha = \widehat{A}_\beta^D \widehat{\mathbf{f}}_\beta$.

(e) $\widehat{B}_\alpha^D \widehat{f}_\alpha = \widehat{B}_\beta^D \widehat{f}_\beta$.

PROOF. (a)

$$\widehat{A}_\alpha^D \widehat{A}_\alpha = [(\alpha A - B)^{-1} A]^D \widehat{A}_\alpha = [(\alpha A - B)^{-1}(\beta A - B)(\beta A - B)^{-1} A]^D \widehat{A}_\alpha$$
$$= [(\alpha \widehat{A}_\beta - \widehat{B}_\beta)\widehat{A}_\beta]^D \widehat{A}_\alpha$$
$$= \widehat{A}_\beta^D (\alpha \widehat{A}_\beta - \widehat{B}_\beta)\widehat{A}_\alpha, \quad \text{by Ex. 4.33,}$$
$$= \widehat{A}_\beta^D [(\beta A - B)^{-1}(\alpha A - B)]\widehat{A}_\alpha$$
$$= \widehat{A}_\beta^D (\beta A - B)^{-1}(\alpha A - B)(\alpha A - B)^{-1} A$$
$$= \widehat{A}_\beta^D \widehat{A}_\beta.$$

The other parts are similarly proved.   □

EX. 30. Complete the proof of Theorem 13.

EX. 31. In Theorems 12–13 consider the special case where B is nonsingular. Then λ can be chosen as zero and $\widehat{A} = B^{-1}A$, $\widehat{B} = I$. Note that Theorem 12 reduces to Example 1.

9. Matrix Volume and the Change-of-Variables Formula in Integration

This section is based on [**73**]. The *change-of-variables formula* in the title is

$$\int_{\mathcal{V}} f(\mathbf{v})\, d\mathbf{v} = \int_{\mathcal{U}} (f \circ \phi)(\mathbf{u})| \det J_\phi(\mathbf{u})|\, d\mathbf{u}, \qquad (122)$$

where $\mathcal{U}, \mathcal{V}$ are sets in $\mathbb{R}^n$, $\phi : \mathcal{U} \to \mathcal{V}$ is a sufficiently well-behaved function, and f is integrable on $\mathcal{V}$. Here $d\mathbf{x}$ denotes the volume element $|dx_1 \wedge dx_2 \wedge \cdots \wedge dx_n|$, and J_ϕ is the *Jacobian matrix* (or *Jacobian*)

$$J_\phi := \left(\frac{\partial \phi_i}{\partial u_j} \right), \quad \text{also denoted} \quad \frac{\partial(v_1, v_2, \ldots, v_n)}{\partial(u_1, u_2, \ldots, u_n)}, \qquad (123)$$

representing the derivative of ϕ. An advantage of (122) is that integration on $\mathcal{V}$ is translated to (perhaps simpler) integration on $\mathcal{U}$.

This formula was given in 1841 by Jacobi [**441**], following Euler (the case $n = 2$) and Lagrange ($n = 3$). It gave prominence to *functional* (or *symbolic*) *determinants*, i.e. (nonnumerical) determinants of matrices including functions or operators as elements.

If $\mathcal{U}$ and $\mathcal{V}$ are in spaces of different dimensions, say $\mathcal{U} \subset \mathbb{R}^n$ and $\mathcal{V} \subset \mathbb{R}^m$ with $n > m$, then the Jacobian J_ϕ is a rectangular matrix and (122) cannot be used in its present form. However, if J_ϕ is of full column rank throughout $\mathcal{U}$, we can replace $|\det J_\phi|$ in (122) by the volume vol J_ϕ of the Jacobian to get

$$\int_{\mathcal{V}} f(\mathbf{v})\, d\mathbf{v} = \int_{\mathcal{U}} (f \circ \phi)(\mathbf{u})\, \text{vol } J_\phi(\mathbf{u})\, d\mathbf{u}. \qquad (124)$$

Recall that the *volume* of an $m \times n$ matrix of rank r is

$$\text{vol } A := \sqrt{\sum_{(I,J)\in\mathcal{N}} \det^2 A_{IJ}}, \qquad (0.89)$$

where A_{IJ} is the submatrix of A with rows I and columns J, and $\mathcal{N}$ is the index set of $r \times r$ nonsingular submatrices of A. If A is of full column rank, its volume is, by the Binet–Cauchy theorem (see Ex. 0.65),

$$\text{vol } A = \sqrt{\det A^T A}. \tag{125}$$

If $m = n$, then $\text{vol } J_\phi = |\det J_\phi|$, and (124) reduces to the classical result (122).

The formula (124) is well known in differential geometry, see, e.g., [**87**, Prop. 6.6.1] and [**270**, §3.2.3]. Although there are elementary accounts of this formula (see, e.g., [**208**, Vol. II, Ch. IV, §4], [**275**, §8.1], and [**730**, §3.4]), it is seldom used in applications.

We illustrate (124) for an elementary calculus example. Let $\mathcal{S}$ be a subset of a surface in $\mathbb{R}^3$ represented by

$$z = g(x, y), \tag{126}$$

and let $f(x, y, z)$ be a function integrable on $\mathcal{S}$. Let $\mathcal{A}$ be the projection of $\mathcal{S}$ on the xy plane. Then $\mathcal{S}$ is the image of $\mathcal{A}$ under a mapping ϕ,

$$\mathcal{S} = \phi(\mathcal{A}), \quad \text{or} \quad \begin{pmatrix} x \\ y \\ z \end{pmatrix} = \begin{pmatrix} x \\ y \\ g(x,y) \end{pmatrix} = \phi\begin{pmatrix} x \\ y \end{pmatrix}, \quad \begin{pmatrix} x \\ y \end{pmatrix} \in \mathcal{A}. \tag{127}$$

The Jacobi matrix of ϕ is the 3×2 matrix

$$J_\phi(x, y) = \frac{\partial(x, y, z)}{\partial(x, y)} = \begin{pmatrix} 1 & 0 \\ 0 & 1 \\ g_x & g_y \end{pmatrix}, \tag{128}$$

where $g_x = \partial g/\partial x$, $g_y = \partial g/\partial y$. The volume of (128) is, by (125),

$$\text{vol } J_\phi(x, y) = \sqrt{1 + g_x^2 + g_y^2}. \tag{129}$$

Substituting (129) in (124) we get the well-known formula

$$\int_{\mathcal{S}} f(x, y, z)\, ds = \int_{\mathcal{A}} f(x, y, g(x, y)) \sqrt{1 + g_x^2 + g_y^2}\; dx\, dy, \tag{130}$$

giving an integral over $\mathcal{S}$ as an integral over its projection in the xy plane.

The simplicity of this approach is not lost in higher dimensions or with different coordinate systems, as demonstrated below by elementary examples. These examples show that the full-rank assumption for J_ϕ is quite natural and presents no real restriction in applications. We see that (124) offers a unified method for a variety of curve and surface integrals, and coordinate systems, without having to construct (and understand) the differential geometry in each application.

A BLANKET ASSUMPTION: All functions are continuously differentiable as needed, all surfaces are smooth, and all curves are rectifiable.

EXAMPLE 2. Let a surface $\mathcal{S}$ in $\mathbb{R}^n$ be given by

$$x_n := g(x_1, x_2, \dots, x_{n-1}), \tag{131}$$

let $\mathcal{V}$ be a subset on $\mathcal{S}$, and let $\mathcal{U}$ be the projection of $\mathcal{V}$ on $\mathbb{R}^{n-1}$, the space of variables $(x_1, \dots, x_{n-1})$. The surface $\mathcal{S}$ is the graph of the mapping

$\phi : \mathcal{U} \to \mathcal{V}$, given by its components $\phi := (\phi_1, \phi_2, \ldots, \phi_n)$,

$$\phi_i(x_1, \ldots, x_{n-1}) := x_i, \quad i = 1, \ldots, n-1,$$
$$\phi_n(x_1, \ldots, x_{n-1}) := g(x_1, \ldots, x_{n-1}).$$

The Jacobi matrix of ϕ is

$$J_\phi = \begin{bmatrix} 1 & 0 & \cdots & 0 & 0 \\ 0 & 1 & \cdots & 0 & 0 \\ 0 & 0 & \ddots & 0 & 0 \\ 0 & 0 & \cdots & 1 & 0 \\ 0 & 0 & \cdots & 0 & 1 \\ \dfrac{\partial g}{\partial x_1} & \dfrac{\partial g}{\partial x_2} & \cdots & \dfrac{\partial g}{\partial x_{n-2}} & \dfrac{\partial g}{\partial x_{n-1}} \end{bmatrix}$$

and its volume is

$$\text{vol } J_\phi = \sqrt{1 + \sum_{i=1}^{n-1} \left(\frac{\partial g}{\partial x_i} \right)^2}. \tag{132}$$

For any function f integrable on $\mathcal{V}$ we therefore have

$$\int_{\mathcal{V}} f(x_1, \ldots, x_{n-1}, x_n) \, dV \tag{133}$$

$$= \int_{\mathcal{U}} f(x_1, \ldots, x_{n-1}, g(x_1, \ldots, x_{n-1})) \sqrt{1 + \sum_{i=1}^{n-1} \left(\frac{\partial g}{\partial x_i} \right)^2} \, dx_1 \cdots dx_{n-1}.$$

In particular, $f \equiv 1$ gives the area of $\mathcal{V}$,

$$\int_{\mathcal{V}} 1 \, dV = \int_{\mathcal{U}} \sqrt{1 + \sum_{i=1}^{n-1} \left(\frac{\partial g}{\partial x_i} \right)^2} \, dx_1 \cdots dx_{n-1}. \tag{134}$$

EXAMPLE 3 (Radon Transform). Let $\mathcal{H}_{\boldsymbol{\xi},p}$ be a hyperplane in $\mathbb{R}^n$ represented by

$$\mathcal{H}_{\boldsymbol{\xi},p} := \left\{ \mathbf{x} \in \mathbb{R}^n : \sum_{i=1}^{n} \xi_i x_i = p \right\} = \{ \mathbf{x} : \langle \boldsymbol{\xi}, \mathbf{x} \rangle = p \}, \tag{135}$$

where $\xi_n \neq 0$ in the normal vector $\boldsymbol{\xi} = (\xi_1, \ldots, \xi_n)$ of $\mathcal{H}_{\boldsymbol{\xi},p}$ (we call such hyperplanes *nonvertical*). Then $\mathcal{H}_{\boldsymbol{\xi},p}$ is given by

$$x_n := \frac{p}{\xi_n} - \sum_{i=1}^{n-1} \frac{\xi_i}{\xi_n} x_i \tag{136}$$

which is of the form (131). The volume (132) is here

$$\text{vol } J_\phi = \sqrt{1 + \sum_{i=1}^{n-1} \left(\frac{\xi_i}{\xi_n} \right)^2} = \frac{\|\boldsymbol{\xi}\|}{|\xi_n|}. \tag{137}$$

The *Radon transform* $(\mathbf{R}f)(\boldsymbol{\xi}, p)$ of a function $f : \mathbb{R}^n \to \mathbb{R}$ is its integral over the hyperplane $\mathcal{H}_{\boldsymbol{\xi},p}$, see [221],

$$(\mathbf{R}f)(\boldsymbol{\xi}, p) := \int_{\{\mathbf{x}:\, \langle \boldsymbol{\xi}, \mathbf{x} \rangle = p\}} f(\mathbf{x})\, d\mathbf{x}. \tag{138}$$

Using (136)–(137), the Radon transform can be computed as an integral in $\mathbb{R}^{n-1}$,

$$(\mathbf{R}f)(\boldsymbol{\xi}, p) = \frac{\|\boldsymbol{\xi}\|}{|\xi_n|} \int_{\mathbb{R}^{n-1}} \widehat{f}(x_1, \dots, x_{n-1}, p)\, dx_1 \cdots dx_{n-1} \tag{139a}$$

where

$$\widehat{f}(x_1, \dots, x_{n-1}, p) = f\left(x_1, \dots, x_{n-1}, \frac{p}{\xi_n} - \sum_{i=1}^{n-1} \frac{\xi_i}{\xi_n} x_i\right). \tag{139b}$$

In tomography applications the Radon transforms are computed by the scanning equipment, and the issue is the *inverse problem*, of reconstructing f from its Radon transforms $(\mathbf{R}f)(\boldsymbol{\xi}, p)$ for all $\boldsymbol{\xi}, p$. The inverse Radon transform is also an integral, see, e.g., [221], [603], and can be expressed analogously to (139), using the method of the next example.

EXAMPLE 4. Consider an integral over $\mathbb{R}^n$,

$$\int_{\mathbb{R}^n} f(\mathbf{x})\, d\mathbf{x} = \int_{\mathbb{R}^n} f(x_1, x_2, \dots, x_n)\, dx_1\, dx_2 \cdots dx_n. \tag{140}$$

Since $\mathbb{R}^n$ is a union of (parallel) hyperplanes,

$$\mathbb{R}^n = \bigcup_{p=-\infty}^{\infty} \{\mathbf{x} : \langle \boldsymbol{\xi}, \mathbf{x} \rangle = p\}, \quad \text{where } \boldsymbol{\xi} \neq \mathbf{0}, \tag{141}$$

we can compute (140) iteratively: an integral over $\mathbb{R}^{n-1}$ (Radon transform), followed by an integral on $\mathbb{R}$,

$$\int_{\mathbb{R}^n} f(\mathbf{x})\, d\mathbf{x} = \int_{-\infty}^{\infty} \frac{dp}{\|\boldsymbol{\xi}\|} (\mathbf{R}f)(\boldsymbol{\xi}, p), \tag{142}$$

where $dp/\|\boldsymbol{\xi}\|$ is the differential of the distance along $\boldsymbol{\xi}$ (i.e., dp times the distance between the parallel hyperplanes $\mathcal{H}_{\boldsymbol{\xi},p}$ and $\mathcal{H}_{\boldsymbol{\xi},p+1}$). Combining (139) and (142) we get the integral of f on $\mathbb{R}^n$,

$$\int_{\mathbb{R}^n} f(\mathbf{x})\, d\mathbf{x}$$
$$= \frac{1}{|\xi_n|} \int_{-\infty}^{\infty} \left\{ \int_{\mathbb{R}^{n-1}} \widehat{f}(x_1, \dots, x_{n-1}, p)\, dx_1 \cdots dx_{n-1} \right\} dp. \tag{143}$$

It is possible to derive (143) directly from the classical change-of-variables formula (122), by changing variables

$$\text{from } \{x_1, \cdots, x_{n-1}, x_n\} \text{ to } \{x_1, \cdots, x_{n-1}, p := \sum_{i=1}^{n} \xi_i x_i\},$$

and using

$$\det\left[\frac{\partial(x_1, \cdots, x_{n-1}, x_n)}{\partial(x_1, \cdots, x_{n-1}, p)} \right] = \frac{1}{\xi_n}.$$

EXAMPLE 5 (Fourier Transform). The *Fourier transform* $(\mathbf{F}f)(\boldsymbol{\xi})$ of f is the integral

$$(\mathbf{F}f)(\boldsymbol{\xi}) := \frac{1}{(2\pi)^{n/2}} \int_{\mathbb{R}^n} e^{-i\langle \boldsymbol{\xi}, \mathbf{x}\rangle} f(\mathbf{x})\, d\mathbf{x} \tag{144}$$

$$= \frac{1}{(2\pi)^{n/2}} \int_{\mathbb{R}^n} f\Big(x_1, \ldots, x_n\Big) \exp\Big\{-i \sum_{k=1}^{n} \xi_k x_k\Big\} dx_1 dx_2 \cdots dx_n.$$

For $\xi_n \neq 0$ we can compute (144) analogously to (143) as

$$(\mathbf{F}f)(\boldsymbol{\xi}) \tag{145}$$

$$= \frac{1}{(2\pi)^{n/2}|\xi_n|} \int_{-\infty}^{\infty} e^{-ip} \Big\{ \int_{\mathbb{R}^{n-1}} \widehat{f}(x_1, \ldots, x_{n-1}, p) \prod_{k=1}^{n-1} dx_k \Big\}\, dp.$$

The Fourier transform of a function of n variables is thus computed as an integral over $\mathbb{R}^{n-1}$ followed by an integral on $\mathbb{R}$. The inverse Fourier transform of a function $g(\boldsymbol{\xi})$ is of the same form as (144),

$$(\mathbf{F}^{-1}g)(\mathbf{x}) := \frac{1}{(2\pi)^{n/2}} \int_{\mathbb{R}^n} e^{i\langle \mathbf{x}, \boldsymbol{\xi}\rangle} g(\boldsymbol{\xi})\, d\boldsymbol{\xi}, \tag{146}$$

and can be computed as in (145).

Exercises

Ex. 32. Let $\phi : \mathbb{R}^2 \to \mathbb{R}^3$ be given by

$$\begin{bmatrix} x \\ y \\ z \end{bmatrix} = \phi\Big(\begin{bmatrix} u \\ v \end{bmatrix}\Big), \quad \text{with Jacobian } J_\phi = \frac{\partial(x,y,z)}{\partial(u,v)} = \begin{bmatrix} x_u & x_v \\ y_u & y_v \\ z_u & z_v \end{bmatrix},$$

assumed of full column rank. Then

$$\text{vol } J_\phi = \sqrt{EG - F^2}, \tag{147}$$

where

$$E = (x_u)^2 + (y_u)^2 + (z_u)^2,$$
$$F = x_u x_v + y_u y_v + z_u z_v,$$
$$G = (x_v)^2 + (y_v)^2 + (z_v)^2.$$

Explanation of the functions E, F, and G: let $\mathbf{r} = \begin{bmatrix} x \\ y \\ z \end{bmatrix}$.

Then:

$$E = \|\mathbf{r}_u\|^2, \quad F = \langle \mathbf{r}_u, \mathbf{r}_v\rangle, \quad G = \|\mathbf{r}_v\|^2,$$

and (147) becomes

$$\text{vol } J_\phi = \sqrt{EG\Big(1 - \frac{F^2}{EG}\Big)} = \sqrt{\|\mathbf{r}_u\|^2 \|\mathbf{r}_v\|^2(1 - \cos^2 \angle\{\mathbf{r}_u, \mathbf{r}_v\})}$$

$$= \|\mathbf{r}_u\| \, \|\mathbf{r}_v\| \, |\sin \angle\{\mathbf{r}_u, \mathbf{r}_v\}| = \text{area of the parallelogram } \Diamond\{\mathbf{r}_u, \mathbf{r}_v\}.$$

Since ϕ maps an area element $\Box\{du, dv\}$ into $\Diamond\{\mathbf{r}_u\, du, \mathbf{r}_v\, dv\}$, we see that E and G measure the stretching of the sides, while F gives the distortion of the angle, under the mapping ϕ. In particular, angles are preserved if $F = 0$, in which case shapes are preserved if $E = G$. See also Ex. 6.21.

Ex. 33. Let C be an arc on a curve in $\mathbb{R}^n$, represented in parametric form as

$$C := \phi([0,1]) = \{(x_1, x_2, \ldots, x_n) : x_i := \phi_i(t),\ 0 \le t \le 1\}. \tag{148}$$

The Jacobi matrix $J_\phi(t) = \partial(x_1, x_2, \cdots, x_n)/\partial t$ is the column matrix $[\phi_i'(t)]$ and its volume is

$$\operatorname{vol} J_\phi = \sqrt{\sum_{i=1}^{n} (\phi_i'(t))^2}.$$

The *line integral* (assuming it exists) of a function f along C, $\int_C f$, is given in terms of the volume of J_ϕ as follows:

$$\int_C f = \int_0^1 f(\phi_1(t), \ldots, \phi_n(t)) \left(\sum_{i=1}^{n} (\phi_i'(t))^2 \right)^{1/2} dt. \tag{149}$$

In particular, $f \equiv 1$ gives

$$\text{arc length } C = \int_0^1 \left(\sum_{i=1}^{n} (\phi_i'(t))^2 \right)^{1/2} dt. \tag{150}$$

If one of the variables, say $a \le x_1 \le b$, is used as parameter, (150) gives the familiar result

$$\text{arc length } C = \int_a^b \left(1 + \sum_{i=2}^{n} \left(\frac{dx_i}{dx_1} \right)^2 \right)^{1/2} dx_1.$$

Ex. 34. Let S be a surface in $\mathbb{R}^3$ represented by

$$z = z(r, \theta) \tag{151}$$

where $\{r, \theta, z\}$ are *cylindrical coordinates*

$$x = r \cos\theta, \tag{152a}$$
$$y = r \sin\theta, \tag{152b}$$
$$z = z. \tag{152c}$$

The Jacobi matrix of the mapping (152a), (152b), and (151) is

$$J_\phi = \frac{\partial(x, y, z)}{\partial(r, \theta)} = \begin{bmatrix} \cos\theta & -r\sin\theta \\ \sin\theta & r\cos\theta \\ \frac{\partial z}{\partial r} & \frac{\partial z}{\partial \theta} \end{bmatrix} \tag{153}$$

and its volume is

$$\operatorname{vol} J_\phi = \sqrt{ r^2 + r^2 \left(\frac{\partial z}{\partial r} \right)^2 + \left(\frac{\partial z}{\partial \theta} \right)^2 } = r \sqrt{ 1 + \left(\frac{\partial z}{\partial r} \right)^2 + \frac{1}{r^2} \left(\frac{\partial z}{\partial \theta} \right)^2 }. \tag{154}$$

An integral over a domain $V \subset S$ is therefore

$$\int_V f(x, y, z)\, dV \tag{155}$$

$$= \int_U f(r \cos\theta,\ r \sin\theta,\ z(r, \theta))\, r \sqrt{ 1 + \left(\frac{\partial z}{\partial r} \right)^2 + \frac{1}{r^2} \left(\frac{\partial z}{\partial \theta} \right)^2 }\, dr\, d\theta.$$

Ex. 35. Let S be a surface in $\mathbb{R}^3$, symmetric about the z-axis. This axial symmetry is expressed in cylindrical coordinates by

$$z = z(r), \quad \text{or} \quad \frac{\partial z}{\partial \theta} = 0, \quad \text{in (153)–(155)}.$$

The volume (154) thus becomes

$$\text{vol } J_\phi = r\sqrt{1 + z'(r)^2} \tag{156}$$

with the axial symmetry "built in." An integral over a domain $\mathcal{V}$ in a z-symmetric surface S is therefore

$$\int_{\mathcal{V}} f(x, y, z)\, dV = \int_{\mathcal{U}} f(r \cos\theta, r \sin\theta, z(r))\, r\sqrt{1 + z'(r)^2}\, dr\, d\theta.$$

Ex. 36. Again let S be a z-symmetric surface in $\mathbb{R}^3$. We use *spherical coordinates*

$$x = \rho \sin\phi \cos\theta, \tag{157a}$$
$$y = \rho \sin\phi \sin\theta, \tag{157b}$$
$$z = \rho \cos\phi. \tag{157c}$$

The axial symmetry is expressed by

$$\rho := \rho(\phi) \tag{158}$$

showing that S is given in terms of the two variables ϕ and θ. The volume of the Jacobi matrix is easily computed

$$\text{vol } \frac{\partial(x, y, z)}{\partial(\phi, \theta)} = \rho\sqrt{\rho^2 + (\rho'(\phi))^2}\, \sin\phi$$

and the change of variables formula gives

$$\int_{\mathcal{V}} f(x, y, z)\, dV = \int_{\mathcal{U}} \widehat{f}(\theta, \phi)\rho(\phi)\sqrt{\rho(\phi)^2 + (\rho'(\phi))^2}\, \sin\phi\, d\phi\, d\theta, \tag{159a}$$

where $\widehat{f}(\theta, \phi)$ is obtained by substitution of (157) and (158) in $f(x, y, z)$,

$$\widehat{f}(\theta, \phi) = f(\rho(\phi) \sin\phi \cos\theta, \rho(\phi) \sin\phi \sin\theta, \rho(\phi) \cos\phi)\rho(\phi). \tag{159b}$$

Ex. 37. *The generalized Pythagorean theorem*, Lin and Lin [**517**], [**95**]. Consider an n-dimensional *simplex*

$$\mathbf{\Delta}_n := \left\{ (x_1, x_2, \ldots, x_n) : \sum_{i=1}^{n} a_i x_i \leq a_0,\ x_i \geq 0,\ i \in \overline{1, n} \right\}, \tag{160}$$

with all $a_j > 0$, $j \in \overline{0, n}$. We denote the $n + 1$ faces of $\mathbf{\Delta}_n$ by

$$\mathcal{F}_0 := \left\{ (x_1, x_2, \ldots, x_n) \in \mathbf{\Delta}_n : \sum_{i=1}^{n} a_i x_i = a_0 \right\}, \tag{161a}$$
$$\mathcal{F}_j := \left\{ (x_1, x_2, \ldots, x_n) \in \mathbf{\Delta}_n : x_j = 0 \right\}, \ j \in \overline{1, n}. \tag{161b}$$

and denote their areas by A_0, A_j, respectively. The *generalized Pythagorean theorem* (see [**517**]) states that

$$A_0^2 = \sum_{j=1}^{n} A_j^2. \tag{162}$$

We prove it here using the change of variables formula (124). For any $j \in \overline{1, n}$ we can represent the (largest) face $\mathcal{F}_0$ as $\mathcal{F}_0 = \phi^{\{j\}}(\mathcal{F}_j)$ where $\phi^{\{j\}} = (\phi_1^{\{j\}}, \dots, \phi_n^{\{j\}})$ is

$$\phi_i^{\{j\}}(x_1, x_2, \dots, x_n) = x_i, \quad i \neq j,$$

$$\phi_j^{\{j\}}(x_1, x_2, \dots, x_n) = \frac{a_0}{a_j} - \sum_{i \neq j} \frac{a_i}{a_j} x_i.$$

The Jacobi matrix of $\phi^{\{j\}}$ is an $n \times (n-1)$ matrix with the ith unit vector in row $i \neq j$ and

$$\left(-\frac{a_1}{a_j}, -\frac{a_2}{a_j}, \dots, -\frac{a_{j-1}}{a_j}, -\frac{a_{j+1}}{a_j}, \dots, -\frac{a_{n-1}}{a_j}, -\frac{a_n}{a_j} \right)$$

in row j. The volume of the Jacobi matrix of $\phi^{\{j\}}$ is computed as

$$\text{vol } J_{\phi^{\{j\}}} = \sqrt{1 + \sum_{i \neq j} \left(\frac{a_i}{a_j} \right)^2} = \sqrt{\frac{\sum_{i=1}^{n} a_i^2}{a_j^2}} = \frac{\|\mathbf{a}\|}{|a_j|}$$

where $\mathbf{a}$ is the vector $(a_1, \dots, a_n)$. Therefore, the area of $\mathcal{F}_0$ is

$$A_0 = \int_{\mathcal{F}_j} \left(\frac{\|\mathbf{a}\|}{|a_j|} \right) \prod_{i \neq j} dx_i = \left(\frac{\|\mathbf{a}\|}{|a_j|} \right) A_j, \quad j \in \overline{1, n}. \tag{163}$$

$$\therefore \quad \frac{\sum_{j=1}^{n} A_j^2}{A_0^2} = \frac{\sum_{j=1}^{n} |a_j|^2}{\|\mathbf{a}\|^2},$$

and (162) reduces to the ordinary Pythagorean theorem

$$\|\mathbf{a}\|^2 = \sum_{j=1}^{n} |a_j|^2.$$

Ex. 38. Let A_0 be the area of largest face $\mathcal{F}_0$ of the regular simplex

$$\mathcal{F}_0 = \left\{ \mathbf{x} : \sum_{i=1}^{n} x_i = 1, \ x_i \geq 0, \ i \in \overline{1, n} \right\}.$$

Then

$$A_0 = \frac{\sqrt{n}}{(n-1)!}.$$

The other n faces have areas

$$A_j = \frac{1}{(n-1)!}, \quad j \in \overline{1, n}.$$

Ex. 39. *Gamma and Beta functions.* We collect here results needed below. The *Gamma function* $\Gamma(p)$ is

$$\Gamma(p) := \int_0^\infty x^{p-1} e^{-x} \, dx. \tag{164}$$

Its properties include

$$\Gamma(1) = 1, \tag{165a}$$

$$\Gamma(p+1) = p\,\Gamma(p), \tag{165b}$$

$$\Gamma\left(\tfrac{1}{2}\right) = \sqrt{\pi}. \tag{165c}$$

The *Beta function* is

$$B(p,q) := \int_0^1 (1-x)^{p-1} x^{q-1}\, dx. \tag{166}$$

Its properties include

$$B(p,q) = \frac{\Gamma(p)\,\Gamma(q)}{\Gamma(p+q)}, \tag{167a}$$

$$\frac{B(p,q+1)}{B(p,q)} = \frac{q}{p+q}, \tag{167b}$$

$$\frac{B(p+1,q)}{B(p,q)} = \frac{p}{p+q}, \tag{167c}$$

where (167b)–(167c) follow from (167a) and (165b).

Ex. 40. Let

$$\mathcal{B}_n(r) := \{\mathbf{x} \in \mathbb{R}^n : \|\mathbf{x}\| \le r\}, \quad \text{the } ball \text{ of radius } r,$$
$$\mathcal{S}_n(r) := \{\mathbf{x} \in \mathbb{R}^n : \|\mathbf{x}\| = r\}, \quad \text{the } sphere \text{ of radius } r,$$

both centered at the origin. Also,

$$v_n(r) := \text{the } volume \text{ of } \mathcal{B}_n(r),$$
$$a_n(r) := \text{the } area \text{ of } \mathcal{S}_n(r),$$

where r is dropped if $r = 1$, so that

$$v_n := \text{the volume of the unit ball } \mathcal{B}_n,$$
$$a_n := \text{the area of the unit sphere } \mathcal{S}_n.$$

Clearly

$$v_n(r) = v_n\, r^n, \quad a_n(r) = a_n\, r^{n-1}, \quad dv_n(r) = v_n'(r)\, dr = a_n(r)\, dr, \tag{168}$$

and it follows that

$$a_n = n\, v_n, \quad n = 2, 3, \ldots. \tag{169}$$

Ex. 41. Integrals on $\mathcal{S}_n$, in particular the area a_n, can be computed using *spherical coordinates*, e.g., [**603**, § VII.2], or the *surface element* of $\mathcal{S}_n$, e.g., [**586**]. An alternative, simpler approach is to use the results of Example 2, representing the "upper hemisphere" as $\phi(\mathcal{B}_{n-1})$, where $\phi = (\phi_1, \phi_2, \ldots, \phi_n)$ is

$$\phi_i(x_1, x_2, \ldots, x_{n-1}) = x_i, \quad i \in \overline{1, n-1},$$

$$\phi_n(x_1, x_2, \ldots, x_{n-1}) = \sqrt{1 - \sum_{i=1}^{n-1} x_i^2}.$$

The Jacobi matrix is

$$J_\phi = \begin{bmatrix} 1 & 0 & \cdots & 0 \\ 0 & 1 & \cdots & 0 \\ \vdots & \vdots & \ddots & \vdots \\ 0 & 0 & \cdots & 1 \\ -\dfrac{x_1}{x_n} & -\dfrac{x_2}{x_n} & \cdots & -\dfrac{x_{n-1}}{x_n} \end{bmatrix}$$

and its volume is easily computed

$$\text{vol } J_\phi = \sqrt{1 + \sum_{i=1}^{n-1}\left(\frac{x_i}{x_n}\right)^2} = \frac{1}{|x_n|} = \frac{1}{\sqrt{1 - \sum_{i=1}^{n-1} x_i^2}}. \tag{170}$$

The area a_n is twice the area of the "upper hemisphere." Therefore

$$a_n = 2\int_{\mathcal{B}_{n-1}} \frac{dx_1\, dx_2 \cdots dx_{n-1}}{\sqrt{1 - \sum_{i=1}^{n-1} x_i^2}}, \quad \text{(by (170))},$$

$$= \frac{2\,\pi^{\frac{n}{2}}}{\Gamma\left(\frac{n}{2}\right)}, \tag{171}$$

where $\Gamma(\cdot)$ is the *Gamma function*, see Ex. 39 above.

PROOF.

$$a_n = 2\int_{\mathcal{B}_{n-1}} \frac{dx_1 \cdots dx_{n-1}}{\sqrt{1 - \sum_{i=1}^{n-1} x_i^2}}, = 2\int_0^1 \frac{dv_{n-1}(r)}{\sqrt{1 - r^2}},$$

using spherical shells of radius r and volume $dv_{n-1}(r)$,

$$= 2\,a_{n-1}\int_0^1 \frac{r^{n-2}}{\sqrt{1 - r^2}}\, dr, \quad \text{by (169)}.$$

$$\therefore \quad \frac{a_n}{a_{n-1}} = \int_0^1 (1-x)^{-1/2}\, x^{(n-3)/2}\, dx\,, \quad \text{using } x = r^2,$$

$$= B\!\left(\frac{(n-1)}{2}, \frac{1}{2}\right) = \frac{\Gamma((n-1)/2)\Gamma(1/2)}{\Gamma(n/2)}$$

and a_n can be computed recursively, beginning with $a_2 = 2\pi$, giving (171). □

EX. 42. *Volume of $\mathcal{B}_n$.* The volume of the unit ball $\mathcal{B}_n$ can be computed by (169) and (171),

$$v_n = \frac{\pi^{n/2}}{\Gamma((n/2) + 1)}, \quad n = 1, 2, \cdots \tag{172}$$

Alternatively, the volume v_n can be computed as the limit of the sum of volumes of cylinders, with base $dx_1 \cdots dx_{n-1}$ and height $2\sqrt{1 - \sum_{k=1}^{n-1} x_k^2}$,

$$v_n = 2\int_{\mathcal{B}_{n-1}} \sqrt{1 - \sum_{k=1}^{n-1} x_k^2}\ dx_1 \cdots dx_{n-1}, \tag{173}$$

a routine integration.

EX. 43. The normal vector $\boldsymbol{\xi}$ of the hyperplane (135) can be normalized and can therefore be assumed a unit vector, see, e.g., [221, Chapter 3] where the Radon transform with respect to a hyperplane

$$\mathcal{H}(\boldsymbol{\xi}^0, p) := \{\mathbf{x} \in \mathbb{R}^n : \langle \boldsymbol{\xi}^0, \mathbf{x}\rangle = p\}, \quad \|\boldsymbol{\xi}^0\| = 1, \tag{174}$$

is represented as

$$(\widehat{\mathbf{R}}f)(\boldsymbol{\xi}^0, p) = \int f(\mathbf{x})\, \delta(p - \langle \boldsymbol{\xi}^0, \mathbf{x}\rangle)\, d\mathbf{x},$$

where $\delta(\cdot)$ is the Dirac delta function. If (135) and (174) represent the same hyperplane, then the correspondence between $(\mathbf{R}f)(\boldsymbol{\xi}, y)$ and $(\widehat{\mathbf{R}}f)(\boldsymbol{\xi}^0, p)$ is given by

$$\xi^0 = \frac{\xi}{\|\xi\|}, \quad p = \frac{y}{\|\xi\|}.$$

EX. 44. The computation of the integrals mentioned above is feasible using available symbolic algebra software (subject to limitations of such software).

As an illustration consider the Radon transform of the function $f(x,y) = e^{-x^2-y^2}$, obtained here by using (139) with the symbolic package DERIVE [**492**]:

Function f	Radon transform $\mathbf{R}f(\xi,p)$, $\xi = \left[\begin{smallmatrix} a \\ b \end{smallmatrix}\right]$
$f(x,y) = \text{EXP}\left[-x^2 - y^2\right]$	$\mathbf{R}f\left(\begin{bmatrix} a \\ b \end{bmatrix}, p\right) = \sqrt{\pi}\ \text{EXP}\left[-\dfrac{p^2}{a^2 + b^2}\right]$
$f(x,y) = \text{EXP}\left[-x^2 - y^2\right]$	$\mathbf{R}f\left(\begin{bmatrix} a \\ 0 \end{bmatrix}, p\right) = \sqrt{\pi}\ \text{EXP}\left[-\dfrac{p^2}{a^2}\right]$ (vertical line)

10. An Application of the Matrix Volume in Probability

This section is based on [**75**]. The abbreviation RV of *random variable* is used throughout. Consider n RVs $(\mathbf{X}_1,\dots,\mathbf{X}_n)$ with a given joint density $f_{\mathbf{X}}(x_1,\dots,x_n)$ and an RV

$$\mathbf{Y} = h(\mathbf{X}_1,\dots,\mathbf{X}_n) \tag{175}$$

defined by the mapping $h : \mathbb{R}^n \to \mathbb{R}$. The density function $f_{\mathbf{Y}}(y)$ of $\mathbf{Y}$ is derived here in two special cases,

h linear:

$$h(\mathbf{X}_1,\cdots,\mathbf{X}_n) = \sum_{i=1}^n \xi_i \mathbf{X}_i, \quad \text{see Corollary 4,} \tag{176}$$

h sum of squares:

$$h(\mathbf{X}_1,\cdots,\mathbf{X}_n) = \sum_{i=1}^n \mathbf{X}_i^2, \quad \text{Corollary 5.} \tag{177}$$

In both cases, the density $f_{\mathbf{Y}}(y)$ is computed as an integral of $f_{\mathbf{X}}$ on the surface

$$\mathcal{V}(y) := \{\mathbf{x} \in \mathbb{R}^n : h(\mathbf{x}) = y\},$$

that is, a hyperplane for (176) and a sphere for (177). These integrals are elementary and computationally feasible, as illustrated in [**75**, Appendix A]. Both results are consequences of Theorem 14, and a comparison between two integrations, one "classical" and the other using the change-of-variables formula (124).

Notation: For a random variable $\mathbf{X}$ we denote:

$\quad F_{\mathbf{X}}(x) := \text{Prob}\{\mathbf{X} \le x\}$, the *distribution function*;

$\quad f_{\mathbf{X}}(x) := \frac{d}{dx}F_{\mathbf{X}}(x)$, the *density function*;

$\quad \text{E}\{\mathbf{X}\}$, the expected value;

$\quad \text{Var}\{\mathbf{X}\}$, the variance;

$\quad \mathbf{X} \sim U(S)$, the fact that $\mathbf{X}$ is uniformly distributed over the set S; and

$\mathbf{X} \sim N(\mu, \sigma)$,   normally distributed with $E\{\mathbf{X}\} = \mu$,
$\text{Var}\{\mathbf{X}\} = \sigma^2$.

BLANKET ASSUMPTION: Throughout this section all random variables are absolutely continuous and the indicated densities exist.

10.1. Probability Densities and Surface Integrals. Let the RV $\mathbf{X} = (\mathbf{X}_1, \ldots, \mathbf{X}_n)$ have joint density $f_{\mathbf{X}}(x_1, \ldots, x_n)$ and let

$$y = h(x_1, \ldots, x_n) \tag{178}$$

where $h : \mathbb{R}^n \to \mathbb{R}$ is sufficiently well-behaved, in particular, $\partial h / \partial x_n \neq 0$, and (178) can be solved for x_n,

$$x_n = h^{-1}(y | x_1, \ldots, x_{n-1}) \tag{179}$$

with $x_1, \ldots, x_{n-1}$ as parameters. By changing variables from $\{x_1, \ldots, x_n\}$ to $\{x_1, \ldots, x_{n-1}, y\}$, and using the fact

$$\det\left[\frac{\partial(x_1, \ldots, x_n)}{\partial(x_1, \ldots, x_{n-1}, y)} \right] = \frac{\partial h^{-1}}{\partial y}, \tag{180}$$

we write the density of $\mathbf{Y} = h(\mathbf{X}_1, \ldots, \mathbf{X}_n)$ as

$$f_{\mathbf{Y}}(y) = \int_{\mathbb{R}^{n-1}} \widehat{f}(y | x_1, \ldots, x_{n-1}) \left| \frac{\partial h^{-1}}{\partial y} \right| dx_1 \cdots dx_{n-1}, \tag{181a}$$

where

$$\widehat{f}(y | x_1, \ldots, x_{n-1}) = f_{\mathbf{X}}(x_1, \ldots, x_{n-1}, h^{-1}(y | x_1, \ldots, x_{n-1})). \tag{181b}$$

Let $\mathcal{V}(y)$ be the surface given by (178), represented as

$$\begin{pmatrix} x_1 \\ \vdots \\ x_{n-1} \\ x_n \end{pmatrix} = \begin{pmatrix} x_1 \\ \vdots \\ x_{n-1} \\ h^{-1}(y | x_1, \ldots, x_{n-1}) \end{pmatrix} = \phi\begin{pmatrix} x_1 \\ \vdots \\ x_{n-1} \end{pmatrix}. \tag{182}$$

Then the surface integral of $f_{\mathbf{X}}$ over $\mathcal{V}(y)$ is given, by (133), as

$$\int_{\mathcal{V}(y)} f_{\mathbf{X}}$$
$$= \int_{\mathbb{R}^{n-1}} \widehat{f}(y | x_1, \ldots, x_{n-1}) \sqrt{1 + \sum_{i=1}^{n-1} \left(\frac{\partial h^{-1}}{\partial x_i} \right)^2} \, dx_1 \cdots dx_{n-1}, \tag{183}$$

with $\widehat{f}(y | x_1, \ldots, x_{n-1})$ given by (181b).

THEOREM 14. *If the ratio*

$$\frac{\dfrac{\partial h^{-1}}{\partial y}}{\sqrt{1 + \sum_{i=1}^{n-1} \left(\frac{\partial h^{-1}}{\partial x_i} \right)^2}} \qquad \textit{does not depend on} \quad x_1, \ldots, x_{n-1}, \tag{184}$$

then

$$f_{\mathbf{Y}}(y) = \frac{\left|\frac{\partial h^{-1}}{\partial y}\right|}{\sqrt{1 + \sum\limits_{i=1}^{n-1}\left(\frac{\partial h^{-1}}{\partial x_i}\right)^2}} \int_{\mathcal{V}(y)} f_{\mathbf{X}}. \tag{185}$$

PROOF. A comparison of (183) and (181a) gives the density $f_{\mathbf{Y}}$ as the surface integral (185). □

Condition (184) holds if $\mathcal{V}(y)$ is a hyperplane (see Section 10.2) or a sphere, see Section 10.3. In these two cases, covering many important probability distributions, the derivation (185) is simpler computationally than classical integration formulas, e.g., [**100**, Theorem 5.1.5], [**377**, Theorem 6-5.4], and transform methods, e.g., [**770**].

10.2. Hyperplanes. Let

$$y = h(x_1, \ldots, x_n) := \sum_{i=1}^{n} \xi_i x_i, \tag{186}$$

where $\boldsymbol{\xi} = (\xi_1, \ldots, \xi_n)$ is a given vector with $\xi_n \neq 0$. Then (179) becomes

$$x_n = h^{-1}(y \,|\, x_1, \ldots, x_{n-1}) := \frac{y}{\xi_n} - \sum_{i=1}^{n-1} \frac{\xi_i}{\xi_n} x_i, \tag{187}$$

with

$$\frac{\partial h^{-1}}{\partial y} = \frac{1}{\xi_n}, \tag{188a}$$

$$\sqrt{1 + \sum_{i=1}^{n-1}\left(\frac{\partial h^{-1}}{\partial x_i}\right)^2} = \sqrt{1 + \sum_{i=1}^{n-1}\left(\frac{\xi_i}{\xi_n}\right)^2} = \frac{\|\boldsymbol{\xi}\|}{|\xi_n|}. \tag{188b}$$

Condition (184) thus holds, and the density of $\sum \xi_i \mathbf{X}_i$ can be expressed as a surface integral of $f_{\mathbf{X}}$ on the hyperplane

$$\mathcal{H}(\boldsymbol{\xi}, y) := \left\{\mathbf{x} \in \mathbb{R}^n : \sum_{i=1}^{n} \xi_i x_i = y\right\},$$

i.e., the *Radon transform* $(\mathbf{R}f_{\mathbf{X}})(\boldsymbol{\xi}, y)$ of $f_{\mathbf{X}}$. Recall that the Radon transform can be computed as an integral on $\mathbb{R}^{n-1}$, see, e.g., (139).

COROLLARY 4. *Let* $\mathbf{X} = (\mathbf{X}_1, \mathbf{X}_2, \ldots, \mathbf{X}_n)$ *be random variables with joint density* $f_{\mathbf{X}}(x_1, x_2, \ldots, x_n)$ *and let* $\mathbf{0} \neq \boldsymbol{\xi} \in \mathbb{R}^n$. *The random variable*

$$\mathbf{Y} := \sum_{i=1}^{n} \xi_i \mathbf{X}_i \tag{189}$$

has the density

$$f_{\mathbf{Y}}(y) = \frac{(\mathbf{R}f_{\mathbf{X}})(\boldsymbol{\xi}, y)}{\|\boldsymbol{\xi}\|}. \tag{190}$$

PROOF. Follows from (185), (137), and (139). □

Explanation of the factor $\|\boldsymbol{\xi}\|$ in (190): the distance between the hyperplanes $\mathcal{H}(\boldsymbol{\xi}, y)$ and $\mathcal{H}(\boldsymbol{\xi}, y + dy)$ is $dy/\|\boldsymbol{\xi}\|$.

10.3. Spheres. Let

$$y = h(x_1, \dots, x_n) := \sum_{i=1}^{n} x_i^2 \tag{191}$$

which has two solutions for x_n, representing the upper and lower hemispheres,

$$x_n = h^{-1}(y \mid x_1, \dots, x_{n-1}) := \pm \sqrt{y - \sum_{i=1}^{n-1} x_i^2}, \tag{192}$$

with

$$\frac{\partial h^{-1}}{\partial y} = \pm \frac{1}{2\sqrt{y - \sum_{i=1}^{n-1} x_i^2}}, \tag{193a}$$

$$\sqrt{1 + \sum_{i=1}^{n-1} \left(\frac{\partial h^{-1}}{\partial x_i}\right)^2} = \frac{\sqrt{y}}{\sqrt{y - \sum_{i=1}^{n-1} x_i^2}}. \tag{193b}$$

Therefore condition (184) holds and the density of $\sum X_i^2$ is, by (185), expressed in terms of the surface integral of f_X on the sphere $\mathcal{S}_n(\sqrt{y})$ of radius $\sqrt{y}$.

COROLLARY 5. *Let $X = (X_1, \dots, X_n)$ have joint density $f_X(x_1, \dots, x_n)$. The density of*

$$Y = \sum_{i=1}^{n} X_i^2 \tag{194}$$

is

$$f_Y(y) = \frac{1}{2\sqrt{y}} \int_{\mathcal{S}_n(\sqrt{y})} f_X, \tag{195}$$

where the integral is over the sphere $\mathcal{S}_n(\sqrt{y})$ of radius $\sqrt{y}$, computed as an integral over the ball $\mathcal{B}_{n-1}(\sqrt{y})$ using (133) with $g = h^{-1}$.

PROOF. Equation (195) follows from (185) and (193a). □

An explanation of the factor $2\sqrt{y}$ in (195): the width of the spherical shell bounded by the two spheres $\mathcal{S}_n(\sqrt{y})$ and $\mathcal{S}_n(\sqrt{y + dy})$ is the difference of radii

$$\sqrt{y + dy} - \sqrt{y} \approx \frac{dy}{2\sqrt{y}}.$$

EXAMPLE 6 (Spherical distribution). If the joint density of $X = (X_1, \dots, X_n)$ is spherical

$$f_X(x_1, \dots, x_n) = p\left(\sum_{i=1}^{n} x_i^2\right), \tag{196}$$

then $Y = \sum_{i=1}^{n} X_i^2$ has the density

$$f_Y(y) = \frac{\pi^{n/2}}{\Gamma(n/2)} \, p(y) \, y^{(n/2)-1}. \tag{197}$$

PROOF. The surface integral of $f_{\mathbf{X}}$ over $\mathcal{S}_n(\sqrt{y})$ is

$$\int_{\mathcal{S}_n(\sqrt{y})} f_{\mathbf{X}} = p(y)\, a_{n-1}(\sqrt{y}) = p(y)\, \frac{2\,\pi^{n/2}}{\Gamma(n/2)}\, \sqrt{y}^{\,n-1}, \quad \text{by (171) and (168).}$$

The proof is completed by (195). $\qquad\qquad\qquad\qquad\qquad\qquad\qquad$ $\square$

Exercises

EX. 45. *Exponential distribution.* Let $\lambda > 0$ be given and let $\mathbf{Y}$ be the mean

$$\mathbf{Y} = \frac{1}{n} \sum_{i=1}^{n} \mathbf{X}_i$$

of n independent RVs, identically distributed with exponential density

$$f_{\mathbf{X}_i}(x_i) = \lambda e^{-\lambda x_i}, \quad x_i \ge 0.$$

Then the density of $\mathbf{Y}$ is

$$f_{\mathbf{Y}}(y) = \frac{(\lambda n)^n}{(n-1)!}\, y^{n-1}\, e^{-\lambda n y}, \quad y \ge 0.$$

PROOF. Use (186) with $\xi_i = 1/n$, (185), and (188) to conclude that

$$f_{\mathbf{Y}}(y) = \frac{1}{\|\boldsymbol{\xi}\|} \int_{\mathcal{F}_0(ny)} f_{\mathbf{X}}\left(x_1, \ldots, x_{n-1}, \frac{y}{\xi_n} - \sum_{i=1}^{n-1} \frac{\xi_i}{\xi_n} x_i\right)$$

$$= \sqrt{n} \int_{\mathcal{F}_0(ny)} f_{\mathbf{X}}\left(x_1, \ldots, x_{n-1}, ny - \sum_{i=1}^{n-1} x_i\right), \qquad (198)$$

where $\mathcal{F}_0(ny) = \left\{\mathbf{x} : \sum_{i=1}^{n} x_i = ny,\ x_i \ge 0,\ i \in \overline{1,n}\right\}$ is the simplex face whose area is, by Ex. 38,

$$A_0(ny) = \frac{\sqrt{n}}{(n-1)!}(ny)^{n-1}.$$

The joint density of $\mathbf{X} = (\mathbf{X}_1, \ldots, \mathbf{X}_n)$ is $f_X(x_1, \ldots, x_n) = \lambda^n e^{-\lambda \sum\limits_{i=1}^{n} x_i}$. Therefore

$$f_{\mathbf{X}}\left(x_1, \ldots, x_{n-1}, ny - \sum_{i=1}^{n-1} x_i\right) = \lambda^n e^{-\lambda n y}$$

and (198) becomes

$$f_{\mathbf{Y}}(y) = \sqrt{n}\, \lambda^n e^{-\lambda n y}\, A_0(ny)$$

$$= \sqrt{n}\, \lambda^n e^{-\lambda n y}\, \frac{\sqrt{n}}{(n-1)!}(ny)^{n-1},$$

completing the proof. $\qquad\qquad\qquad\qquad\qquad\qquad\qquad\qquad\qquad$ $\square$

EX. 46. *Bivariate normal distribution.* Let $(\mathbf{X}_1, \mathbf{X}_2)$ have the bivariate normal distribution, with zero means and unit variances,

$$f_{\mathbf{X}}(x_1, x_2) = \frac{1}{2\pi\sqrt{1-\rho^2}} \exp\left\{-\frac{x_1^2 - 2\rho x_1 x_2 + x_2^2}{2(1-\rho^2)}\right\} \qquad (199)$$

and let $\mathbf{Y} := a\mathbf{X}_1 + b\mathbf{X}_2$. The density of $\mathbf{Y}$ is, by (190),

$$f_{\mathbf{Y}}(y) = \frac{1}{\sqrt{a^2 + b^2}}\, (\mathbf{R} f_{\mathbf{X}})((a,b), y)$$

$$= \frac{1}{\sqrt{2\pi}\,\sqrt{a^2 + 2ab\rho + b^2}} \exp\left\{-\frac{y^2}{2\,(a^2 + 2ab\rho + b^2)}\right\}. \qquad (200)$$

Therefore $a\mathbf{X}_1 + b\mathbf{X}_2 \sim N(0, \sqrt{a^2 + 2ab\rho + b^2})$. In particular, $\mathbf{X}_1 + \mathbf{X}_2 \sim N(0, \sqrt{2(1+\rho)})$ and $\mathbf{X}_1 - \mathbf{X}_2 \sim N(0, \sqrt{2(1-\rho)})$.

EX. 47. *Uniform distribution.* Let $(\mathbf{X}_1, \mathbf{X}_2)$ be independent and uniformly distributed on $[0,1]$. Their joint density is

$$f_{\mathbf{X}}(x_1, x_2) = \begin{cases} 1, & \text{if } 0 \le x_1, x_2 \le 1, \\ 0, & \text{otherwise,} \end{cases}$$

and the density of $a\mathbf{X}_1 + b\mathbf{X}_2$ is, by (190),

$$f_{a\mathbf{X}_1 + b\mathbf{X}_2}(y) = \frac{1}{\sqrt{a^2 + b^2}} \left(\mathbf{R}f_{\mathbf{X}}\right)\left(\begin{bmatrix} a \\ b \end{bmatrix}, y\right)$$

$$= \frac{|y - a - b| - |y - a| - |y - b| + |y|}{2ab}. \tag{201}$$

In particular,

$$f_{\mathbf{X}_1 + \mathbf{X}_2}(y) = \begin{cases} y, & \text{if } 0 \le y < 1, \\ 2 - y, & \text{if } 1 \le y \le 2, \\ 0, & \text{otherwise,} \end{cases}$$

a symmetric triangular distribution on $[0,2]$.

EX. 48. χ^2 *distribution.* If $\mathbf{X}_i \sim N(0,1)$ and are independent, $i \in \overline{1,n}$, their joint density is of the form (196),

$$f_{\mathbf{X}}(x_1, \dots, x_n) = (2\pi)^{-n/2} \exp\left\{-\frac{\sum\limits_{i=1}^{n} x_i^2}{2}\right\}$$

and (197) gives

$$f_{\mathbf{Y}}(y) = \frac{1}{2^{n/2}\,\Gamma(n/2)}\, y^{(n/2)-1} \exp\left\{-\frac{y}{2}\right\}, \tag{202}$$

the χ^2 *distribution* with n degrees of freedom.

Suggested Further Reading

SECTION 2. Butler and Morley [**143**], Duffin and Morley [**242**], [**243**], Hartwig [**388**], Hartwig and Shoaf [**394**], Mitra [**560**], Mitra and Prasad [**566**], Mitra and Puri [**567**], [**568**], Piziak, Odell, and Hahn [**643**], Taurian and Löwdin [**797**].

SECTION 3. Akdeniz [**5**], Albert [**9**], [**10**], [**11**], Albert and Sittler [**12**], Baksalary and Kala [**33**], Baksalary and Mathew [**34**], Bapat [**44**], [**43**], Bhimasankaram and SahaRay [**98**], Bjerhammar [**104**], Björck [**108**], Campbell and Meyer [**159**, pp. 104–115], Chipman [**186**], [**187**], [**189**], Chipman and Rao [**190**], Corradi [**206**], De Pierro and Wei [**220**], Drygas [**236**], [**237**], [**238**], [**239**], [**240**], Goldman and Zelen [**301**], Golub and Styan [**310**], Graybill [**316**], [**317**], [**318**], Greville [**325**], Groß [**344**], [**345**], Groß and Puntanen [**351**], Groß and Trenkler [**353**], [**354**], Haberman [**363**], Hall and Meyer [**364**], Harville [**397**], Hawkins and Bradu [**401**], Kala [**456**], Khatri [**474**], Kreijger and Neudecker [**484**], Kruskal [**485**], [**486**], Lewis and Odell [**514**], Lowerre [**526**], [**527**], Mäkeläinen [**530**], Marquardt [**536**], Meyer and Painter [**547**], Mitra [**559**], Morley [**579**], Mitra and Rao [**569**], [**570**], Price [**650**], Pringle and Rayner [**651**], [**652**], Puntanen and Scott [**654**], Puntanen and Styan [**655**], Puntanen, Styan, and Werner [**656**], Rao [**669**], [**670**], [**671**], [**672**], [**673**], [**674**], [**675**], [**676**], [**677**], Rao and Mitra [**678**], [**679**], Rayner and Pringle [**680**], [**681**], Rohde [**704**], [**706**], Rohde and Harvey [**707**], von Rosen [**842**], Sakallıoğlu and Akdeniz [**722**], Schönfeld and Werner [**729**], Searle [**737**], [**738**], [**739**], Seely [**740**], [**741**], Seely and Zyskind [**742**], Stahlecker and Trenkler [772], Tan [794], Trenkler [813], Wang [850], Werner [863], [865], Werner and Yapar [867], [866], and Zyskind [898], [899], [900], Zyskind and Martin [901].

OTHER APPLICATIONS IN STATISTICS. Banerjee [41], Banerjee and Federer [42], Chernoff [185], [305]), Good [314], Graybill and Marsaglia [319], J.A. John [448], P.W.M. John [449], Zacks [883].

APPLICATIONS TO THE KALMAN FILTER. Catlin [174], Duncan and Horn [245], Kalman [458], [459], [460], [461].

SECTION 4. Goldstein and Smith [303], Hoerl and Kennard [423], [425], Hoerl [426], Leamer [506], Marquardt [536], McDonald [539], Obenchain [618], Smith and Campbell [764], Sundberg [789], Trenkler [814], Vinod [836].

SECTION 5. Allgower, Böhmer, Hoy, and Janovský [13], Allgower and Georg [14], Altman [19], [20], Ben-Israel [59], [60], [66], Burmeister [139], Chen, Nashed and Qi [179], Deuflhard and Heindl [227], Fletcher [276], [278], Golub and Pereyra [307], Leach [505], Levin, Nediak, and Ben-Israel [510], Nashed and Chen [596], Rheinboldt [688], especially Theorem 3.5, Tapia [795], [796], Tsuji and Yang [821], Vaarmann [826], [827].

SECTION 6. APPLICATIONS TO SYSTEM THEORY. Balakrishnan [40], Barnett [51], Ho and Kalman [422], Kalman [458], [459], [460], [461], Kalman, Ho and Narendra [462], Kishi [476], Kuo and Mazda [490], Minamide and Nakamura [556], [557], Porter [645], [646], Porter and Williams [648], [647], Sontag [769], Tokarzewski [811], [812], Wahba and Nashed [844], and Zadeh and Desoer [884].

SECTION 7. Lamond and Puterman [493], Meyer [544], [545], [546], Meyer and Shoaf [548], Rising [692], Sonin and Thornton [767], Zarnowski [886].

SECTION 8. Campbell [150].

APPLICATIONS TO SINGULAR LINEAR DIFFERENTIAL EQUATIONS AND BOUNDARY VALUE PROBLEMS. Boyarintsev [128], [129], [130], [131], Boyarintsev and Korsukov [132], Campbell [147], [149], [152], [156], [157], Campbell, Meyer and Rose [160], Campbell and Rakowski [161], Campbell and Rose [162], [163], [164].

APPLICATIONS TO OPTIMAL CONTROL. Campbell [146], [148], [151], [154], [158].

Generalized Inverses of Linear Operators between Hilbert Spaces

1. Introduction

The observation that generalized inverses are like prose ("Good Heavens! For more than forty years I have been speaking prose without knowing it" – Molière, *Le Bourgeois Gentilhomme*) is nowhere truer than in the literature of linear operators. In fact, generalized inverses of integral and differential operators were studied by Fredholm, Hilbert, Schmidt, Bounitzky, Hurwitz, and others, before E.H. Moore introduced generalized inverses in an algebraic setting; see, e.g., the historic survey in Reid [685].

This chapter is a brief and biased introduction to generalized inverses of linear operators between Hilbert spaces, with special emphasis on the similarities to the finite-dimensional case. Thus the spectral theory of such operators is omitted.

Following the preliminaries in Section 2, generalized inverses are introduced in Section 3. Applications to integral and differential operators are sampled in Exs. 18–36. The minimization properties of generalized inverses are studied in Section 6. Integral and series representations of generalized inverses, and iterative methods for their computation, are given in Section 7.

This chapter requires familiarity with the basic concepts of linear functional analysis, in particular, the theory of linear operators in Hilbert space.

2. Hilbert Spaces and Operators: Preliminaries and Notation

In this section we have collected, for convenience, some preliminary results that can be found in the form stated here or in a more general form, in the standard texts on functional analysis; see, e.g., Taylor [800] and Yosida [882].

(A)   Our Hilbert spaces will be denoted by $\mathcal{H}, \mathcal{H}_1, \mathcal{H}_2$, etc. In each space, the *inner product* of two vectors $\mathbf{x}$ and $\mathbf{y}$ is denoted by $\langle \mathbf{x}, \mathbf{y} \rangle$ and the *norm* is denoted by $\| \ \|$. The *closure* of a subset L of $\mathcal{H}$ will be denoted by $\overline{L}$ and its *orthogonal complement* by $L^\perp$. $L^\perp$ is a closed subspace of $\mathcal{H}$ and

$$L^\perp = \overline{L}^\perp.$$

The *sum*, $L + M$, of two subsets $L, M \subset \mathcal{H}$ is

$$L + M = \{\mathbf{x} + \mathbf{y} : \mathbf{x} \in L, \ \mathbf{y} \in M\}.$$

If L, M are subspaces of $\mathcal{H}$ and $L \cap M = \{\mathbf{0}\}$, then $L + M$ is called the *direct sum* of L and M, and denoted by $L \oplus M$. If, in addition, $L \subset M^\perp$ we denote their sum by $L \overset{\perp}{\oplus} M$ and call it the *orthogonal direct sum* of L and M. Even if the subspaces L, M are closed, their sum $L + M$ need not be closed; see, e.g., Ex. 1. An orthogonal direct sum of two closed subspaces is closed. Conversely, if L, M are closed subspaces of $\mathcal{H}$ and $M \subset L$, then

$$L = M \overset{\perp}{\oplus} (L \cap M^\perp). \tag{1}$$

If (1) holds for two subspaces $M \subset L$, we say that L is *decomposable with respect to* M. See Exs. 5–6.

(B)   The (*Cartesian*) *product* of $\mathcal{H}_1, \mathcal{H}_2$ will be denoted by

$$\mathcal{H}_{1,2} = \mathcal{H}_1 \times \mathcal{H}_2 = \{\{\mathbf{x}, \mathbf{y}\} : \mathbf{x} \in \mathcal{H}_1, \mathbf{y} \in \mathcal{H}_2\}$$

where $\{\mathbf{x}, \mathbf{y}\}$ is an ordered pair. $\mathcal{H}_{1,2}$ is a Hilbert space with inner product

$$\langle \{\mathbf{x}_1, \mathbf{y}_1\}, \{\mathbf{x}_2, \mathbf{y}_2\} \rangle = \langle \mathbf{x}_1, \mathbf{y}_1 \rangle + \langle \mathbf{x}_2, \mathbf{y}_2 \rangle.$$

Let $J_i : \mathcal{H}_i \to \mathcal{H}_{1,2}$, $i = 1, 2$, be defined by

$$J_1 \mathbf{x} = \{\mathbf{x}, \mathbf{0}\}, \quad \text{for all } \mathbf{x} \in \mathcal{H}_1,$$

and

$$J_2 \mathbf{y} = \{\mathbf{0}, \mathbf{y}\}, \quad \text{for all } \mathbf{y} \in \mathcal{H}_2.$$

The transformations J_1 and J_2 are isometric isomorphisms, mapping $\mathcal{H}_1$ and $\mathcal{H}_2$ onto

$$\mathcal{H}_{1,0} = J_1 \mathcal{H}_1 = \mathcal{H}_1 \times \{\mathbf{0}\}$$

and

$$\mathcal{H}_{0,2} = J_2 \mathcal{H}_2 = \{\mathbf{0}\} \times \mathcal{H}_2,$$

respectively. Here $\{\mathbf{0}\}$ is an appropriate zero space.

(C)   Let $\mathcal{L}(\mathcal{H}_1, \mathcal{H}_2)$ denote the *class of linear operators from* $\mathcal{H}_1$ *to* $\mathcal{H}_2$. In what follows we will use *operator* to mean a linear operator. For any $T \in \mathcal{L}(\mathcal{H}_1, \mathcal{H}_2)$ we denote the *domain* of T by $D(T)$, the *range* of T by $R(T)$, the *null space* of T by $N(T)$, and the *carrier* of T by $C(T)$, where

$$C(T) = D(T) \cap N(T)^\perp. \tag{2}$$

The *graph*, $G(T)$, of a $T \in \mathcal{L}(\mathcal{H}_1, \mathcal{H}_2)$ is

$$G(T) = \{\{\mathbf{x}, T\mathbf{x}\} : \mathbf{x} \in D(T)\}.$$

Clearly, $G(T)$ is a subspace of $\mathcal{H}_{1,2}$ and $G(T) \cap \mathcal{H}_{0,2} = \{\mathbf{0}, \mathbf{0}\}$. Conversely, if G is a subspace of $\mathcal{H}_{1,2}$ and $G(T) \cap \mathcal{H}_{0,2} = \{\mathbf{0}, \mathbf{0}\}$, then G is the graph of a unique $T \in \mathcal{L}(\mathcal{H}_1, \mathcal{H}_2)$, defined for any point $\mathbf{x}$ in its domain

$$D(T) = J_1^{-1} P_{\mathcal{H}_{1,0}} G(T)$$

by

$$T\mathbf{x} = \mathbf{y},$$

where $\mathbf{y}$ is the unique vector in $\mathcal{H}_2$ such that $\{\mathbf{x}, \mathbf{y}\} \in G$ and $P_{\mathcal{H}_{1,0}}$ is the orthogonal projector: $\mathcal{H}_{1,2} \to \mathcal{A}_{1,0}$, see (L) below.

Similarly, for any $T \in \mathcal{L}(\mathcal{H}_2, \mathcal{H}_1)$, the *inverse graph* of T, $G^{-1}(T)$, is defined by

$$G^{-1}(T) = \{\{T\mathbf{y}, \mathbf{y}\} : \mathbf{y} \in D(T)\}.$$

A subspace G in $\mathcal{H}_{1,2}$ is an inverse graph of some $T \in \mathcal{L}(\mathcal{H}_2, \mathcal{H}_1)$ if and only if $G \cap \mathcal{H}_{1,0} = \{\mathbf{0}, \mathbf{0}\}$, in which case T is uniquely determined by G (von Neumann [**840**]).

(D)   An operator $T \in \mathcal{L}(\mathcal{H}_1, \mathcal{H}_2)$ is called *closed* if $G(T)$ is a closed subspace of $\mathcal{H}_{1,2}$. Equivalently, T is closed if

$$\mathbf{x}_n \in D(T), \ \mathbf{x}_n \to \mathbf{x}_0, \ T\mathbf{x}_n \to \mathbf{y}_0 \quad \Longrightarrow \quad \mathbf{x}_0 \in D(T) \text{ and } T\mathbf{x}_0 = \mathbf{y}_0,$$

where $\to$ denotes strong convergence. A closed operator has a closed null space. The subclass of *closed operators* in $\mathcal{L}(\mathcal{H}_1, \mathcal{H}_2)$ is denoted by $\mathcal{C}(\mathcal{H}_1, \mathcal{H}_2)$.

(E)   An operator $T \in \mathcal{L}(\mathcal{H}_1, \mathcal{H}_2)$ is called *bounded* if its norm $\|T\|$ is finite, where

$$\|T\| = \sup_{\mathbf{0} \neq \mathbf{x} \in \mathcal{H}_1} \frac{\|T\mathbf{x}\|}{\|\mathbf{x}\|}.$$

The subclass of *bounded operators* is denoted by $\mathcal{B}(\mathcal{H}_1, \mathcal{H}_2)$. If $T \in \mathcal{B}(\mathcal{H}_1, \mathcal{H}_2)$, then it may be assumed, without loss of generality, that $D(T)$ is closed or even that $D(T) = \mathcal{H}_1$. A bounded $T \in \mathcal{B}(\mathcal{H}_1, \mathcal{H}_2)$ is closed if and only if $D(T)$ is closed. Thus we may write $\mathcal{B}(\mathcal{H}_1, \mathcal{H}_2) \subset \mathcal{C}(\mathcal{H}_1, \mathcal{H}_2)$. Conversely, a closed $T \in \mathcal{C}(\mathcal{H}_1, \mathcal{H}_2)$ is bounded if $D(T) = \mathcal{H}_1$. This statement is the *closed graph theorem*.

(F)   Let $T_1, T_2 \in \mathcal{L}(\mathcal{H}_1, \mathcal{H}_2)$ with $D(T_1) \subset D(T_2)$. If $T_2 \mathbf{x} = T_1 \mathbf{x}$ for all $\mathbf{x} \in D(T_1)$, then T_2 is called an *extension* of T_1 and T_1 is called a *restriction* of T_2. These relations are denoted by

$$T_1 \subset T_2$$

or by

$$T_1 = (T_2)_{[D(T_1)]}.$$

Let $T \in \mathcal{L}(\mathcal{H}_1, \mathcal{H}_2)$ and let the restriction of T to $C(T)$ be denoted by T_0

$$T_0 = T_{[C(T)]}.$$

Then

$$G(T_0) = \{\{\mathbf{x}, T\mathbf{x}\} : \mathbf{x} \in C(T)\}$$

satisfies

$$G(T_0) \cap \mathcal{H}_{1,0} = \{\mathbf{0}, \mathbf{0}\}$$

and hence is the inverse graph of an operator $S \in \mathcal{L}(\mathcal{H}_2, \mathcal{H}_1)$ with

$$D(S) = R(T_0).$$

Clearly,

$$ST\mathbf{x} = \mathbf{x}, \quad \text{for all } \mathbf{x} \in C(T),$$

and

$$TS\mathbf{y} = \mathbf{y}, \quad \text{for all } \mathbf{y} \in R(T_0).$$

Thus, if T_0 is considered as an operator in $\mathcal{L}(\overline{C(T)}, \overline{R(T_0)})$, then T_0 is invertible in its domain. The inverse T_0^{-1} is closed if and only if T_0 is closed. For $T \in \mathcal{L}(\mathcal{H}_1, \mathcal{H}_2)$, both $C(T)$ and T_0 may be trivial; see, e.g., Exs. 2 and 4.

(G)   An operator $T \in \mathcal{L}(\mathcal{H}_1, \mathcal{H}_2)$ is called *dense* (or *densely defined*) if $\overline{D(T)} = \mathcal{H}_1$. Since any $T \in \mathcal{L}(\mathcal{H}_1, \mathcal{H}_2)$ can be viewed as an element of $T \in \mathcal{L}(\overline{D(T)}, \mathcal{H}_2)$, any operator can be assumed to be dense without loss of generality.

For any $T \in \mathcal{L}(\mathcal{H}_1, \mathcal{H}_2)$, the condition $\overline{D(T)} = \mathcal{H}_1$ is equivalent to

$$G(T)^\perp \cap \mathcal{H}_{1,0} = \{\mathbf{0}, \mathbf{0}\},$$

where

$$G(T)^\perp = \{\{\mathbf{y}, \mathbf{z}\} : \langle \mathbf{y}, \mathbf{x} \rangle + \langle \mathbf{z}, T\mathbf{x} \rangle = 0 \text{ for all } \mathbf{x} \in D(T)\} \subset \mathcal{H}_{1,2}.$$

Thus, for any dense $T \in \mathcal{L}(\mathcal{H}_1, \mathcal{H}_2)$, $G(T)^\perp$ is the inverse graph of a unique operator in $\mathcal{C}(\mathcal{H}_1, \mathcal{H}_2)$. This operator is $-T^*$ where T^*, the *adjoint* of T, satisfies

$$\langle T^*\mathbf{y}, \mathbf{x} \rangle = \langle \mathbf{y}, T\mathbf{x} \rangle, \quad \text{for all } \mathbf{x} \in D(T).$$

(H)   For any dense $T \in \mathcal{L}(\mathcal{H}_1, \mathcal{H}_2)$,

$$\overline{N(T)} = R(T^*)^\perp, \quad N(T^*) = R(T)^\perp. \tag{3}$$

In particular, $T\,[T^*]$ has a dense range if and only if $T^*\,[T]$ is one-to-one.

(I)   Let $T \in \mathcal{L}(\mathcal{H}_1, \mathcal{H}_2)$ be dense.
If both T and T^* have inverses, then $(T^{-1})^* = (T^*)^{-1}$.
T has a bounded inverse if and only if $R(T^*) = \mathcal{H}_1$.
T^* has a bounded inverse if $R(T) = \mathcal{H}_2$. The converse holds if T is closed.
T^* has a bounded inverse and $R(T^*) = \mathcal{H}_1$ if and only if T has a bounded inverse and $\overline{R(T)} = \mathcal{H}_1$ (Taylor [**800**], Goldberg [**300**]).

(J)   An operator $T \in \mathcal{L}(\mathcal{H}_1, \mathcal{H}_2)$ is called *closable* (or *preclosed*) if T has a closed extension. Equivalently, T is closable if

$$\overline{G(T)} \cap \mathcal{H}_{0,2} = \{\mathbf{0}, \mathbf{0}\},$$

in which case $\overline{G(T)}$ is the graph of an operator $\overline{T}$, called the *closure* of T. $\overline{T}$ is the minimal closed extension of T.

Since $G(T)^{\perp\perp} = \overline{G(T)}$ it follows that for a dense T, T^{**} is defined only if T is closable, in which case

$$T \subset T^{**} = \overline{T}$$

and

$$T = T^{**}$$

if and only if T is closed.

(K) A dense operator $T \in \mathcal{L}(\mathcal{H}, \mathcal{H})$ is called *symmetric* if

$$T \subset T^*$$

and *self-adjoint* if

$$T = T^*,$$

in which case it is called *nonnegative*, and denoted by $T \succcurlyeq O$, if

$$\langle T\mathbf{x}, \mathbf{x} \rangle \geq 0, \quad \text{for all } \mathbf{x} \in D(T).$$

If $T \in \mathcal{C}(\mathcal{H}_1, \mathcal{H}_2)$ is dense, then T^*T and TT^* are nonnegative, and $I + TT^*$ and $I + T^*T$ have bounded inverses (von Neumann [**838**]).

(L)   An operator $T \in \mathcal{B}(\mathcal{H}, \mathcal{H})$ is an *orthogonal projector* if

$$P = P^* = P^2,$$

in which case $R(P)$ is closed and

$$\mathcal{H} = R(P) \overset{\perp}{\oplus} N(P).$$

Conversely, if L is a closed subspace of $\mathcal{H}$, then there is a unique orthogonal projector P_L such that

$$L = R(P_L) \quad \text{and} \quad L^\perp = N(P_L).$$

(M)   An operator $T \in \mathcal{C}(\mathcal{H}_1, \mathcal{H}_2)$ is called *normally solvable* if $R(T)$ is closed which, by (3), is equivalent to the following condition: The equation

$$T\mathbf{x} = \mathbf{y}$$

is consistent if and only if $\mathbf{y}$ is orthogonal to any solution $\mathbf{u}$ of

$$T^*\mathbf{u} = \mathbf{0}.$$

This condition accounts for the name "normally solvable."
     For any $T \in \mathcal{C}(\mathcal{H}_1, \mathcal{H}_2)$, the following statements are equivalent:

(a) T is normally solvable.
(b) The restriction $T_0 = T_{[C(T)]}$ has a bounded inverse.
(c) The nonnegative number

$$\gamma(T) = \inf\left\{ \frac{\|T\mathbf{x}\|}{\|\mathbf{x}\|} : \mathbf{0} \neq \mathbf{x} \in C(T) \right\} \tag{4}$$

is positive (Hestenes [**416**, Theorem 3.3]).

Exercises

EX. 1. *A nonclosed sum of closed subspaces.* Let $T \in \mathcal{B}(\mathcal{H}_1, \mathcal{H}_2)$ and let

$$D = J_1 D(T) = \{\{\mathbf{x}, \mathbf{0}\} : \mathbf{x} \in D(T)\}.$$

Without loss of generality we assume that $D(T)$ is closed. Then D is closed. Also $G(T)$ is closed since T is bounded. But

$$G(T) + D$$

is nonclosed if $R(T)$ is nonclosed, since

$$\{\mathbf{x}, \mathbf{y}\} \in G(T) + D \quad \Longleftrightarrow \quad \mathbf{y} \in R(T) \quad \text{(Halmos [**366**, p. 26])}.$$

EX. 2. *Unbounded linear functionals.* Let T be an unbounded linear functional on $\mathcal{H}$. Then $N(T)$ is dense in $\mathcal{H}$ and, consequently, $N(T)^{\perp} = \{0\}$, $C(T) = \{0\}$.

An example of such a functional on $L^2[0, \infty]$ is

$$T\mathbf{x} = \int_0^{\infty} t\mathbf{x}(t)\, dt.$$

To show that $N(T)$ is dense, let $\mathbf{x}_0 \in L^2[0, \infty]$ with $T\mathbf{x}_0 = \alpha$. Then a sequence $\{\mathbf{x}_n\} \subset N(T)$ converging to $\mathbf{x}_0$ is

$$\mathbf{x}_n(t) = \begin{cases} \mathbf{x}_0(t), & \text{if } t < 1 \text{ or } t > n+1, \\ \mathbf{x}_0(t) - \alpha/nt, & \text{if } 1 \le t \le n+1. \end{cases}$$

Indeed,

$$\|\mathbf{x}_n - \mathbf{x}_0\|^2 = \int_1^{n+1} \frac{\alpha^2}{(nt)^2}\, dt = \frac{\alpha^2}{n(n+1)} \to 0.$$

EX. 3. Let D be a dense subspace of $\mathcal{H}$ and let F be a closed subspace such that $F^{\perp}$ is finite dimensional. Then

$$\overline{D \cap F} = F \quad \text{(Erdelyi and Ben-Israel [265, Lemma 5.1])}.$$

EX. 4. *An operator with trivial carrier.* Let D be any proper dense subspace of $\mathcal{H}$ and choose $\mathbf{x} \notin D$. Let $F = [\mathbf{x}]^{\perp}$, where $[\mathbf{x}]$ is the line generated by $\mathbf{x}$. Then $\overline{D \cap F} = F$, by Ex. 3. However, $D \not\subset F$, so we can choose a subspace $A \ne \{0\}$ in D such that

$$D = A \oplus (D \cap F).$$

Define $T \in \mathcal{L}(\mathcal{H}, \mathcal{H})$ by

$$D(T) = D$$

and

$$T(\mathbf{y} + \mathbf{z}) = \mathbf{y}, \quad \text{if } \mathbf{y} \in A, \mathbf{z} \in D \cap F.$$

Then

$$N(T) = D \cap F,$$
$$\overline{N(T)} = \overline{D \cap F} = F,$$
$$N(T)^{\perp} = F^{\perp} = [\mathbf{x}],$$
$$C(T) = D(T) \cap N(T)^{\perp} = \{0\}.$$

EX. 5. (Arghiriade [25]). Let L, M be subspaces of $\mathcal{H}$ and let $M \subset L$. Then

$$L = M \overset{\perp}{\oplus} (L \cap M^{\perp}) \tag{1}$$

if and only if

$$P_{\overline{M}}\mathbf{x} \in M, \quad \text{for all } \mathbf{x} \in L.$$

In particular, a space is decomposable with respect to any closed subspace.

EX. 6. Let L, M, N be subspaces of $\mathcal{H}$ such that

$$L = M \overset{\perp}{\oplus} N.$$

Then

$$M = L \cap N^{\perp}, \quad N = L \cap M^{\perp}.$$

Thus an orthogonal direct sum is decomposable with respect to each summand.

Ex. 7. *A bounded operator with nonclosed range.* Let ℓ^2 denote the Hilbert space of square summable sequences and let $T \in \mathcal{B}(\ell^2, \ell^2)$ be defined, for some $0 < k < 1$, by

$$T(\alpha_0, \alpha_1, \alpha_2, \ldots, \alpha_n, \ldots) = (\alpha_0, k\alpha_1, k^2\alpha_2, \ldots, k^n\alpha_n, \ldots).$$

Consider the sequence

$$\mathbf{x}_n = \left(1, \frac{1}{2k}, \frac{1}{3k^2}, \ldots, \frac{1}{nk^{n-1}}, 0, 0, \ldots\right),$$

and the vector

$$\mathbf{y} = \lim_{n \to \infty} T\mathbf{x}_n = \left(1, \frac{1}{2}, \frac{1}{3}, \ldots, \frac{1}{n}, \ldots\right).$$

Then,

$$\mathbf{y} \in \overline{R(T)}, \quad \mathbf{y} \notin R(T).$$

Ex. 8. *Linear integral operators.* Let $L^2 = L^2[a, b]$, the Lebesgue square integrable functions on the finite interval $[a, b]$. Let $K(s, t)$ be an L^2–*kernel* on $a \leq s, t, \leq b$, meaning that the Lebesgue integral

$$\int_a^b \int_a^b |K(s, t)|^2 \, ds \, dt$$

exists and is finite; see, e.g., Smithies [**765**, Section 1.6].

Consider the two operators $T_1, T_2 \in \mathcal{B}(L^2, L^2)$ defined by

$$(T_1\mathbf{x})(s) = \int_a^b K(s, t)\mathbf{x}(t) \, dt, \quad a \leq s \leq b,$$

$$(T_2\mathbf{x})(s) = \mathbf{x}(s) - \int_a^b K(s, t)\mathbf{x}(t) \, dt, \quad a \leq s \leq b,$$

called *Fredholm integral operators of the first kind* and *the second kind,* respectively. Then:

(a) $R(T_2)$ is closed.

(b) $R(T_1)$ is nonclosed unless it is finite dimensional.

More generally, if $T \in \mathcal{L}(\mathcal{H}_1, \mathcal{H}_2)$ is completely continuous, then $R(T)$ is nonclosed unless it is finite dimensional (Kammerer and Nashed [**465**, Prop. 2.5]).

Ex. 9. Let $T \in \mathcal{C}(\mathcal{H}_1, \mathcal{H}_2)$. Then T is normally solvable if and only if T^* is. Also, T is normally solvable if and only if TT^* or T^*T is.

3. Generalized Inverses of Linear Operators Between Hilbert Spaces

A natural definition of generalized inverses in $\mathcal{L}(\mathcal{H}_1, \mathcal{H}_2)$ is the following one due to Tseng [**817**].

DEFINITION 1. Let $T \in \mathcal{L}(\mathcal{H}_1, \mathcal{H}_2)$. Then an operator $T^g \in \mathcal{L}(\mathcal{H}_2, \mathcal{H}_1)$ is a *Tseng inverse* of T if

$$R(T) \subset D(T^g), \tag{5}$$

$$R(T^g) \subset D(T), \tag{6}$$

$$T^g T \mathbf{x} = P_{\overline{R(T^g)}} \mathbf{x}, \quad \text{for all } \mathbf{x} \in D(T), \tag{7}$$

$$TT^g \mathbf{y} = P_{\overline{R(T)}} \mathbf{y}, \quad \text{for all } \mathbf{y} \in D(T^g). \tag{8}$$

This definition is symmetric in T and T^g, thus T is a Tseng inverse of T^g.

An operator $T \in \mathcal{L}(\mathcal{H}_1, \mathcal{H}_2)$ may have a unique Tseng inverse, or infinitely many Tseng inverses or it may have none. We will show in Theorem 1 that T has a Tseng inverse if and only if its domain is decomposable with respect to its null space,

$$D(T) = N(T) \overset{\perp}{\oplus} (D(T) \cap N(T)^{\perp})$$
$$= N(T) \overset{\perp}{\oplus} C(T). \tag{9}$$

By Ex. 5, this condition is satisfied if $N(T)$ is closed. Thus it holds for all closed operators and, in particular, for bounded operators. If T has Tseng inverses, then it has a maximal Tseng inverse, some of whose properties are collected in Theorem 2. For bounded operators with closed range, the maximal Tseng inverse coincides with the Moore–Penrose inverse and will likewise be denoted by $T^\dagger$. See Theorem 3.

For operators T without Tseng inverses, the maximal Tseng inverse $T^\dagger$ can be "approximated" in several ways, with the objective of retaining as many of its useful properties as possible. One such approach, due to Erdélyi [264], is described in Definition 3 and Theorem 4.

Some properties of Tseng inverses, when they exist, are given in the following three lemmas, due to Arghiriade [25], which are needed later.

LEMMA 1. *If $T^g \in \mathcal{L}(\mathcal{H}_2, \mathcal{H}_1)$ is a Tseng inverse of $T \in \mathcal{L}(\mathcal{H}_1, \mathcal{H}_2)$, then $D(T)$ is decomposable with respect to $R(T^g)$.*

PROOF. Follows from Ex. 5 since, for any $\mathbf{x} \in D(T)$,

$$P_{\overline{R(T^g)}} \mathbf{x} = T^g T \mathbf{x}, \quad \text{by (7)}. \qquad \square$$

LEMMA 2. *If $T^g \in \mathcal{L}(\mathcal{H}_2, \mathcal{H}_1)$ is a Tseng inverse of $T \in \mathcal{L}(\mathcal{H}_1, \mathcal{H}_2)$, then T is a one-to-one mapping of $R(T^g)$ onto $R(T)$.*

PROOF. Let $\mathbf{y} \in R(T)$. Then

$$\mathbf{y} = P_{\overline{R(T)}} \mathbf{y} = T T^g \mathbf{y}, \quad \text{by (8)},$$

proving that $T(R(T^g)) = R(T)$.

Now we prove that T is one-to-one on $R(T^g)$. Let $\mathbf{x}_1, \mathbf{x}_2 \in R(T^g)$ satisfy

$$T\mathbf{x}_1 = T\mathbf{x}_2.$$

Then

$$\mathbf{x}_1 = P_{\overline{R(T^g)}} \mathbf{x}_1 = T^g T \mathbf{x}_1 = T^g T \mathbf{x}_2 = P_{\overline{R(T^g)}} \mathbf{x}_2 = \mathbf{x}_2. \qquad \square$$

LEMMA 3. *If $T^g \in \mathcal{L}(\mathcal{H}_2, \mathcal{H}_1)$ is a Tseng inverse of $T \in \mathcal{L}(\mathcal{H}_1, \mathcal{H}_2)$, then*

$$N(T) = D(T) \cap R(T^g)^{\perp} \tag{10}$$

and

$$C(T) = R(T^g). \tag{11}$$

PROOF. Let $\mathbf{x} \in D(T)$. Then, by Lemma 1,

$$\mathbf{x} = \mathbf{x}_1 + \mathbf{x}_2, \quad \mathbf{x}_1 \in R(T^g), \quad \mathbf{x}_2 \in D(T) \cap R(T^g)^\perp, \quad \mathbf{x}_1 \perp \mathbf{x}_2. \qquad (12)$$

Now

$$\mathbf{x}_1 = P_{\overline{R(T^g)}} \mathbf{x} = T^g T(\mathbf{x}_1 + \mathbf{x}_2) = T^g T \mathbf{x}_1$$

and, therefore,

$$T^g T \mathbf{x}_2 = \mathbf{0},$$

which, by Lemma 2 with T and T^g interchanged, implies that

$$T \mathbf{x}_2 = \mathbf{0}, \qquad (13)$$

hence

$$D(T) \cap R(T^g)^\perp \subset N(T).$$

Conversely, let $\mathbf{x} \in N(T)$ be decomposed as in (12). Then

$$0 = T\mathbf{x} = T(\mathbf{x}_1 + \mathbf{x}_2)$$
$$= T\mathbf{x}_1, \quad \text{by (13)},$$

which, by Lemma 2, implies that $\mathbf{x}_1 = \mathbf{0}$ and, therefore,

$$N(T) \subset D(T) \cap R(T^g)^\perp,$$

completing the proof of (10). Now

$$D(T) = R(T^g) \overset{\perp}{\oplus} (D(T) \cap R(T^g)^\perp), \quad \text{by Lemma 1},$$
$$= R(T^g) \overset{\perp}{\oplus} N(T),$$

which, by Ex. 6, implies that

$$R(T^g) = D(T) \cap N(T)^\perp,$$

proving (11). $\qquad \qquad \square$

The existence of Tseng inverses is settled in the following theorem announced, without proof, by Tseng [817]. Our proof follows that of Arghiriade [25].

THEOREM 1. *Let $T \in \mathcal{L}(\mathcal{H}_1, \mathcal{H}_2)$. Then T has a Tseng inverse if and only if*

$$D(T) = N(T) \overset{\perp}{\oplus} C(T), \qquad (9)$$

in which case, for any subspace $L \subset R(T)^\perp$, there is a Tseng inverse T_L^g of T, with

$$D(T_L^g) = R(T) \overset{\perp}{\oplus} L \qquad (14)$$

and

$$N(T_L^g) = L. \qquad (15)$$

PROOF. If T has a Tseng inverse, then (9) follows from Lemmas 1 and 3.
Conversely, suppose that (9) holds. Then

$$R(T) = T(D(T)) = T(C(T)) = R(T_0), \tag{16}$$

where $T_0 = T_{[C(T)]}$ is the restriction of T to $C(T)$. The inverse T_0^{-1} exists, by
Section 2(F), and satisfies

$$R(T_0^{-1}) = C(T)$$

and, by (16),

$$D(T_0^{-1}) = R(T).$$

For any subspace $L \subset R(T)^\perp$, consider the extension T_L^g of T_0^{-1} with

$$\text{domain} \quad D(T_L^g) = R(T) \overset{\perp}{\oplus} L \tag{14}$$

and

$$\text{null space} \quad N(T_L^g) = L. \tag{15}$$

From its definition, it follows that T_L^g satisfies

$$D(T_L^g) \supset R(T)$$

and

$$R(T_L^g) = R(T_0^{-1}) = C(T) \subset D(T). \tag{17}$$

For any $\mathbf{x} \in D(T)$,

$$\begin{aligned}
T_L^g T \mathbf{x} &= T_L^g T P_{\overline{C(T)}} \mathbf{x}, \quad \text{by (9),} \\
&= T_0^{-1} T_0 P_{\overline{C(T)}} \mathbf{x}, \quad \text{by Ex. 5,} \\
&= P_{\overline{R(T_L^g)}} \mathbf{x}, \quad \text{by (17).}
\end{aligned}$$

Finally, any $\mathbf{y} \in D(T_L^g)$ can be written, by (14), as

$$\mathbf{y} = \mathbf{y}_1 + \mathbf{y}_2, \quad \mathbf{y}_1 \in R(T), \quad \mathbf{y}_2 \in L, \quad \mathbf{y}_1 \perp \mathbf{y}_2,$$

and, therefore,

$$\begin{aligned}
T T_L^g \mathbf{y} &= T T_L^g \mathbf{y}_1, \quad \text{by (15),} \\
&= T_0 T_0^{-1} \mathbf{y}_1 = \mathbf{y}_1 \\
&= P_{\overline{R(T)}} \mathbf{y}.
\end{aligned}$$

Thus T_L^g is a Tseng inverse of T. □

The Tseng inverse T_L^g is uniquely determined by its domain (14) and
null space (15); see Ex. 10.

The maximal choice of the subspace L in (14) and (15) is $L = R(T)^\perp$.
For this choice we have the following:

DEFINITION 2. Let $T \in \mathcal{L}(\mathcal{H}_1, \mathcal{H}_2)$ satisfy (9). Then the *maximal
Tseng inverse* of T, denoted by $T^\dagger$, is the Tseng inverse of T with domain

$$D(T^\dagger) = R(T) \overset{\perp}{\oplus} R(T)^\perp \tag{18}$$

and null space

$$N(T^\dagger) = R(T)^\perp. \tag{19}$$

By Ex. 10, the Tseng inverse $T^\dagger$ so defined is unique. It is maximal in the sense that any other Tseng inverse of T is a restriction of $T^\dagger$.

Moreover, $T^\dagger$ is dense, by (18), and has a closed null space, by (19). Choosing L as a dense subspace of $R(T)^\perp$ shows that an operator T may have infinitely many dense Tseng inverses T_L^g. Also, T may have infinitely many Tseng inverses T_L^g with closed null space, each obtained by choosing L as a closed subspace of $R(T)^\perp$. However, $T^\dagger$ is the unique dense Tseng inverse with closed null space; see Ex. 11.

For closed operators, the maximal Tseng inverse can be alternatively defined, by means of the following construction, due to Hestenes [416]; see also Landesman [498]:

Let $T \in \mathcal{C}(\mathcal{H}_1, \mathcal{H}_2)$ be dense. Since $N(T)$ is closed, it follows, from Ex. 5, that

$$D(T) = N(T) \overset{\perp}{\oplus} C(T), \tag{9}$$

and, therefore,

$$G(T) = N \overset{\perp}{\oplus} C, \tag{20}$$

where, using the notation of Section 2(B), (C), and (F),

$$N = J_1 N(T) = G(T) \cap \mathcal{H}_{1,0}, \tag{21a}$$
$$C = \{\{\mathbf{x}, T\mathbf{x}\} : \mathbf{x} \in C(T)\}. \tag{21b}$$

Similarly, since T^* is closed, it follows, from Section 2(G), that

$$G(T)^\perp = N^* \overset{\perp}{\oplus} C^* \tag{22}$$

with

$$N^* = J_2 N(T^*) = G(T)^\perp \cap \mathcal{H}_{0,2}, \tag{23a}$$
$$C^* = \{\{-T^*\mathbf{y}, \mathbf{y}\} : \mathbf{y} \in C(T^*)\}. \tag{23b}$$

Now

$$\mathcal{H}_{1,2} = G(T) \overset{\perp}{\oplus} G(T)^\perp, \quad \text{since } T \text{ is closed,}$$
$$= (N \overset{\perp}{\oplus} C) \overset{\perp}{\oplus} (N^* \overset{\perp}{\oplus} C^*), \quad \text{by (20) and (22),}$$
$$= (C \overset{\perp}{\oplus} N^*) \overset{\perp}{\oplus} (C^* \overset{\perp}{\oplus} N)$$
$$= G^\dagger \overset{\perp}{\oplus} G^{\dagger *}, \tag{24}$$

where

$$G^\dagger = C \overset{\perp}{\oplus} N^*, \tag{25a}$$

$$G^{\dagger *} = C^* \overset{\perp}{\oplus} N. \tag{25b}$$

Since

$$G^\dagger \cap \mathcal{H}_{1,0} = \{\mathbf{0}, \mathbf{0}\}, \quad \text{by Section 2(F),}$$

it follows that $G^\dagger$ is the inverse graph of an operator $T^\dagger \in C(\mathcal{H}_2, \mathcal{H}_1)$, with domain

$$J_2^{-1} P_{\mathcal{H}_{0,2}} G^\dagger = T(C(T)) \overset{\perp}{\oplus} N(T^*)$$

$$= R(T) \overset{\perp}{\oplus} R(T)^\perp, \quad \text{by (16) and (3)},$$

and null space

$$J_2^{-1} M^* = N(T^*) = R(T)^\perp$$

and such that

$$T^\dagger T \mathbf{x} = P_{\overline{C(T)}} \mathbf{x}, \quad \text{for any } \mathbf{x} \in N(T) \overset{\perp}{\oplus} C(T),$$

and

$$TT^\dagger \mathbf{y} = P_{\overline{R(T)}} \mathbf{y}, \quad \text{for any } \mathbf{y} \in R(T) \overset{\perp}{\oplus} R(T)^\perp.$$

Thus $T^\dagger$ is the maximal Tseng inverse of Definition 2.

Similarly, $G^{\dagger *}$ is the graph of the operator $-T^{*\dagger} \in C(\mathcal{H}_1, \mathcal{H}_2)$, which is the maximal Tseng inverse of $-T^*$.

This elegant construction makes obvious the properties of the maximal Tseng inverse, collected in the following:

THEOREM 2 (Hestenes [**416**]). *Let $T \in C(\mathcal{H}_1, \mathcal{H}_2)$ be dense. Then:*

(a) $T^\dagger \in C(\mathcal{H}_2, \mathcal{H}_1)$;

(b) $D(T^\dagger) = R(T) \overset{\perp}{\oplus} N(T^*)$; $N(T^\dagger) = N(T^*)$;

(c) $R(T^\dagger) = C(T)$;

(d) $T^\dagger T \mathbf{x} = P_{\overline{R(T^\dagger)}} \mathbf{x}$ *for any* $\mathbf{x} \in D(T)$;

(e) $TT^\dagger \mathbf{y} = P_{\overline{R(T)}} \mathbf{y}$ *for any* $\mathbf{y} \in D(T^\dagger)$;

(f) $T^{\dagger\dagger} = T$;

(g) $T^{*\dagger} = T^{\dagger *}$;

(h) $N(T^{*\dagger}) = N(T)$;

(i) T^*T *and* $T^\dagger T^{*\dagger}$ *are nonnegative and*

$$(T^*T)^\dagger = T^\dagger T^{*\dagger}, \quad N(T^*T) = N(T); \quad and$$

(j) TT^* *and* $T^{*\dagger}T^\dagger$ *are nonnegative and*

$$(TT^*)^\dagger = T^{*\dagger}T^\dagger, \quad N(TT^*) = N(T^*). \qquad \square$$

For bounded operators with closed range, various characterizations of the maximal Tseng inverse are collected in the following:

THEOREM 3 (Petryshyn [**641**]). *If $T \in \mathcal{B}(\mathcal{H}_1, \mathcal{H}_2)$ and $R(T)$ is closed, then $T^\dagger$ is characterized as the unique solution X of the following equivalent systems:*

(a) $TXT = T$, $XTX = X$, $(TX)^* = TX$, $(XT)^* = XT$;

(b) $TX = P_{R(T)}$, $N(X^*) = N(T)$;

(c) $TX = P_{R(T)}$, $XT = P_{R(T^*)}$, $XTX = X$;

(d) $XTT^* = T^*$, $XX^*T^* = X$;

(e) $XT\mathbf{x} = \mathbf{x}$ *for all* $\mathbf{x} \in R(T^*)$;
      $X\mathbf{y} = \mathbf{0}$ *for all* $\mathbf{y} \in N(T^*)$;

(f) $XT = P_{R(T^*)}$, $N(X) = N(T^*)$; and

(g) $TX = P_{R(T)}$, $XT = P_{R(X)}$. $\square$

The notation $T^\dagger$ is justified by Theorem 3(a), which lists the four *Penrose equations* (1.1)–(1.4).

If $T \in \mathcal{L}(\mathcal{H}_1, \mathcal{H}_2)$ does not satisfy (9), then it has no Tseng inverse, by Theorem 1. In this case one can still approximate $T^\dagger$ by an operator that has some properties of $T^\dagger$, and reduces to it if $T^\dagger$ exists. Such an approach, due to Erdélyi [**264**], is described in the following:

DEFINITION 3. Let $T \in \mathcal{L}(\mathcal{H}_1, \mathcal{H}_2)$ and let T_r be the restriction of T defined by

$$D(T_r) = N(T) \overset{\perp}{\oplus} C(T), \quad N(T_r) = N(T). \tag{26}$$

The *Erdélyi inverse* of T is defined as $T_r^\dagger$, which exists since T_r satisfies (9).

The inverse graph of $T_r^\dagger$ is

$$G^{-1}(T_r) = \{\{\mathbf{x}, T\mathbf{x} + \mathbf{z}\} : \mathbf{x} \in C(T), \ \mathbf{z} \in (T(C(T)))^\perp\}, \tag{27}$$

from which the following properties of $T_r^\dagger$ can be easily deduced:

THEOREM 4 (Erdélyi [**264**]). *Let* $T \in \mathcal{L}(\mathcal{H}_1, \mathcal{H}_2)$ *and let its restriction* T_r *be defined by* (26). *Then:*

(a) $T_r^\dagger = T^\dagger$ *if* $T^\dagger$ *exists;*

(b) $D(T_r^\dagger) = T(C(T)) \overset{\perp}{\oplus} T(C(T))^\perp$ *and, in general,* $R(T) \not\subset D(T_r^\dagger)$;

(c) $R(T_r^\dagger) = C(T)$, $\overline{R(T_r^\dagger)} = N(T)^\perp$;

(d) $T_r^\dagger T\mathbf{x} = P_{\overline{R(T_r^\dagger)}}\mathbf{x}$ *for all* $\mathbf{x} \in D(T_r)$;

(e) $TT_r^\dagger \mathbf{y} = P_{\overline{R(T)}}\mathbf{y}$ *for all* $\mathbf{y} \in D(T_r^\dagger)$;

(f) $D((T_r^\dagger)_r^\dagger) = \overline{N(T)} \overset{\perp}{\oplus} C(T)$;

(g) $R((T_r^\dagger)_r^\dagger) = \overline{T(C(T))}$;

(h) $N((T_r^\dagger)_r^\dagger) = \overline{N(T)}$;

(i) $T \subset (T_r^\dagger)_r^\dagger$ *if* (9) *holds;*

(j) $T = (T_r^\dagger)_r^\dagger$ *if and only if* $N(T)$ *is closed; and*

(k) $T_r^{\dagger*} \subset (T^*)_r^\dagger$ *if* T *is dense and closable.* $\square$

See also Ex. 15.

Exercises

Ex. 10. Let $T \in \mathcal{L}(\mathcal{H}_1, \mathcal{H}_2)$ have Tseng inverses and let L be a subspace of $R(T)^\perp$. Then the conditions

$$D(T_L^g) = R(T) \overset{\perp}{\oplus} L, \tag{14}$$

$$N(T_L^g) = L, \tag{15}$$

determine a unique Tseng inverse, which is thus equal to T_L^g as constructed in the proof of Theorem 1.

PROOF. Let T^g be a Tseng inverse of T satisfying (14) and (15) and let $\mathbf{y} \in D(T^g)$ be written as

$$\mathbf{y} = \mathbf{y}_1 + \mathbf{y}_2, \quad \mathbf{y}_1 \in R(T), \quad \mathbf{y}_2 \in L.$$

Then

$$T^g\mathbf{y} = T^g\mathbf{y}_1, \quad \text{by (15)},$$
$$= T^g T\mathbf{x}_1, \quad \text{for some } \mathbf{x}_1 \in D(T),$$
$$= P_{\overline{R(T^g)}}\mathbf{x}_1, \quad \text{by (7)},$$
$$= P_{\overline{C(T)}}\mathbf{x}_1, \quad \text{by (11)}.$$

We claim that this determines T^g uniquely. For, suppose there is an $\mathbf{x}_2 \in D(T)$ with $\mathbf{y}_1 = T\mathbf{x}_2$. Then, as above,

$$T^g\mathbf{y} = P_{\overline{C(T)}}\mathbf{x}_2$$

and, therefore,

$$P_{\overline{C(T)}}\mathbf{x}_1 - P_{\overline{C(T)}}\mathbf{x}_2 = P_{\overline{C(T)}}(\mathbf{x}_1 - \mathbf{x}_2)$$
$$= \mathbf{0}, \quad \text{since } \mathbf{x}_1 - \mathbf{x}_2 \in N(T). \qquad \square$$

EX. 11. Let $T \in \mathcal{L}(\mathcal{H}_1, \mathcal{H}_2)$ have Tseng inverses. Then $T^\dagger$ is the unique dense Tseng inverse with closed null space.

PROOF. Let T^g be any dense Tseng inverse with closed null space. Then

$$D(T^g) = N(T^g) \overset{\perp}{\oplus} C(T^g), \quad \text{by Theorem 1},$$
$$= N(T^g) \overset{\perp}{\oplus} R(T), \quad \text{by (11)},$$

which, together with the assumptions $\overline{D(T^g)} = \mathcal{H}_2$ and $N(T^g) = \overline{N(T^g)}$, implies that

$$N(T^g) = R(T)^\perp.$$

Thus, T^g has the same domain and null space as $T^\dagger$ and, therefore, $T^g = T^\dagger$, by Ex. 10. $\qquad \square$

EX. 12. (Kurepa [491]). Let $T \in \mathcal{B}(\mathcal{H}_1, \mathcal{H}_2)$ have a closed range $R(T)$ and let $T_1 \in \mathcal{B}(\mathcal{H}_1, R(T))$ be defined by

$$T_1\mathbf{x} = T\mathbf{x}, \quad \text{for all } \mathbf{x} \in \mathcal{H}_1.$$

Then:

    (a) T_1^* is the restriction of T^* to $R(T)$.
    (b)   The operator $T_1 T_1^* \in \mathcal{B}(R(T), R(T))$ is invertible.
    (c) $T^\dagger = P_{R(T^*)} T_1^* (T_1 T_1^*)^{-1} P_{R(T)}$.

EX. 13. (Landesman [498]). Let $T \in \mathcal{C}(\mathcal{H}_1, \mathcal{H}_2)$. Then $R(T)$ is closed if and only if $T^\dagger$ is bounded.

PROOF. Follows from Section 2(M). $\qquad \square$

EX. 14. (Desoer and Whalen [226]). Let $T \in \mathcal{B}(\mathcal{H}_1, \mathcal{H}_2)$ have closed range. Then

$$T^\dagger = (T^*T)^\dagger T^* = T^*(TT^*)^\dagger.$$

EX. 15. For arbitrary $T \in \mathcal{L}(\mathcal{H}_1, \mathcal{H}_2)$ consider its extension $\widetilde{T}$ with

$$D(\widetilde{T}) = D(T) + \overline{N(T)}, \quad N(\widetilde{T}) = \overline{N(T)}, \quad \widetilde{T} = T \quad \text{on } D(T), \qquad (28)$$

which coincides with T if $N(T)$ is closed. Since $D(\widetilde{T})$ is decomposable with respect to $N(\widetilde{T})$, it might seem that $\widetilde{T}$ can be used to obtain $\widetilde{T}^\dagger$, a substitute for (possibly nonexisting) $T^\dagger$.

Show that $\widetilde{T}$ is not well defined by (28) if

$$D(T) \cap \overline{N(T)} \neq N(T) \quad \text{and} \quad N(\widetilde{T}) \neq D(\widetilde{T}), \tag{29}$$

which is the only case of interest since otherwise $D(T)$ is decomposable with respect to $N(T)$ or $\widetilde{T}$ is identically O in its domain.

PROOF. By (29) there exist $\mathbf{x}_0$ and $\mathbf{y}$ such that

$$\mathbf{x}_0 \in D(T) \cap \overline{N(T)}, \quad \mathbf{x}_0 \notin N(T),$$

and

$$\mathbf{y} \in D(T), \quad \mathbf{y} \notin \overline{N(T)}.$$

Then

$$\widetilde{T}(\mathbf{x}_0 + \mathbf{y}) = \widetilde{T}\mathbf{y}, \quad \text{since } \mathbf{x}_0 \in N(\widetilde{T}),$$

and, on the other hand,

$$\widetilde{T}(\mathbf{x}_0 + \mathbf{y}) = T(\mathbf{x}_0 + \mathbf{y}), \quad \text{since } \mathbf{x}_0, \mathbf{y} \in D(T),$$
$$\neq T\mathbf{y}, \quad \text{since } \mathbf{x}_0 \notin N(T). \qquad \square$$

Ex. 16. (Petryshyn [**641**, Lemma 2]). Let $T \in \mathcal{B}(\mathcal{H}_1, \mathcal{H}_2)$ have closed range. Then

$$\|T^\dagger\| = \frac{1}{\gamma(T)},$$

where $\gamma(T)$ is defined in (4).

Ex. 17. (Holmes [**427**, p. 223]). Let $F \in \mathcal{B}(\mathcal{H}_3, \mathcal{H}_2)$ and $G \in \mathcal{B}(\mathcal{H}_1, \mathcal{H}_3)$ with $R(G) = \mathcal{H}_3 = R(F^*)$ and define $A \in \mathcal{B}(\mathcal{H}_1, \mathcal{H}_2)$ by $A = FG$. Then

$$A^\dagger = G^*(GG^*)^{-1}(F^*F)^{-1}F^*$$
$$= G^\dagger F^\dagger.$$

Compare with Theorem 1.5 and Ex. 1.17.

4. Generalized Inverses of Linear Integral Operators

Consider the *Fredholm integral equation of the second kind*

$$x(s) - \lambda \int_a^b K(s,t) x(t) \, dt = y(s), \quad a \leq s \leq b, \tag{30}$$

written for short as

$$(I - \lambda K) \mathbf{x} = \mathbf{y},$$

where all functions are complex, $[a, b]$ is a bounded interval, λ is a complex scalar, and $K(s, t)$ is a L^2-*kernel* on $[a, b] \times [a, b]$; see Ex. 8. Writing L^2 for $L^2[a, b]$, we need the following facts from the Fredholm theory of integral equations; see, e.g., Smithies [**765**]. For any λ, K as above:

(a) $(I - \lambda K) \in \mathcal{B}(L^2, L^2)$.

(b) $(I - \lambda K)^* = I - \overline{\lambda} K^*$, where $K^*(s,t) = \overline{K(t,s)}$.

(c) The null spaces $N(I - \lambda K)$ and $N(I - \overline{\lambda} K^*)$ have equal finite dimensions,

$$\dim N(I - \lambda K) = \dim N(I - \overline{\lambda} K^*) = n(\lambda), \quad \text{say.} \tag{31}$$

(d)   A scalar λ is called a *regular value* of K if $n(\lambda) = 0$, in which case the operator $I - \lambda K$ has an inverse $(I - \lambda K)^{-1} \in \mathcal{B}(L^2, L^2)$ written as

$$(I - \lambda K)^{-1} = I + \lambda R, \tag{32}$$

where $R = R(s, t; \lambda)$ is an L^2-kernel called the *resolvent* of K.

(e)   A scalar λ is called an *eigenvalue* of K if $n(\lambda) > 0$, in which case any nonzero $\mathbf{x} \in N(I - \lambda K)$ is called an *eigenfunction* of K corresponding to λ.

For any λ and, in particular, for any eigenvalue λ, both range spaces $R(I - \lambda K)$ and $R(I - \overline{\lambda} K^*)$ are closed and, by (3),

$$R(I - \lambda K) = N(I - \overline{\lambda} K^*)^\perp, \quad R(I - \overline{\lambda} K^*) = N(I - \lambda K)^\perp. \tag{33}$$

Thus, if λ is a regular value of K, then (30) has, for any $\mathbf{y} \in L^2$, a unique solution given by

$$\mathbf{x} = (I + \lambda R)\,\mathbf{y},$$

that is,

$$x(s) = y(s) + \lambda \int_a^b R(s, t, \lambda) y(t)\, dt, \quad a \le s \le b. \tag{34}$$

If λ is an eigenvalue of K, then (30) is consistent if and only if $\mathbf{y}$ is orthogonal to every $\mathbf{u} \in N(I - \overline{\lambda} K^*)$, in which case the general solution of (30) is

$$\mathbf{x} = \mathbf{x}_0 + \sum_{i=1}^{n(\lambda)} c_i \mathbf{x}_i, \quad c_i \text{ arbitrary scalars}, \tag{35}$$

where $\mathbf{x}_0$ is a particular solution of (30) and $\{\mathbf{x}_1, \ldots, \mathbf{x}_{n(\lambda)}\}$ is a basis of $N(I - \lambda K)$.

Exercises

Ex. 18. *Pseudoresolvents.* Let λ be an eigenvalue of K. Following Hurwitz [435], an L^2-kernel $R = R(s, t, \lambda)$ is called a *pseudoresolvent* of K if for any $\mathbf{y} \in R(I - \lambda K)$, the function

$$x(s) = y(s) + \lambda \int_a^b R(s, t, \lambda)\, y(t)\, dt \tag{34}$$

is a solution of (30).

A pseudoresolvent was constructed by Hurwitz as follows:

Let λ_0 be an eigenvalue of K and let $\{\mathbf{x}_1, \ldots, \mathbf{x}_n\}$ and $\{\mathbf{u}_1, \ldots, \mathbf{u}_n\}$ be o.n. bases of $N(I - \lambda_0 K)$ and $N(I - \overline{\lambda_0} K^*)$, respectively. Then λ_0 is a regular value of the kernel

$$K_0(s, t) = K(s, t) - \frac{1}{\lambda_0} \sum_{i=1}^n u_i(s)\, \overline{x_i(t)}, \tag{36}$$

written for short as

$$K_0 = K - \frac{1}{\lambda_0} \sum_{i=1}^n \mathbf{u}_i \mathbf{x}_i^*$$

and the resolvent R_0 of K_0 is a pseudoresolvent of K satisfying

$$(I + \lambda_0 R_0)(I - \lambda_0 K)\mathbf{x} = \mathbf{x}, \quad \text{for all } \mathbf{x} \in R(I - \overline{\lambda_0}K^*),$$
$$(I - \lambda_0 K)(I + \lambda_0 R_0)\mathbf{y} = \mathbf{y}, \quad \text{for all } \mathbf{y} \in R(I - \lambda_0 K),$$
$$(I + \lambda_0 R_0)\mathbf{u}_i = \mathbf{x}_i, \quad i = 1, \dots, n.$$

PROOF. Follows from the matrix case, Ex. 2.53.   □

EX. 19. (Hurwitz [**435**]). A comparison with Theorem 2.2 shows that $I + \lambda R$ is a {1}-inverse of $I - \lambda K$, if R is a pseudoresolvent of K. As with {1}-inverses, the pseudoresolvent is nonunique. Indeed, for $R_0, \mathbf{u}_i, \mathbf{x}_i$ as above, the kernel

$$R_0 + \sum_{i,j=1}^{n} c_{ij}\mathbf{x}_i\mathbf{u}_j^* \tag{37}$$

is a pseudoresolvent of K for any choice of scalars c_{ij}.

The pseudoresolvent constructed by Fredholm [**290**], who called the resulting operator $I + \lambda R$ a *pseudoinverse* of $I - \lambda K$, is the first explicit application, known to us, of a generalized inverse.

The class of all pseudoresolvents of a given kernel K is characterized as follows:

Let K be an L^2-kernel, let λ_0 be an eigenvalue of K, and let $\{\mathbf{x}_1, \dots, \mathbf{x}_n\}$ and $\{\mathbf{u}_1, \dots, \mathbf{u}_n\}$ be o.n. bases of $N(I - \lambda_0 K)$ and $N(I - \overline{\lambda_0}K^*)$, respectively. An L^2-kernel R is a pseudoresolvent of K if and only if

$$R = K + \lambda_0 KR - \frac{1}{\lambda_0}\sum_{i=1}^{n} \boldsymbol{\beta}_i \mathbf{u}_i^*, \tag{38a}$$

$$R = K + \lambda_0 RK - \frac{1}{\lambda_0}\sum_{i=1}^{n} \mathbf{x}_i \boldsymbol{\alpha}_i^*, \tag{38b}$$

where $\boldsymbol{\alpha}_i, \boldsymbol{\beta}_i \in L^2$ satisfy

$$\langle \boldsymbol{\alpha}_i, \mathbf{x}_j \rangle = \delta_{ij}, \quad \langle \boldsymbol{\beta}_i, \mathbf{u}_j \rangle = \delta_{ij}, \quad i, j = 1, \dots, n. \tag{39}$$

Here KR stands for the kernel $KR(s,t) = \int_a^b K(s,u)R(u,t)\,du$, etc.

If λ is a regular value of K, then (38) reduces to

$$R = K + \lambda KR, \quad R = K + \lambda RK, \tag{40}$$

which uniquely determines the resolvent $R(s,t,\lambda)$.

EX. 20. Let $K, \lambda_0, \mathbf{x}_i, \mathbf{u}_i$, and R_0 be as above. Then the maximal Tseng inverse of $I - \lambda_0 K$ is

$$(I - \lambda_0 K)^\dagger = I + \lambda_0 R_0 - \sum_{i=1}^{n} \mathbf{x}_i \mathbf{u}_i^*, \tag{41}$$

corresponding to the pseudoresolvent

$$R = R_0 - \frac{1}{\lambda_0}\sum_{i=1}^{n} \mathbf{x}_i \mathbf{u}_i^*. \tag{42}$$

EX. 21. Let $K(s,t) = u(s)\overline{v(t)}$, where

$$\int_a^b u(s)\overline{v(s)}\,ds = 0.$$

Then every scalar λ is a regular value of K.

Ex. 22. *Degenerate kernels.* A kernel $K(s,t)$ is called *degenerate* if it is a finite sum of products of L^2 functions, as follows:

$$K(s,t) = \sum_{i=1}^{m} f_i(s)\,\overline{g_i(t)}. \tag{43}$$

Degenerate kernels are convenient because they reduce the integral equation (30) to a finite system of linear equations. Also, any L^2-kernel can be approximated, arbitrarily close, by a degenerate kernel; see, e.g., Smithies [**765**, p. 40], and Halmos [**366**, Problem 137].

Let $K(s,t)$ be given by (43). Then:

(a) The scalar λ is an eigenvalue of (43) if and only if $1/\lambda$ is an eigenvalue of the $m \times m$ matrix

$$B = [b_{ij}], \quad \text{where } b_{ij} = \int_a^b f_j(s)\,\overline{g_i(s)}\,ds.$$

(b) Any eigenfunction of K $[K^*]$ corresponding to an eigenvalue λ $[\overline{\lambda}]$ is a linear combination of the m functions $f_1, \ldots, f_m$ $[g_1, \ldots, g_m]$.

(c) If λ is a regular value of (43), then the resolvent at λ is

$$R(s,t,;\lambda) = \frac{\det\begin{bmatrix} 0 & \vdots & f_1(s) & \cdots & f_m(s) \\ \cdots & \cdots & \cdots & \cdots & \cdots \\ -g_1(t) & \vdots & & & \\ \vdots & \vdots & & I - \lambda B & \\ -g_m(t) & \vdots & & & \end{bmatrix}}{\det(I - \lambda B)}.$$

See also Kantorovich and Krylov [**468**, Chapter II].

Ex. 23. Consider the equation

$$x(s) - \lambda \int_{-1}^{1} (1 + 3st)x(t)\,dt = y(s) \tag{44}$$

with $K(s,t) = 1 + 3st$. The resolvent is

$$R(s,t;\lambda) = \frac{1 + 3st}{1 - 2\lambda}.$$

K has a single eigenvalue $\lambda = \frac{1}{2}$ and an o.n. basis of $N(I - \frac{1}{2}K)$ is

$$\left\{ x_1(s) = \frac{1}{\sqrt{2}}, \ x_2(s) = \frac{\sqrt{3}}{\sqrt{2}}\,s \right\}$$

which, by symmetry, is also an o.n. basis of $N(I - \frac{1}{2}K^*)$. From (36) we get

$$K_0(s,t) = K(s,t) - \frac{1}{\lambda_0} \sum u_i(s)\overline{x_i(t)}$$

$$= (1 + 3st) - 2\left(\frac{1}{\sqrt{2}}\frac{1}{\sqrt{2}} + \frac{\sqrt{3}}{\sqrt{2}}s\frac{\sqrt{3}}{\sqrt{2}}t \right)$$

$$= 0,$$

and the resolvent of $K_0(s,t)$ is, therefore,

$$R_0(s,t;\lambda) = 0.$$

If $\lambda \neq \frac{1}{2}$, then, for each $y \in L^2[-1,1]$, equation (44) has a unique solution

$$x(s) = y(s) + \lambda \int_{-1}^{1} \frac{1 + 3st}{1 - 2\lambda} y(t)\, dt.$$

If $\lambda = \frac{1}{2}$, then (44) is consistent if and only if

$$\int_{-1}^{1} y(t)\, dt = 0, \qquad \int_{-1}^{1} ty(t)\, dt = 0,$$

in which case the general solution is

$$x(s) = y(s) + c_1 + c_2 s, \qquad c_1, c_2 \text{ arbitrary.}$$

Ex. 24. Let

$$K(s,t) = 1 + s + 3st, \qquad -1 \leq s, t \leq 1.$$

Then $\lambda = \frac{1}{2}$ is the only eigenvalue and

$$\dim N(I - \tfrac{1}{2}K) = 1.$$

An o.n. basis of $N(I - \frac{1}{2}K)$ is the single vector

$$x_1(s) = \frac{\sqrt{3}}{\sqrt{2}}\, s, \qquad -1 \leq s \leq 1.$$

An o.n. basis of $N(I - \frac{1}{2}K^*)$ is

$$u_1(s) = \frac{1}{\sqrt{2}}, \qquad -1 \leq s \leq 1.$$

The Hurwitz kernel (36) is

$$K_0(s,t) = (1 + s + 3st) - 2\left(\frac{1}{\sqrt{2}} \frac{\sqrt{3}}{\sqrt{2}} t\right)$$

$$= 1 + s - \sqrt{3}t + 3st, \qquad -1 \leq s, t \leq 1.$$

Compute the resolvent R_0 of K_0, which is a pseudoresolvent of K.

5. Generalized Inverses of Linear Differential Operators

This section deals with generalized inverses of closed dense operators $L \in \mathcal{C}(\mathcal{S}_1, \mathcal{S}_2)$ with $\overline{D(L)} = \mathcal{S}_1$, where:

(i) $\mathcal{S}_1, \mathcal{S}_2$ are spaces of (scalar or vector) functions which are either the Hilbert space $L^2[a,b]$ or the *space of continuous functions* $C[a,b]$, where $[a,b]$ is a given finite real interval. Since $C[a,b]$ is a dense subspace of $L^2[a,b]$, a closed dense linear operator mapping $C[a,b]$ into $\mathcal{S}_2$ may be considered as a dense operator in $\mathcal{C}(L^2[a,b], \mathcal{S}_2)$.

(ii) L is defined for all $\mathbf{x}$ in its domain $D(L)$ by

$$L\mathbf{x} = \ell\mathbf{x}, \qquad (45)$$

where ℓ is a *differential expression*, for example, in the vector case,

$$\ell\mathbf{x} = A_1(t)\frac{d}{dt}\mathbf{x} + A_0(t)\mathbf{x}, \qquad (46)$$

where $A_0(t), A_1(t)$ are $n \times n$ matrix coefficients, with suitable regularity conditions; see, e.g., Ex. 30 below.

(iii) The domain of L consists of those functions in S_1 for which ℓ makes sense and $\ell x \in S_2$, and which satisfy certain conditions, such as *initial* or *boundary conditions*.

If a differential operator L is invertible and there is a kernel (function, or matrix in the vector case)

$$G(s,t), \quad a \le s, t \le b,$$

such that, for all $\mathbf{y} \in R(L)$,

$$(L^{-1}\mathbf{y})(s) = \int_a^b G(s,t)\mathbf{y}(t)\,dt, \quad a \le s \le b,$$

then $G(s,t)$ is called the *Green function* (or *matrix*) of L. In this case, for any $\mathbf{y} \in R(L)$, the unique solution of

$$L\mathbf{x} = \mathbf{y} \tag{47}$$

is given by

$$\mathbf{x}(s) \doteq \int_a^b G(s,t)\mathbf{y}(t)\,dt, \quad a \le s \le b. \tag{48}$$

If L is not invertible, but there is a kernel $G(s,t)$ such that, for any $\mathbf{y} \in R(L)$, a particular solution of (47) is given by (48), then $G(s,t)$ is called a *generalized Green function* (or *matrix*) of L. A generalized Green function of L is therefore a kernel of an integral operator which is a generalized inverse of L.

Generalized Green functions were introduced by Hilbert [**418**] in 1904 and, consequently studied by Myller, Westfall, and Bounitzky [**124**], Elliott [**251**], [**252**], and Reid [**682**]; see, e.g., the historical survey in [**685**].

Exercises

Ex. 25. *Derivatives* (Hestenes [**416**, Example 1]). Let:

S = the real space $L^2[0,\pi]$ of real valued functions;
S^1 = the absolutely continuous functions $\mathbf{x}(t)$, $0 \le t \le \pi$, whose derivatives $\mathbf{x}'$ are in S; and
$S^2 = \{\mathbf{x} \in S^1 : \mathbf{x}' \in S^1\}$;

and let L be the differential operator d/dt with

$$D(L) = \{\mathbf{x} \in S^1 : \mathbf{x}(0) = \mathbf{x}(\pi) = 0\}.$$

Then:

(a) $L \in \mathcal{C}(S,S)$, $\overline{D(L)} = S$, $C(L) = D(L)$,

$$R(L) = \left\{ \mathbf{y} \in S : \int_0^\pi \mathbf{y}(t)\,dt = 0 \right\} = \overline{R(L)}.$$

(b)   The adjoint L^* is the operator $-d/dt$ with

$$D(L^*) = S^1, \quad C(L^*) = S^1 \cap R(L), \quad R(L^*) = S.$$

(c) $L^*L = -d^2/dt^2$ with $D(L^*L) = \{\mathbf{x} \in S^2 : \mathbf{x}(0) = \mathbf{x}(\pi) = 0\}$ and $R(L^*L) = S$.

(d) $LL^* = -d^2/dt^2$ with $D(LL^*) = \{\mathbf{x} \in S^2 : \mathbf{x}'(0) = \mathbf{x}'(\pi) = 0\}$ and $R(LL^*) = R(L)$.

(e) $L^\dagger$ is defined on $D(L^\dagger) = S$ by

$$(L^\dagger \mathbf{y})(t) = \int_0^t \mathbf{y}(s)\, ds - \frac{t}{\pi} \int_0^\pi \mathbf{y}(s)\, ds, \quad 0 \le t \le \pi.$$

Ex. 26. For L of Ex. 25, determine which of the following equations hold and interpret your results:

(a) $L^{\dagger *} = L^{*\dagger}$;

(b) $L^\dagger = (L^* L)^\dagger L^* = L^* (L L^*)^\dagger$; and

(c) $L^{\dagger\dagger} = L$.

Ex. 27. *Gradients* (Landesman [**498**, Section 5]). Let:

$S = $ the real space $L^2([0,\pi] \times [0,\pi])$ of real valued functions $\mathbf{x}(t_1, t_2)$, $0 \le t_1, t_2 \le \pi$; and

$S^1 = $ the subclass of S with the properties:

(i) $\mathbf{x}(t_1, t_2)$ is absolutely continuous in $t_1[t_2]$ for almost all $t_2[t_1]$, $0 \le t_1, t_2 \le \pi$; and

(ii)   the partial derivatives $\partial \mathbf{x} / \partial t_1$, $\partial \mathbf{x} / \partial t_2$ which exist almost everywhere are in S;

and let L be the gradient operator

$$\ell \mathbf{x} = \begin{bmatrix} \frac{\partial \mathbf{x}}{\partial t_1} \\ \frac{\partial \mathbf{x}}{\partial t_2} \end{bmatrix}$$

with domain

$$D(L) = \left\{ \mathbf{x} \in S^1 : \begin{array}{l} \mathbf{x}(0, t_2) = \mathbf{x}(\pi, t_2) = 0 \text{ for almost all } t_2, \\ \mathbf{x}(t_1, 0) = \mathbf{x}(t_1, \pi) = 0 \text{ for almost all } t_1, \end{array} \quad 0 \le t_1, t_2 \le \pi \right\}.$$

Then:

(a) $L \in C(S, S \times S)$, $\overline{D(L)} = S$.

(b)   The adjoint L^* is the negative of the divergence operator

$$\ell^* \mathbf{y} = \ell^* \begin{bmatrix} y_1 \\ y_2 \end{bmatrix} = -\frac{\partial y_1}{\partial t_1} - \frac{\partial y_2}{\partial t_2}$$

with

$$D(L^*) = \{ \mathbf{y} \in S \times S : \mathbf{y} \in C^1 \}.$$

(c) $L^* L$ is the negative of the Laplacian operator

$$L^* L = - \left[\frac{\partial^2}{\partial t_1^2} + \frac{\partial^2}{\partial t_2^2} \right].$$

(d)   The Green function of $L^* L$ is

$$G(s_1, s_2, t_1, t_2)$$

$$= \frac{4}{\pi^2} \sum_{m,n=1}^{\infty} \frac{1}{m^2 + n^2} \sin(m s_1) \sin(n s_2) \sin(m t_1) \sin(n t_2), \quad 0 \le s_i, t_j \le \pi.$$

(e)   If

$$\mathbf{y} = \begin{bmatrix} y_1 \\ y_2 \end{bmatrix} \in S \times S,$$

then

$$(L^\dagger \mathbf{y})(t_1, t_2) = \sum_{j=1}^{2} \int_0^\pi \int_0^\pi \frac{\partial}{\partial s_j} G(s_1, s_2, t_1, t_2)\, y_j(s_1, s_2)\, ds_1\, ds_2.$$

Ex. 28. *Ordinary linear differential equations with homogeneous boundary conditions.* Let

$S =$ the real space $L^2[a, b]$ of real valued functions;

$C^k[a, b] =$ the real valued functions on $[a, b]$ with k derivatives and

$$\mathbf{x}^{(k)} = \frac{d^k \mathbf{x}}{dt^k} \in C[a, b];$$

$S^k = \{\mathbf{x} \in C^{k-1}[a, b] : \mathbf{x}^{(k-1)} \text{ absolutely continuous, } \mathbf{x}^{(k)} \in S\};$
and let L be the operator

$$\ell = \sum_{i=0}^{n} a_i(t) \left(\frac{d}{dt}\right)^i, \quad a_i \in C^i[a, b], \ i = 0, 1, \dots, n, \tag{49}$$

$$a_n(t) \neq 0, \ a \leq t \leq b,$$

with domain $D(L)$ consisting of all $\mathbf{x} \in S^n$ which satisfy

$$M\widehat{\mathbf{x}} = 0, \tag{50}$$

where $M \in \mathbb{R}_m^{m \times 2n}$ is a matrix with a specific null space $N(M)$ and $\widehat{\mathbf{x}} \in \mathbb{R}^{2n}$ is the boundary vector

$$\widehat{\mathbf{x}}^T = [\mathbf{x}(a), \mathbf{x}'(a), \dots, \mathbf{x}^{(n-1)}(a); \ \mathbf{x}(b), \mathbf{x}'(b), \dots, \mathbf{x}^{(n-1)}(b)].$$

Finally let $\widetilde{L}$ be the operator ℓ of (49) with $D(\widetilde{L}) = S^n$. Then:

   (a) $L \in C(S, S)$, $\overline{D(L)} = S$.
   (b) $\dim N(\widetilde{L}) = n = \dim N(\widetilde{L}^*)$.
   (c) $N(L) \subset N(\widetilde{L})$, $N(L^*) \subset N(\widetilde{L}^*)$, hence $\dim N(L) \leq n$ and $\dim N(L^*) \leq n$.
   (d) $R(L)$ is closed.
   (e) The restriction $L_0 = L_{[C(L)]}$ of L to its carrier is a one-to-one mapping of $C(L)$ onto $R(L)$;

$$L_0 \in C(C(L), R(L)).$$

   (f) $L_0^{-1} \in B(R(L), C(L))$.
   (g) $L^\dagger$, the extension of L_0^{-1} to all of S with $N(L^\dagger) = R(L)^\perp$ is bounded and satisfies

$$LL^\dagger \mathbf{y} = P_{R(L)} \mathbf{y}, \quad \text{for all } \mathbf{y} \in S,$$

$$L^\dagger L \mathbf{x} = P_{N(L)^\perp} \mathbf{x}, \quad \text{for all } \mathbf{x} \in D(L).$$

For proofs of (a) and (d), see Halperin [**369**] and Schwartz [**733**]. The proof of (e) is contained in Section 2(F), and (f) follows from the closed graph theorem (Locker [**522**]).

Ex. 29. For L as in Ex. 28, find the generalized Green function which corresponds to $L^\dagger$, i.e., find the kernel $L^\dagger(s, t)$ such that

$$(L^\dagger \mathbf{y})(s) = \int_a^b L^\dagger(s, t) \mathbf{y}(t) \, dt, \quad \text{for all } \mathbf{y} \in D(L^\dagger) = S.$$

SOLUTION. A generalized Green function of $\widetilde{L}$ is (see Coddington and Levinson [**204**, Theorem 6.4]),

$$\widetilde{G}(s, t) = \begin{cases} \displaystyle\sum_{j=1}^{n} \frac{\mathbf{x}_j(s) \det(X_j(t))}{a_n(t) \det(X(t))}, & a \leq t \leq s \leq b, \\ 0, & a \leq s \leq t \leq b, \end{cases} \tag{51}$$

where:

$\{\mathbf{x}_1, \dots, \mathbf{x}_k\}$ is an o.n. basis of $N(L)$;

$\{\mathbf{x}_1, \dots, \mathbf{x}_k, \mathbf{x}_{k+1}, \dots, \mathbf{x}_n\}$ is an o.n. basis of $N(\widetilde{L})$;

$$X(t) = [\mathbf{x}_j^{(i-1)}(t)], \quad i, j = 1, \dots, n;$$

$X_j(t)$ is the matrix obtained from $X(t)$ by replacing the jth column by $[0, 0, \dots, 0, 1]^T$.

Since $R(L) \subset R(\widetilde{L})$ it follows, for any $\mathbf{y} \in R(L)$, that the general solution of

$$L\mathbf{x} = \mathbf{y}$$

is

$$\mathbf{x}(s) = \int_a^b \widetilde{G}(s, t)\mathbf{y}(t)\, dt + \sum_{i=1}^n c_i\, \mathbf{x}_i(s), \quad c_i \text{ arbitrary.} \tag{52}$$

Writing the particular solution $L^\dagger \mathbf{y}$ in the form (52),

$$L^\dagger \mathbf{y} = \mathbf{x}_0 + \sum_{i=1}^n c_i\, \mathbf{x}_i, \tag{53}$$

$$\mathbf{x}_0(s) = \int_a^b \widetilde{G}(s, t)\mathbf{y}(t)\, dt,$$

we determine its coefficients $\{c_1, \dots, c_n\}$ as follows:

(a) The coefficients $\{c_1, \dots, c_k\}$ are determined by $L^\dagger \mathbf{y} \in N(L)^\perp$, since, by (53),

$$\langle L^\dagger \mathbf{y}, \mathbf{x}_j \rangle = 0 \quad \Longrightarrow \quad c_j = -\langle \mathbf{x}_0, \mathbf{x}_j \rangle, \quad j = 1, \dots, k.$$

(b) The remaining coefficients $\{c_{k+1}, \dots, c_n\}$ are determined by the boundary condition (50). Indeed, writing (53) as

$$L^\dagger \mathbf{y} = \mathbf{x}_0 + X\mathbf{c}, \quad \mathbf{c}^T = [c_1, \dots, c_n],$$

it follows from (50) that

$$M\widehat{\mathbf{x}}_0 + M\widehat{X}\mathbf{c} = \mathbf{0}, \quad \text{where } \widehat{X} = \begin{bmatrix} X(a) \\ X(b) \end{bmatrix}. \tag{54}$$

A solution of (54) is

$$\mathbf{c} = -(M\widehat{X})^{(1)} M\widehat{\mathbf{x}}_0, \tag{55}$$

where $(M\widehat{X})^{(1)} \in \mathbb{R}^{n \times m}$ is any $\{1\}$-inverse of $M\widehat{X} \in \mathbb{R}^{m \times n}$. Now $\{\mathbf{x}_1, \dots, \mathbf{x}_k\} \subset D(L)$ and, therefore,

$$M\widehat{X} = [O \;\; B], \quad B \in \mathbb{R}_{n-k}^{m \times (n-k)}.$$

Thus, we may use in (55),

$$(M\widehat{X})^{(1)} = \begin{bmatrix} O \\ B^{(1)} \end{bmatrix}, \quad \text{for any } B^{(1)} \in B\{1\},$$

obtaining

$$\mathbf{c} = -\begin{bmatrix} O \\ B^{(1)} \end{bmatrix} M\widehat{\mathbf{x}}_0,$$

which uniquely determines $\{c_{k+1}, \dots, c_n\}$.

Substituting these coefficients $\{c_1, \dots, c_n\}$ in (52) finally gives $L^\dagger(s, t)$ (Locker [522]). $\qquad\square$

Ex. 30. *The vector case* (Reid [685] and [686, Chapter III]). Let $\mathcal{S}_n$ and $\mathcal{S}_n^k$ denote the spaces of n-dimensional vector functions whose components belong to $\mathcal{S}$ and $\mathcal{S}^k$, respectively, of Ex. 28. Let L be the differential operator

$$\ell\mathbf{x} = A_1(t)\frac{d\mathbf{x}}{dt} + A_0(t)\mathbf{x}, \quad a \le t \le b, \tag{46}$$

where A_0, A_1 are $n \times n$ matrix functions satisfying[1]:

(i) $A_0(t)$ is continuous on $[a, b]$;
(ii) $A_1(t)$ is continuously differentiable and nonsingular on $[a, b]$;

with domain $D(L)$ consisting of those vector functions $\mathbf{x} \in \mathcal{S}_n^1$ which satisfy

$$M\widehat{\mathbf{x}} = \mathbf{0}, \tag{50}$$

where $M \in \mathbb{R}_m^{m \times 2n}$ is a matrix with a specified null space $N(M)$ and $\widehat{\mathbf{x}} \in \mathbb{R}^{2n}$ is the boundary vector

$$\widehat{\mathbf{x}} = \begin{bmatrix} \mathbf{x}(a) \\ \mathbf{x}(b) \end{bmatrix}. \tag{56}$$

Let $\widetilde{L}$ be the differential operator (46) with domain $D(\widetilde{L}) = \mathcal{S}_n^1$. Then:

(a) $L \in \mathcal{C}(\mathcal{S}_n, \mathcal{S}_n)$, $\overline{D(L)} = \mathcal{S}_n$.
(b)   The adjoint of L is the operator L^* defined by

$$\ell^*\mathbf{y} = -\frac{d}{dt}(A_1^*(t)\mathbf{y}) + A_0^*(t)\mathbf{y} \tag{57}$$

on its domain

$$D(L^*) = \{\mathbf{y} \in \mathcal{S}_n^1 : \mathbf{y}^*(b)\mathbf{x}(b) - \mathbf{y}^*(a)\mathbf{x}(a) = 0 \text{ for all } \mathbf{x} \in D(L)\} \tag{58}$$

$$= \left\{\mathbf{y} \in \mathcal{S}_n^1 : P^* \begin{bmatrix} I & O \\ O & -I \end{bmatrix} \widehat{\mathbf{y}} = \mathbf{0} \text{ for any } P \in \mathbb{R}_{2n-m}^{(2n-m) \times 2n} \text{ with } MP = O \right\}.$$

(c) $\dim N(\widetilde{L}) = n$.
(d)   Let

$$k = \dim N(L) \quad \text{and} \quad k^* = \dim N(L^*).$$

Then

$$\max\{0, n - m\} \le k \le \min\{n, 2n - m\}$$

and

$$k + m = k^* + n.$$

(e) $R(L) = N(L^*)^\perp$, $R(L^*) = N(L)^\perp$,
        hence both $R(L)$ and $R(L^*)$ are closed.
(f)   Let

$$X(t) = [\mathbf{x}_1(t), \ldots, \mathbf{x}_n(t)]$$

be a *fundamental matrix* of $\widetilde{L}$, i.e., let the vectors $\{\mathbf{x}_1, \ldots, \mathbf{x}_n\}$ form a basis of $N(\widetilde{L})$. Then

$$\widetilde{G}(s, t) = \tfrac{1}{2}\operatorname{sign}(s - t)X(s)X(t)^{-1} \tag{59}$$

is a generalized Green matrix of $\widetilde{L}$.

---

[1]Weaker regularity conditions will do; see, e.g., Reid [684] and [686, Chapter III].

(g)   Let $(M\widehat{X})^{(1)}$ be any $\{1\}$-inverse of $M\widehat{X}$ where $\widehat{X} = \begin{bmatrix} X(a) \\ X(b) \end{bmatrix}$. Then

$$G(s,t) = \tfrac{1}{2}X(s)\left(\operatorname{sign}(s-t)I - (M\widehat{X})^{(1)}M\begin{bmatrix} I & O \\ O & -I \end{bmatrix}\widehat{X}\right)X(t)^{-1} \qquad (60)$$

is a generalized Green matrix of L.

PROOF OF (g). For any $y \in R(L)$, the general solution of

$$Lx = y \qquad (47)$$

is

$$x(s) = \int_a^b \widetilde{G}(s,t)y(t)\,dt + \sum_{i=1}^n c_i x_i(s) \qquad (52)$$

or

$$x = x_0 + Xc, \quad c^T = [c_1,\ldots,c_n],$$

and from (50) it follows that

$$c = -(M\widehat{X})^{(1)}M\widehat{x}_0 \qquad (55)$$

and (60) follows by substituting (55) in (52).   □

EX. 31. The differential expression

$$\ell x = \sum_{i=1}^n a_i(t)\frac{d^i x}{dt^i}, \quad x \text{ scalar function}, \qquad (49)$$

is a special case of

$$\ell x = A_1(t)\frac{dx}{dt} + A_0(t)x, \quad x \text{ vector function}. \qquad (46)$$

EX. 32. *The class of all generalized Green functions* (Reid [684]). Let L be as in Ex. 30 and let $X_0(t)$ and $Y_0(t)$ be $n \times k$ and $n \times k^*$ matrix functions whose columns are bases of $N(L)$ and $N(L^*)$, respectively. Then a kernel $H(s,t)$ is a generalized Green matrix of L if and only if

$$H(s,t) = G(s,t) + X_0(s)A^*(t) + B(s)Y_0^*(t), \qquad (61)$$

where $G(s,t)$ is any generalized Green matrix of L (in particular (60)), and $A(t)$ and $B(s)$ are $n \times k$ and $n \times k^*$ matrix functions which are Lebesgue measurable and essentially bounded.

EX. 33. (Reid [684]). Let $X_0(t)$ and $Y_0(t)$ be as in Ex. 32. If $\Theta(t)$ and $\Psi(t)$ are Lebesgue measurable and essentially bounded matrix functions such that the matrices

$$\int_a^b \Theta^*(t)X_0(t)\,dt, \quad \int_a^b Y_0^*(t)\Psi(t)\,dt,$$

are nonsingular, then L has a unique generalized Green function $G_{\Theta,\Psi}$ such that

$$\int_a^b \Theta^*(s)\,G(s,t)\,ds = O, \quad \int_a^b G(s,t)\,\Psi(t)\,dt = O, \quad a \le s,t \le b. \qquad (62)$$

Thus the generalized inverse determined by $G_{\Theta,\Psi}$ has null space spanned by the columns of Ψ and range which is the orthogonal complement of the columns of Θ. Compare with Section 2.6.

Ex. 34. *Existence and properties of* $L^\dagger$ (Loud [**524**], [**525**]). If in Ex. 33 we take

$$\Theta = X_0, \quad \Psi = Y_0,$$

then we get a generalized inverse of L which has the same range and null space as L^*. This generalized inverse is the analog of the *Moore–Penrose inverse* of L and will likewise be denoted by $L^\dagger$.

Show that $L^\dagger$ satisfies the four *Penrose equations* (1.1)–(1.4) as far as can be expected:

(a) $LL^\dagger L = L$;

(b) $L^\dagger LL^\dagger = L^\dagger$;

(c) $LL^\dagger = P_{R(L)}$;
       $(LL^\dagger)^* = P_{R(L)}$ on $D(L^*)$;

(d) $L^\dagger L = P_{R(L^*)}$ on $D(L)$; and
       $(L^\dagger L)^* = P_{R(L^*)}$.

Ex. 35. *Loud's construction of* $L^\dagger$ (Loud [**525**]). Just as in the matrix case (see Theorem 2.12(c) and Ex. 2.38) it follows here that

$$L^\dagger = P_{R(L^*)} G P_{R(L)}, \tag{63}$$

where G is any generalized Green matrix.

In computing $P_{R(L^*)}$ and $P_{R(L)}$ we use Ex. 30(e) to obtain

$$P_{R(L^*)} = I - P_{N(L)}, \quad P_{R(L)} = I - P_{N(L^*)}. \tag{64}$$

Here $P_{N(L)}$ and $P_{N(L^*)}$ are integral operators of the first kind with kernels

$$K_{N(L)} = X_0(s) \left(\int_a^b X_0^*(u) X_0(u) \, du \right)^{-1} X_0^*(t) \tag{65}$$

and

$$K_{N(L^*)} = Y_0(s) \left(\int_a^b Y_0^*(u) Y_0(u) \, du \right)^{-1} Y_0^*(t), \tag{66}$$

respectively, where X_0 and Y_0 are as in Ex. 32.

Thus, for any generalized Green matrix $G(s,t)$, $L^\dagger$ has the kernel

$$L^\dagger(s,t) = G(s,t) - \int_a^b K_{N(L)}(s,u) G(u,t) \, du - \int_a^b G(s,u) K_{N(L^*)}(u,t) \, du$$

$$+ \int_a^b \int_a^b K_{N(L)}(s,u) G(u,v) K_{N(L^*)}(v,t) \, du \, dv. \tag{67}$$

Ex. 36. (Loud [**525**, pp. 201–202]). Let L be the differential operator given by

$$\ell \mathbf{x} = \mathbf{x}' - B(t)\mathbf{x}, \quad 0 \le t \le 1,$$

with boundary conditions

$$\mathbf{x}(0) = \mathbf{x}(1) = \mathbf{0}.$$

Then the adjoint L^* is given by

$$\ell^* \mathbf{y} = -\mathbf{y}' - B(t)^* \mathbf{y}$$

with no boundary conditions.

Let $X(t)$ be a fundamental matrix for

$$\ell \mathbf{x} = \mathbf{0}.$$

Then $X(t)^{*-1}$ is a fundamental matrix for

$$\ell^* \mathbf{y} = \mathbf{0}.$$

Now $N(L) = \{\mathbf{0}\}$ and therefore $K_{N(L)} = O$. Also, $N(L^*)$ is spanned by the columns of $X(t)^{*-1}$ so, by (66),

$$K_{N(L^*)}(s,t) = X(s)^{*-1}\left(\int_0^1 X(u)X(u)^{*-1}\,du\right)X(t)^{-1}. \tag{68}$$

A generalized Green matrix for L is

$$G(s,t) = \begin{cases} X(s)X(t)^{-1}, & 0 \le s < t \le 1, \\ O, & 0 \le t < s \le 1. \end{cases} \tag{69}$$

Finally, by (67),

$$L^\dagger(s,t) = G(s,t) - \int_0^1 G(s,u)K_{N(L^*)}(u,t)\,du,$$

with G and $K_{N(L^*)}$ given by (69) and (68), respectively.

6. Minimal Properties of Generalized Inverses

In this section, which is based on Erdélyi and Ben-Israel [265], we develop certain minimal properties of generalized inverses of operators between Hilbert spaces, analogous to the matrix case studied in Chapter 3.

DEFINITION 4. Let $T \in \mathcal{L}(\mathcal{H}_1, \mathcal{H}_2)$ and consider the linear equation

$$T\mathbf{x} = \mathbf{y}. \tag{70}$$

If the infimum

$$\|T\mathbf{x}' - \mathbf{y}\| = \inf_{\mathbf{x} \in D(T)} \|T\mathbf{x} - \mathbf{y}\| \tag{71}$$

is attained by a vector $\mathbf{x}' \in D(T)$, then $\mathbf{x}'$ is called an *extremal solution* of (70). Among the extremal solutions there may exist a unique vector $\mathbf{x}_0$ of least norm

$$\|\mathbf{x}_0\| < \|\mathbf{x}'\|,$$

for all extremal solutions $\mathbf{x}' \neq \mathbf{x}_0$. Then $\mathbf{x}_0$ is called the *least extremal solution*.

Other names for extremal solutions are *virtual solutions* (Tseng [820]) and *approximate solutions*.

Example 37 shows that extremal solutions need not exist. Their existence is characterized in the following theorem:

THEOREM 5. *Let $T \in \mathcal{L}(\mathcal{H}_1, \mathcal{H}_2)$. Then*

$$T\mathbf{x} = \mathbf{y} \tag{70}$$

has an extremal solution if and only if

$$P_{\overline{R(T)}}\mathbf{y} \in R(T). \tag{72}$$

PROOF. For every $\mathbf{x} \in D(T)$,

$$\|T\mathbf{x} - \mathbf{y}\|^2 = \|P_{\overline{R(T)}}(T\mathbf{x} - \mathbf{y})\|^2 + \|P_{R(T)^\perp}(T\mathbf{x} - \mathbf{y})\|^2$$
$$= \|P_{\overline{R(T)}}(T\mathbf{x} - \mathbf{y})\|^2 + \|P_{R(T)^\perp}\mathbf{y}\|^2.$$

Thus

$$\|T\mathbf{x} - \mathbf{y}\| \geq \|P_{R(T)^\perp}\mathbf{y}\|, \quad \text{for all } \mathbf{x} \in D(T),$$

with equality if and only if

$$T\mathbf{x} = P_{\overline{R(T)}}\mathbf{y}. \tag{73}$$

Clearly,

$$\inf_{\mathbf{x} \in D(T)} \|T\mathbf{x} - \mathbf{y}\| = P_{\overline{R(T)}}\mathbf{y}, \tag{74}$$

which is attained if and only if (73) is satisfied for some $\mathbf{x} \in D(T)$. $\quad\square$
See also Ex. 44.

The existence of extremal solutions does not guarantee the existence of a least extremal solution; see, e.g., Ex. 39. Before settling this issue we require

LEMMA 4. *Let $\mathbf{x}'$ and $\mathbf{x}''$ be extremal solutions of* (70). *Then:*

(a) $P_{N(T)^\perp}\mathbf{x}' = P_{N(T)^\perp}\mathbf{x}''$; *and*
(b) $P_{\overline{N(T)}}\mathbf{x}' \in N(T)$ *if and only if* $P_{\overline{N(T)}}\mathbf{x}'' \in N(T)$.

PROOF. (a) From (73),

$$T\mathbf{x}' = T\mathbf{x}'' = P_{\overline{R(T)}}\mathbf{y}$$

and, hence,

$$T(\mathbf{x}' - \mathbf{x}'') = \mathbf{0}, \tag{75}$$

proving (a).

(b) From (75),

$$\mathbf{x}' - \mathbf{x}'' = P_{\overline{N(T)}}(\mathbf{x}' - \mathbf{x}'')$$

and then

$$P_{\overline{N(T)}}\mathbf{x}' = P_{\overline{N(T)}}\mathbf{x}'' + (\mathbf{x}' - \mathbf{x}''),$$

proving (b). $\quad\square$

The existence of the least extremal solution is characterized in the following:

THEOREM 6 (Erdélyi and Ben-Israel [**265**]). *Let $\mathbf{x}$ be an extremal solution of* (70). *There exists a least extremal solution if and only if*

$$P_{\overline{N(T)}}\mathbf{x} \in N(T), \tag{76}$$

in which case, the least extremal solution is

$$\mathbf{x}_0 = P_{N(T)^\perp}\mathbf{x}. \tag{77}$$

PROOF. Let $\mathbf{x}'$ be an extremal solution of (70). Then

$$\|\mathbf{x}'\|^2 = \|P_{\overline{N(T)}}\mathbf{x}'\|^2 + \|P_{N(T)^\perp}\mathbf{x}'\|^2$$
$$= \|P_{\overline{N(T)}}\mathbf{x}'\|^2 + \|P_{N(T)^\perp}\mathbf{x}\|^2, \quad \text{by Lemma 4,}$$

proving that

$$\|\mathbf{x}'\| \geq \|P_{N(T)^\perp}\mathbf{x}\|$$

with equality if and only if

$$P_{\overline{N(T)}}\mathbf{x}' = \mathbf{0}. \tag{78}$$

If: Let condition (76) be satisfied and define

$$\mathbf{x}_0 = \mathbf{x} - P_{\overline{N(T)}}\mathbf{x}.$$

Then $\mathbf{x}_0$ is an extremal solution since

$$T\mathbf{x}_0 = T\mathbf{x}.$$

Also

$$P_{\overline{N(T)}}\mathbf{x}_0 = \mathbf{0},$$

which, by (78), proves that $\mathbf{x}_0$ is the least extremal solution.

Only if: Let $\mathbf{x}_0$ be the least extremal solution of (70). Then, by (78),

$$\mathbf{x}_0 = P_{\overline{N(T)}}\mathbf{x}_0 + P_{N(T)^\perp}\mathbf{x}_0 = P_{N(T)^\perp}\mathbf{x},$$

and hence

$$\mathbf{x}_0 = \mathbf{x} - P_{\overline{N(T)}}\mathbf{x}.$$

But

$$T\mathbf{x}_0 = T\mathbf{x},$$

since both $\mathbf{x}_0$ and $\mathbf{x}$ are extremal solutions and, therefore,

$$TP_{\overline{N(T)}} = \mathbf{0},$$

proving (76). $\qquad\square$

As in the matrix case (see Corollary 3.3, p. 109), here too a unique generalized inverse is characterized by the property that it gives the least extremal solution whenever it exists. We define this inverse as follows:

DEFINITION 5. Let $T \in \mathcal{L}(\mathcal{H}_1, \mathcal{H}_2)$, let

$$C(T) = D(T) \cap N(T)^\perp, \tag{2}$$

$$B(T) = D(T) \cap \overline{N(T)}, \tag{79}$$

and let $A(T)$ be a subspace satisfying

$$D(T) = A(T) \oplus \left(B(T) \overset{\perp}{\oplus} C(T)\right). \tag{80}$$

(Examples 42 and 43 below show that, in the general case, this complicated decomposition cannot be avoided.) Let

$$G_0 = \{\{\mathbf{x}, T\mathbf{x}\} : \mathbf{x} \in C(T)\}, \quad G_1 = G(T)^\perp \cap \mathcal{H}_{0,2} = J_2 R(T)^\perp.$$

The *extremal inverse* of T, denoted by $T_e^\dagger$, is defined by its inverse graph

$$G_0 + G_1 = \{\{\mathbf{x}, T\mathbf{x} + \mathbf{z}\} : \mathbf{x} \in C(T), \mathbf{z} \in R(T)^\perp\}.$$

The following properties of $T_e^\dagger$ are easy consequences of the above construction:

THEOREM 7 (Erdélyi and Ben-Israel [**265**]). *Let* $T \in \mathcal{L}(\mathcal{H}_1, \mathcal{H}_2)$. *Then*:

(a) $D(T_e^\dagger) = T(C(T)) \overset{\perp}{\oplus} R(T)^\perp$ *and, in general,* $R(T) \not\subset D(T_e^\dagger)$.
(b) $R(T_e^\dagger) = C(T)$.
(c) $N(T_e^\dagger) = R(T)^\perp$.
(d) $TT_e^\dagger \mathbf{y} = P_{\overline{R(T)}}\mathbf{y}$, *for all* $\mathbf{y} \in D(T_e^\dagger)$.

(e) $T_e^\dagger T \mathbf{x} = P_{\overline{R(T_e^\dagger)}} \mathbf{x}$, *for all* $\mathbf{x} \in N(T) \overset{\perp}{\oplus} C(T)$. $\square$

See also Exs. 40–41 below.

The extremal inverse $T_e^\dagger$ is characterized in terms of the least extremal solution, as follows:

THEOREM 8 (Erdélyi and Ben-Israel [**265**]). *The least extremal solution* $\mathbf{x}_0$ *of* (70) *exists if and only if*

$$\mathbf{y} \in D(T_e^\dagger), \tag{81}$$

in which case

$$\mathbf{x}_0 = T_e^\dagger \mathbf{y}. \tag{82}$$

PROOF. Assume (81). By Theorem 7(a),

$$P_{\overline{R(T)}} \mathbf{y} = \mathbf{y}_0 \in T(C(T)) \subset R(T),$$

and, by Theorem 5, extremal solutions do exist. Let $\mathbf{x}_0$ be the unique vector in $C(T)$ such that

$$P_{\overline{R(T)}} \mathbf{y} = \mathbf{y}_0 = T\mathbf{x}_0.$$

Then, by Theorem 3(a), (c), and (e),

$$T_e^\dagger \mathbf{y} = T_e^\dagger \mathbf{y}_0 = T_e^\dagger T \mathbf{x}_0 = \mathbf{x}_0,$$

and, by Theorem 3(d),

$$\|T\mathbf{x}_0 - \mathbf{y}\| = \|TT_e^\dagger \mathbf{y} - \mathbf{y}\| = \|P_{\overline{R(T)}} \mathbf{y} - \mathbf{y}\| = \|P_{R(T)^\perp} \mathbf{y}\|,$$

which, by (74), shows that $\mathbf{x}_0$ is an extremal solution. Since

$$\mathbf{x}_0 \in R(T_e^\dagger) \subset N(T)^\perp,$$

it follows, from Lemma 4, that

$$\mathbf{x}_0 = P_{N(T)^\perp} \mathbf{x}$$

for any extremal solution $\mathbf{x}$ of (70). By Theorem 6, $\mathbf{x}_0$ is the least extremal solution.

Conversely, let $\mathbf{x}_0$ be the least extremal solution whose existence we assume. By Theorem 2, $\mathbf{x}_0 \in C(T)$, and, by Theorem 3(e),

$$T_e^\dagger T \mathbf{x}_0 = \mathbf{x}_0.$$

Since $\mathbf{x}_0$ is an extremal solution, it follows from (73) that

$$T\mathbf{x}_0 = P_{\overline{R(T)}} \mathbf{y} \in T(C(T))$$

and, therefore,

$$\mathbf{x}_0 = T_e^\dagger T \mathbf{x} = T_e^\dagger P_{\overline{R(T)}} \mathbf{y}$$
$$= T_e^\dagger \mathbf{y}. \qquad \square$$

If $N(T)$ is closed then $T_e^\dagger$ coincides with the maximal Tseng inverse $T^\dagger$. Thus, for closed operators and, in particular, for bounded operators, $T_e^\dagger$ should be replaced by $T^\dagger$ in the statement of Theorem 8.

Exercises

Ex. 37. *A linear equation without extremal solution.* Let T and $\mathbf{y}$ be as in Ex. 7. Then

$$T\mathbf{x} = \mathbf{y}$$

has no extremal solutions.

Ex. 38. It was noted in Ex. 8 that, in general, the Fredholm integral operator of the first kind has a nonclosed range. Consider the kernel

$$G(s,t) = \begin{cases} s(1-t), & 0 \le s \le t \le 1, \\ t(1-s), & 0 \le t \le s \le 1, \end{cases}$$

which is a generalized Green function of the operator

$$-\frac{d^2}{dt^2}, \quad 0 \le t \le 1.$$

Let $T \in \mathcal{B}(L^2[0,1], L^2[0,1])$ be defined by

$$(T\mathbf{x})(s) = \int_0^1 G(s,t)\mathbf{x}(t)\,dt.$$

Show that there exists a $\mathbf{y} \in L^2[0,1]$ for which

$$T\mathbf{x} = \mathbf{y}$$

has no extremal solution.

Ex. 39. *An equation without a least extremal solution.* Consider the unbounded functional on $L^2[0,\infty]$,

$$T\mathbf{x} = \int_0^\infty t\mathbf{x}(t)\,dt$$

discussed in Ex. 2. Then the equation

$$T\mathbf{x} = 1$$

is consistent, and each of the functions

$$\mathbf{x}_n(t) = \begin{cases} 1/nt, & 1 \le t \le n+1, \\ 0, & \text{otherwise}, \end{cases}$$

is a solution, $n = 1, 2, \ldots$. Since

$$\|\mathbf{x}_n\|^2 = \int_1^{n+1} \frac{1}{(nt)^2}\,dt = \frac{1}{n(n+1)} \to 0,$$

there is no extremal solution of least norm.

Ex. 40. *Properties of* $(T_e^\dagger)^\dagger$. By Theorem 7(a) and (c), it follows that $D(T_e^\dagger)$ is decomposable with respect to $N(T_e^\dagger)$. Thus $T_e^\dagger$ has a maximal Tseng inverse, denoted by $T_e^{\dagger\dagger}$. Some of its properties are listed below:

(a) $G(T_e^{\dagger\dagger}) = \{\{\mathbf{x}+\mathbf{z}, T\mathbf{x}\} : \mathbf{x} \in C(T), \mathbf{z} \in C(T)^\perp\}$.

(b) $D(T_e^{\dagger\dagger}) = C(T) \overset{\perp}{\oplus} C(T)^\perp$.

(c) $R(T_e^{\dagger\dagger}) = T(C(T))$.

(d) $N(T_e^{\dagger\dagger}) = C(T)^\perp$.

Ex. 41. Let $T \in \mathcal{L}(\mathcal{H}_1, \mathcal{H}_2)$ and let

$$D_0(T) = N(T) \overset{\perp}{\oplus} C(T).$$

Then:

(a) $D(T_e^{\dagger\dagger}) = C(T) \overset{\perp}{\oplus} \overline{N(T)} \overset{\perp}{\oplus} D_0(T)^\perp$, a refinement of Ex. 40(b).

(b) $D_0(T) \subset D(T) \cap D(T_e^{\dagger\dagger})$ and $T_{[D_0(T)]} = (T_e^{\dagger\dagger})_{[D_0(T)]}$.

(c) $T_e^{\dagger\dagger}$ is an extension of T if and only if $D(T)$ is decomposable with respect to $N(T)$, in which case $T_e^{\dagger\dagger}$ is an extension by zero to $\overline{N(T)} \stackrel{\perp}{\oplus} D(T)^{\perp}$.

Ex. 42. *An example of* $A(T) \neq \{0\}$, $A(T) \subset D(T_e^{\dagger\dagger})$. Let T be the operator defined in Ex. 4. Then,

$$B(T) = D(T) \cap \overline{N(T)} = D \cap (\overline{D \cap F}) = D \cap F$$
$$= N(T),$$

and $C(T) = \{0\}$, showing that

$$A(T) \neq \{0\}, \quad \text{by (80)}.$$

Thus

$$A(T) = A, \quad \text{of Ex. 4,}$$

and

$$D(T_e^{\dagger}) = A^{\perp} = N(T_e^{\dagger}).$$

Finally, from $C(T)^{\perp} = \mathcal{H}$,

$$D(T_e^{\dagger\dagger}) = \mathcal{H} \supset A$$

with

$$N(T_e^{\dagger\dagger}) = \mathcal{H}.$$

Ex. 43. *An example of* $A(T) \neq \{0\}$, $A(T) \cap D(T_e^{\dagger\dagger}) = \{0\}$. Let $\mathcal{H}$ be a Hilbert space and let M, N be subspaces of $\mathcal{H}$ such that

$$M \neq \overline{M}, \quad N \neq \overline{N} \subset M^{\perp}.$$

Let $\mathbf{x} \in (X \backslash Y)$ denote $\mathbf{x} \in X$, $\mathbf{x} \notin Y$. Choose

$$\mathbf{y} \in \overline{M} \backslash M \quad \text{and} \quad \mathbf{z} \in M^{\perp} \backslash (N \stackrel{\perp}{\oplus} (N^{\perp} \cap M^{\perp})),$$

let

$$\mathbf{x} = \mathbf{y} + \mathbf{z}$$

and

$$D = M \oplus N \oplus [\mathbf{x}]$$

where $[\mathbf{x}]$ is the line spanned by $\mathbf{x}$. Define $T \in \mathcal{L}(\mathcal{H}, \mathcal{H})$ on $D(T) = D$ by

$$T(\mathbf{u} + \mathbf{v} + \alpha\mathbf{x}) = \mathbf{v} + \alpha\mathbf{x}, \quad \mathbf{u} \in M, \quad \mathbf{v} \in N, \quad \alpha\mathbf{x} \in [\mathbf{x}].$$

Then

$$C(T) = N, \quad N(T) = M, \quad A(T) = [\mathbf{x}],$$

and

$$\mathbf{x} \notin D(T_e^{\dagger\dagger}).$$

Ex. 44. (Tseng [820]; see also Holmes [427, Section 35]). Let $T \in \mathcal{B}(\mathcal{H}_1, \mathcal{H}_2)$. Then

$$Tx = y \tag{70}$$

has an extremal solution if and only if there is a positive scalar β such that

$$|\langle y, z \rangle|^2 \le \beta \langle z, AA^* z \rangle, \quad \text{for every } z \in N(AA^*)^{\perp}.$$

Ex. 45. (Minamide and Nakamura [557]). Let $T \in \mathcal{B}(\mathcal{H}_1, \mathcal{H}_2)$, $S \in \mathcal{B}(\mathcal{H}_1, \mathcal{H}_3)$ be normally solvable, and let

$$T_S = T_{[N(S)]}$$

denote the restriction of T to $N(S)$. If T_S is also normally solvable, then $T_S^{\dagger}$ is called the $N(S)$-*restricted pseudoinverse of* T. It is the unique solution X of the following five equations

$$SX = O,$$
$$XTX = X,$$
$$(TX)^* = TX,$$
$$TXT = T \quad \text{on } N(S),$$
$$P_{N(S)}(XT)^* = XT \quad \text{on } N(S).$$

Ex. 46. (Minamide and Nakamura [557]). Let T, S, and $T_S^{\dagger}$ be as in Ex. 45. Then, for any $y_0 \in \mathcal{H}_2$ and $z_0 \in R(S)$, the least extremal solution of

$$Tx = y_0$$

subject to

$$Sx = z_0$$

is given by

$$x_0 = T_S^{\dagger}(y_0 - TS^{\dagger}z_0) + S^{\dagger}z_0.$$

Ex. 47. (Porter and Williams [647]). Let $\mathcal{H}_1, \mathcal{H}_2, \mathcal{H}_3$ be Hilbert spaces, let $T \in \mathcal{B}(\mathcal{H}_1, \mathcal{H}_2)$ with $R(T) = \mathcal{H}_2$, and let $S \in \mathcal{B}(\mathcal{H}_1, \mathcal{H}_3)$. For any $y \in \mathcal{H}_2$, there is a unique $x_0 \in \mathcal{H}_1$ satisfying

$$Tx = y \tag{70}$$

and which minimizes the functional

$$\|Sx\|^2 + \|x\|^2$$

over all solutions of (70). This x_0 is given by

$$x_0 = (I + S^*S)^{-1}T^{\dagger}y_0$$

where y_0 is the unique vector in $\mathcal{H}_2$ satisfying

$$y = T(I + S^*S)^{-1}T^{\dagger}y_0.$$

EX. 48. (Porter and Williams [**647**]). Let $\mathcal{H}_1, \mathcal{H}_2, \mathcal{H}_3, T$, and S be as above. Then, for any $\mathbf{y} \in \mathcal{H}_2, \mathbf{x}_1 \in \mathcal{H}_1$, and $\mathbf{y}_1 \in \mathcal{H}_2$, there is a unique $\mathbf{x}_0 \in \mathcal{H}_1$ which is a solution of

$$T\mathbf{x} = \mathbf{y} \tag{70}$$

and which minimizes

$$\|S\mathbf{x} - \mathbf{y}_1\|^2 + \|\mathbf{x} - \mathbf{x}_1\|^2$$

from among all solutions of (70). This $\mathbf{x}_0$ is given by

$$\mathbf{x}_0 = (I + S^*S)^{-1}(T^\dagger \mathbf{y}_0 + \mathbf{x}_0 + S^*\mathbf{y}_1)$$

where $\mathbf{y}_0$ is the unique vector in $\mathcal{H}_2$ satisfying

$$\mathbf{y} = T(I + S^*S)^{-1}(T^\dagger \mathbf{y}_0 + \mathbf{x}_1 + S^*\mathbf{y}_1).$$

7. Series and Integral Representations and Iterative Computation of Generalized Inverses

Direct computational methods, in which the exact solution requires a finite number of steps (such as the elimination methods of Sections 7.2–7.4) cannot be used, in general, for the computation of generalized inverses of operators. The exceptions are operators with nice algebraic properties, such as the integral and differential operators of Exs. 18–36 with their finite-dimensional null spaces. In the general case, the only computable representations of generalized inverses involve infinite series, or integrals, approximated by suitable iterative methods. Such representations and methods are sampled in this section, based on Showalter and Ben-Israel [**756**], where the proofs, omitted here, can be found.

To motivate the idea behind our development consider the problem of minimizing

$$f(\mathbf{x}) = \langle A\mathbf{x} - \mathbf{y}, A\mathbf{x} - \mathbf{y} \rangle, \tag{83}$$

where $A \in \mathcal{B}(\mathcal{H}_1, \mathcal{H}_2)$ and $\mathcal{H}_1, \mathcal{H}_2$ are Hilbert spaces.

Treating $\mathbf{x}$ as a function $\mathbf{x}(t)$, $t \geq 0$, with $\mathbf{x}(0) = \mathbf{0}$, we differentiate (83):

$$\frac{d}{dt} f(\mathbf{x}) = 2\Re\langle A\mathbf{x} - \mathbf{y}, A\dot{\mathbf{x}} \rangle, \quad \dot{\mathbf{x}} = \frac{d}{dt}\mathbf{x},$$
$$= 2\Re\langle A^*(A\mathbf{x} - \mathbf{y}), \dot{\mathbf{x}} \rangle \tag{84}$$

and setting

$$\dot{\mathbf{x}} = -A^*(A\mathbf{x} - \mathbf{y}), \tag{85}$$

it follows from (84) that

$$\frac{d}{dt} f(\mathbf{x}) = -2\|A^*(A\mathbf{x} - \mathbf{y})\|^2 < 0. \tag{86}$$

This version of the steepest descent method, given in Rosenbloom [**713**], results in $f(\mathbf{x}(t))$ being a monotone decreasing function of t, asymptotically approaching its infimum as $t \to \infty$. We expect $\mathbf{x}(t)$ to approach asymptotically $A^\dagger \mathbf{y}$ so, by solving (85),

$$\mathbf{x}(t) = \int_0^t \exp\{-A^*A(t-s)\}A^*\mathbf{y}\,ds \tag{87}$$

and observing that $\mathbf{y}$ is arbitrary we get

$$A^\dagger = \lim_{t\to\infty} \int_0^t \exp\{-A^*A(t-s)\}A^*\,ds \tag{88}$$

which is the essence of Theorem 9 below.

Here as elsewhere in this section, the convergence is in the strong operator topology. Thus the limiting expression

$$A^\dagger = \lim_{t\to\infty} B(t) \quad \text{or} \quad B(t) \to A^\dagger \text{ or } t \to \infty \tag{89}$$

means that, for all $\mathbf{y} \in D(A^\dagger)$,

$$A^\dagger\mathbf{y} = \lim_{t\to\infty} B(t)\mathbf{y}$$

in the sense that

$$\lim_{t\to\infty} \|(A^\dagger - B(t))\mathbf{y}\| = 0. \tag{90}$$

A numerical integration of (85) with a suitably chosen step size similarly results in

$$A^\dagger = \sum_{k=0}^{\infty} (I - \alpha A^*A)^k \alpha A^*, \tag{91}$$

where

$$0 < \alpha < \frac{2}{\|A\|^2}, \tag{92}$$

which is the essence of Theorem 10 below.

In statements like (90) it is necessary to distinguish between points $\mathbf{y} \in \mathcal{H}_2$ relative to the given $A \in \mathcal{B}(\mathcal{H}_1, \mathcal{H}_2)$. Indeed, the three cases where $P_{\overline{R(A)}}\mathbf{y}$ lies in $R(AA^*)$, $(R(A)\backslash R(AA^*))$, or $(\overline{R(A)}\backslash R(A))$, have different rates of convergence in (90). We abbreviate these cases as follows:

$$
\begin{aligned}
(\mathbf{y} \in \text{I}) \quad &\text{means} \quad P_{\overline{R(A)}}\mathbf{y} \in R(AA^*), \\
(\mathbf{y} \in \text{II}) \quad &\text{means} \quad P_{\overline{R(A)}}\mathbf{y} \in (R(A)\backslash R(AA^*)), \\
(\mathbf{y} \in \text{III}) \quad &\text{means} \quad P_{\overline{R(A)}}\mathbf{y} \in (\overline{R(A)}\backslash R(A)).
\end{aligned}
\tag{93}
$$

We note that $A^\dagger\mathbf{y}$ is not defined for $(\mathbf{y} \in \text{III})$, a case which is impossible if $R(A)$ is closed.

THEOREM 9 (Showalter and Ben-Israel [**756**]). *Let $A \in \mathcal{B}(\mathcal{H}_1, \mathcal{H}_2)$ and define, for $t \geq 0$,*

$$L_1(t) = \int_0^t \exp\{-A^*A(t-s)\}\,ds,$$

$$L_2(t) = \int_0^t \exp\{-AA^*(t-s)\}\,ds, \tag{94}$$

$$B(t) = L_1(t)A^* = A^*L_2(t).$$

Then:

(a) $\|(A^\dagger - B(t))\mathbf{y}\|^2 \leq \dfrac{\|A^\dagger \mathbf{y}\|^2 \|(AA^*)^\dagger \mathbf{y}\|^2}{\|(AA^*)^\dagger \mathbf{y}\|^2 + 2\|A^\dagger \mathbf{y}\|^2 t}$
  if $(\mathbf{y} \in \mathrm{I})$ *and* $t \geq 0$.

(b) $\|(A^\dagger - B(t))\mathbf{y}\|^2$ *is a decreasing function of* $t \geq 0$,
  with limit zero as $t \to \infty$, *if* $(\mathbf{y} \in \mathrm{II})$.

(c) $\|(P_{\overline{R(A)}} - AB(t))\mathbf{y}\|^2 \leq \dfrac{\|\mathbf{y}\|^2 \|A^\dagger \mathbf{y}\|^2}{\|A^\dagger \mathbf{y}\|^2 + 2\|\mathbf{y}\|^2 t}$
  if $(\mathbf{y} \in \mathrm{I})$ *or* $(\mathbf{y} \in \mathrm{II})$, *and* $t \geq 0$.

(d) $\|(P_{\overline{R(A)}} - AB(t))\mathbf{y}\|^2$ *is a decreasing function of* $t \geq 0$,
  with limit zero as $t \to \infty$, *if* $(\mathbf{y} \in \mathrm{III})$. □

Note that even though $A^\dagger \mathbf{y}$ is not defined for $(\mathbf{y} \in \mathrm{III})$, still

$$AB(t) \to P_{\overline{R(A)}}, \quad \text{as } t \to \infty.$$

The discrete version of Theorem 9 is the following theorem:

THEOREM 10 (Showalter and Ben-Israel [**756**]). *Let* $A \in \mathcal{B}(\mathcal{H}_1, \mathcal{H}_2)$, *let* c *be a real number,* $0 < c < 2$, *and let*

$$\alpha = \frac{c}{\|A\|^2}.$$

For any $\mathbf{y} \in \mathcal{H}_2$ *define*

$$\mathbf{x} = T^\dagger \mathbf{y}, \quad \text{if } (\mathbf{y} \in \mathrm{I}) \text{ or } (\mathbf{y} \in \mathrm{II}),$$

and define the sequence

$$\mathbf{y}_0 = \mathbf{0}, \quad \mathbf{x}_0 = \mathbf{0},$$
$$(\mathbf{y} - \mathbf{y}_{N+1}) = (I - \alpha AA^*)(\mathbf{y} - \mathbf{y}_N), \quad \text{if } (\mathbf{y} \in \mathrm{I}) \text{ or } (\mathbf{y} \in \mathrm{II}) \text{ or } (\mathbf{y} \in \mathrm{III}),$$
$$(\mathbf{x} - \mathbf{x}_{N+1}) = (I - \alpha A^*A)(\mathbf{x} - \mathbf{x}_N), \quad \text{if } (\mathbf{y} \in \mathrm{I}) \text{ or } (\mathbf{y} \in \mathrm{II}),$$
$$N = 1, 2, \ldots$$

Then the sequence

$$B_N = \sum_{k=0}^{N} (I - \alpha A^*A)^k \alpha A^*, \quad N = 0, 1, \ldots \tag{95}$$

converges to $A^\dagger$ *as follows:*

(a) $\|(A^\dagger - B_N)\mathbf{y}\|^2 \leq \dfrac{\|A^\dagger \mathbf{y}\|^2 \|(AA^*)^\dagger \mathbf{y}\|^2}{\|(AA^*)^\dagger \mathbf{y}\|^2 + N[(2-c)c/\|A\|^2]\|A^\dagger \mathbf{y}\|^2}$
  if $(\mathbf{y} \in \mathrm{I})$ *and* $N = 1, 2, \ldots$.

(b) $\|(A^\dagger - B_N)\mathbf{y}\|^2 = \|\mathbf{x} - \mathbf{x}_N\|^2$
  converges monotonically to zero if $(\mathbf{y} \in \mathrm{II})$.

(c) $\|(P_{\overline{R(A)}} - AB_N}\mathbf{y}\|^2 \leq \dfrac{\|\mathbf{y}\|^2 \|A^\dagger \mathbf{y}\|^2}{\|A^\dagger \mathbf{y}\|^2 + N[(2-c)c/\|A\|^2]\|\mathbf{y}\|^2}$
  if $(\mathbf{y} \in \mathrm{I})$ *or* $(\mathbf{y} \in \mathrm{II})$ *and* $N = 1, 2, \ldots$.

(d) $\|(P_{\overline{R(A)} - AB_N}\mathbf{y}\|^2 = \|\mathbf{y} - \mathbf{y}_N\|^2$ *converges monotonically to zero*
  if $(\mathbf{y} \in \mathrm{III})$. □

The convergence $B_N \to A^\dagger$, in the uniform operator topology, was established by Petryshyn [**641**], restricting A to have closed range.

As in the matrix case, studied in Section 7.7, higher-order iterative methods are more efficient means of summing the series (91) than the first-order method (95). Two such methods, of order $p \geq 2$, are given in the following:

THEOREM 11 (Showalter and Ben-Israel [**756**]). *Let A, α, and $\{B_N : N = 0, 1, \ldots\}$ be as in Theorem 10. Let p be an integer*

$$p \geq 2$$

and define the sequences $\{C_{N,p}\}$ and $\{D_{N,p}\}$ as follows:

$$C_{0,p} = \alpha A^*, \quad C_{N+1,p} = C_{N,p} \sum_{k=0}^{p-1} (I - AC_{N,p})^k, \tag{96}$$

$$D_{0,p} = \alpha A^*, \quad D_{N+1,p} = D_{N,p} \sum_{k=0}^{p} \binom{p}{k} (-AD_{N,p})^{k-1}. \tag{97}$$

Then, for all $N = 0, 1, \ldots$,

$$B_{(p^{N+1}-1)} = C_{N+1,p} = D_{N+1,p}. \tag{98}$$

$\square$

Consequently $\{C_{N,p}\}$ and $\{D_{N,p}\}$ are p^{th}-order iterative methods for computing $A^\dagger$, with the convergence rates established in Theorem 10; e.g.,

$$\|(A^\dagger - C_{N,p})\mathbf{y}\|^2 \leq \frac{\|A^\dagger \mathbf{y}\|^2 \|(AA^*)^\dagger \mathbf{y}\|^2}{\|(AA^*)^\dagger \mathbf{y}\|^2 + (p^N - 1)[(2-c)c/\|A\|^2]\|A^\dagger \mathbf{y}\|^2},$$
$$\text{if } (\mathbf{y} \in \mathrm{I}) \text{ and } N = 1, 2, \ldots.$$

The series (96) is somewhat simpler to use if the term $(I - AC_{N,p})^k$ can be evaluated by only $k-1$ operator multiplications, e.g., for matrices. The form (97) is preferable otherwise, e.g., for integral operators.

Exercises

EX. 49. (Zlobec [**892**]). Let $A \in \mathcal{B}(\mathcal{H}_1, \mathcal{H}_2)$ have closed range, let $\mathbf{b} \in \mathcal{H}_2$, and let[2] $B \in R(A^*, A^*)$. Then the sequence

$$\mathbf{x}_{k+1} = \mathbf{x}_k - B(A\mathbf{x}_k - \mathbf{b}), \quad k = 0, 1, \ldots, \tag{99}$$

converges to $A^\dagger \mathbf{b}$ for all $\mathbf{x}_0 \in R(A^*)$ if

$$\rho(P_{R(A^*)} - BA) < 1,$$

where $\rho(T)$ denotes the spectral radius of T; see, e.g., Taylor [**800**, p. 262].

The choice $B = \alpha A^*$ in (99) reduces it to the iterative method (95). Other choices of B are given in the following exercise:

EX. 50. *Splitting methods* (Zlobec [**892**], Berman and Neumann [**89**], Berman and Plemmons [**90**]). Let A be as in Ex. 49, and write

$$A = M + N, \tag{100}$$

where $M \in \mathcal{B}(\mathcal{H}_1, \mathcal{H}_2)$ has closed range and $N(A) = N(M)$. Choosing

$$B = w M^\dagger, \quad w \neq 0,$$

---

[2]For $S, T \in \mathcal{B}(\mathcal{H}_1, \mathcal{H}_2)$ with closed ranges, $R(S, T) = \{Z : Z = SWR \text{ for some } W \in \mathcal{B}(\mathcal{H}_2, \mathcal{H}_1)\}$.

in (99) gives

$$\mathbf{x}_{k+1} = [(1 - w)I - wM^\dagger N]\mathbf{x}_k + wM^\dagger \mathbf{b}, \quad \mathbf{x}_0 \in R(A^*), \tag{101}$$

in particular, for $w = 1$,

$$\mathbf{x}_{k+1} = -M^\dagger N \mathbf{x}_k + M^\dagger \mathbf{b}, \quad \mathbf{x}_0 \in R(A^*). \tag{102}$$

8. Frames

This section is based on Christensen [**194**] and Daubechies [**213**, Chapter 3].

Let $\mathcal{H}$ be a separable Hilbert space. A *basis* of $\mathcal{H}$ is a set $\{\mathbf{f}_n\} \subset \mathcal{H}$ such that every $\mathbf{f} \in \mathcal{H}$ is represented as

$$\mathbf{f} = \sum_n c_n \mathbf{f}_n \tag{103}$$

with unique scalar coefficients $\{c_n\}$. A basis is *unconditional* if (103) converges unconditionally for all $\mathbf{f} \in \mathcal{H}$. Unconditional bases are generalized as follows:

DEFINITION 6. (Duffin and Schaeffer [**244**, p. 358]). A sequence $\{\mathbf{f}_n\} \subset \mathcal{H}$ is:

(a) a *frame* for $\mathcal{H}$, if there exist constants $A, B > 0$ such that

$$A\|\mathbf{f}\|^2 \leq \sum |\langle \mathbf{f}, \mathbf{f}_n \rangle|^2 \leq B\|\mathbf{f}\|^2, \quad \text{for all } \mathbf{f} \in \mathcal{H}; \tag{104}$$

(b) a *Bessel sequence* if there is a $B > 0$ such that the upper bound holds in (104).

Consider the mapping $T : \ell^2 \to \mathcal{H}$ given by

$$T : \{c_n\} \to \sum_n c_n \mathbf{f}_n. \tag{105}$$

$\{\mathbf{f}_n\}$ is a Bessel sequence if and only if T is a well-defined operator from ℓ^2 into $\mathcal{H}$, in which case T is bounded, and its adjoint is

$$T^* : \mathcal{H} \to \ell^2, \quad T^* \mathbf{f} = \{\langle \mathbf{f}, \mathbf{f}_n \rangle\}, \quad [\mathbf{194}, \text{Lemma 2.2}]. \tag{106}$$

If $\{\mathbf{f}_n\}$ is a frame, its *frame operator* $S : \mathcal{H} \to \mathcal{H}$ is $S = TT^*$, or

$$Sf = TT^* \mathbf{f} = \sum \langle \mathbf{f}, \mathbf{f}_n \rangle \mathbf{f}_n. \tag{107}$$

S is bounded and surjective, [**411**], allowing the representation

$$\mathbf{f} = SS^{-1}\mathbf{f} = \sum \langle \mathbf{f}, S^{-1}\mathbf{f}_n \rangle \mathbf{f}_n, \quad \forall \mathbf{f} \in \mathcal{H}, \tag{108}$$

see [**194**, Theorem 2.4]. The coefficients $\langle \mathbf{f}, S^{-1}\mathbf{f}_n \rangle$ are not unique, however (108) converges unconditionally, showing frames to be generalizations of unconditional bases (uniqueness lost).

The following proof uses generalized inverses:

THEOREM 12 (Christensen [**194**], Theorem 2.5). *A sequence* $\{\mathbf{f}_n\} \subset \mathcal{H}$ *is a frame if and only if* T *is a well defined operator from* ℓ^2 *onto* $\mathcal{H}$.

PROOF. *Only if*: Let $\{\mathbf{f}_n\}$ be a frame. Then T is a bounded operator from ℓ^2 into $\mathcal{H}$ (since $\{\mathbf{f}_n\}$ is a Bessel sequence) and T is surjective since S is.

If: Let T be a well-defined operator from ℓ^2 onto $\mathcal{H}$. Therefore $\{\mathbf{f}_n\}$ is a Bessel sequence. Let $N(T)^{\perp}$ be the orthogonal complement of $N(T)$, the kernel of T, and let $\widetilde{T} = T_{[N(T)^{\perp}]} : N(T)^{\perp} \to \mathcal{H}$ be the restriction of T to $N(T)^{\perp}$. $\widetilde{T}$ is clearly bounded and bijective and, therefore, has a bounded inverse $T^{\dagger} = \widetilde{T}^{-1} : \mathcal{H} \to N(T)^{\perp}$. Writing a decomposition of $T^{\dagger}\mathbf{f}, \mathbf{f} \in \mathcal{H}$, as $T^{\dagger}\mathbf{f} = \{(T^{\dagger}\mathbf{f})_n\}$, we have

$$\mathbf{f} = TT^{\dagger}\mathbf{f} = \sum (T^{\dagger}\mathbf{f})_n \mathbf{f}_n. \tag{109}$$

$$\therefore \ \|\mathbf{f}\|^4 = \langle \mathbf{f}, \mathbf{f} \rangle^2 = |\langle \sum (T^{\dagger}\mathbf{f})_n \mathbf{f}_n, \mathbf{f} \rangle|^2$$

$$\leq \sum |(T^{\dagger}\mathbf{f})_n|^2 \sum |\langle \mathbf{f}, \mathbf{f}_n \rangle|^2 \leq \|T^{\dagger}\|^2 \|\mathbf{f}\|^2 \sum |\langle \mathbf{f}, \mathbf{f}_n \rangle|^2.$$

$$\therefore \ \sum |\langle \mathbf{f}, \mathbf{f}_n \rangle|^2 \geq \frac{\|\mathbf{f}\|^2}{\|T^{\dagger}\|^2}, \quad \forall \mathbf{f} \in \mathcal{H},$$

establishing the lower bound

$$A = \frac{1}{\|T^{\dagger}\|^2} = \frac{1}{\|S^{-1}\|}, \tag{110}$$

needed in (104) to make $\{\mathbf{f}_n\}$ a frame. $\qquad\square$

The bound (110) was shown in [**192**] to be optimal.

Suggested Further Reading

SECTION 3. The annotated bibliography of Nashed and Rall [**597**] is an essential source.

For alternative or more general treatments of generalized inverses of operators see F.V. Atkinson [**27**], [**28**], Beutler [**92**], [**93**], Davis and Robinson [**215**], Deutsch [**230**], Hamburger [**370**], Hansen and Robinson [**373**], Hestenes [**417**], Holmes [**427**], Leach [**505**], Nashed [**592**]–[**595**], Nashed and Votruba [**598**], [**599**], [**600**], Pietsch [**642**], Porter and Williams [**647**], [**648**], Przeworska-Rolewicz and Rolewicz [**653**], Sengupta [**744**], Sheffield [**748**], Votruba [**843**], Wyler [**880**], Zarantonello [**885**].

SENSITIVITY ANALYSIS OF THE MOORE–PENROSE INVERSE. Koliha [**478**], Moore and Nashed [**577**], [**578**], Nashed [**594**], Rakočević [**664**], Roch and Silbermann [**702**].

INTEGRAL EQUATIONS. K.E. Atkinson [**29**], Courant and Hilbert [**209**], Kammerer and Nashed [**464**], [**465**], Korganoff and Pavel-Parvu [**481**], Lonseth [**523**], Rall [**667**].

REPRODUCING KERNEL HILBERT SPACES. Alpay, Ball, and Bolotnikov [**15**, p. 286], Alpay, Bolotnikov, and Loubaton [**16**, p. 84], Alpay, Bolotnikov, and Rodman [**17**, p. 258], [**18**, p. 33], Aronszajn [**26**], Hilgers [**419**], Moore [**574**], Nashed and Wahba [**601**], Saitoh [**717**], [**718**], [**719**].

GENERALIZED INVERSES OF NONLINEAR MAPPINGS: Ben-Israel [**74**], Saitoh [**720**].

DRAZIN INVERSES OF INFINITE MATRICES AND LINEAR OPERATORS: Campbell [**145**], [**153**], Castro González [**169**], Castro González and Koliha [**170**], [**171**], Koliha [**477**], Kuang [**487**], Nashed and Zhao [**602**].

SENSITIVITY ANALYSIS OF THE DRAZIN INVERSE. Castro González and Koliha [**171**], Castro González, Koliha, and Straškraba [**172**], Castro González, Koliha, and Wei [**173**], Hartwig and Shoaf [**393**], Koliha [**479**], Koliha and Rakočević [**480**], Rakočević [**665**], Rakočević and Wei [**666**].

SECTION 5. For applications to differential operators and related areas see also Bradley [133], [134], Courant and Hilbert [209], Greub and Rheinboldt [322], Kallina [457], Kunkel and Mehrmann [489], Lanczos [497], Lent [507], Locker [521], Stakgold [773], Tucker [822], Van Hamme [828], [829], and Wyler [881].

SECTION 6. Groetsch [331], Ivanov and Kudrinskii [439], Tseng [819], [820], Wahba and Nashed [844].

REGULARIZATION plays an important role in the approximate solution of operator equations, see Engl, Hanke, and Neubauer [253], Groetsch [335], Groetsch and Hanke [338], Groetsch and King [340], Groetsch and Neubauer [341], Gulliksson, Wedin, and Wei [360], Hanke [371], Hansen [374], Hilgers [419], [420], [421], Meleško [541], Morozov [581], [582], Nashed and Wahba [601], Neubauer [605], [606], Neumaier [607], Tikhonov [806], [805], [807], [808], Tikhonov and Arsenin [809].

APPLICATIONS TO APPROXIMATION AND SPLINE APPROXIMATION. Groetsch [334], [337], Herring [413], Izumino [440], Jerome and Schumaker [445], Laurent [502], [503], Weinberger [862].

SECTION 7. Groetsch ([332], [333], [336]), Groetsch and Jacobs [339], Kammerer and Nashed [464], [463], [465], [466], Lardy ([500], [501]), Showalter [755], Zlobec [892].

SECTION 8. Christensen [192], [193], [194], Daubechies [213], Heil and Walnut [411], and their references.

The Moore of the Moore–Penrose Inverse

1. Introduction

This Appendix is based on [**76**].

E.H. Moore (1862–1932) introduced and studied the *general reciprocal* during the decade 1910–1920. He stated the objective as follows:

> "The effectiveness of the reciprocal of a non–singular finite matrix in the study of properties of such matrices makes it desirable to define if possible an analogous matrix to be associated with each finite matrix κ^{12} even if κ^{12} is not square or, if square, is not necessarily non–singular." [**576**, p. 197].

Moore constructed the general reciprocal, established its uniqueness and main properties, and justified its application to linear equations. This work appears in [**575**], [**576**, Part 1, pp. 197–209].

The general reciprocal was rediscovered by R. Penrose [**635**] in 1955 and is, nowadays, called the *Moore–Penrose inverse*. It had to be rediscovered because Moore's work was sinking into oblivion even during his lifetime: it was much too idiosyncratic and used unnecessarily complicated notation, making it illegible for all but very dedicated readers.

Much of Moore's work is today of interest only for historians. One of the exceptions is his work on the general reciprocal, that may still interest, and benefit, mathematical researchers. It is summarized below, and – where necessary – restated in plain English and modern notation.

To illustrate the difficulty of reading the original Moore, and the need for translation, here is a theorem from [**576**, Part 1, p. 202].

(29.3) **Theorem.**

$$\mathfrak{U}^C \, \mathfrak{B}^1 \, {}^{\text{II}} \, \mathfrak{B}^2 \, {}^{\text{II}} \, \kappa^{12} .)\, \cdot$$
$$\exists \,|\, \lambda^{21} \text{ type } \mathfrak{M}^2_{\kappa*} \, \overline{\mathfrak{M}}^1_{\kappa} \, \ni \, \cdot \, S^2 \, \kappa^{12} \, \lambda^{21} = \delta^{11}_{\mathfrak{M}^1_{\kappa}} \, \cdot \, S^1 \, \lambda^{21} \, \kappa^{12} = \delta^{22}_{\mathfrak{M}^2_{\kappa*}} .$$

One symbol needs explanation: $\mathfrak{U}$ stands for the *number system* used throughout and $\mathfrak{U}^C$ denotes a number system of *type C*, that is, a *quasi-field* with a *conjugate* and an *order relation*, see [**576**, Part 1, p. 174] for details. All results below are for type C number systems, so this assumption will not be repeated. The rest of the theorem, in plain English, is:

(29.3) **Theorem.** *For every matrix A there exists a unique matrix X : $R(A) \to R(A^H)$ such that*

$$AX = P_{R(A)}, \; XA = P_{R(A^H)}. \qquad \square$$

The plan of this appendix:

Section 2 summarizes the results of Moore's lecture to the American Mathematical Society in 1920 [**575**].

Section 3 is a translation of the main results in [**576**, Part 1, pp. 197–209].

2. The 1920 Lecture to the American Mathematical Society

This is an abstract of a lecture given by E.H. Moore at the Fourteenth Western Meeting of the American Mathematical Society, held at the University of Chicago in April 9–10, 1920. There were 19 lectures in two afternoons; only the abstracts, written by Arnold Dresden (Secretary of the Chicago Section) appear in the *Bulletin.* Dresden writes

> "In this paper Professor Moore calls attention to a useful extension of the classical notion of the reciprocal of a nonsingular square matrix." [**575**, p. 394].

The details: Let A be any $m \times n$ complex matrix. Then there exists a unique $n \times m$ matrix $A^\dagger$, the *reciprocal* of A, such that:

(1) the columns of $A^\dagger$ are linear combinations of the conjugate of the rows of A;

(2) the rows of $A^\dagger$ are linear combinations of the conjugate of the columns of A; and

(3) $AA^\dagger A = A$.

The relation between A and $A^\dagger$ is mutual: A is the reciprocal of $A^\dagger$, viz.:

(4) the columns of A are linear combinations of the conjugates of rows of $A^\dagger$;

(5) the rows of A are linear combinations of the conjugates of columns of $A^\dagger$; and

(6) $A^\dagger A A^\dagger = A^\dagger$.

If A is of rank r, then $A^\dagger$ is given explicitly as follows:

$(r \geq 2)$:

$$A^\dagger[j_1, i_1] = \frac{\displaystyle\sum_{\substack{i_2 < \cdots < i_r \\ j_2 < \cdots < j_r}} A\begin{pmatrix} i_2 \cdots i_r \\ j_2 \cdots j_r \end{pmatrix} \overline{A\begin{pmatrix} i_1\ i_2 \cdots i_r \\ j_1\ j_2 \cdots j_r \end{pmatrix}}}{\displaystyle\sum_{\substack{k_1 < \cdots < k_r \\ \ell_1 < \cdots < \ell_r}} A\begin{pmatrix} k_1 \cdots k_r \\ \ell_1 \cdots \ell_r \end{pmatrix} \overline{A\begin{pmatrix} k_1 \cdots k_r \\ \ell_1 \cdots \ell_r \end{pmatrix}}} \, ;$$

$(r = 1)$:

$$A^\dagger[j, i] = \frac{\overline{A[i, j]}}{\sum_{k\ell} A[k, \ell]\overline{A[k, \ell]}} \, ;$$

$(r = 0)$:

$$A^\dagger[j, i] = 0 \, ;$$

where $A \begin{pmatrix} g_1 \cdots g_k \\ h_1 \cdots h_k \end{pmatrix}$ denotes the determinant of the $k \times k$ matrix with elements $A[g_i, h_j]$ and $\overline{x}$ denotes the conjugate of x.

The linear combinations of the columns of A $(A^\dagger)$ are the linear combinations of the rows of $A^\dagger$ (A) and constitute the m–dimensional vectors $\mathbf{y}$ (n–dimensional vectors $\mathbf{x}$) of an r–dimensional subspace M (N) of $\mathbb{C}^m$ ($\mathbb{C}^n$). Let $\overline{M}$ $(\overline{N})$ denote the conjugate space of the conjugate vectors $\overline{\mathbf{y}}$ $(\overline{\mathbf{x}})$. Then the matrices $A, A^\dagger$ establish one-to-one linear vector correspondences between the spaces $M, \overline{M}$ and the respective subspaces $N, \overline{N}$; $\mathbf{y} = A\mathbf{x}$ is equivalent to $\mathbf{x} = A^\dagger \mathbf{y}$ and $\overline{\mathbf{x}} = \overline{\mathbf{y}}A$ is equivalent to $\overline{\mathbf{u}} = \overline{\mathbf{v}}A^\dagger$.

3. The General Reciprocal in *General Analysis*

The centerpiece of Moore's work on the general reciprocal is Section 29 of [**576**], his treatise on *General Analysis*, edited by R.W. Barnard and published posthumously. These results were since rediscovered, some more than once.

For a matrix A denote:

A^H the conjugate transpose of A; and

$R(A)$ the range of A.

For index sets I, J:

A_{I*} or $A[I, *]$ the submatrix of rows indexed by I;

A_{*J} or $A[*, J]$ the submatrix of columns indexed by J; and

A_{IJ} the submatrix of A with rows in I and columns in J.

If A is nonsingular, its inverse A^{-1} satisfies,

$$AX = I, \quad XA = I.$$

Moore begins by introducing *generalized identity matrices* (orthogonal projectors) to replace the identity matrices above. This is done in Lemma (29.1)(5) and (6), and Theorem (29.2). The *general reciprocal* is then constructed in Theorems (29.3) and (29.4), and its properties are studied in the sequel.

(29.1) **Lemma.** *Let A be a non–zero $m \times n$ matrix, and let A_{IJ} be a maximal nonsingular submatrix of A.*

(1) $A_{*J}^H A_{*J}$ *is Hermitian and PD*[1].

(2) $\left(A_{*J}^H A_{*J}\right)^{-1}$ *is Hermitian and PD.*

(3) $A_{I*} A_{I*}^H$ *is Hermitian and PD.*

(4) $\left(A_{I*} A_{I*}^H\right)^{-1}$ *is Hermitian and PD.*

(5) $P_{R(A)} := A_{*J} \left(A_{*J}^H A_{*J}\right)^{-1} A_{*J}^H$,
        the generalized identity on $R(A)$.

(6) $P_{R(A^H)} := A_{I*}^H \left(A_{I*} A_{I*}^H\right)^{-1} A_{I*}$,
        the generalized identity on $R(A^H)$.

(7) $P_{R(A)}\mathbf{x} = \mathbf{x}$ *for all $\mathbf{x} \in R(A)$.*

(8) $\mathbf{x}^H P_{R(A)} = \mathbf{x}^H$ *for all $\mathbf{x} \in R(A)$.*

(9) $P_{R(A^H)}\mathbf{x} = \mathbf{x}$ *for all $\mathbf{x} \in R(A^H)$.*

---

[1]Moore calls it *proper* (i.e., the determinants of all principal minors are nonzero), *positive* (i.e., the corresponding quadratic form is nonnegative) and Hermitian.

(10) $\mathbf{x}^H P_{R(A^H)} = \mathbf{x}^H$ for all $\mathbf{x} \in R(A^H)$.

(11) Let $X := A_{I*}^H \left(A_{I*} A_{I*}^H \right)^{-1} A_{IJ} \left(A_{*J}^H A_{*J} \right)^{-1} A_{*J}^H$

$\qquad = A_{I*}^H \left(A_{I*}^H A_{I*} \right)^{-1} P_{R(A)}[I, *]$

$\qquad = P_{R(A^*)}[*, J] \left(A_{*J}^H A_{*J} \right)^{-1} A_{*J}^H,$

   the general reciprocal *of* A.

(12) X maps $R(A^H)$ onto $R(A)$.

(13) $AX = P_{R(A)}$.

(14) $XA = P_{R(A^H)}$.   □

(29.2) Theorem. *Let M be a finite-dimensional subspace.*

(1) *There exists a unique linear operator*[2] P_M *such that*

$$P_M \mathbf{x} = \mathbf{x}, \ \mathbf{x}^H P_M = \mathbf{x}^H, \quad \text{for all } \mathbf{x} \in M.$$

(2) P_M *is Hermitian, PSD, and idempotent.*

(3) $M = R(P_M)$.

(4) *For all* $\mathbf{x}$: $P_M \mathbf{x} \in M$, $(\mathbf{x} - P_M \mathbf{x}) \in M^\perp$.

(5) $\mathbf{x} \perp M \iff P_M \mathbf{x} = \mathbf{0}$.

(6) *For any matrix* A,

$$A = P_M \iff \begin{cases} A\mathbf{x} = \mathbf{x}, & \text{for all } \mathbf{x} \in M, \\ R(A^H) \subset M, \end{cases}$$

$$\iff \begin{cases} A\mathbf{x} = \mathbf{x}, & \text{for all } \mathbf{x} \in M, \\ A\mathbf{x} = \mathbf{0}, & \text{for all } \mathbf{x} \in M^\perp. \end{cases}$$ □

(29.3) Theorem. *For every matrix A there exists a unique matrix* $X : R(A) \to R(A^H)$ *such that*

$$AX = P_{R(A)}, \quad XA = P_{R(A^H)}.$$

We call X the *general reciprocal* and denote it by $A^\dagger$.

(29.4) Theorem. *For every matrix A the general reciprocal $A^\dagger$ satisfies:*

(1) $A^\dagger A A^\dagger = A^\dagger$, $AA^\dagger A = A$.

(2) $\operatorname{rank} A = \operatorname{rank} A^\dagger$.

(3) $R(A) = R(A^{\dagger H})$, $R(A^H) = R(A^\dagger)$.

(4) $A^{\dagger H} = (A^H)^\dagger$, $A = (A^\dagger)^\dagger$.   □

(29.45) Corollary. *If $A[I, J]$ is a maximal nonsingular submatrix of A, then:*

(1) $A^\dagger = P_{R(A^H)}[*, J] A_{IJ}^{-1} P_{R(A)}[I, *]$.

(2) $\mathbf{x}^H A^\dagger \mathbf{y} = \mathbf{x}_I^H A_{IJ}^{-1} \mathbf{y}_J$.   □

(29.5) Theorem. *For any matrix A, the following statements on a matrix X are equivalent:*

(a) $X = A^\dagger$.

(b) $R(X) \subset R(A^H)$, $AX = P_{R(A)}$.

(c) $R(X) \subset R(A^H)$, $R(X^H) \subset R(A)$, $AXA = A$.   □

(29.55) Corollary. *If* $A = \begin{bmatrix} B & O \\ O & C \end{bmatrix}$, *then* $A^\dagger = \begin{bmatrix} B^\dagger & O \\ O & C^\dagger \end{bmatrix}$.   □

---

[2]The *generalized identity matrix* for the subspace M, denoted by δ_M [**576**, p. 199].

(29.6) **Theorem.** *Let the matrix A be Hermitian. Then:*
  (1) *$A^\dagger$ is Hermitian.*
  (2) *If A is PSD, then so is $A^\dagger$.*   □

Consider a square matrix A. Then for any principal submatrix A_{II},

$$A_{II} = A_{II} A_{II}^\dagger A_{II}$$

More can be said if A is Hermitian and PSD:

(29.7) **Theorem.** *Let A be Hermitian and PSD. Then, for any principal submatrix A_{II},*
  (1) $A_{II} A_{II}^\dagger A_{I*} = A_{I*}$.
  (2) $A_{*I} A_{II}^\dagger A_{II} = A_{*I}$.   □

(29.8) **Theorem.** *Let A be Hermitian and PSD. Then the following statements, about a vector $\mathbf{x}$, are equivalent:*
  (a) $\mathbf{x}^H A \mathbf{x} = 0$;
  (b) $\mathbf{x} \perp R(A)$;
  (c) $\mathbf{x} \perp R(A^\dagger)$; and
  (d) $\mathbf{x}^H A^\dagger \mathbf{x} = 0$.   □

The general reciprocal can be used to solve linear equations

$$A\mathbf{x} = \mathbf{b},$$

that are assumed consistent, i.e., $\mathbf{b} \in R(A)$, or the way Moore expresses consistency: $\operatorname{rank} A = \operatorname{rank}[A \quad \mathbf{b}]$.

(29.9) **Theorem.** *Let A be a matrix and $\mathbf{b}$ a vector in $R(A)$. Then the general solution of $A\mathbf{x} = \mathbf{b}$ is*

$$A^\dagger \mathbf{b} + \{\mathbf{y} : \mathbf{y} \perp R(A^H)\}.$$ □

REMARK. Moore avoids the concept of null space, and the equivalent form of the general solution, $A^\dagger \mathbf{b} + N(A)$. Also, Moore does not consider the case where $A\mathbf{x} = \mathbf{b}$ is inconsistent. A. Bjerhammar [102], R. Penrose [636], and Yuan–Yung Tseng[3] [820] would later use $A^\dagger$ to obtain least–squares solutions. This has become the major application of the Moore–Penrose inverse.

---

[3]Tseng, a student of Barnard at Chicago (1933), extended the Moore–Penrose inverse to linear operators, see Definition 9.1, page 336.

Bibliography

1. N. N. Abdelmalek, *On the solutions of the linear least squares problems and pseudo-inverses*, Computing **13** (1974), no. 3-4, 215–228.

2. S. N. Afriat, *On the latent vectors and characteristic values of products of pairs of symmetric idempotents*, Quart. J. Math. Oxford Ser. (2) **7** (1956), 76–78.

3. _____, *Orthogonal and oblique projectors and the characteristics of pairs of vector spaces*, Proc. Cambridge Philos. Soc. **53** (1957), 800–816.

4. A. C. Aitken, *On least squares and linear combinations of observations*, Proc. Roy. Soc. Edinburgh, Sec A **55** (1934), 42–47.

5. F. Akdeniz, *A note concerning the Gauss–Markov theorem*, J. Fac. Sci. Karadeniz Tech. Univ. **1** (1977), 129–133.

6. I. S. Alalouf and G. P. H. Styan, *Characterizations of estimability in the general linear model*, Ann. Statist. **7** (1979), no. 1, 194–200.

7. _____, *Estimability and testability in restricted linear models*, Math. Operations-forsch. Statist. Ser. Statist. **10** (1979), no. 2, 189–201.

8. A. Albert, *Conditions for positive and nonnegative definiteness in terms of pseudo-inverses*, SIAM J. Appl. Math. **17** (1969), 434–440.

9. _____, *Regression and the Moore–Penrose Pseudoinverse*, Academic Press, New York, 1972.

10. _____, *The Gauss–Markov theorem for regression models with possibly singular covariances*, SIAM J. Appl. Math. **24** (1973), 182–187.

11. _____, *Statistical applications of the pseudo inverse*, In Nashed [**593**], pp. 525–548.

12. A. Albert and R. W. Sittler, *A method for computing least squares estimators that keep up with the data*, SIAM J. Control **3** (1965), 384–417.

13. E. L. Allgower, K. Böhmer, A. Hoy, and V. Janovský, *Direct methods for solving singular nonlinear equations*, Z. Angew. Math. Mech. **79** (1999), 219–231.

14. E. L. Allgower and K. Georg, *Numerical path following*, Handbook of Numerical Analysis, Vol. V, North-Holland, Amsterdam, 1997, pp. 3–207.

15. D. Alpay, J. A. Ball, and V. Bolotnikov, *On the bitangential interpolation problem for contractive valued functions in the polydisk*, J. Operator Theory **44** (2000), no. 2, 277–301.

16. D. Alpay, V. Bolotnikov, and Ph. Loubaton, *One two-sided residue interpolation for matrix-valued H_2-functions with symmetries*, J. Math. Anal. Appl. **200** (1996), no. 1, 76–105.

17. D. Alpay, V. Bolotnikov, and L. Rodman, *One-sided tangential interpolation for operator-valued Hardy functions on polydisks*, Integral Equations Operator Theory **35** (1999), no. 3, 253–270.

18. _____, *Two-sided tangential interpolation for Hilbert-Schmidt operator functions on polydisks*, Operator Theory: Advances and Applications **124** (2001), 21–62.

19. M. Altman, *A generalization of Newton's method*, Bull. Acad. Polon. Sci. Sér. Sci. Math. Astronom. Phys. **3** (1955), 189–193.

20. _____, *On a generalization of Newton's method*, Bull. Acad. Polon. Sci. Sér. Sci. Math. Astronom. Phys. **5** (1957), 789–795.

21. W. N. Anderson, Jr. and R. J. Duffin, *Series and parallel addition of matrices*, SIAM J. Appl. Math. **26** (1969), 576–594, (see [**483**]).

22. T. Ando, *Generalized Schur complements*, Linear Algebra Appl. **27** (1979), 173–186.

23. A. C. Antoulas, *Approximation of linear operators in the 2-norm*, Linear Algebra Appl. **278** (1998), no. 1-3, 309–316.

24. E. Arghiriade, *Sur les matrices qui sont permutables avec leur inverse généralisée*, Atti Accad. Naz. Lincei Rend. Cl. Sci. Fis. Mat. Natur. (8) **35** (1963), 244–251.

25. ———, *Sur l'inverse généralisée d'un operateur lineaire dans les espaces de Hilbert*, Atti Accad. Naz. Lincei Rend. Cl. Sci. Fis. Mat. Natur. (8) **45** (1968), 471–477.

26. N. Aronszajn, *Theory of reproducing kernels*, Trans. Amer. Math. Soc. **68** (1950), 337–404.

27. F. V. Atkinson, *The normal solvability of linear equations in normed spaces* (Russian), Mat. Sb. (N.S.) **28(70)** (1951), 3–14.

28. ———, *On relatively regular operators*, Acta Sci. Math. Szeged **15** (1953), 38–56.

29. K. E. Atkinson, *The solution of non-unique linear integral equations*, Numer. Math. **10** (1967), 117–124, (see also [**578**]).

30. L. Autonne, *Sur les matrices hypohermitiennes et sur les matrices unitaires*, Ann. Univ. Lyon, Nouvelle Sér. I **38** (1915), 1–77, (see history of SVD in [**783**]).

31. J. K. Baksalary, *A relationship between the star and minus orderings*, Linear Algebra Appl. **82** (1986), 163–167.

32. J. K. Baksalary and J. Hauke, *Partial orderings of matrices referring to singular values or eigenvalues*, Linear Algebra Appl. **96** (1987), 17–26.

33. J. K. Baksalary and R. Kala, *On equalities between BLUEs, WLSEs, and SLSEs*, Canad. J. Statist. **11** (1983), no. 2, 119–123, (extension of [**526**]).

34. J. K. Baksalary and T. Mathew, *Rank invariance criterion and its application to the unified theory of least squares*, Linear Algebra Appl. **127** (1990), 393–401.

35. J. K. Baksalary and S. K. Mitra, *Left-star and right-star partial orderings*, Linear Algebra Appl. **149** (1991), 73–89.

36. J. K. Baksalary and F. Pukelsheim, *On the Löwner, minus, and star partial orderings of nonnegative definite matrices and their squares*, Linear Algebra Appl. **151** (1991), 135–141.

37. J. K. Baksalary, F. Pukelsheim, and G. P. H. Styan, *Some properties of matrix partial orderings*, Linear Algebra Appl. **119** (1989), 57–85.

38. J. K. Baksalary, S. Puntanen, and H. Yanai, *Canonical correlations associated with symmetric reflexive generalized inverses of the dispersion matrix*, Linear Algebra Appl. **176** (1992), 61–74.

39. J. K. Baksalary and G. P. H. Styan, *Generalized inverses of partitioned matrices in Banachiewicz-Schur form*, Linear Algebra Appl. **354** (2002), 41–47, Ninth special issue on linear algebra and statistics.

40. A. V. Balakrishnan, *An operator theoretic formulation of a class of control problems and a steepest descent method of solution*, J. Soc. Indust. Appl. Math. Ser. A: Control **1** (1963), 109–127.

41. K. S. Banerjee, *Singularity in Hotelling's weighing designs and generalized inverses*, Ann. Math. Statist. **37** (1966), 1021–1032, (corrections: *ibid* **40**(1969), 710).

42. K. S. Banerjee and W. T. Federer, *On the structure and analysis of singular fractional replicates*, Ann. Math. Statist. **39** (1968), 657–663.

43. R. B. Bapat, *Linear Algebra and Linear Models*, 2nd ed., Hindustan Book Agency, New Delhi, 1999.

44. ———, *Linear estimation in models based on a graph*, Linear Algebra Appl. **302/303** (1999), 223–230.

45. ———, *Moore–Penrose inverse of set inclusion matrices*, Linear Algebra Appl. **318** (2000), no. 1-3, 35–44.

46. ———, *Outer inverses: Jacobi type identities and nullities of submatrices*, Linear Algebra Appl. **361** (2003), 107–120.

47. R. B. Bapat and A. Ben-Israel, *Singular values and maximum rank minors of generalized inverses*, Linear and Multilinear Algebra **40** (1995), no. 2, 153–161.

48. R. B. Bapat, S. K. Jain, and L. E. Snyder, *Nonnegative idempotent matrices and the minus partial order*, Linear Algebra Appl. **261** (1997), 143–154.

49. R. B. Bapat, K. P. S. Bhaskara Rao, and K. M. Prasad, *Generalized inverses over integral domains*, Linear Algebra Appl. **140** (1990), 181–196.

50. R. B. Bapat and B. Zheng, *Generalized inverses of bordered matrices*, Electron. J. Lin. Algeb. **10** (2003), 16–30.

51. S. Barnett, *Matrices in Control Theory*, Van Nostrand Reinhold, London, 1971.

52. D. Batigne, *Integral generalized inverses of integral matrices*, Linear Algebra Appl. **22** (1978), 125–134.

53. D. R. Batigne, F. J. Hall, and I. J. Katz, *Further results on integral generalized inverses of integral matrices*, Linear and Multilinear Algebra **6** (1978/79), no. 3, 233–241.

54. F. L. Bauer, *Theory of norms*, Technical Report CS 75, Stanford University, Computer Science Department, Stanford, CA, 1967.

55. E. F. Beckenbach and R. Bellman, *Inequalities*, 3rd ed., Springer-Verlag, New York, 1971.

56. R. Bellman, *Introduction to Matrix Analysis*, 2nd ed., McGraw-Hill, New York, 1970.

57. E. Beltrami, *Sulle funzioni bilineari*, Giornale di Matematiche ad Uso degli Studenti Delle Universita **11** (1873), 98–106, (an English translation by D. Boley is available as University of Minnesota, Department of Computer Science, Technical Report 90–37, 1990 (see history of SVD in [**783**])).

58. E. J. Beltrami, *A constructive proof of the Kuhn–Tucker multiplier rule*, J. Math. Anal. Appl. **26** (1969), 297–306.

59. A. Ben-Israel, *A modified Newton–Raphson method for the solution of systems of equations*, Israel J. Math. **3** (1965), 94–98.

60. _____, *A Newton–Raphson method for the solution of systems of equations*, J. Math. Anal. Appl. **15** (1966), 243–252.

61. _____, *A note on an iterative method for generalized inversion of matrices*, Math. Comp. **20** (1966), 439–440.

62. _____, *A note on the Cayley transform*, Notices Amer. Math. Soc. **13** (1966), 599.

63. _____, *On error bounds for generalized inverses*, SIAM J. Numer. Anal. **3** (1966), 585–592, (see also [**779**]).

64. _____, *On iterative methods for solving nonlinear least squares problems over convex sets*, Israel J. Math. **5** (1967), 211–214.

65. _____, *On the geometry of subspaces in Euclidean n-spaces*, SIAM J. Appl. Math. **15** (1967), 1184–1198.

66. _____, *On applications of generalized inverses in nonlinear analysis*, In Boullion and Odell [**121**], pp. 183–202.

67. _____, *On decompositions of matrix spaces with applications to matrix equations*, Atti Accad. Naz. Lincei Rend. Cl. Sci. Fis. Mat. Natur. (8) **45** (1968), 122–128.

68. _____, *On optimal solutions of 2-person 0-sum games*, Atti Accad. Naz. Lincei Rend. Cl. Sci. Fis. Mat. Natur. (8) **44** (1968), 512–516.

69. _____, *A note on partitioned matrices and equations*, SIAM Rev. **11** (1969), 247–250.

70. _____, *On matrices of index zero or one*, SIAM J. Appl. Math. **17** (1969), 1118–1121, (see [**543**], [**714**]).

71. _____, *A Cramer rule for least-norm solutions of consistent linear equations*, Linear Algebra Appl. **43** (1982), 223–226, (see [**835**], [**181**], [**183**], [**182**], [**787**], [**847**], [**849**], [**864**]).

72. _____, *A volume associated with m×n matrices*, Linear Algebra Appl. **167** (1992), 87–111, (this concept was introduced by Good [**315**]).

73. _____, *The change-of-variables formula using matrix volume*, SIAM J. Matrix Anal. Appl. **21** (1999), no. 1, 300–312 (electronic).

74. _____, *A local inverse for nonlinear mappings*, Numer. Algorithms **25** (2000), no. 1-4, 37–46, Mathematical Journey through Analysis, Matrix Theory and Scientific Computation (Kent, OH, 1999).

75. _____, *An application of the matrix volume in probability*, Linear Algebra Appl. **321** (2001), 9–25.

76. _____, *The Moore of the Moore–Penrose inverse*, Electron. J. Lin. Algeb. **9** (2002), 150–157.

77. A. Ben-Israel and A. Charnes, *Contributions to the theory of generalized inverses*, J. Soc. Indust. Appl. Math. **11** (1963), 667–699.

78. _____, *Generalized inverses and the Bott-Duffin network analysis*, J. Math. Anal. Appl. **7** (1963), 428–435, (corrections: *ibid* **18**(1967), 393).

79. _____, *An explicit solution of a special class of linear programming problems*, Oper. Res. **16** (1968), 1166–1175, (see [**88**], [**99**], [**694**], [**723**]).

80. A. Ben-Israel, A. Charnes, and P. D. Robers, *On generalized inverses and interval linear programming*, In Boullion and Odell [**121**], pp. 53–70.

81. A. Ben-Israel and D. Cohen, *On iterative computation of generalized inverses and associated projections*, SIAM J. Numer. Anal. **3** (1966), 410–419.

82. A. Ben-Israel and M. J. L. Kirby, *A characterization of equilibrium points of bimatrix games*, Atti Accad. Naz. Lincei Rend. Cl. Sci. Fis. Mat. Natur. (8) **46** (1969), 402–407.

83. A. Ben-Israel and S. J. Wersan, *An elimination method for computing the generalized inverse of an arbitrary complex matrix*, J. Assoc. Comput. Mach. **10** (1963), 532–537.

84. A. Ben-Tal and M. Teboulle, *A geometric property of the least squares solution of linear equations*, Linear Algebra Appl. **139** (1990), 165–170, (see [**86**], [**231**], [**72**], [**85**], [**114**], [**279**]).

85. _____, Addenda: *A geometric property of the least squares solution of linear equations* [Linear Algebra Appl. **139** (1990), 165–170], Linear Algebra Appl. **180** (1993), 5.

86. L. Berg, *Three results in connection with inverse matrices*, Linear Algebra Appl. **84** (1986), 63–77, (see also [**84**]).

87. M. Berger and B. Gostiaux, *Differential Geometry: Manifolds, Curves and Surfaces*, Graduate Texts in Mathematics, No. 115, Springer-Verlag, New York, 1988, (translated by S. Levy).

88. A. Berman, *Generalized interval programming*, Bull. Calcutta Math. Soc. **71** (1979), no. 3, 169–176.

89. A. Berman and M. Neumann, *Proper splittings of rectangular matrices*, SIAM J. Appl. Math. **31** (1976), no. 2, 307–312.

90. A. Berman and R. J. Plemmons, *Monotonicity and the generalized inverse*, SIAM J. Appl. Math. **22** (1972), 155–161.

91. _____, *Nonnegative Matrices in the Mathematical Sciences*, Society for Industrial and Applied Mathematics (SIAM), Philadelphia, PA, 1994, (revised reprint of the 1979 original).

92. F. J. Beutler, *The operator theory of the pseudo-inverse. I. Bounded operators*, J. Math. Anal. Appl. **10** (1965), 451–470.

93. _____, *The operator theory of the pseudo-inverse. II. Unbounded operators with arbitrary range*, J. Math. Anal. Appl. **10** (1965), 471–493.

94. K. P. S. Bhaskarà Rao, *The Theory of Generalized Inverses over Commutative Rings*, Taylor & Francis, London, 2002.

95. R. Bhatia, Letter to the editor: *The n-dimensional Pythagorean theorem*, Linear and Multilinear Algebra **30** (1991), no. 1-2, 155, (see [**517**]).

96. _____, *Matrix Analysis*, Springer-Verlag, New York, 1997.

97. P. Bhimasankaram, *Rank factorization of a matrix and its applications*, Math. Sci. **13** (1988), no. 1, 4–14, (see [**687**]).

98. P. Bhimasankaram and R. SahaRay, *On a partitioned linear model and some associated reduced models*, Linear Algebra Appl. **264** (1997), 329–339.

99. M. Bilodeau, *Sur une représentation explicite des solutions optimales d'un programme linéaire*, Canad. Math. Bull. **29** (1986), no. 4, 419–425.

100. Z. W. Birnbaum, *Introduction to Probability and Mathematical Statistics*, Harper & Brothers, New York, 1962.

101. A. Bjerhammar, *Application of calculus of matrices to method of least squares with special reference to geodetic calculations*, Trans. Roy. Inst. Tech. Stockholm (1951), no. 49, 86 pp. (2 plates).

102. _____, *Rectangular reciprocal matrices, with special reference to geodetic calculations*, Bull. Géodésique (1951), 188–220.

103. _____, *A generalized matrix algebra*, Trans. Roy. Inst. Tech. Stockholm **1958** (1958), no. 124, 32 pp.

104. _____, *Theory of Errors and Generalized Matrix Inverses*, Elsevier Scientific, Amsterdam, 1973.

105. Å. Björck, *Iterative refinement of linear least squares solutions I*, BIT **7** (1967), 257–278.

106. _____, *Solving linear least squares problems by Gram–Schmidt orthogonalization*, BIT **7** (1967), 1–21.

107. _____, *Iterative refinement of linear least squares solutions II*, BIT **8** (1968), 8–30.

108. _____, *A uniform numerical method for linear estimation from general Gauss–Markov models*, Proceedings of the First Symposium on Comutational Statistics (COMPSTAT), (G. Bruckmann, F. Ferschl and L. Schmetterer, editors), Physica Verlag, Vienna, 1974, pp. 131–140.

109. _____, *Numerical Methods for Least Squares Problems*, Society for Industrial and Applied Mathematics (SIAM), Philadelphia, PA, 1994.

110. Å. Björck and C. Bowie, *An iterative algorithm for computing the best estimate of an orthogonal matrix*, SIAM J. Numer. Anal. **8** (1971), no. 2, 358–364.

111. Å. Björck and G. H. Golub, *Iterative refinement of linear least squares solutions by householder transformation*, BIT **7** (1967), 322–337.

112. _____, *Numerical methods for computing angles between linear subspaces*, Math. Comp. **27** (1973), 579–594.

113. J. W. Blattner, *Bordered matrices*, J. Soc. Indust. Appl. Math. **10** (1962), 528–536.

114. E. Y. Bobrovnikova and S. A. Vavasis, *A norm bound for projections with complex weights*, Linear Algebra Appl. **307** (2000), no. 1-3, 69–75, (a complex version of the bounds in [**782**], [**810**]).

115. F. Bohnenblust, *A characterization of complex Hilbert spaces*, Portugal. Math. **3** (1942), 103–109.

116. T. Bonnesen and W. Fenchel, *Theorie der konvexen Körper*, Springer-Verlag, Berlin, 1934.

117. C. de Boor, *The Method of Projections as Applied to the Numerical Solution of Two Point Boundary Value Problems Using Cubic Splines*, Doctoral dissertation in mathematics, University of Michigan, Ann Arbor, MI, 1966.

118. N. K. Bose and S. K. Mitra, *Generalized inverse of polynomial matrices*, IEEE Trans. Autom. Control **23** (1978), no. 3, 491–493, (see also [**768**]).

119. R. C. Bose, *The fundamental theorem of linear estimation* (abstract), Proc. 31st Indian Sci. Congress (1944), 2–3.

120. R. Bott and R. J. Duffin, *On the algebra of networks*, Trans. Amer. Math. Soc. **74** (1953), 99–109.

121. T. L. Boullion and P. L. Odell (eds.), *Proceedings of the Symposium on Theory and Applications of Generalized Inverses of Matrices*, Lubbock, Texas Tech. Press, 1968.

122. _____, *A note on the Scroggs–Odell pseudoinverse*, SIAM J. Appl. Math. **17** (1969), 7–10, (correction of [**736**, Theorem 6]).

123. _____, *Generalized Inverse Matrices*, John Wiley & Sons, New York, 1971.

124. E. Bounitzky, *Sur la fonction de Green des équations differentielles linéaires ordinaires*, J. Math. Pures Appl. **5** (1909), no. 6, 65–125.

125. N. Bourbaki, *Espaces Vectoriels Topologiques*, Eléments de Mathématiques. Livre V, Hermann & Cie, Paris, 1953.

126. _____, *Algèbre*, Eléments de Mathématiques. Livre II, Hermann & Cie, Paris, 1958.

127. V. J. Bowman and C.-A. Burdet, *On the general solution to systems of mixed-integer linear equations*, SIAM J. Appl. Math. **26** (1974), 120–125.

128. Yu. E. Boyarintsev, *General solutions of boundary value problems for singular systems of ordinary differential equations*, Čisl. Metody Meh. Splošn. Sredy **8** (1977), no. 7, 12–21.

129. _____, *Regulyarnye i Singulyarnye Sistemy Lineinykh Obyknovennykh Differentsialnykh Uravnenii*, Nauka Sibirsk. Otdel., Novosibirsk, 1980.

130. _____, *Methods of Solving Singular Systems of Ordinary Differential Equations*, Wiley, Chichester, 1992, (translation of the 1988 Russian original).

131. _____, *A resolving transformation of unknowns in an implicit system of ordinary differential equations*, Algebrodifferential Systems and Methods for their Solution (Russian), Nauka, Novosibirsk, 1993, pp. 4–19, 90.

132. Yu. E. Boyarintsev and V. M. Korsukov, *The structure of a general continuously differentiable solution of a boundary value problem for a singular system of ordinary differential equations*, Problems in Applied Mathematics (Russian), Sibirsk. Ènerget. Inst., Akad. Nauk SSSR Sibirsk. Otdel., Irkutsk, 1977, pp. 73–93.

133. J. S. Bradley, *Adjoint quasi-differential operators of Euler type*, Pacific J. Math. **16** (1966), 213–237.

134. _____, *Generalized Green's matrices for compatible differential systems*, Michigan Math. J. **13** (1966), 97–108.

135. L. Brand, *The solution of linear algebraic equations*, Math. Gaz. **46** (1962), 203–237.

136. C. G. den Broeder Jr. and A. Charnes, *Contributions to the theory of generalized inverses for matrices*, Technical Report, Purdue University, Department of Mathematics, Lafayette, IN, 1957, (reprinted as ONR Res. Memo. No. 39, Northwestern University, Evanston, IL, 1962).

137. R. A. Brualdi and H. Schneider, *Determinantal identities: Gauss, Schur, Cauchy, Sylvester, Kronecker, Jacobi, Binet, Laplace, Muir and Cayley*, Linear Algebra Appl. **52** (1983), 769–791.

138. J. T. Bruening, *A new formula for the Moore–Penrose inverse*, Current Trends in Matrix Theory (Auburn, AL, 1986), North-Holland, New York, 1987, pp. 65–74.

139. W. Burmeister, *Inversionfreie Verfahren zur Lösung nichtlinearer Operatorgleichungen*, Z. Angew. Math. Mech. **52** (1972), 101–110.

140. F. Burns, D. Carlson, E. V. Haynsworth, and T. Markham, *Generalized inverse formulas using the Schur complement*, SIAM J. Appl. Math. **26** (1974), 254–259.

141. P. A. Businger and G. H. Golub, *Linear least squares by Householder transformations*, Numer. Math. **7** (1965), 269–276, (republished, pp. 111–118 in [**875**]).

142. _____, *Algorithm 358: Singular value decomposition of a complex matrix*, Comm. ACM **12** (1969), 564–565.

143. C. A. Butler and T. D. Morley, *A note on the shorted operator*, SIAM J. Matrix Anal. Appl. **9** (1988), no. 2, 147–155.

144. _____, *Six generalized Schur complements*, Linear Algebra Appl. **106** (1988), 259–269.

145. S. L. Campbell, *The Drazin inverse of an infinite matrix*, SIAM J. Appl. Math. **31** (1976), no. 3, 492–503, (see [**153**]).

146. _____, *Optimal control of autonomous linear processes with singular matrices in the quadratic cost functional*, SIAM J. Control Optim. **14** (1976), no. 6, 1092–1106.

147. _____, *Linear systems of differential equations with singular coefficients*, SIAM J. Math. Anal. **8** (1977), no. 6, 1057–1066.

148. _____, *Optimal control of discrete linear processes with quadratic cost*, Internat. J. Systems Sci. **9** (1978), no. 8, 841–847.

149. _____, *Singular perturbation of autonomous linear systems. II*, J. Differential Equations **29** (1978), no. 3, 362–373.

150. _____, *Limit behavior of solutions of singular difference equations*, Linear Algebra Appl. **23** (1979), 167–178.

151. _____, *On a singularly perturbed autonomous linear control problem*, IEEE Trans. Automat. Control **24** (1979), no. 1, 115–117.

152. _____, *Singular Systems of Differential Equations*, Pitman (Advanced Publishing Program), Boston, MA, 1980.

153. _____, *The Drazin inverse of an operator*, [**155**], pp. 250–260.

154. _____, *On positive controllers and linear quadratic optimal control problems*, Internat. J. Control **36** (1982), no. 5, 885–888.

155. S. L. Campbell (ed.), *Recent Applications of Generalized Inverses*, Boston, MA, Pitman (Advanced Publishing Program), 1982.

156. _____, *Singular Systems of Differential Equations*. II, Pitman (Advanced Publishing Program), Boston, MA, 1982.

157. _____, *Index two linear time-varying singular systems of differential equations*, SIAM J. Algebraic Discrete Methods **4** (1983), no. 2, 237–243.

158. _____, *Control problem structure and the numerical solution of linear singular systems*, Math. Control Signals Systems **1** (1988), no. 1, 73–87.

159. S. L. Campbell and C. D. Meyer, Jr., *Generalized Inverses of Linear Transformations*, Pitman (Advanced Publishing Program), Boston, MA, 1979, (reprinted by Dover, 1991).

160. S. L. Campbell, C. D. Meyer, Jr., and N. J. Rose, *Applications of the Drazin inverse to linear systems of differential equations with singular constant coefficients*, SIAM J. Appl. Math. **31** (1976), no. 3, 411–425, (see [**627**], [**195**], [**358**], [**874**]).

161. S. L. Campbell and M. Rakowski, *Explicit formulae for completions of linear time varying singular systems of differential equations*, Circuits Systems Signal Process. **13** (1994), no. 2-3, 185–199.

162. S. L. Campbell and N. J. Rose, *Singular perturbation of autonomous linear systems*. III, Houston J. Math. **4** (1978), no. 4, 527–539.

163. _____, *Singular perturbation of autonomous linear systems*, SIAM J. Math. Anal. **10** (1979), no. 3, 542–551.

164. _____, *A second order singular linear system arising in electric power systems analysis*, Internat. J. Systems Sci. **13** (1982), no. 1, 101–108.

165. D. Carlson, *Matrix decompositions involving the Schur complement*, SIAM J. Appl. Math. **28** (1975), 577–587.

166. _____, *What are Schur complements, anyway?*, Linear Algebra Appl. **74** (1986), 257–275.

167. _____, *Generalized inverse invariance, partial orders, and rank-minimization problems for matrices*, Current Trends in Matrix Theory (Auburn, AL, 1986), North-Holland, New York, 1987, pp. 81–87.

168. D. Carlson, E. V. Haynsworth, and T. Markham, *A generalization of the Schur complement by means of the Moore–Penrose inverse*, SIAM J. Appl. Math. **26** (1974), 169–175.

169. N. Castro González, *On the convergence of semi-iterative methods to the Drazin inverse solution of linear equations in Banach spaces*, Collect. Math. **46** (1995), no. 3, 303–314.

170. N. Castro González and J. J. Koliha, *Semi-iterative methods for the Drazin inverse solution of linear equations in Banach spaces*, Numer. Funct. Anal. Optim. **20** (1999), no. 5-6, 405–418.

171. _____, *Perturbation of the Drazin inverse for closed linear operators*, Integral Equations Operator Theory **36** (2000), no. 1, 92–106.

172. N. Castro González, J. J. Koliha, and I. Straškraba, *Perturbation of the Drazin inverse*, Soochow J. Math. **27** (2001), no. 2, 201–211.

173. N. Castro González, J. J. Koliha, and Y. Wei, *Perturbation of the Drazin inverse for matrices with equal eigenprojections at zero*, Linear Algebra Appl. **312** (2000), no. 1-3, 181–189.

174. D. E. Catlin, *Estimation, Control, and the Discrete Kalman Filter*, Springer–Verlag, New York, 1989, (see, in particular, pp. 100–113).

175. A. Charnes and W. W. Cooper, *Structural sensitivity analysis in linear programming and an exact product form left inverse*, Naval Res. Logist. Quart. **15** (1968), 517–522.

176. A. Charnes, W. W. Cooper, and G. L. Thompson, *Constrained generalized medians and hypermedians as deterministic equivalents for two-stage linear programs under uncertainty*, Management Sci. **12** (1965), 83–112.

177. A. Charnes and F. Granot, *Existence and representation of Diophantine and mixed Diophantine solutions to linear equations and inequalities*, Technical Report, The University of Texas, Center for Cybernetic Studies, Austin, TX, 1973.

178. A. Charnes and M. J. L. Kirby, *Modular design, generalized inverses and convex programming*, Oper. Res. **13** (1965), 836–847.

179. X.-J. Chen, M. Z. Nashed, and L. Qi, *Convergence of Newton's method for singular smooth and nonsmooth equations using adaptive outer inverses*, SIAM J. Optim. **7** (1997), 445–462.

180. Y.-L. Chen, *The generalized Bott-Duffin inverse and its applications*, Linear Algebra Appl. **134** (1990), 71–91.

181. _____, *A Cramer rule for solution of the general restricted linear equation*, Linear and Multilinear Algebra **34** (1993), no. 2, 177–186.

182. _____, *An explicit representation of the general solution to a system of constrained linear equations and Cramer's rule*, Gaoxiao Yingyong Shuxue Xuebao **8** (1993), no. 1, Ser. A, 61–70.

183. _____, *Representations and Cramer rules for the solution of a restricted matrix equation*, Linear and Multilinear Algebra **35** (1993), no. 3-4, 339–354.

184. E. W. Cheney, *Introduction to Approximation Theory*, McGraw–Hill, New York, 1966.

185. H. Chernoff, *Locally optimal designs for estimating parameters*, Ann. Math. Statist. **24** (1953), 586–602.

186. J. S. Chipman, *On least squares with insufficient observations*, J. Amer. Statist. Assoc. **54** (1964), 1078–1111, (see [**851**]).

187. _____, *Specification problems in regression analysis*, In Boullion and Odell [**121**], pp. 114–176.

188. _____, *"Proofs" and proofs of the Eckart–Young theorem*, Stochastic Processes and Functional Analysis (Riverside, CA, 1994), Dekker, New York, 1997, pp. 71–83.

189. _____, *Linear restrictions, rank reduction, and biased estimation in linear regression*, Linear Algebra Appl. **289** (1999), no. 1-3, 55–74.

190. J. S. Chipman and M. M. Rao, *On the treatment of linear restrictions in regression analysis*, Econometrica **32** (1964), 198–209.

191. _____, *Projections, generalized inverses and quadratic forms*, J. Math. Anal. Appl. **9** (1964), 1–11.

192. O. Christensen, *Frames and pseudo-inverses*, J. Math. Anal. Appl. **195** (1995), no. 2, 401–414.

193. _____, *Operators with closed range, pseudo-inverses, and perturbation of frames for a subspace*, Canad. Math. Bull. **42** (1999), no. 1, 37–45.

194. _____, *Frames, Riesz bases, and discrete Gabor/Wavelet expansions*, Bull. Amer. Math. Soc. **38** (2001), no. 3, 273–291.

195. M. A. Christodoulou and P. N. Paraskevopoulos, *Solvability, controllability, and observability of singular systems*, J. Optim. Theory Appl. **45** (1985), no. 1, 53–72.

196. M. T. Chu, R. E. Funderlic, and G. H. Golub, *On a variational formulation of the generalized singular value decomposition*, SIAM J. Matrix Anal. Appl. **18** (1997), no. 4, 1082–1092.

197. K.-L. Chung, *Elementary Probability Theory with Stochastic Processes*, Springer–Verlag, New York, 1974.

198. G. Cimmino, *Cramer's rule without the notion of determinant*, Atti Accad. Sci. Istit. Bologna Cl. Sci. Fis. Rend. (14) **3** (1985/86), 115–138 (1987).

199. J. A. Clarkson, *Uniformly convex spaces*, Trans. Amer. Math. Soc. **40** (1936), 396–414.

200. R. E. Cline, *Representations for the generalized inverse of a partitioned matrix*, J. Soc. Indust. Appl. Math. **12** (1964), 588–600.

201. _____, *Inverses of rank invariant powers of a matrix*, SIAM J. Numer. Anal. **5** (1968), 182–197.

202. R. E. Cline and T. N. E. Greville, *An extension of the generalized inverse of a matrix*, SIAM J. Appl. Math. **19** (1970), 682–688.

203. R. E. Cline and L. D. Pyle, *The generalized inverse in linear programming. An intersecton projection method and the solution of a class of structured linear programming problems*, SIAM J. Appl. Math. **24** (1973), 338–351.

204. E. A. Coddington and N. Levinson, *Theory of Ordinary Differential Equations*, McGraw–Hill, New York, 1955.

205. G. Corach, A. Maestripieri, and D. Stojanoff, *Generalized Schur complements and oblique projections*, Linear Algebra Appl. **341** (2002), 259–272.

206. C. Corradi, *Computing methods for restricted estimation in linear models*, Statistica (Bologna) **42** (1982), no. 1, 55–68, (see [**295**]).

207. R. W. Cottle, *Manifestations of the Schur complement*, Linear Algebra and Appl. **8** (1974), 189–211.

208. R. Courant, *Differential and Integral Calculus*, Interscience, New York, 1936, (translated by E.J. McShane).

209. R. Courant and D. Hilbert, *Methods of Mathematical Physics*. Vol. I, Interscience, New York, 1953, (first published in German 1924).

210. D. E. Crabtree and E. V. Haynsworth, *An identity for the Schur complement of a matrix*, Proc. Amer. Math. Soc. **22** (1969), 364–366.

211. D. F. Cudia, *Rotundity*, Convexity, Proc. Sympos. Pure Math., Vol. VII (V. Klee, editor), Amer. Math. Soc., Providence, RI, 1963, pp. 73–97.

212. C. G. Cullen and K. J. Gale, *A functional definition of the determinant*, Amer. Math. Monthly **72** (1965), 403–406.

213. I. Daubechies, *Ten Lectures on Wavelets*, CBMS-NSF Regional Conference Series in Applied Mathematics, vol. 61, SIAM, Philadelphia, PA, 1992.

214. C. Davis and W. M. Kahan, *The rotation of eigenvectors by a perturbation. III*, SIAM J. Numer. Anal. **7** (1970), 1–46.

215. D. L. Davis and D. W. Robinson, *Generalized inverses of morphisms*, Linear Algebra Appl. **5** (1972), 329–338.

216. L. De Lathauwer, B. De Moor, and J. Vanderwalle, *A multilinear singular value decomposition*, SIAM J. Matrix Anal. Appl. **21** (2000), no. 4, 1253–1278 (electronic).

217. _____, *On the best rank-1 and rank-$(R_1, R_2, \ldots, R_N)$ approximation of higher-order tensors*, SIAM J. Matrix Anal. Appl. **21** (2000), no. 4, 1324–1342 (electronic).

218. R. de Meersman, *Geometric meaning of the method of Golub and Kahan for the calculation of the singular values of a matrix*, Bull. Soc. Math. Belg. **22** (1970), 146–154.

219. B. De Moor, *Structured total least squares and l_2 approximation problems*, Linear Algebra Appl. **188/189** (1993), 163–205.

220. A. R. De Pierro and M.-S. Wei, *Some new properties of the equality constrained and weighted least squares problem*, Linear Algebra Appl. **320** (2000), no. 1-3, 145–165.

221. S. R. Deans, *The Radon Transform and some of its Applications* (revised reprint of the 1983 original), Robert E. Krieger, Malabar, FL, 1993.

222. H. P. Decell, Jr., *An alternate form of the generalized inverse of an arbitrary complex matrix*, SIAM Rev. **7** (1965), 356–358, (see [**891**]).

223. _____, *An application of the Cayley–Hamilton theorem to generalized matrix inversion*, SIAM Rev. **7** (1965), 526–528, (extended in [**848**]).

224. _____, *On the derivative of the generalized inverse of a matrix*, Linear and Multilinear Algebra **1** (1973/74), 357–359.

225. J. B. Dennis, *Mathematical Programming and Electrical Networks*, MIT Press, Cambridge, MA, 1959.

226. C. A. Desoer and B. H. Whalen, *A note on pseudoinverses*, J. Soc. Indust. Appl. Math. **11** (1963), 442–447.

227. P. Deuflhard and G. Heindl, *Affine invariant convergence theorems for Newton's method and extensions to related methods*, SIAM J. Numer. Anal. **16** (1979), 1–10.

228. E. Deutsch, *Semi-inverses, reflexive semi-inverses, and pseudo-inverses of an arbitrary linear transformation*, Linear Algebra Appl. **4** (1971), 313–322.

229. F. R. Deutsch, *The angle between subspaces of a Hilbert space*, Approximation Theory, Wavelets and Applications (Maratea, 1994), Kluwer Academic, Dordrecht, 1995, pp. 107–130.

230. _____, *Best Approximation in Inner Product Spaces*, Springer–Verlag, New York, 2001.

231. I. I. Dikin, *On the speed of an iterative process*, Upravlyaemye Sistemi **12** (1974), 54–60, (see [**833**]).

232. D. S. Djordjević, *Unified approach to the reverse order rule for generalized inverses*, Acta Sci. Math. (Szeged) **67** (2001), no. 3-4, 761–776.

233. M. P. Drazin, *Pseudo inverses in associative rings and semigroups*, Amer. Math. Monthly **65** (1958), 506–514.

234. _____, *Natural structures on semigroups with involution*, Bull. Amer. Math. Soc. **84** (1978), no. 1, 139–141.

235. _____, *A partial order in completely regular semigroups*, J. Algebra **98** (1986), 362–374.

236. H. Drygas, *The Coordinate-Free Approach to Gauss–Markov Estimation*, Springer–Verlag, Berlin, 1970.

237. _____, *Consistency of the least squares and Gauss–Markov estimators in regression models*, Z. Wahrsch. Verw. Gebiete **17** (1971), 309–326.

238. _____, *Estimation and prediction for linear models in general spaces*, Math. Operationsforsch. Statist. **6** (1975), no. 2, 301–324.

239. _____, *Gauss–Markov estimation for multivariate linear models with missing observations*, Ann. Statist. **4** (1976), no. 4, 779–787.

240. _____, *On the unified theory of least squares*, Probab. Math. Statist. **5** (1985), no. 2, 177–186.

241. R. J. Duffin, *Network models*, SIAM–AMS Proc., vol. III, Amer. Math. Soc., Providence, RI, 1971, pp. 65–91.

242. R. J. Duffin and T. D. Morley, *Inequalities induced by network connections. II. Hybrid connections*, J. Math. Anal. Appl. **67** (1979), no. 1, 215–231.

243. _____, *Inequalities induced by network connections*, In Campbell [**155**], pp. 27–49.

244. R. J. Duffin and A. C. Schaeffer, *A class of nonharmonic Fourier series*, Trans. Amer. Math. Soc. **72** (1952), 341–366.

245. D. B. Duncan and S. D. Horn, *Linear dynamic recursive estimation from the viewpoint of regression analysis*, J. Amer. Statist. Assoc. **67** (1972), 815–821, (connection between Kalman filter and least-squares regression)).

246. N. Dunford and J. T. Schwartz, *Linear Operators. Part I*, Interscience, New York, 1957.

247. C. Eckart and G. Young, *The approximation of one matrix by another of lower rank*, Psychometrika **1** (1936), 211–218.

248. _____, *A principal axis transformation for non-Hermitian matrices*, Bull. Amer. Math. Soc. **45** (1939), 118–121.

249. L. Eldén, *Perturbation theory for the least squares problem with linear equality constraints*, SIAM J. Numer. Anal. **17** (1980), 338–350.

250. _____, *A weighted pseudoinverse, generalized singular values, and constrained least squares problems*, BIT **22** (1983), 487–502.

251. W. W. Elliott, *Generalized Green's functions for compatible differential systems*, Amer. J. Math. **50** (1928), 243–258.

252. _____, *Green's functions for differential systems containing a parameter*, Amer. J. Math. **51** (1929), 397–416.

253. H. W. Engl, M. Hanke, and A. Neubauer, *Regularization of Inverse Problems*, Kluwer Academic, Dordrecht, 1996.

254. M. J. Englefield, *The commuting inverses of a square matrix*, Proc. Cambridge Philos. Soc. **62** (1966), 667–671.

255. P. J. Erdelsky, *Projections in a normed linear space and a generalization of the pseudo–inverse*, Doctoral dissertation in mathematics, California Inst. Tech., Pasadena, CA, 1969.

256. I. Erdélyi, *On partial isometries in finite dimensional Euclidean spaces*, SIAM J. Appl. Math. **14** (1966), 453–467.

257. _____, *On the "reversed order law" related to the generalized inverse of matrix products*, J. Assoc. Comput. Mach. **13** (1966), 439–443.

258. _____, *On the matrix equation $Ax = \lambda Bx$*, J. Math. Anal. Appl. **17** (1967), 119–132.

259. _____, *The quasi-commuting inverse for a square matrix*, Atti Accad. Naz. Lincei Rend. Cl. Sci. Fis. Mat. Natur. (8) **42** (1967), 626–633.

260. _____, *Normal partial isometries closed under multiplication on unitary spaces*, Atti Accad. Naz. Lincei Rend. Cl. Sci. Fis. Mat. Natur. (8) **43** (1968), 186–190.

261. _____, *Partial isometries and generalized inverses*, In Boullion and Odell [**121**], pp. 203–217.

262. _____, *Partial isometries closed under multiplication on Hilbert spaces*, J. Math. Anal. Appl. **22** (1968), 546–551.

263. _____, *Partial isometries defined by a spectral property on unitary spaces*, Atti Accad. Naz. Lincei Rend. Cl. Sci. Fis. Mat. Natur. (8) **44** (1968), 741–747.

264. _____, *A generalized inverse for arbitrary operators between hilbert spaces*, Proc. Cambridge Philos. Soc. **71** (1972), 43–50.

265. I. Erdélyi and A. Ben-Israel, *Extremal solutions of linear equations and generalized inversion between Hilbert spaces*, J. Math. Anal. Appl. **39** (1972), 298–313.

266. I. Erdélyi and F. R. Miller, *Decomposition theorems for partial isometries*, J. Math. Anal. Appl. **30** (1970), 665–679.

267. Ky Fan and A. J. Hoffman, *Some metric inequalities in the space of matrices*, Proc. Amer. Math. Soc. **6** (1955), 111–116.

268. L. Fantappiè, *Le calcul des matrices*, C. R. Acad. Sci. Paris **186** (1928), 619–621, (see [**691**]).

269. R. W. Farebrother, *Relations among set estimators: A bibliographical note*, Technometrics **27** (1985), 85–86.

270. H. Federer, *Geometric Measure Theory*, Springer-Verlag, New York, 1969.

271. W. Feller, *An Introduction to Probability Theory and Applications*, Vol. 1, Wiley, New York, 1950.

272. P. A. Filmore and J. P. Williams, *On operator ranges*, Adv. in Math. **7** (1971), 254–281.

273. M. Finzel, *Plücker–Grassmann coordinates and elementary vectors*, Approximation Theory VIII, Vol. 1 (College Station, TX, 1995), World Scientific, River Edge, NJ, 1995, pp. 207–214.

274. A. G. Fisher, *On construction and properties of the generalized inverse*, SIAM J. Appl. Math. **15** (1967), 269–272.

275. W. Fleming, *Functions of Several Variables* (2nd edition), Springer-Verlag, New York, 1977.

276. R. Fletcher, *Generalized inverse methods for the best least squares solution of systems of non-linear equations*, Comput. J. **10** (1968), 392–399.

277. _____, *A technique for orthogonalization*, J. Inst. Math. Appl. **5** (1969), 162–166.

278. _____, *Generalized inverses for nonlinear equations and optimization*, Numerical Methods for Nonlinear Algebraic Equations (P. Rabinowitz, editor), pp. 75–85, Gordon and Breach, London, 1970.

279. A. Forsgren, *On linear least-squares problems with diagonally dominant weight matrices*, SIAM J. Matrix Anal. Appl. **17** (1996), no. 4, 763–788.

280. A. Forsgren and G. Sporre, *On weighted linear least-squares problems related to interior methods for convex quadratic programming*, SIAM J. Matrix Anal. Appl. **23** (2001), 42–56.

281. G. E. Forsythe, *The maximum and minimum of a positive definite quadratic polynomial on a sphere are convex functions of the radius*, SIAM J. Appl. Math. **19** (1970), 551–554.

282. G. E. Forsythe and G. H. Golub, *On the stationary values of a second-degree polynomial on the unit sphere*, SIAM J. Appl. Math. **13** (1965), 1050–1068.

283. M. Foster, *An application of the Wiener–Kolmogorov smoothing theory to matrix inversion*, J. Soc. Indust. Appl. Math. **9** (1961), 387–392.

284. D. J. Foulis, *Relative inverses in Baer *-semigroups*, Michigan Math. J. **10** (1963), 65–84.

285. J. S. Frame, *Matrix functions and applications. I. Matrix operations and generalized inverses*, IEEE Spectrum **1** (1964), 209–220.

286. _____, *Matrix functions and applications. II. Functions of a matrix*, IEEE Spectrum **1** (1964), no. 4, 102–108.

287. _____, *Matrix functions and applications. IV. Matrix functions and constituent matrices*, IEEE Spectrum **1** (1964), no. 6, 123–131.

288. _____, *Matrix functions and applications. V. Similarity reductions by rational or orthogonal matrices*, IEEE Spectrum **1** (1964), no. 7, 103–109.

289. P. Franck, *Sur la distance minimale d'une matrice régulière donnée au lieu des matrices singulières*, Deux. Congr. Assoc. Franc. Calcul. Trait. Inform. Paris 1961, Gauthiers–Villars, Paris, 1962, pp. 55–60.

290. I. Fredholm, *Sur une classe d'équations fonctionnelles*, Acta Math. **27** (1903), 365–390.

291. R. Gabriel, *Extensions of generalized algebraic complement to arbitrary matrices* (Romanian), Stud. Cerc. Mat. **17** (1965), 1567–1581.

292. _____, *Das verallgemeinerte Inverse einer Matrix deren Elemente über einem beliebigen Körper angehören*, J. Reine Angew. Math. **234** (1969), 107–122.

293. _____, *Das verallgemeinerte Inverse einer Matrix über einem beliebigen Körperanalytisch betrachtet*, J. Reine Angew. Math. **244** (1970), 83–93.

294. J. Gaches, J.-L. Rigal, and X. Rousset de Pina, *Distance euclidienne d'une application linéaire σ au lieu des applications de rang r donné*, C. R. Acad. Sci. Paris **260** (1965), 5672–5674.

295. A. R. Gallant and T. M. Gerig, *Computations for constrained linear models*, J. Econometrics **12** (1980), no. 1, 59–84, (see [**206**]).

296. F. R. Gantmacher, *The Theory of Matrices*, vol. I, II, Chelsea, New York, 1959.

297. J. M. Garnett III, A. Ben-Israel, and S. S. Yau, *A hyperpower iterative method for computing matrix products involving the generalized inverse*, SIAM J. Numer. Anal. **8** (1971), 104–109.

298. B. Germain-Bonne, *Calcul de pseuodo–inverses*, Rev. Francaise Informat. Rech. Opérationelle **3** (1969), 3–14.

299. I. M. Glazman and Ju. I. Ljubich, *Finite Dimensional Linear Analysis*, Nauka, Moscow, 1969, (English translation published by MIT Press).

300. S. Goldberg, *Unbounded Linear Operators*, McGraw-Hill, New York, 1966.

301. A. J. Goldman and M. Zelen, *Weak generalized inverses and minimum variance linear unbiased estimation*, J. Res. Nat. Bur. Standards Sect. B **68B** (1964), 151–172.

302. A. A. Goldstein, *Constructive Real Analysis*, Harper and Row, New York, 1967.

303. M. Goldstein and A. F. M. Smith, *Ridge-type estimators for regression analysis*, J. Roy. Statist. Soc. Ser. B **36** (1974), 284–291.

304. G. H. Golub, *Least squares, singular values and matrix approximations*, Apl. Mat. **13** (1968), 44–51.

305. _____, *Matrix decompositions and statistical calculations*, Technical Report STAN-CS-124, Stanford University, Stanford, CA, March 1969.

306. G. H. Golub and W. Kahan, *Calculating the singular values and pseudo-inverse of a matrix*, J. Soc. Indust. Appl. Math. Ser. B Numer. Anal. **2** (1965), 205–224, (see [**218**]).

307. G. H. Golub and V. Pereyra, *The differentiation of pseudoinverses and nonlinear least squares problems whose variables separate*, SIAM J. Numer. Anal. **10** (1973), 413–432.

308. _____, *Differentiation of pseudoinverses, separable nonlinear least squares problems and other tales*, In Nashed [**593**], pp. 303–324.

309. G. H. Golub and C. Reinsch, *Singular value decompositions and least squares solutions*, Numer. Math. **14** (1970), 403–420, (republished, pp. 134–151 in [**875**]).

310. G. H. Golub and G. P. H. Styan, *Numerical computations for univariate linear models*, J. Statist. Comput. Simulation **2** (1973), 253–274.

311. G. H. Golub and C. F. Van Loan, *Matrix Computations*, 3rd ed., Johns Hopkins University Press, Baltimore, MD, 1996.

312. G. H. Golub and J. H. Wilkinson, *Note on the iterative refinement of least squares solutions*, Numer. Math. **9** (1966), 139–148.

313. C. C. Gonzaga and H. J. Lara, *A note on properties of condition numbers*, Linear Algebra Appl. **261** (1997), 269–273, (see [**437**]).

314. I. J. Good, *Some applications of the singular decomposition of a matrix*, Technometrics **11** (1969), 823–831.

315. _____, *Generalized determinants and generalized generalized variance*, J. Statist. Comput. Simulation **12** (1980/81), no. 3-4, 311–315, (see [**72**]).

316. F. A. Graybill, *An Introduction to Linear Statistical Models*. Vol I, McGraw-Hill, New York, 1961.

317. _____, *Theory and Application of the Linear Model*, Duxbury Press, North Scituate, MA, 1976.

318. _____, *Matrices with Applications in Statistics*, 2nd ed., Wadsworth Advanced Books and Software, Belmont, Calif., 1983.

319. F. A. Graybill and G. Marsaglia, *Idempotent matrices and quadratic forms in the general linear hypothesis*, Ann. Math. Statist. **28** (1957), 678–686.

320. F. A. Graybill, C. D. Meyer, Jr., and R. J. Painter, *Note on the computation of the generalized inverse of a matrix*, SIAM Rev. **8** (1966), 522–524.

321. B. Green, *The orthogonal approximation of an oblique structure in factor analysis*, Psychometrika **17** (1952), 429–440.

322. W. Greub and W. C. Rheinboldt, *Non self-adjoint boundary value problems in ordinary differential equations*, J. Res. Nat. Bur. Standards Sect. B **1960** (64B), 83–90.

323. T. N. E. Greville, *On smoothing a finite table: A matrix approach*, J. Soc. Indust. Appl. Math. **5** (1957), 137–154.

324. _____, *The pseudoinverse of a rectangular matrix and its application to the solution of systems of linear equations*, SIAM Rev. **1** (1959), 38–43.

325. _____, *Some applications of the pseudoinverse of a matrix*, SIAM Rev. **2** (1960), 15–22.

326. _____, *Note on fitting functions of several independent variables*, J. Soc. Indust. Appl. Math. **9** (1961), 109–115, (Erratum, *ibid* **9** (1961), 317).

327. _____, *Note on the generalized inverse of a matrix product*, J. Soc. Indust. Appl. Math. **9** (1966), 109–115.

328. _____, *Spectral generalized inverses of square matrices*, Math. Research Center Technical Summary Report 823, University of Wisconsin, Madison, WI, October 1967.

329. _____, *Some new generalized inverses with spectral properties*, In Boullion and Odell [**121**], pp. 26–46.

330. _____, *Solutions of the matrix equation $XAX = X$ and relations between oblique and orthogonal projectors*, SIAM J. Appl. Math. **26** (1974), 828–832.

331. C. W. Groetsch, *Steepest descent and least squares solvability*, Canad. Math. Bull. **17** (1974), 275–276.

332. _____, *Generalized Inverses of Linear Operators: Representation and Approximation*, Marcel Dekker, New York, 1977.

333. _____, *On rates of convergence for approximations to the generalized inverse*, Numer. Funct. Anal. Optim. **1** (1979), no. 2, 195–201.

334. _____, *Generalized inverses and generalized splines*, Numer. Funct. Anal. Optim. **2** (1980), no. 1, 93–97, (connection between generalized inverses and generalized splines, see [**724**]).

335. _____, *The Theory of Tikhonov Regularization for Fredholm Equations of the First Kind*, Pitman, London, 1984.

336. _____, *Inclusions for the Moore–Penrose inverse with applications to computational methods*, Contributions in Numerical Mathematics, World Scientific, River Edge, NJ, 1993, pp. 203–211.

337. _____, *Dykstra's algorithm and a representation of the Moore-Penrose inverse*, J. Approx. Theory **117** (2002), no. 1, 179–184.

338. C. W. Groetsch and M. Hanke, *A general framework for regularized evaluation of unstable operators*, J. Math. Anal. Appl. **203** (1996), no. 2, 451–463.

339. C. W. Groetsch and B. J. Jacobs, *Iterative methods for generalized inverses based on functional interpolation*, In Campbell [**155**], pp. 220–232.

340. C. W. Groetsch and J. T. King, *Extrapolation and the method of regularization for generalized inverses*, J. Approx. Theory **25** (1979), no. 3, 233–247.

341. C. W. Groetsch and A. Neubauer, *Regularization of ill-posed problems: Optimal parameter choice in finite dimensions*, J. Approx. Theory **58** (1989), no. 2, 184–200.

342. J. Groß, *A note on a partial ordering in the set of Hermitian matrices*, SIAM J. Matrix Anal. Appl. **18** (1997), no. 4, 887–892.

343. _____, *Some remarks concerning the reverse order law*, Discuss. Math. Algebra Stochastic Methods **17** (1997), no. 2, 135–141.

344. _____, *Special generalized inverse matrices connected with the theory of unified least squares*, Linear Algebra Appl. **264** (1997), 325–327.

345. _____, *On contractions in linear regression*, J. Statist. Plann. Inference **74** (1998), no. 2, 343–351.

346. _____, *A note on the rank-subtractivity ordering*, Linear Algebra Appl. **289** (1999), no. 1-3, 151–160.

347. _____, *On oblique projection, rank additivity and the Moore–Penrose inverse of the sum of two matrices*, Linear and Multilinear Algebra **46** (1999), no. 4, 265–275.

348. _____, *Löwner partial ordering and space preordering of Hermitian non-negative definite matrices*, Linear Algebra Appl. **326** (2001), no. 1-3, 215–223.

349. J. Groß, J. Hauke, and A. Markiewicz, *Some comments on matrix partial orderings*, Discuss. Math. Algebra Stochastic Methods **17** (1997), no. 2, 203–214.

350. _____, *Partial orderings, preorderings, and the polar decomposition of matrices*, Linear Algebra Appl. **289** (1999), no. 1-3, 161–168.

351. J. Groß and S. Puntanen, *Estimation under a general partitioned linear model*, Linear Algebra Appl. **321** (2000), no. 1-3, 131–144.

352. J. Groß and G. Trenkler, *On the least squares distance between affine subspaces*, Linear Algebra Appl. **237/238** (1996), 269–276.

353. _____, *On the equality of usual and Amemiya's partially generalized least squares estimator*, Comm. Statist. A – Theory and Meth. **26** (1997), no. 9, 2075–2086.

354. _____, *Restrictions and projections in linear regression*, Internat. J. Math. Ed. Sci. Tech. **28** (1997), no. 3, 465–468.

355. _____, *On the product of oblique projectors*, Linear and Multilinear Algebra **44** (1998), no. 3, 247–259.

356. _____, *Nonsingularity of the difference of two oblique projectors*, SIAM J. Matrix Anal. Appl. **21** (1999), no. 2, 390–395 (electronic).

357. J. Groß and S. -O. Troschke, *Some remarks on partial orderings of nonnegative definite matrices*, Linear Algebra Appl. **264** (1997), 457–461.

358. Z.-H. Guan, X.-C. Wen, and Y.-Q. Liu, *Variation of the parameters formula and the problem of BIBO for singular measure differential systems with impulse effect*, Appl. Math. Comput. **60** (1994), no. 2-3, 153–169.

359. E. A. Guillemin, *Theory of Linear Physical Systems*, Wiley, New York, 1963.

360. M. E. Gulliksson, P.-Å. Wedin, and Y. Wei, *Perturbation identities for regularized Tikhonov inverses and weighted pseudoinverses*, BIT **40** (2000), no. 3, 513–523.

361. A. Guterman, *Linear preservers for Drazin star partial order*, Comm. in Algebra (2001).

362. _____, *Linear preservers for matrix inequalities and partial orderings*, Linear Algebra Appl. **331** (2001), 75–87.

363. S. J. Haberman, *How much do Gauss–Markov and least square estimates differ? A coordinate-free approach*, Ann. Statist. **3** (1975), no. 4, 982–990, (extension of [**486**].

364. F. J. Hall and C. D. Meyer, Jr., *Generalized inverses of the fundamental bordered matrix used in linear estimation*, Sankhyā Ser. A **37** (1975), no. 3, 428–438, (corrections: *ibid* **40**(1980), p. 399).

365. P. R. Halmos, *Finite-Dimensional Vector Spaces*, 2nd ed., D. Van Nostrand, Princeton, NJ, 1958.

366. _____, *A Hilbert Space Problem Book*, D. Van Nostrand, Princeton, NJ, 1967.

367. P. R. Halmos and J. E. McLaughlin, *Partial isometries*, Pacific J. Math. **13** (1963), 585–596.

368. P. R. Halmos and L. J. Wallen, *Powers of partial isometries*, J. Math. Mech. **19** (1970), 657–663.

369. I. Halperin, *Closures and adjoints of linear differential operators*, Ann. of Math. (1937), 880–919.

370. H. Hamburger, *Non-symmetric operators in Hilbert space*, Proceedings Symposium on Spectral Theory and Differential Problems, Oklahoma A& M College, Stillwater, OK, 1951, pp. 67–112.

371. M. Hanke, *Regularization with differential operators: An iterative approach*, Numer. Funct. Anal. Optim. **13** (1992), no. 5-6, 523–540.

372. M. Hanke and M. Neumann, *The geometry of the set of scaled projections*, Linear Algebra Appl. **190** (1993), 137–148.

373. G. W. Hansen and D. W. Robinson, *On the existence of generalized inverses*, Linear Algebra Appl. **8** (1974), 95–104.

374. P. C. Hansen, *The truncated SVD as a method for regularization*, BIT **27** (1987), 534–553.

375. R. J. Hanson, *A numerical method for solving Fredholm integral equations of the first kind using singular values*, SIAM J. Numer. Anal. **8** (1971), 616–622.

376. R. J. Hanson and M. J. Norris, *Analysis of measurements based on the singular value decomposition*, SIAM J. Sci. Statist. Comput. **2** (1981), no. 3, 363–373.

377. B. Harris, *Theory of Probability*, Addison-Wesley, Reading, MA, 1966.

378. R. E. Hartwig, *1-2 inverses and the invariance of BA^+C*, Linear Algebra Appl. **11** (1975), no. 3, 271–275.

379. _____, *Block generalized inverses*, Arch. Rational Mech. Anal. **61** (1976), no. 3, 197–251.

380. _____, *Rank factorization and Moore–Penrose inversion*, Indust. Math. **26** (1976), no. 1, 49–63.

381. _____, *Singular value decomposition and the Moore–Penrose inverse of bordered matrices*, SIAM J. Appl. Math. **31** (1976), no. 1, 31–41.

382. _____, *A note on the partial ordering of positive semidefinite matrices*, Linear and Multilinear Algebra **6** (1978), 223–226.

383. _____, *Schur's theorem and the Drazin inverse*, Pacific J. Math. **78** (1978), no. 1, 133–138.

384. _____, *How to partially order regular elements*, Math. Japon. **25** (1980), no. 1, 1–13.

385. _____, *Drazin inverses and canonical forms in $M_n(\mathbf{z}/h)$*, Linear Algebra Appl. **37** (1981), 205–233.

386. _____, *A method for calculating A^D*, Math. Japon. **26** (1981), no. 1, 37–43.

387. _____, *The reverse order law revisited*, Linear Algebra Appl. **76** (1986), 241–246.

388. _____, *A remark on the characterization of the parallel sum of two matrices*, Linear and Multilinear Algebra **22** (1987), no. 2, 193–197.

389. R. E. Hartwig and M. P. Drazin, *Lattice properties of the * order for matrices*, J. Math. Anal. Appl. **86** (1982), 359–378.

390. R. E. Hartwig and F. J. Hall, *Applications of the Drazin inverse to Cesàro–Neumann iterations*, In Campbell [155], pp. 145–195.

391. R. E. Hartwig and J. Levine, *Applications of the Drazin inverse to the Hill cryptographic system. III*, Cryptologia **5** (1981), no. 2, 67–77.

392. _____, *Applications of the Drazin inverse to the Hill cryptographic system. IV*, Cryptologia **5** (1981), no. 4, 213–228.

393. R. E. Hartwig and J. M. Shoaf, *On the derivative of the Drazin inverse of a complex matrix*, SIAM J. Math. Anal. **10** (1979), no. 1, 207–216.

394. _____, *Invariance, group inverses and parallel sums*, Rev. Roumaine Math. Pures Appl. **25** (1980), no. 1, 33–42.

395. R. E. Hartwig and G. P. H. Styan, *On some characterizations of the star partial orderings and rank subtractivity*, Linear Algebra Appl. **82** (1986), 145–161.

396. R. E. Hartwig, G.-R. Wang, and Y. Wei, *Some additive results on Drazin inverse*, Linear Algebra Appl. **322** (2001), no. 1-3, 207–217.

397. D. A. Harville, *Extension of the Gauss–Markov theorem to include the estimation of random effects*, Ann. Statist. **4** (1976), no. 2, 384–395.

398. W. R. Harwood, V. Lovass-Nagy, and D. L. Powers, *A note on the generalized inverses of some partitioned matrices*, SIAM J. Appl. Math. **19** (1970), 555–559.

399. J. Hauke and A. Markiewicz, *On partial orderings on the set of rectangular matrices*, Linear Algebra Appl. **219** (1995), 187–193, (see [**535**]).

400. _____, *On partial orderings on the set of rectangular matrices and their properties*, Discuss. Math. Algebra Stochastic Methods **15** (1995), no. 1, 5–10.

401. D. M. Hawkins and D. Bradu, *Application of the Moore–Penrose inverse of a data matrix in multiple regression*, Linear Algebra Appl. **127** (1990), 403–425.

402. J. B. Hawkins and A. Ben-Israel, *On generalized matrix functions*, Linear and Multilinear Algebra **1** (1973), no. 2, 163–171.

403. E. V. Haynsworth, *Applications of an inequality for the Schur complement*, Proc. Amer. Math. Soc. **24** (1970), 512–516.

404. E. V. Haynsworth and J. R. Wall, *Group inverses of certain nonnegative matrices*, Linear Algebra Appl. **25** (1979), 271–288.

405. _____, *Group inverses of certain positive operators*, Linear Algebra Appl. **40** (1981), 143–159.

406. J. Z. Hearon, *On the singularity of a certain bordered matrix*, SIAM J. Appl. Math. **15** (1967), 1413–1421.

407. _____, *Partially isometric matrices*, J. Res. Nat. Bur. Standards Sect. B **71B** (1967), 225–228.

408. _____, *Polar factorization of a matrix*, J. Res. Nat. Bur. Standards Sect. B **71B** (1967), 65–67.

409. _____, *Generalized inverses and solutions of linear systems*, J. Res. Nat. Bur. Standards Sect. B **72B** (1968), 303–308.

410. J. Z. Hearon and J. W. Evans, *On spaces and maps of generalized inverses*, J. Res. Nat. Bur. Standards Sect. B **72B** (1968), 103–107.

411. C. Heil and D. Walnut, *Continuous and discrete wavelet transforms*, SIAM Rev. **31** (1989), 628–666.

412. G. Heinig, *The group inverse of the transformation $S(X) = AX - XB$*, Linear Algebra Appl. **257** (1997), 321–342.

413. G. P. Herring, *A note on generalized interpolation and the pseudoinverse*, SIAM J. Numer. Anal. **4** (1967), 548–556.

414. M. R. Hestenes, *Inversion of matrices by biorthogonalization and related results*, J. Soc. Indust. Appl. Math. **6** (1958), 51–90.

415. _____, *Relative Hermitian matrices*, Pacific J. Math. **11** (1961), 225–245.

416. _____, *Relative self-adjoint operators in Hilbert space*, Pacific J. Math. **11** (1961), 1315–1357.

417. _____, *A ternary algelbra with applications to matrices and linear transformations*, Arch. Rational Mech. Anal. **11** (1962), 138–194.

418. D. Hilbert, *Grundzüge einer algemeinen Theorie der linearen Integralgleichungen*, Teubner, Leipzig, 1912, (reprint of six articles which appeared originally in the *Götingen Nachrichten* (1904), 49–51; (1904), 213–259; (1905), 307–338; (1906), 157–227; (1906), 439–480; (1910), 355–417).

419. J. W. Hilgers, *On the equivalence of regularization and certain reproducing kernel Hilbert space approaches for solving first kind problems*, SIAM J. Numer. Anal. **13** (1976), no. 2, 172–184, (corrections: ibid **15**(1978), no. 6, p. 1301).

420. _____, *A note on estimating the optimal regularization parameter*, SIAM J. Numer. Anal. **17** (1980), no. 3, 472–473.

421. _____, *A theory for optimal regularization in the finite-dimensional case*, Linear Algebra Appl. **48** (1982), 359–379.

422. B. L. Ho and R. E. Kalman, *Effective construction of linear state–variables models from input/output functions*, Regelungstechnik **14** (1966), 545–548.

423. A. E. Hoerl and R. W. Kennard, *Ridge regression: Biased estimation of nonorthogonal problems*, Technometrics **12** (1970), 55–67.

424. _____, *A note on least squares estimates*, Comm. Statist. B – Simulation and Comput. **9** (1980), 315–317, (generalization of [**785**]).

425. _____, *Ridge regression—1980. Advances, algorithms, and applications*, Amer. J. Math. Management Sci. **1** (1981), no. 1, 5–83.

426. R. W. Hoerl, *Ridge analysis 25 years later*, Amer. Statist. **39** (1985), no. 3, 186–192.

427. R. B. Holmes, *A Course on Optimization and Best Approximation*, Springer-Verlag, Berlin, 1972.

428. R. A. Horn and C. R. Johnson, *Matrix Analysis*, Cambridge University Press, Cambridge, 1985, (corrected reprint Cambridge University Press, Cambridge, 1990).

429. _____, *Topics in Matrix Analysis*, Cambridge University Press, Cambridge, 1991.

430. A. Höskuldsson, *Data analysis, matrix decompositions, and generalized inverse*, SIAM J. Sci. Comput. **15** (1994), no. 2, 239–262.

431. H. Hotelling, *Relation between two sets of variates*, Biometrika **28** (1936), 322–377.

432. A. S. Householder, *The Theory of Matrices in Numerical Analysis*, Blaisdell, New York, 1964.

433. A. S. Householder and G. Young, *Matrix approximation and latent roots*, Amer. Math. Monthly **45** (1938), 165–171.

434. M. F. Hurt and C. Waid, *A generalized inverse which gives all the integral solutions to a system of linear equations*, SIAM J. Appl. Math. **19** (1970), 547–550.

435. W. A. Hurwitz, *On the pseudo–resolvent to the kernel of an integral equation*, Trans. Amer. Math. Soc. **13** (1912), 405–418.

436. Y. Ijiri, *On the generalized inverse of an incidence matrix*, J. Soc. Indust. Appl. Math. **13** (1965), 941–945.

437. Kh. D. Ikramov, *An algebraic proof of a result by Gonzaga and Lara*, Linear Algebra Appl. **299** (1999), no. 1-3, 191–194.

438. I. C. F. Ipsen and C. D. Meyer, Jr., *The angle between complementary subspaces*, Amer. Math. Monthly **102** (1995), no. 10, 904–911.

439. V. V. Ivanov and V. Yu. Kudrinskii, *Approximate solution of linear operator equations in Hilbert space by the method of least squares. I*, Zh. Vychisl. Mat. i Mat. Fiz. **6** (1966), no. 5, 831–841.

440. S. Izumino, *Convergence of generalized inverses and spline projectors*, J. Approx. Theory **38** (1983), no. 3, 269–278.

441. C. G. J. Jacobi, *De determinantibus functionalibus*, Crelle Journal für die reine und angewandte Mathematik **22** (1841), 319–359, (reprinted in *C.G.J. Jacobi's Gesammelte Werke* (K. Weierstrass, editor), Vol. 3, 393–438, Berlin 1884, and Chelsea, New York, 1969).

442. _____, *De formatione et proprietatibus determinantium*, Crelle Journal für die reine und angewandte Mathematik **22** (1841), 285–318, (reprinted in *C.G.J. Jacobi's Gesammelte Werke* (K. Weierstrass, editor), Vol. 3, 355–392, Berlin 1884, and Chelsea, New York, 1969).

443. S. K. Jain, S. K. Mitra, and H.-J. Werner, *Extensions of G-based matrix partial orders*, SIAM J. Matrix Anal. Appl. **17** (1996), no. 4, 834–850.

444. D. James, *Implicit nullspace iterative methods for constrained least squares problems*, SIAM J. Matrix. Anal. Appl. **13** (1992), 962–978.

445. J. W. Jerome and L. L. Schumaker, *A note on obtaining natural spline functions by the abstract approach of Atteia and Laurent*, SIAM J. Numer. Anal. **5** (1968), 657–663.

446. E. P. Jiang and M. W. Berry, *Solving total least-squares problems in information retrieval*, Linear Algebra Appl. **316** (2000), no. 1-3, 137–156.

447. S. Jiang, *Angles between Euclidean subspaces*, Geom. Dedicata **63** (1996), no. 2, 113–121.

448. J. A. John, *Use of generalized inverse matrices in MANOVA*, J. Roy. Statist. Soc. Ser. B **32** (1970), 137–143.

449. P. W. M. John, *Pseudo-inverses in the analysis of variance*, Ann. Math. Statist. **35** (1964), 895–896.

450. J. Jones, Jr., *Solution of certain matrix equations*, Proc. Amer. Math. Soc. **31** (1972), 333–339.

451. Jon Jones, N. P. Karampetakis, and A. C. Pugh, *The computation and application of the generalized inverse via Maple*, J. Symbolic Comput. **25** (1998), no. 1, 99–124.

452. C. Jordan, *Mémoires sur les formes bilinéaires*, J. Math. Pures Appl. Deuxième Sér. **19** (1874), 35–54, (see history of SVD in [**783**]).

453. _____, *Sur la réduction des formes bilinéaires*, C. R. Acad. Sci. Paris **78** (1874), 614–617, (see history of SVD in [**783**]).

454. S. Kaczmarz, *Angenäherte Auflösung von Systemen linearer Gleichungen*, Bull. Acad. Polon. Sci. **A35** (1937), 355–357.

455. S. Kakutani, *Some characterizations of Euclidean spaces*, Japan J. Math. **16** (1939), 93–97.

456. R. Kala, *Projectors and linear estimation in general linear models*, Comm. Statist. A – Theory and Meth. **10** (1981), no. 9, 849–873.

457. C. Kallina, *A Green's function approach to perturbations of periodic solutions*, Pacific J. Math. **29** (1969), 325–334.

458. R. E. Kalman, *Contributions to the theory of optimal control*, Bol. Soc. Mat. Mexicana **5** (1960), no. 2, 102–119.

459. _____, *A new approach to linear filtering and prediction problems*, Trans. ASME Ser. D. J. Basic Eng. **82** (1960), 35–45.

460. _____, *New results in linear filtering and prediction problems*, Trans. ASME Ser. D. J. Basic Eng. **83** (1961), 95–107.

461. _____, *Mathematical description of linear dynamical systems*, SIAM J. Control **1** (1963), 152–192.

462. R. E. Kalman, Y. C. Ho, and K. S. Narendra, *Controllability of linear dynamical systems*, Contributions to Differential Equations, Vol. I, Interscience, New York, 1963, pp. 189–213.

463. W. J. Kammerer and M. Z. Nashed, *A generalization of a matrix iterative method of G. Cimmino to best approximate solution of linear integral equations of the first kind*, Atti Accad. Naz. Lincei Rend. Cl. Sci. Fis. Mat. Natur. (8) **51** (1971), 20–25.

464. _____, *Steepest descent for singular linear operators with nonclosed range*, Applicable Anal. **1** (1971), no. 2, 143–159.

465. _____, *Iterative methods for best approximate solutions of linear integral equations of the first and second kinds*, J. Math. Anal. Appl. **40** (1972), 547–573.

466. _____, *On the convergence of the conjugate gradient method for singular linear operator equations*, SIAM J. Numer. Anal. **9** (1972), 165–181.

467. L. V. Kantorovich and G. P. Akilov, *Functional Analysis in Normed Spaces*, Pergamon Press, Oxford, 1964, (translated from Russian).

468. L. V. Kantorovich and V. I. Krylov, *Approximate Methods of Higher Analysis*, Interscience, New York, 1958, (translated from Russian).

469. N. P. Karampetakis, *Computation of the generalized inverse of a polynomial matrix and applications*, Linear Algebra Appl. **252** (1997), 35–60.

470. _____, *Generalized inverses of two-variable polynomial matrices and applications*, Circuits Systems Signal Process. **16** (1997), no. 4, 439–453.

471. I. J. Katz, *Wiegmann type theorems for EP_r matrices*, Duke Math. J. **32** (1965), 423–427.

472. _____, *Remarks on a paper of Ben-Israel*, SIAM J. Appl. Math. **18** (1970), 511–513.

473. I. J. Katz and M. H. Pearl, *On EPr and normal EPr matrices*, J. Res. Nat. Bur. Standards Sect. B **70B** (1966), 47–77.

474. C. G. Khatri, *A representation of a matrix and its use in the Gauss–Markoff model*, J. Indian Statist. Assoc. **20** (1982), 89–98.

475. M. J. L. Kirby, *Generalized Inverses and Chance Constrained Programming*, Applied mathematics, Northwestern University, Evanston, IL, June 1965.

476. F. H. Kishi, *On line computer control techniques and their application to re-entry aerospace vehicle control*, Advances in Control Systems Theory and Applications (C. T. Leondes, editor), Academic Press, New York, 1964, pp. 245–257.

477. J. J. Koliha, *A generalized Drazin inverse*, Glasgow Math. J. **38** (1996), no. 3, 367–381.

478. _____, *Continuity and differentiability of the Moore–Penrose inverse in C^*-algebras*, Math. Scand. **88** (2001), no. 1, 154–160.

479. _____, *Error bounds for a general perturbation of the Drazin inverse*, Appl. Math. Comput. **126** (2002), no. 2-3, 61–65.

480. J. J. Koliha and V. Rakočević, *Continuity of the Drazin inverse. II*, Studia Math. **131** (1998), no. 2, 167–177.

481. A. Korganoff and M. Pavel-Parvu, *Méthodes de calcul numérique. Tome II: Éléments de théorie des matrices carrées et rectangles en analyse numérique*, Dunod, Paris, 1967.

482. S. Kourouklis and C. C. Paige, *A constrained least squares approach to the general Gauss–Markov linear model*, J. Amer. Statist. Assoc. **76** (1981), no. 375, 620–625.

483. O. Krafft, *An arithmetic-harmonic-mean inequality for nonnegative definite matrices*, Linear Algebra Appl. **268** (1998), 243–246.

484. R. G. Kreijger and H. Neudecker, *Exact linear restrictions on parameters in the general linear model with a singular covariance matrix*, J. Amer. Statist. Assoc. **72** (1977), no. 358, 430–432.

485. W. Kruskal, *The coordinate-free approach to Gauss–Markov estimation, and its application to missing and extra observations*, Proc. 4th Berkeley Sympos. Math. Statist. and Prob., Vol. I, University of California Press, Berkeley, CA, 1961, pp. 435–451.

486. ———, *When are Gauss–Markov and least squares estimators identical? A coordinate-free approach*, Ann. Math. Statist **39** (1968), 70–75, (see [**363**]).

487. J.-X. Kuang, *The representation and approximation of Drazin inverses of linear operators*, Numer. Math. J. Chinese Univ. **4** (1982), no. 2, 97–106.

488. V. N. Kublanovskaya, *On the calculation of generalized inverses and projections* (Russian), Z. Vycisl. Mat. i Mat. Fiz. **6** (1966), 326–332.

489. P. Kunkel and V. Mehrmann, *Generalized inverses of differential-algebraic operators*, SIAM J. Matrix Anal. Appl. **17** (1996), no. 2, 426–442.

490. M. C. Y. Kuo and L. F. Mazda, *Mimimum energy problems in Hilbert function space*, J. Franklin Inst. **283** (1967), 38–54.

491. S. Kurepa, *Generalized inverse of an operator with a closed range*, Glas. Mat. **3** (1968), no. 23, 207–214.

492. B. Kutzler and V. Kokol-Voljc, *Introduction to Derive 5*, Texas Instruments, Dallas, TX 75265, 2000.

493. B. F. Lamond and M. L. Puterman, *Generalized inverses in discrete time Markov decision processes*, SIAM J. Matrix Anal. Appl. **10** (1989), no. 1, 118–134.

494. P. Lancaster, *Explicit solutions of linear matrix equations*, SIAM Rev. **12** (1970), 544–566.

495. P. Lancaster and M. Tismenetsky, *The Theory of Matrices* (2nd edition), Academic Press, San Diego, 1985.

496. C. Lanczos, *Linear systems in self-adjoint form*, Amer. Math. Monthly **65** (1958), 665–679.

497. ———, *Linear Differential Operators*, D. Van Nostrand, Co., Princeton, 1961.

498. E. M. Landesman, *Hilbert–space methods in elliptic partial differential equations*, Pacific J. Math. **21** (1967), 113–131.

499. C. E. Langenhop, *On generalized inverses of matrices*, SIAM J. Appl. Math. **15** (1967), 1239–1246.

500. L. J. Lardy, *A series representation for the generalized inverse of a closed linear operator*, Atti Accad. Naz. Lincei Rend. Cl. Sci. Fis. Mat. Natur. (8) **58** (1975), no. 2, 152–157.

501. ———, *Some iterative methods for linear operator equations with applications to generalized inverses*, SIAM J. Appl. Math. **32** (1977), no. 3, 610–618.

502. P.-J. Laurent, *Approximation et Optimisation. Collection Enseignement des Sciences*, No. 13, Hermann, Paris, 1972.

503. ———, *Quadratic convex analysis and splines*, Methods of Functional Analysis in Approximation Theory (Bombay, 1985), Birkhäuser, Basel, 1986, pp. 17–43.

504. C. L. Lawson and R. J. Hanson, *Solving Least Squares Problems*, Prentice-Hall, Englewood Cliffs, NJ, 1974, (reprinted, *Classics in Applied Mathematics*, No. 15. Society for Industrial and Applied Mathematics (SIAM), Philadelphia, PA, 1995).

505. E. B. Leach, *A note on inverse function theorems*, Proc. Amer. Math. Soc. **12** (1961), 694–697.

506. E. E. Leamer, *Coordinate-free ridge regression bounds*, J. Amer. Statist. Assoc. **76** (1981), no. 376, 842–849.

507. A. H. Lent, *Wiener–Hopf Operators and Factorizations*, Doctoral Dissertation in Applied Mathematics, Northwestern University, Evanston, IL, June 1971.

508. Ö. Leringe and P.-Å. Wedin, *A comparison between different methods to compute a vector* **x** *which minimizes* $\|A\mathbf{x} - \mathbf{b}\|_2$ *when* $G\mathbf{x} = \mathbf{h}$, Technical Report, Lund University, Department of Computer Science, Lund, Sweden, March 1970.

509. Y. Levin and A. Ben-Israel, *A Newton method for systems of m equations in n variables*, Nonlinear Anal. **47** (2001), 1961–1971.

510. Y. Levin, M. Nediak, and A. Ben-Israel, *A direct approach to calculus of variations via Newton–Raphson method*, Comput. Appl. Math. **139** (2001), 197–213.

511. J. Levine and R. E. Hartwig, *Applications of the Drazin inverse to the Hill cryptographic system*. I, Cryptologia **4** (1980), no. 2, 71–85.

512. _____, *Applications of the Drazin inverse to the Hill cryptographic system*. II, Cryptologia **4** (1980), no. 3, 150–168.

513. B. C. Levy, *A note on the hyperbolic singular value decomposition*, Linear Algebra Appl. **277** (1998), no. 1-3, 135–142.

514. T. O. Lewis and P. L. Odell, *A generalization of the Gauss–Markov theorem*, J. Amer. Statist. Assoc. **61** (1966), 1063–1066.

515. C.-K. Li and R. Mathias, *Extremal characterizations of the Schur complement and resulting inequalities*, SIAM Rev. **42** (2000), no. 2, 233–246 (electronic).

516. X. Li and Y. Wei, *An improvement on the perturbation of the group inverse and oblique projection*, Linear Algebra Appl. **338** (2001), 53–66.

517. S.-Y. T. Lin and Y.-F. Lin, *The n-dimensional Pythagorean theorem*, Linear and Multilinear Algebra **26** (1990), no. 1-2, 9–13, (see [**95**]).

518. E. P. Liski, *On Löwner-ordering antitonicity of matrix inversion*, Acta Math. Appl. Sinica (English Ser.) **12** (1996), no. 4, 435–442.

519. E. P. Liski and S.-G. Wang, *On the {2}-inverse and some ordering properties of nonnegative definite matrices*, Acta Math. Appl. Sinica (English Ser.) **12** (1996), no. 1, 22–27.

520. J.-Z. Liu and J. Wang, *Some inequalities for Schur complements*, Linear Algebra Appl. **293** (1999), no. 1-3, 233–241.

521. J. Locker, *An existence analysis for nonlinear equations in Hilbert space*, Trans. Amer. Math. Soc. **128** (1967), 403–413.

522. _____, *An existence analysis for nonlinear boundary value problems*, SIAM J. Appl. Math. **19** (1970), 199–207.

523. A. T. Lonseth, *Approximate solutions of Fredholm-type integral equations*, Bull. Amer. Math. Soc. **60** (1954), 415–430.

524. W. S. Loud, *Generalized inverses and generalized Green's functions*, SIAM J. Appl. Math. **14** (1966), 342–369.

525. _____, *Some examples of generalized Green's functions and generalized Green's matrices*, SIAM Rev. **12** (1970), 194–210.

526. J. M. Lowerre, *Some relationships between BLUEs, WISEs and SLSEs*, J. Amer. Statist. Assoc. **69** (1974), 223–225, (see [**33**]).

527. _____, *Some simplifying results on BLUEs*, J. Amer. Statist. Assoc. **72** (1977), no. 358, 433–437.

528. C. C. MacDuffee, *The Theory of Matrices*, Chelsea, New York, 1956.

529. J. R. Magnus and H. Neudecker, *Matrix Differential Calculus with Applications in Statistics and Econometrics*, Wiley, Chichester, 1999, (revised reprint of the 1988 original).

530. T. Mäkeläinen, *Projections and generalized inverses in the general linear model*, Soc. Sci. Fenn. Comment. Phys.-Math. **38** (1970), 13–25.

531. M. Marcus, *Finite Dimensional Multilinear Algebra*, Part 1, Marcel Dekker, New York, 1973.

532. _____, *Finite Dimensional Multilinear Algebra*, Part 2, Marcel Dekker, New York, 1975.

533. _____, *A unified exposition of some classical matrix theorems*, Linear and Multilinear Algebra **25** (1989), 137–147.

534. M. Marcus and H. Minc, *A Survey of Matrix Theory and Matrix Inequalities*, Allyn & Bacon, Boston, MA, 1964.

535. A. Markiewicz, *Simultaneous polar decomposition of rectangular complex matrices*, Linear Algebra Appl. **289** (1999), no. 1-3, 279–284, (application of [**399**]).

536. M. D. Marquardt, *Generalized inverses, ridge regression, biased linear estimation, and nonlinear regression*, Technometrics **12** (1970), 591–613.

537. J. C. Maxwell, *Treatise of Electricity and Magnetism*, 3rd ed., vol. I, Oxford University Press, Oxford, 1892.

538. N. H. McCoy, *Generalized regular rings*, Bull. Amer. Math. Soc. **45** (1939), 175–178.

539. G. C. McDonald, *Some algebraic properties of ridge coefficients*, J. Roy. Statist. Soc. Ser. B **42** (1980), no. 1, 31–34.

540. M. Meicler, *Chebyshev solution of an inconsistent system of $n+1$ linear equations in n unknowns in terms of its least squares solution*, SIAM Rev. **10** (1968), 373–375.

541. V. I. Meleško, *An investigation of stable L-pseudo-inverses of unbounded closed operators by the regularization method*, Differentsial'nye Uravneniya **15** (1979), no. 5, 921–935, 958–959, (English translation: Differential Equations 15 (1979), no. 5, 653–664).

542. C. D. Meyer, Jr., *Representations for (1)- and (1, 2)-inverses for partitioned matrices*, Linear Algebra Appl. **4** (1971), 221–232.

543. _____, *Limits and the index of a square matrix*, SIAM J. Appl. Math. **26** (1974), 469–478, (see [**714**]).

544. _____, *The role of the group generalized inverse in the theory of finite Markov chains*, SIAM Rev. **17** (1975), 443–464.

545. _____, *Analysis of finite Markov chains by group inversion techniques*, In Campbell [**155**], pp. 50–81.

546. _____, *The character of a finite Markov chain*, Linear Algebra, Markov Chains, and Queueing Models (Minneapolis, MN, 1992), Springer-Verlag, New York, 1993, pp. 47–58.

547. C. D. Meyer, Jr. and R. J. Painter, *Note on a least squares inverse for a matrix*, J. Assoc. Comput. Mach. **17** (1970), 110–112.

548. C. D. Meyer, Jr. and J. M. Shoaf, *Updating finite Markov chains by using techniques of group matrix inversion*, J. Statist. Comput. Simulation **11** (1980), no. 3-4, 163–181.

549. J.-M. Miao and A. Ben-Israel, *On principal angles between subspaces in R^n*, Linear Algebra Appl. **171** (1992), 81–98.

550. _____, *Minors of the Moore–Penrose inverse*, Linear Algebra Appl. **195** (1993), 191–207.

551. _____, *On l_p-approximate solutions of linear equations*, Linear Algebra Appl. **199** (1994), 305–327.

552. _____, *The geometry of basic, approximate, and minimum-norm solutions of linear equations*, Linear Algebra Appl. **216** (1995), 25–41.

553. _____, *Product cosines of angles between subspaces*, Linear Algebra Appl. **237/238** (1996), 71–81.

554. L. Mihályffy, *A note on the matrix inversion by the partitioning technique*, Studia Sci. Math. Hungar. **5** (1970), 127–135.

555. R. D. Milne, *An oblique matrix pseudoinverse*, SIAM J. Appl. Math. **16** (1968), 931–944, (see [**851**]).

556. N. Minamide and K. Nakamura, *Minimum error control problem in Banach space*, Technical Report 16, Nagoya University, Nagoya, Japan, 1969.

557. _____, *A restricted pseudoinverse and its applications to constrained minima*, SIAM J. Appl. Math. **19** (1970), 167–177.

558. L. Mirsky, *Symmetric gauge functions and unitarily invariant norms*, Quart. J. Math. Oxford **11** (1960), 50–59.

559. S. K. Mitra, *On a generalized inverse of a matrix and applications*, Sankhyā Ser. A **30** (1968), 107–114.

560. _____, *Shorted operators and the identification problem*, IEEE Trans. Circuits and Systems **29** (1982), no. 8, 581–583.

561. _____, *The minus partial order and the shorted matrix*, Linear Algebra Appl. **83** (1986), 1–27.

562. _____, *On group inverses and the sharp order*, Linear Algebra Appl. **92** (1987), 17–37.

563. _____, *Shorted matrices in star and related orderings*, Circuits Systems Signal Process. **9** (1990), no. 2, 197–212.

564. _____, *Matrix partial order through generalized inverses: unified theory*, Linear Algebra Appl. **148** (1991), 237–263.

565. S. K. Mitra and R. E. Hartwig, *Partial orders based on outer inverses*, Linear Algebra Appl. **176** (1992), 3–20.

566. S. K. Mitra and K. M. Prasad, *The nonunique parallel sum*, Linear Algebra Appl. **259** (1997), 77–99.

567. S. K. Mitra and M. L. Puri, *Shorted operators and generalized inverses of matrices*, Linear Algebra Appl. **25** (1979), 45–56.

568. _____, *Shorted matrices—an extended concept and some applications*, Linear Algebra Appl. **42** (1982), 57–79.

569. S. K. Mitra and C. R. Rao, *Some results in estimation and tests of linear hypotheses under the Gauss–Markov model*, Sankhyā Ser. A **30** (1968), 281–290.

570. _____, *Conditions for optimality and validity of simple least squares theory*, Ann. Math. Statist. **40** (1969), 1617–1624.

571. V. J. Mizel and M. M. Rao, *Nonsymmetric projections in Hilbert space*, Pacific J. Math. **12** (1962), 343–357.

572. C. B. Moler, *Iterative refinement in floating point*, J. Assoc. Comput. Mach. **14** (1967), 316–321.

573. C. B. Moler and C. F. Van Loan, *Nineteen dubious ways to compute the exponential of a matrix*, SIAM Rev. **20** (1978), 801–836.

574. E. H. Moore, *On properly positive Hermitian matrices*, Bull. Amer. Math. Soc. **23** (1916), 59.

575. _____, *On the reciprocal of the general algebraic matrix*, Bull. Amer. Math. Soc. **26** (1920), 394–395, (see [**76**]).

576. E. H. Moore and R. W. Barnard, *General Analysis*, Memoirs of the American Philosophical Society, I, American Philosophical Society, Philadelphia, PA, 1935, (specially Part 1, pp. 197–209, see [**76**]).

577. R. H. Moore and M. Z. Nashed, *Approximations of generalized inverses of linear operators in Banach spaces*, Approximation Theory (Proc. Internat. Sympos., Univ. Texas, Austin, TX, 1973), Academic Press, New York, 1973, pp. 425–428.

578. _____, *Approximations to generalized inverses of linear operators*, SIAM J. Appl. Math. **27** (1974), 1–16.

579. T. D. Morley, *A Gauss–Markov theorem for infinite-dimensional regression models with possibly singular covariance*, SIAM J. Appl. Math. **37** (1979), no. 2, 257–260.

580. _____, *Parallel summation, Maxwell's principle and the infimum of projections*, J. Math. Anal. Appl. **70** (1979), no. 1, 33–41.

581. V. A. Morozov, *The principle of disparity in solving operator equations by the method of regularization*, Zh. Vychisl. Mat. i Mat. Fiz. **8** (1968), 295–309.

582. _____, *The optimal regularization of operator equations*, Zh. Vychisl. Mat. i Mat. Fiz. **10** (1970), 818–829.

583. G. L. Morris and P. L. Odell, *Common solutions for n matrix equations with applications*, J. Assoc. Comput. Mach. **15** (1968), 272–274.

584. G. D. Mostow and J. H. Sampson, *Linear Algebra*, McGraw–Hill, New York, 1969.

585. G. D. Mostow, J. H. Sampson, and J. -P. Meyer, *Fundamental Structures of Algebra*, McGraw–Hill, New York, 1963.

586. C. Müller, *Spherical Harmonics*, Lecture Notes in Mathematics, No. 17, Springer-Verlag, New York, 1966.

587. W. D. Munn, *Pseudoinverses in semigroups*, Proc. Cambridge Philos. Soc. **57** (1961), 247–250.

588. W. D. Munn and R. Penrose, *A note on inverse semigroups*, Proc. Cambridge Philos. Soc. **51** (1955), 396–399.

589. F. J. Murray and J. von Neumann, *On rings of operators*, Ann. of Math. **37** (1936), 116–229.

590. V. C. Nanda, *A generalization of Cayley's theorem*, Math. Z. **101** (1967), 331–334.

591. M. Z. Nashed, *Steepest descent for singular linear operator equations*, SIAM J. Numer. Anal. **7** (1970), 358–362.

592. _____, *Generalized inverses, normal solvability, and iteration for singular operator equations*, Nonlinear Functional Anal. and Appl. (Proc. Advanced Sem., Math. Res. Center, University of Wisconsin, Madison, WI, 1970), Academic Press, New York, 1971, pp. 311–359.

593. M. Z. Nashed (ed.), *Generalized Inverses and Applications* (Proc. Sem., Math. Res. Center, University of Wisconsin, Madison, WI, 1973), New York, Academic Press, 1976.

594. _____, *Perturbations and approximations for generalized inverses and linear operator equations*, [**593**], pp. 325–396.

595. _____, *On generalized inverses and operator ranges*, Functional Analysis and Approximation (Oberwolfach, 1980), Birkhäuser, Basel, 1981, pp. 85–96.

596. M. Z. Nashed and X. Chen, *Convergence of Newton-like methods for singular operator equations using outer inverses*, Numer. Math. **66** (1993), 235–257.

597. M. Z. Nashed and L. B. Rall, *Annotated bibliography on generalized inverses and applications*, In Nashed [**593**], pp. 771–1041.

598. M. Z. Nashed and G. F. Votruba, *A unified approach to generalized inverses of linear operators. I. Algebraic, topological and projectional properties*, Bull. Amer. Math. Soc. **80** (1974), 825–830.

599. _____, *A unified approach to generalized inverses of linear operators. II. Extremal and proximal properties*, Bull. Amer. Math. Soc. **80** (1974), 831–835.

600. _____, *A unified operator theory of generalized inverses*, In Nashed [**593**], pp. 1–109.

601. M. Z. Nashed and G. Wahba, *Generalized inverses in reproducing kernel spaces: An approach to regularization of linear operator equations*, SIAM J. Math. Anal. **5** (1974), 974–987.

602. M. Z. Nashed and Y.-G. Zhao, *The Drazin inverse for singular evolution equations and partial differential operators*, Recent Trends in Differential Equations, World Scientific, River Edge, NJ, 1992, pp. 441–456.

603. F. Natterer, *The Mathematics of Computerized Tomography*, Wiley, New York, 1986.

604. D. L. Nelson, T. O. Lewis, and T. L. Boullion, *A quadratic programming technique using matrix pseudoinverses*, Indust. Math. **21** (1971), 1–21.

605. A. Neubauer, *Tikhonov-regularization of ill-posed linear operator equations on closed convex sets*, J. Approx. Theory **53** (1988), no. 3, 304–320.

606. _____, *On converse and saturation results for Tikhonov regularization of linear ill-posed problems*, SIAM J. Numer. Anal. **34** (1997), no. 2, 517–527.

607. A. Neumaier, *Solving ill-conditioned and singular linear systems: A tutorial on regularization*, SIAM Rev. **40** (1998), no. 3, 636–666 (electronic).

608. M. Neumann, *On the Schur complement and the LU-factorization of a matrix*, Linear and Multilinear Algebra **9** (1980/81), no. 4, 241–254.

609. T. G. Newman and P. L. Odell, *On the concept of a p − q generalized inverse of a matrix*, SIAM J. Appl. Math. **17** (1969), 520–525.

610. Y. Nievergelt, *Total least squares: State-of-the-art regression in numerical analysis*, SIAM Rev. **36** (1994), no. 2, 258–264.

611. _____, *Schmidt-Mirsky matrix approximation with linearly constrained singular values*, Linear Algebra Appl. **261** (1997), 207–219.

612. _____, *A tutorial history of least squares with applications to astronomy and geodesy*, J. Comput. Appl. Math. **121** (2000), no. 1-2, 37–72.

613. X.-W. Niu, *Plücker coordinates representation for relations and operations between linear subspaces*, J. Math. Res. Exposition **21** (2001), no. 1, 143–147.

614. B. Noble, *A method for computing the generalized inverse of a matrix*, SIAM J. Numer. Anal. **3** (1966), 582–584.

615. _____, *Applied Linear Algebra*, Prentice-Hall, Englewood Cliffs, NJ, 1969.

616. _____, *Methods for computing the Moore–Penrose generalized inverse, and related matters*, In Nashed [**593**], pp. 245–301.

617. K. Nordström, *Some further aspects of the Löwner-ordering antitonicity of the Moore–Penrose inverse*, Comm. Statist. A – Theory and Meth. **18** (1989), no. 12, 4471–4489 (1990).

618. R. L. Obenchain, *Good and optimal ridge estimators*, Ann. Statist. **6** (1978), no. 5, 1111–1121.

619. D. P. O'Leary, *On bounds for scaled projections and pseudo-inverses*, Linear Algebra Appl. **132** (1990), 115–117, (Answer to a question in [**782**]).

620. J. M. Ortega and W. C. Rheinboldt, *Iterative Solution of Nonlinear Equations in Several Variables*, Academic Press, New York, 1970.

621. E. E. Osborne, *On least squares solutions of linear equations*, J. Assoc. Comput. Mach. **8** (1961), 628–636.

622. _____, *Smallest least squares solutions of linear equations*, SIAM J. Numer. Anal. **2** (1965), 300–307.

623. A. Ostrowski, *A new proof of Haynsworth's quotient formula for Schur complements.*, Linear Algebra Appl. **4** (1971), 389–392.

624. D. V. Ouellette, *Schur complements and statistics*, Linear Algebra Appl. **36** (1981), 187–295.

625. C. C. Paige and M. A. Saunders, *Towards a generalized singular value decomposition*, SIAM J. Numer. Anal. **18** (1981), 398–405.

626. C.-T. Pan and K. Sigmon, *A bottom-up inductive proof of the singular value decomposition*, SIAM J. Matrix Anal. Appl. **15** (1994), no. 1, 59–61.

627. L. Pandolfi, *Controllability and stabilization for linear systems of algebraic and differential equations*, J. Optim. Theory Appl. **30** (1980), no. 4, 601–620.

628. B. N. Parlett, *The LU and QR algorithms*, In Ralston and Wilf [**668**], pp. 116–130, Vol. II.

629. M. H. Pearl, *On Cayley's parametrization*, Canad. J. Math. **9** (1957), 553–562.

630. _____, *A further extension of Cayley's parametrization*, Canad. J. Math. **11** (1959), 48–50.

631. _____, *On normal and EP_r matrices*, Michigan Math. J. **6** (1959), 1–5.

632. _____, *On normal EP_r matrices*, Michigan Math. J. **8** (1961), 33–37.

633. _____, *On generalized inverses of matrices*, Proc. Cambridge Philos. Soc. **62** (1966), 673–677.

634. _____, *Generalized inverses of matrices with entries taken from an arbitrary field*, Linear Algebra Appl. **1** (1968), 571–587, (see also [**690**]).

635. R. Penrose, *A generalized inverse for matrices*, Proc. Cambridge Philos. Soc. **51** (1955), 406–413.

636. _____, *On best approximate solutions of linear matrix equations*, Proc. Cambridge Philos. Soc. **52** (1956), 17–19.

637. V. Pereyra, *Iterative methods for solving nonlinear least squares problems*, SIAM J. Numer. Anal. **4** (1967), 27–36.

638. _____, *Stability of general systems of linear equations*, Aequationes Math. **2** (1969), 194–206.

639. V. Pereyra and J. B. Rosen, *Computation of the pseudoinverse of a matrix of unknown rank*, Technical Report CS 13, Department of Computer Science, Stanford University, Stanford, CA, Sept. 1964, (Comp. Rev. **6**(1965), 259 #7948).

640. G. Peters and J. H. Wilkinson, *The least squares problem and pseudo-inverses*, Comput. J. **13** (1970), 309–316.

641. W. V. Petryshyn, *On generalized inverses and on the uniform convergence of $(I - \beta K)^n$ with application to iterative methods*, J. Math. Anal. Appl. **18** (1967), 417–439.

642. A. Pietsch, *Zur Theorie der σ-Transformationen in lokalkonvexen Vektorräumen*, Math. Nachr. **21** (1960), 347–369.

643. R. Piziak, P. L. Odell, and R. Hahn, *Constructing projections on sums and intersections*, Comput. Math. Appl. **37** (1999), no. 1, 67–74.

644. G. D. Poole and T. L. Boullion, *Weak spectral inverses which are partial isometries*, SIAM J. Appl. Math. **23** (1972), 171–172.

645. W. A. Porter, *Modern Foundations of System Engineering*, Macmillan, New York, 1966.

646. _____, *A basic optimization problem in linear systems*, Math. Systems Theory **5** (1971), 20–44.

647. W. A. Porter and J. P. Williams, *Extension of the minimum effort control problem*, J. Math. Anal. Appl. **13** (1966), 536–549.

648. ———, *A note on the minimum effort control problem*, J. Math. Anal. Appl. **13** (1966), 251–264.

649. K. M. Prasad, K. P. S. Bhaskara Rao, and R. B. Bapat, *Generalized inverses over integral domains. II. Group inverses and Drazin inverses*, Linear Algebra Appl. **146** (1991), 31–47.

650. C. M. Price, *The matrix pseudoinverse and minimal variance estimates*, SIAM Rev. **6** (1964), 115–120.

651. R. M. Pringle and A. A. Rayner, *Expressions for generalized inverses of a bordered matrix with application to the theory of constrained linear models*, SIAM Rev. **12** (1970), 107–115.

652. ———, *Generalized Inverse Matrices with Applications to Statistics*. Griffin's Statistical Monographs and Courses, No. 28, Hafner, New York, 1971.

653. D. Przeworska-Rolewicz and S. Rolewicz, *Equations in Linear Spaces*, Polska Akad. Nauk Monog. Mat., vol. 47, PWN Polish Scientific Publishers, Warsaw, 1968.

654. S. Puntanen and A. J. Scott, *Some further remarks on the singular linear model*, Linear Algebra Appl. **237/238** (1996), 313–327.

655. S. Puntanen and G. P. H. Styan, *The equality of the ordinary least squares estimator and the best linear unbiased estimator* (with comments by O. Kempthorne and S. R. Searle and a reply by the authors), Amer. Statist. **43** (1989), no. 3, 153–164.

656. S. Puntanen, G. P. H. Styan, and H.-J. Werner, *Two matrix-based proofs that the linear estimator Gy is the best linear unbiased estimator*, J. Statist. Plann. Inference **88** (2000), no. 2, 173–179.

657. M. L. Puri, C. T. Russell, and T. Mathew, *Convergence of generalized inverses with applications to asymptotic hypothesis testing*, Sankhyā Ser. A **46** (1984), no. 2, 277–286.

658. R. Puystjens and R. E. Hartwig, *The group inverse of a companion matrix*, Linear and Multilinear Algebra **43** (1997), no. 1-3, 137–150.

659. L. D. Pyle, *Generalized inverse computations using the gradient projection method*, J. Assoc. Comput. Mach. **11** (1964), 422–428.

660. ———, *A generalized inverse ε-algorithm for constructing intersection projection matrices, with applications*, Numer. Math. **10** (1967), 86–102.

661. ———, *The generalized inverse in linear programming. Basic structure*, SIAM J. Appl. Math. **22** (1972), 335–355.

662. G. Rabson, *The generalized inverse in set theory and matrix theory*, Technical Report, Deptartment of Mathematics, Clarkson College of Technology, Potsdam, NY, 1969.

663. R. Rado, *Note on generalized inverses of matrices*, Proc. Cambridge Philos. Soc. **52** (1956), 600–601.

664. V. Rakočević, *On the continuity of the Moore–Penrose inverse in Banach algebras*, Facta Univ. Ser. Math. Inform. (1991), no. 6, 133–138.

665. ———, *Continuity of the Drazin inverse*, J. Operator Theory **41** (1999), no. 1, 55–68.

666. V. Rakočević and Y. Wei, *The perturbation theory for the Drazin inverse and its applications* II, J. Austral. Math. Soc. **70** (2001), no. 2, 189–197.

667. L. B. Rall, *Computational Solution of Nonlinear Operator Equations*, Wiley, New York, 1969.

668. A. Ralston and H. Wilf (eds.), *Mathematical Methods for Digital Computers*, New York, Wiley, 1967.

669. C. R. Rao, *Markoff's theorem with linear restrictions on parameters*, Sankhyā Ser. A **7** (1945), 16–20.

670. ———, *A note on a generalized inverse of a matrix with applications to problems in mathematical statistics*, J. Roy. Statist. Soc. Ser. B **24** (1962), 152–158.

671. ———, *Linear Statistical Inference and its Applications*, Wiley, New York, 1965, (second edition, 1973).

672. _____, *Generalized inverse for matrices and its applications in mathematical statistics*, Research Papers in Statistics (Festschrift J. Neyman), Wiley, London, 1966, pp. 263–279.

673. _____, *Calculus of generalized inverses of matrices. Part I: General theory*, Sankhyā Ser. A **29** (1967), 317–342.

674. _____, *Unified theory of linear estimation*, Sankhyā Ser. A **33** (1971), 371–394, (corrections: *ibid* **34** (1972), p. 194, and Sankhyā Ser. A **34** (1972), p. 477).

675. _____, *Projectors, generalized inverses and the BLUEs*, J. Roy. Statist. Soc. Ser. B **36** (1974), 442–448.

676. _____, *Estimation of parameters in a linear model (first 1975 Wald memorial lecture)*, Ann. Statist. **4** (1976), no. 6, 1023–1037, (corrections: *ibid* **7**(1979), no. 3, p. 696).

677. _____, *Choice of best linear estimators in the Gauss–Markoff model with a singular dispersion matrix*, Comm. Statist. A – Theory and Meth. **7** (1978), no. 13, 1199–1208.

678. C. R. Rao and S. K. Mitra, *Generalized Inverse of Matrices and its Applications*, Wiley, New York, 1971.

679. _____, *Theory and application of constrained inverse of matrices*, SIAM J. Appl. Math. **24** (1973), 473–488.

680. A. A. Rayner and R. M. Pringle, *A note on generalized inverses in the linear hypothesis not of full rank*, Ann. Math. Statist. **38** (1967), 271–273.

681. _____, *Some aspects of the solution of singular normal equations with the use of linear restrictions*, SIAM J. Appl. Math. **31** (1976), no. 3, 449–460, (corrections: *ibid* **47**(1987), p. 1130).

682. W. T. Reid, *Generalized Green's matrices for compatible systems of differential equations*, Amer. J. Math. **53** (1931), 443–459.

683. _____, *Principal solutions of non-oscillatory linear differential systems*, J. Math. Anal. Appl. **9** (1964), 397–423.

684. _____, *Generalized Green's matrices for two–point boundary problems*, SIAM J. Appl. Math. **15** (1967), 856–870.

685. _____, *Generalized inverses of differential and integral operators*, In Boullion and Odell [**121**], pp. 1–25.

686. _____, *Ordinary Differential Equations*, Wiley-Interscience, New York, 1970.

687. B. C. Rennie, Letter to the editor: *Rank factorization of a matrix and its applications* [Math. Sci. **13** (1988), no. 1, 4–14; MR 90a:15009a] by P. Bhimasankaram, Math. Sci. **13** (1988), no. 2, 152, (see [**97**]).

688. W. C. Rheinboldt, *A unified convergence theory for a class of iterative processes*, SIAM J. Numer. Anal. **5** (1968), 42–63.

689. J. Rice, *Experiments on Gram–Schmidt orthogonalization*, Math. Comp. **20** (1966), 325–328.

690. M. Q. Rieck, *Totally isotropic subspaces, complementary subspaces, and generalized inverses*, Linear Algebra Appl. **251** (1997), 239–248, (extension of a result of [**634**]).

691. R. F. Rinehart, *The equivalence of definitions of a matric function*, Amer. Math. Monthly **62** (1955), 395–414.

692. W. Rising, *Applications of generalized inverses to Markov chains*, Adv. in Appl. Probab. **23** (1991), 293–302.

693. P. D. Robers and A. Ben-Israel, *An interval programming algorithm for discrete linear L_1 approximation problems*, J. Approx. Theory **2** (1969), 323–336.

694. _____, *A suboptimization method for interval linear programming: A new method for linear programming*, Linear Algebra Appl. **3** (1970), 383–405.

695. P. Robert, *On the group-inverse of a linear transformation*, J. Math. Anal. Appl. **22** (1968), 658–669.

696. D. W. Robinson, *A proof of the composite function theorem for matric functions*, Amer. Math. Monthly **64** (1957), 34–35.

697. _____, *On the genralized inverse of an arbitrary linear transformation*, Amer. Math. Monthly **69** (1962), 412–416.

698. _____, *Gauss and generalized inverses*, Historia Math. **7** (1980), 118–125.

699. _____, *Separation of subspaces by volume*, Amer. Math. Monthly **105** (1998), no. 1, 22–27.

700. S. M. Robinson, *A short proof of Cramer's rule*, Math. Mag. **43** (1977), 94–95, (reprinted in *Selected Papers on Algebra* (S. Montgomery et al., editors), Mathematical Association of America, 1977, pp. 313–314).

701. S. Roch and B. Silbermann, *Index calculus for approximation methods and singular value decomposition*, J. Math. Anal. Appl. **225** (1998), no. 2, 401–426.

702. _____, *Continuity of generalized inverses in Banach algebras*, Studia Math. **136** (1999), no. 3, 197–227.

703. R. T. Rockafellar, *Convex Analysis*, Princeton University Press, Princeton, NJ, 1970.

704. C. A. Rohde, *Contributions to the Theory, Computation and Application of Generalized Inverses*, Ph.D., University of North Carolina, Raleigh, N.C., May 1964.

705. _____, *Generalized inverses of partitioned matrices*, J. Soc. Indust. Appl. Math. **13** (1965), 1033–1035.

706. _____, *Special applications of the theory of generalized matrix inversion to statistics*, In Boullion and Odell [**121**], pp. 239–266.

707. C. A. Rohde and J. R. Harvey, *Unified least squares analysis*, J. Amer. Statist. Assoc. **60** (1965), 523–527.

708. N. J. Rose, *A note on computing the Drazin inverse*, Linear Algebra Appl. **15** (1976), no. 2, 95–98.

709. _____, *The Laurent expansion of a generalized resolvent with some applications*, SIAM J. Math. Anal. **9** (1978), no. 4, 751–758.

710. J. B. Rosen, *The gradient projection method for nonlinear programming*. Part I: *Linear Constraints*, J. Soc. Indust. Appl. Math. **8** (1960), 181–217.

711. _____, *The gradient projection method for nonlinear programming*. Part II: *Nonlinear Constraints*, J. Soc. Indust. Appl. Math. **9** (1961), 514–532.

712. _____, *Minimum and basic solutions to singular linear systems*, J. Soc. Indust. Appl. Math. **12** (1964), 156–162.

713. P. C. Rosenbloom, *The method of steepest descent*, Numerical Analysis. Proceedings of the Sixth Symposium in Applied Mathematics, McGraw–Hill, New York, 1956, pp. 127–176.

714. U. G. Rothblum, *A representation of the Drazin inverse and characterizations of the index*, SIAM J. Appl. Math. **31** (1976), no. 4, 646–648.

715. _____, *Resolvent expansions of matrices and applications*, Linear Algebra Appl. **38** (1981), 33–49.

716. B. Rust, W. R. Burrus, and C. Schneeberger, *A simple algorithm for computing the generalized inverse of a matrix*, Comm. ACM **9** (1966), 381–385, 387.

717. S. Saitoh, *Positive definite Hermitian matrices and reproducing kernels*, Linear Algebra Appl. **48** (1982), 119–130.

718. _____, *One approach to some general integral transforms and its applications*, Integral Transform. Spec. Funct. **3** (1995), no. 1, 49–84.

719. _____, *Integral Transforms, Reproducing Kernels and their Applications*, Longman, Harlow, 1997.

720. _____, *Representations of inverse functions*, Proc. Amer. Math. Soc. **125** (1997), no. 12, 3633–3639.

721. S. Sakallıoğlu and F. Akdeniz, *Computation of the Moore–Penrose inverse of a matrix using the full rank factorization*, Pure Appl. Math. Sci. **39** (1994), no. 1-2, 79–84.

722. _____, *Generalized inverse estimator and comparison with least squares estimator*, Turkish J. Math. **22** (1998), no. 1, 77–84.

723. G. Salinetti, *The generalized inverse in parametric programming*, Calcolo **11** (1974), 351–363 (1975).

724. A. Sard, *Approximation based on nonscalar observations*, J. Approx. Theory **8** (1973), 315–334, (see [**334**]).

725. K. Scharnhorst, *Angles in complex vector spaces*, Acta Appl. Math. **69** (2001), 95–103.

726. H. Scheffé, *The Analysis of Variance*, Wiley, New York, 1959.

727. E. Schmidt, *Zur Theorie der linearen und nichlinearen Integralgleichungen, I. Entwicklung willküricher Funktionen nach Systemen vorgeschriebener*, Math. Ann. **63** (1907), 433–476, (see SVD history in [**783**])).

728. P. H. Schönemann, *A generalized solution of the orthogonal Procrustes problem*, Psychmoetrika **31** (1966), 1–10.

729. P. Schönfeld and H.-J. Werner, *A note on C. R. Rao's wider definition BLUE in the general Gauss-Markov model*, Sankhyā Ser. B **49** (1987), no. 1, 1–8.

730. M. Schreiber, *Differential Forms: A Heuristic Introduction*, Springer-Verlag, New York, 1977.

731. G. Schulz, *Iterative Berechnung der Reziproken Matrix*, Z. Angew. Math. Mech. **13** (1933), 57–59.

732. I. Schur, *Potenzreihen im Innern des Einheitskreises*, J. Reine Angew. Math. **147** (1917), 205–232.

733. J. T. Schwartz, *Perturbations of spectral operators, and applications*, Pacific J. Math. **4** (1954), 415–458.

734. H. Schwerdtfeger, *Les Fonctions de Matrices I. Les Fonctions Univalents*, Actualités Scientifiques et Industrielles, No. 649, Herman, Paris, 1938.

735. _____, *Introduction to Linear Algebra and the Theory of Matrices*, Noordhoff, Groningen, 1950.

736. J. E. Scroggs and P. L. Odell, *An alternate definition of a pseudoinverse of a matrix*, SIAM J. Appl. Math. **14** (1966), 796–810, (see [**122**]).

737. S. R. Searle, *Additional results concerning estimable functions and generalized inverse matrices*, J. Roy. Statist. Soc. Ser. B **27** (1965), 486–490.

738. _____, *Linear Models*, Wiley, New York, 1971.

739. _____, *Extending some results and proofs for the singular linear model*, Linear Algebra Appl. **210** (1994), 139–151.

740. J. Seely, *Linear spaces and unbiased estimation*, Ann. Math. Statist. **41** (1970), 1725–1734.

741. _____, *Linear spaces and unbiased estimation—Application to the mixed linear model*, Ann. Math. Statist. **41** (1970), 1735–1748.

742. J. Seely and G. Zyskind, *Linear spaces and minimum variance unbiased estimation*, Ann. Math. Statist. **42** (1971), 691–703.

743. J. Seidel, *Angles and distances in n-dimensional Euclidean and non-Euclidean geometry. I, II, III*, Nederl. Akad. Wetensch. Proc. Ser. A. **58** = Indag. Math. **17** (1955), 329–335, 336–340, 535–541.

744. A. Sengupta, *Multifunction and generalized inverse*, J. Inverse Ill-Posed Probl. **5** (1997), 265–285.

745. G. E. Sharpe and G. P. H. Styan, *Circuit duality and the general network inverse*, IEEE Trans. Circuit Theory **12** (1965), 22–27.

746. _____, *A note on the general network inverse*, IEEE Trans. Circuit Theory **12** (1965), 632–633.

747. _____, *Circuit duality and the general network inverse*, Proc. IEEE **55** (1967), 1226–1227.

748. R. D. Sheffield, *On pseudo-inverses of linear transformations in Banach space*, Technical Report 2133, Oak Ridge National Laboratory, 1956.

749. _____, *A general theory for linear systems*, Amer. Math. Monthly **65** (1958), 109–111.

750. O. B. Sheynin, *R. J. Boscovich's work on probability*, Arch. Hist. Exact Sci. **9** (1972/73), no. 4-5, 306–324.

751. N. Shinozaki and M. Sibuya, *The reverse order law $(AB)^- = B^- A^-$*, Linear Algebra Appl. **9** (1974), 29–40.

752. _____, *Further results on the reverse-order law*, Linear Algebra Appl. **27** (1979), 9–16.

753. N. Shinozaki, M. Sibuya, and K. Tanabe, *Numerical algorithms for the Moore-Penrose inverse of a matrix: Direct methods*, Ann. Inst. Statist. Math. **24** (1972), 193–203.

754. _____, *Numerical algorithms for the Moore-Penrose inverse of a matrix: Iterative methods*, Ann. Inst. Statist. Math. **24** (1972), 621–629.

755. D. W. Showalter, *Representation and computation of the pseudoinverse*, Proc. Amer. Math. Soc. **18** (1967), 584–586.

756. D. W. Showalter and A. Ben-Israel, *Representation and computation of the generalized inverse of a bounded linear operator between Hilbert spaces*, Atti Accad. Naz. Lincei Rend. Cl. Sci. Fis. Mat. Natur. (8) **48** (1970), 184–194.

757. M. Sibuya, *Subclasses of generalized inverses of matrices*, Ann. Inst. Statist. Math. **22** (1970), 543–556.

758. _____, *The Azumaya-Drazin pseudoinverse and the spectral inverses of a matrix*, Sankhyā Ser. A **35** (1973), 95–102.

759. A. Sidi, *Development of iterative techniques and extrapolation methods for Drazin inverse solution of consistent or inconsistent singular linear systems*, Linear Algebra Appl. **167** (1992), 171–203.

760. _____, *A unified approach to Krylov subspace methods for the Drazin-inverse solution of singular nonsymmetric linear systems*, Linear Algebra Appl. **298** (1999), no. 1-3, 99–113.

761. A. Sidi and V. Kluzner, *A Bi-CG type iterative method for Drazin-inverse solution of singular inconsistent nonsymmetric linear systems of arbitrary index*, Electron. J. Linear Algebra **6** (1999/00), 72–94 (electronic).

762. C. L. Siegel, *Über die analytische Theorie der quadratischen Formen III*, Ann. of Math. **38** (1937), 212–291, (see, in particular, pp. 217–229).

763. I. Singer, *Best Approximation in Normed Linear Spaces by Elements of Linear Subspaces*, Springer–Verlag, Berlin, 1970.

764. G. Smith and F. Campbell, *A critique of some ridge regression methods* (with comments by R. A. Thisted, et al, and with a reply by the authors, J. Amer. Statist. Assoc. **75** (1980), no. 369, 74–103.

765. F. Smithies, *Integral Equations*, Cambridge University Press, Cambridge, 1958.

766. T. Söderstörm and G. W. Stewart, *On the numerical proerties of an iterative method for computing the Moore–Penrose generalized inverse*, SIAM J. Numer. Anal. **11** (1974), 61–74.

767. I. Sonin and J. Thornton, *Recursive algorithm for the fundamental/group inverse matrix of a Markov chain from an explicit formula*, SIAM J. Matrix Anal. Appl. **23** (2001), no. 1, 209–224.

768. E. D. Sontag, *On generalized inverses of polynomial and other matrices*, IEEE Trans. Autom. Control **25** (1980), no. 3, 514–517, (extension of [**118**]).

769. _____, *Mathematical Control Theory: Deterministic Finite Dimensional Systems*, Speinger-Verlag, New York, 1990.

770. M. D. Springer, *The Algebra of Random Variables*, Wiley, New York, 1979.

771. V. P. Sreedharan, *Least squares algorithms for finding solutions of overdetermined linear equations which minimize error in an abstract norm*, Numer. Math. **17** (1971), 387–401.

772. P. Stahlecker and G. Trenkler, *Linear and ellipsoidal restrictions in linear regression*, Statistics **22** (1991), no. 2, 163–176.

773. I. Stakgold, *Branching of solutions of nonlinear equations*, SIAM Rev. **13** (1971), 289–332, (corrections: *ibid* **14**(1972), p. 492).

774. W. T. Stallings and T. L. Boullion, *Computation of pseudoinverse matrices using residue arithmetic*, SIAM Rev. **14** (1972), 152–163.

775. P. S. Stanimirović and D. S. Djordjević, *Full-rank and determinantal representation of the Drazin inverse*, Linear Algebra Appl. **311** (2000), no. 1-3, 131–151.

776. P. Stein, *Some general theorems on iterants*, J. Res. Nat. Bur. Standards **48** (1952), 82–82.

777. T. E. Stern, *Extremum relations in nonlinear networks and their applications to mathematical programming*, Journées d'Études sur le Contrôle Optimum el les Systèmes Nonlinéaires, Institut National des Sciences et Techniques Nucleaires, Saclay, France, pp. 135–156.

778. _____, *Theory of Nonlinear Networks and Systems*, Addison–Wesley, Reading, MA, 1965.

779. G. W. Stewart, *On the continuity of the generalized inverse*, SIAM J. Appl. Math. **17** (1969), 33–45, (see [**657**]).

780. _____, *Projectors and generalized inverses*, Technical Report TNN-97, University of Texas at Austin Computation Center, October 1969.

781. _____, *On the perturbation of pseudo-inverses, projections, and linear least squares problems*, SIAM Rev. **19** (1977), 634–662.

782. _____, *On scaled projections and pseudo-inverses*, Linear Algebra Appl. **112** (1989), 189–194, (see [**619**], [**810**], [**313**], [**114**]).

783. _____, *On the early history of the singular value decomposition*, SIAM Rev. **35** (1993), 551–566, (see [**57**], [**429**, Chapter 3], [**452**]–[**453**], [**727**], [**790**]–[**792**], [**868**]).

784. G. W. Stewart and J.-G. Sun, *Matrix Perturbation Theory*, Academic Press, Boston, MA, 1990.

785. M. Subrahmanyan, *A property of simple least squares estimates*, Sankhyā Ser. B **34** (1972), 355–356, (see [**424**]).

786. J.-G. Sun, *Perturbation of angles between linear subspaces*, J. Comput. Math. **5** (1987), no. 1, 58–61.

787. W. Sun, *Cramer rules for weighted systems*, Nanjing Daxue Xuebao Shuxue Bannian Kan **3** (1986), no. 2, 117–121.

788. W. Sun and Y. Wei, *Triple reverse-order law for weighted generalized inverses*, Appl. Math. Comput. **125** (2002), no. 2-3, 221–229.

789. R. Sundberg, *Continuum regression and ridge regression*, J. Roy. Statist. Soc. Ser. B **55** (1993), no. 3, 653–659.

790. J. J. Sylvester, *A new proof that a general quadratic may be reduced to its canonical form (that is, a linear function of squares) by means of real orthogonal substitution*, Messenger of Math. **19** (1889), 1–5, (see history of SVD in [**783**]).

791. _____, *On the reduction of a bilinear quantic of the nth order to the form of a sum of n products by a double orthogonal substitution*, Messenger of Math. **19** (1889), 42–46, (see history of SVD in [**783**]).

792. _____, *Sur la réduction biorthogonale d'une form linéaire à sa forme cannonique*, C. R. Acad. Sci. Paris **108** (1889), 651–653, (see history of SVD in [**783**]).

793. Y. Takane and H. Yanai, *On oblique projectors*, Linear Algebra Appl. **289** (1999), no. 1-3, 297–310.

794. W. Y. Tan, *Note on an extension of the Gauss–Markov theorems to multivariate linear regression models*, SIAM J. Appl. Math. **20** (1971), 24–29.

795. R. A. Tapia, *An application of a Newton-like method to the Euler-Lagrange equation.*, Pacific J. Math. **29** (1969), 235–246.

796. _____, *The weak Newton method and boundary value problems*, SIAM J. Numer. Anal. **6** (1969), 539–550.

797. O. E. Taurian and P.-O. Löwdin, *Some remarks on the projector associated with the intersection of two linear manifolds*, Acta Phys. Acad. Sci. Hungar. **51** (1981), no. 1-2, 5–12 (1982).

798. O. Taussky, *Note on the condition of matrices*, Math. Tables Aids Comput. **4** (1950), 111–112.

799. _____, *Matrices C with $C^n \to O$*, J. Algebra **1** (1964), 5–10.

800. A. E. Taylor, *Introduction to Functional Analysis*, Wiley, New York, 1958.

801. R. P. Tewarson, *A direct method for generalized matrix inversion*, SIAM J. Numer. Anal. **4** (1967), 499–507.

802. _____, *On some representations of generalized inverses*, SIAM Rev. **11** (1969), 272–276.

803. _____, *On two direct methods for computing generalized inverses*, Computing **7** (1971), 236–239.

804. W. Thomson (Lord Kelvin), Cambridge and Dublin Math. J. (1848), 84–87.

805. A. N. Tikhonov, *On the regularization of ill-posed problems*, Dokl. Akad. Nauk SSSR **153** (1963), 49–52.

806. _____, *On the solution of ill-posed problems and the method of regularization*, Dokl. Akad. Nauk SSSR **151** (1963), 501–504.

807. _____, *On the stability of algorithms for the solution of degenerate systems of linear algebraic equations*, Zh. Vychisl. Mat. i Mat. Fiz. **5** (1965), 718–722.

808. _____, *On the problems with approximately specified information*, Ill-Posed Problems in the Natural Sciences, Mir, Moscow, 1987, pp. 13–20.

809. A. N. Tikhonov and V. Y. Arsenin, *Solutions of Ill-Posed Problems*, V. H. Winston & Sons, Washington, D.C.: Wiley, New York, 1977, (translated from the Russian, Preface by translation editor Fritz John, Scripta Series in Mathematics).

810. M. J. Todd, *A Dantzig–Wolfe-like variant of Karmarkar's interior-point linear programming algorithm*, Oper. Res. **38** (1990), no. 6, 1006–1018, (see [**782**]).

811. J. Tokarzewski, *Geometric characterization of system zeros and zero directions by the Moore–Penrose inverse of the first non-zero Markov parameter*, Arch. Control Sci. **5** (1996), no. 3-4(41), 245–264.

812. _____, *Zeros in Linear Systems: A Geometric Approach*, Oficyna Wydawnicza Politehniki Warszawskiej, Warsaw, 2002.

813. G. Trenkler, *Biased Estimators in the Linear Regression Model*, Athenäum, Königstein, 1981.

814. _____, *Generalizing Mallows' C_L and optimal ridge constants*, VII. Symposium on Operations Research, Sektionen 4–9 (St. Gallen, 1982), Athenäum, Königstein, 1983, pp. 157–166.

815. _____, *Characterizations of oblique and orthogonal projectors*, Proceedings of the International Conference on Linear Statistical Inference LINSTAT '93 (Poznań, 1993) (Dordrecht), Kluwer Academic, 1994, pp. 255–270.

816. Y.-Y. Tseng, *The Characteristic Value Problem of Hermitian Functional Operators in a Non–Hilbert Space*, Ph.d. in mathematics, University of Chicago, Chicago, 1933, (published by the University of Chicago Libraries, 1936).

817. _____, *Generalized inverses of unbounded operators between two unitary spaces*, Doklady Akad. Nauk SSSR (N.S.) **67** (1949), 431–434.

818. _____, *Properties and classification of generalized inverses of closed operators*, Doklady Akad. Nauk SSSR (N.S.) **67** (1949), 607–610.

819. _____, *Sur les solutions des équations opératrices fonctionnelles entre les espaces unitaires. Solutions extrémales. Solutions virtuelles*, C. R. Acad. Sci. Paris **228** (1949), 640–641.

820. _____, *Virtual solutions and general inversions*, Uspehi Mat. Nauk (N.S.) **11** (1956), no. 6(72), 213–215.

821. T. Tsuji and Z.-J. Yang, *Pseudo inverse in Banach space for Newton method*, Proceedings of the Fifth International Colloquium on Differential Equations (Plovdiv, 1994) (Utrecht), VSP, 1995, pp. 341–344.

822. D. H. Tucker, *Boundary value problems for linear differential systems*, SIAM J. Appl. Math. **17** (1969), 769–783.

823. F. E. Udwadia and R. E. Kalaba, *An alternative proof of the Greville formula*, J. Optim. Theory Appl. **94** (1997), no. 1, 23–28.

824. N. S. Urquhart, *Computation of generalized inverse matrices which satisfy specified conditions*, SIAM Rev. **10** (1968), 216–218.

825. _____, *The nature of the lack of uniqueness of generalized inverse matrices*, SIAM Rev. **11** (1969), 268–271.

826. O. Vaarmann, *The application of generalized inverse operators and their approximations to the solution of nonlinear equations*, Eesti NSV Tead. Akad. Toimetised Füüs.-Mat. **19** (1970), 265–274.

827. _____, *Approximations of pseudo-inverse operators as applied to the solution of nonlinear equations*, Eesti NSV Tead. Akad. Toimetised Füüs.-Mat. **20** (1971), 386–394.

828. J. Van Hamme, *Generalized inverses of linear operators in Hilbert spaces*, Bull. Soc. Math. Belg. Sér. B **41** (1989), no. 1, 83–93.

829. _____, *On the generalized inverse of a matrix partial differential operator*, Simon Stevin **66** (1992), no. 1-2, 185–194.

830. S. Van Huffel and J. Vanderwalle, *The Total Least Squares Problem: Computational Aspects and Analysis*, SIAM, Philadelphia, PA, 1991.

831. C. F. Van Loan, *Generalizing the singular value decomposition*, SIAM J. Numer. Anal. **13** (1976), no. 1, 76–83.

832. _____, *Computing the CS and the generalized singular value decompositions*, Numer. Math. **479–491** (1985), 479–491.

833. R. J. Vanderbei and J. C. Lagarias, *I. I. Dikin's convergence result for the affine-scaling algorithm*, Mathematical Developments arising from Linear Programming (Brunswick, ME, 1988), Amer. Math. Soc., Providence, RI, 1990, pp. 109–119.

834. S. A. Vavasis and Y. Ye, *Condition numbers for polyhedra with real number data*, Oper. Res. Lett. **17** (1995), no. 5, 209–214.

835. G. C. Verghese, *A "Cramer rule" for the least-norm, least-squared-error solution of inconsistent linear equations*, Linear Algebra Appl. **48** (1982), 315–316, (extension of [**71**]).

836. H. D. Vinod, *A survey of ridge regression and related techniques for improvements over ordinary least squares*, Rev. Econom. Statist. **60** (1978), no. 1, 121–131.

837. A. Vogt, *On the linearity of form isometries*, SIAM J. Appl. Math. **22** (1972), 553–560.

838. J. von Neumann, *Über adjungierte Funktionaloperatoren*, Ann. of Math. **33** (1932), 294–310.

839. _____, *Some matrix inequalities and metrization of matric-space*, Tomsk Univ. Rev. **1** (1937), 286–300, (republished in *John von Neumann Collected Works*, Macmillan, New York, Vol IV, pp. 205–219).

840. _____, *Functional Operators. Vol II: The Geometry of Orthogonal Spaces*, Annals of Math. Studies, vol. 29, Princeton University Press, Princeton, 1950.

841. _____, *Continuous Geometry*, Princeton University Press, Princeton, 1960.

842. D. von Rosen, *A matrix formula for testing linear hypotheses in linear models*, Linear Algebra Appl. **127** (1990), 457–461.

843. G. Votruba, *Generalized Inverses and Singular Equations in Functional Analysis*, Doctoral Dissertation in Mathematics, University of Michigan, Ann Arbor, MI, 1963.

844. G. Wahba and M. Z. Nashed, *The approximate solution of a class of constrained control problems*, Proceedings of the Sixth Hawaii International Conference on Systems Sciences (Hawaii), 1973.

845. B.-Y. Wang, X. Zhang, and F. Zhang, *Some inequalities on generalized Schur complements*, Linear Algebra Appl. **302/303** (1999), 163–172.

846. G.-R. Wang, *A new proof of Greville's method for computing M–P inverse* (Chinese), J. Shanghai Teachers University **14** (1985), no. 3, 32–38.

847. _____, *A Cramer rule for minimum-norm (T) least-squares (S) solution of inconsistent linear equations*, Linear Algebra Appl. **74** (1986), 213–218, (see [**71**], [**835**]).

848. _____, *A finite algorithm for computing the weighted Moore–Penrose inverse A^+_{MN}*, Appl. Math. Comput. **23** (1987), no. 4, 277–289.

849. _____, *A Cramer rule for finding the solution of a class of singular equations*, Linear Algebra Appl. **116** (1989), 27–34.

850. S.-G. Wang, *On biased linear estimators in models with arbitrary rank*, Comm. Statist. A – Theory and Meth. **11** (1982), no. 14, 1571–1581.

851. J. F. Ward, Jr., T. L. Boullion, and T. O. Lewis, *A note on the oblique matrix pseudoinverse*, SIAM J. Appl. Math. **20** (1971), 173–175, (proof of equivalence of weighted inverses [**186**] and oblique inverses [**555**]).

852. _____, *Weak spectral inverses*, SIAM J. Appl. Math. **22** (1972), 514–518.

853. J. H. M. Wedderburn, *Lectures on Matrices*, Colloq. Publ., vol. XVII, Amer. Math. Soc., Providence, RI, 1934.

854. P.-Å. Wedin, *Perturbation bounds in connection with singular value decomposition*, BIT **12** (1972), 99–111.

855. _____, *Pertubation theory for pseudo-inverses*, BIT **13** (1973), 217–232.

856. _____, *On angles between subspaces*, Matrix Pencils, (B. Kågström and A. Ruhe, editors), Springer, New York, 1983, pp. 263–285.

857. M.-S. Wei, *Upper bounds and stability of scaled pseudoinverses*, Numer. Math. **72** (1995), 285–293.

858. Y. Wei, *The reverse order rule of weighted Moore–Penrose of matrix product* (Chinese), J. Shanghai Teachers Univ. **24** (1994), no. 3, 19–23.

859. ———, *On the perturbation of the group inverse and oblique projection*, Appl. Math. Comput. **98** (1999), no. 1, 29–42.

860. Y. Wei and G.-R. Wang, *On continuity of the generalized inverse* $A_{T,S}^{(2)}$, Appl. Math. Comput. **136** (2003), no. 2-3, 289–295.

861. Y. Wei and H. Wu, *The perturbation of the Drazin inverse and oblique projection*, Appl. Math. Lett. **13** (2000), no. 3, 77–83.

862. H. F. Weinberger, *On optimal numerical solution of partial differential equations*, SIAM J. Numer. Anal. **9** (1972), 182–198.

863. H.-J. Werner, *More on BLIMB-estimation*, Contributions to Operations Research and Mathematical Economics, Vol. II, Athenäum/Hain/Hanstein, Königstein, 1984, pp. 629–638.

864. ———, *On extensions of Cramer's rule for solutions of restricted linear systems*, Linear and Multilinear Algebra **15** (1984), no. 3-4, 319–330.

865. ———, *More on BLU estimation in regression models with possibly singular covariances*, Linear Algebra Appl. **67** (1985), 207–214.

866. H.-J. Werner and C. Yapar, *A BLUE decomposition in the general linear regression model*, Linear Algebra Appl. **237/238** (1996), 395–404.

867. ———, *On inequality constrained generalized least squares selections in the general possibly singular Gauss–Markov model: a projector theoretical approach*, Linear Algebra Appl. **237/238** (1996), 359–393.

868. H. Weyl, *Das asymptotische Verteilingsgesetz der Eigenwert linearer partieller Differentialgleichungen (mit einer Anwendung auf die Theorie der Hohlraumstrahlung)*, Math. Ann. **71** (1912), 441–479, (see SVD history in [**783**]).

869. ———, *Inequalities between the two kinds of eigenvalues of a linear transformation*, Proc. Nat. Acad. Sci. U.S.A. **35** (1949), 408–411.

870. T. M. Whitney and R. K. Meany, *Two algorithms related to the method of steepest descent*, SIAM J. Numer. Anal. **4** (1967), 109–118.

871. E. A. Wibker, R. B. Howe, and J. D. Gilbert, *Explicit solutions to the reverse order law* $(AB)^+ = B_{mr}^- A_{lr}^-$, Linear Algebra Appl. **25** (1979), 107–114.

872. J. H. Wilkinson, *The Algebraic Eigenvalue Problem*, Oxford University Press, London, 1965.

873. ———, *The solution of ill-conditioned linear equations*, In Ralston and Wilf [**668**], pp. 65–93, Vol. II.

874. ———, *Note on the practical significance of the Drazin inverse*, In Campbell [**155**], pp. 82–99.

875. J. H. Wilkinson and C. Reinsch (eds.), *Handbook for Automatic Computation*, Vol. II: *Linear Algebra*, Springer-Verlag, Berlin, 1971.

876. J. Williamson, *A polar representation of singular matrices*, Bull. Amer. Math. Soc. **41** (1935), 118–123.

877. L. B. Willner, *An elimination method for computing the generalized inverse*, Math. Comp. **21** (1967), 227–229.

878. H. K. Wimmer, *Generalized singular values and interlacing inequalities*, J. Math. Anal. Appl. **137** (1989), no. 1, 181–184.

879. H. K. Wimmer and A. D. Ziebur, *Solving the equation* $\sum_{\rho=1}^{r} f_\rho(A) X g_\rho(B) = C$, SIAM Rev. **14** (1972), 318–323.

880. O. Wyler, *Green's operators*, Ann. Mat. Pura Appl. (4) **66** (1964), 252–263, (see [**895**]).

881. ———, *On two-point boundary-value problems*, Ann. Mat. Pura Appl. (4) **67** (1965), 127–142.

882. K. Yosida, *Functional Analysis*, 2nd ed., Springer–Verlag, Berlin, 1958.

883. S. Zacks, *Generalized least squares estimators for randomized replication designs*, Ann. Math. Statist. **35** (1964), 696–704.

884. L. A. Zadeh and C. A. Desoer, *Linear Syetem Theory*, McGraw–Hill, New York, 1963.

885. E. H. Zarantonello, *Differentoids*, Adv. in Math. **2** (1968), 187–306.

886. R. E. Zarnowski, *Generalized inverses and the total stopping times of Collatz sequences*, Linear and Multilinear Algebra **49** (2001), no. 2, 115–130.

887. F. Zhang, *Schur complements and matrix inequalities in the Löwner ordering*, Linear Algebra Appl. **321** (2000), 399–410.

888. K. Ziętak, *Orthant-monotonic norms and overdetermined linear systems*, J. Approx. Theory **88** (1997), no. 2, 209–227.

889. S. Zlobec, *On computing the generalized inverse of a linear operator*, Glas. Mat. Ser. III **2 (22)** (1967), 265–271.

890. _____ , *Contributions to mathematical programming and generalized inversion*, Applied math., Northwestern University, Evanston, IL, 1970.

891. _____ , *An explicit form of the Moore–Penrose inverse of an arbitrary complex matrix*, SIAM Rev. **12** (1970), 132–134.

892. _____ , *On computing the best least squares solutions in Hilbert space*, Rend. Circ. Mat. Palermo (2) **25** (1976), no. 3, 256–270 (1977).

893. S. Zlobec and A. Ben-Israel, *On explicit solutions of interval linear programs*, Israel J. Math. **8** (1970), 265–271.

894. _____ , *Explicit solutions of interval linear programs*, Oper. Res. **21** (1973), 390–393.

895. V. M. Zubov, *Some properties of a generalized inverse operator in a vector space*, Izv. Vyssh. Uchebn. Zaved. Mat. (1983), no. 12, 67–69.

896. E. L. Žukovskiĭ, *The method of least squares for degenerate and ill-conditioned systems of linear algebraic equations*, Zh. Vychisl. Mat. i Mat. Fiz. **17** (1977), no. 4, 814–827, 1091.

897. E. L. Žukovskiĭ and R. Š. Lipcer, *A recurrence method for computing the normal solutions of linear algebraic equations*, Zh. Vychisl. Mat. i Mat. Fiz. **12** (1972), 843–857, 1084.

898. G. Zyskind, *A note on residue analysis*, J. Amer. Statist. Soc. **58** (1963), 1125–1132.

899. _____ , *On canonical forms, nonnegative covariance matrices and best and simple least squares linear estimators in linear models*, Ann. Math. Statist. **38** (1967), 1092–1109.

900. _____ , *Error structures, projections and conditional inverses in linear model theory*, A Survey of Statistical Design and Linear Models (Proc. Internat. Sympos., Colorado State University, Ft. Collins, CO, 1973), North-Holland, Amsterdam, 1975, pp. 647–663.

901. G. Zyskind and F. B. Martin, *On best linear estimation and a general Gauss–Markov theorem in linear models with arbitrary nonnegative covariance structure*, SIAM J. Appl. Math. **17** (1969), 1190–1202.

Subject Index

α-β generalized inverse, 134
α-approximate solution, 136
ℓ_p-norm, 9, 141
λ-vector, 34
ϕ-metric projector, 132
$\{1, 2, 3\}$-inverse, 46
  computation, 47, 179, 260
$\{1, 2, 4\}$-inverse, 46
  computation, 47, 179
$\{1, 2\}$-inverse, 45, 208
  computation, 45, 46, 179, 208, 258
  weighted, 119, 121, 255
$\{1, 3\}$-inverse, 104, 111, 208
  computation, 208
$\{1, 4\}$-inverse, 111, 208
  computation, 208
$\{1\}$-inverse, 41, 42, 208
  computation, 41, 208, 258, 259
$\{2\}$-inverse, 295, 296, 301
  computation, 50
$\{i, j, \dots, k\}$-inverse, 40

Absorbing
  chain, 304
  state, 304
acute
  matrices, 239
  perturbation, 239
adjoint, 12
admittance matrix, 102
affine set, 182
algebraic multiplicity, 35
angle, 8
aperiodic state, 304

B-SVD, 251
B-singular values, 251
basic
  solution, 122
  subspaces, 236
best linear unbiased estimator, *see* BLUE,
  5
best rank-k approximation, 213
Beta function, 321
bias, 293
Binet–Cauchy formula, 29
BLUE, 5, 285

Bott–Duffin inverse, 92, 98, 148
branch currents, 100

Canonical angles, 232
carrier, 331
Cauchy–Schwartz inequality, 7, 141, 234
  generalized, 141
Cesaro mean, 306
characteristic polynomial, 35
Cholesky factorization, 119
Cline's method, 166, 262
closed set of states, 304
closure, 330
coefficient of inclination, 230
communicating states, 303
complementary subspaces, 6
compound matrix, 32
computation
  $\{1, 2, 3\}$-inverse, 47, 179, 260
  $\{1, 2, 4\}$-inverse, 47, 179
  $\{1, 2\}$-inverse, 45, 46, 179, 208, 258
  $\{1, 3\}$-inverse, 208
  $\{1, 4\}$-inverse, 208
  $\{1\}$-inverse, 41, 208, 258
  $\{2\}$-inverse, 50
  basis for null space, 25
  basis for range space, 25
  Drazin inverse, 164–168, 261, 262
  group inverse, 181, 182, 262
  Hermite normal form, 24, 26
  Moore–Penrose inverse, 48, 179, 207,
    208, 250, 261–263, 272, 277
  rank factorization, 26
  Smith normal form, 38
condition number, 204
  spectral, 204
consistent norms, 19
constrained
  inverse, 92
  least-squares solution, 108
  minimum-norm least-squares solution,
    113, 255
contraction, 223
convergent matrix, 21
convex
  body, 140
  function, 131

set
  rotund, 142
  smooth, 142
coordinates
  cylindrical, 318
  Plücker, 32, 210, 237
  spherical, 319
covariance matrix, 284
Cramer's rule, 30, 78, 124, 197, 200
current, 100
cylindrical coordinates, 318

Decomposable, 331
density function, 323
derivative, 300
determinant, 28
diagonable matrix, 60, 62, 153, 155
difference equation
  consistent initial solution, 310
  homogeneous, 310
  tractable, 310
differentiable, 300
dimension of inclination, 230
direct sum, 331
discriminant, 94
distance between subspaces, 233
distribution
  χ^2, 328
  bivariate normal, 327
  exponential, 327
  function, 323
  spherical, 326
  uniform, 328
domain, 331
Drazin inverse, 156, 163, 164
  Cline's method, 166, 262
  computation, 164–168, 261, 262
  limit form, 168
dual
  function, 138
  norms, 140
  set, 140
  vectors, 141

Eigenfunction, 345
eigenspace, 13
eigenvalue, 13, 345
eigenvector, 13
electrical network, 99, 149
  currents, 99
  dual transfer matrix, 101
  transfer matrix, 101
  voltages, 99
elementary
  matrices, 22
  operations, 38
  row operations, 22
EP matrix, 157
EP_r matrix, 157

equilibrated convex body, 140
equivalent
  matrices, 18
  norms, 9
  over $\mathbb{Z}$, 38
Erdélyi inverse, 342
ergodic chain, 304
ergodic state, 304
e.s.c., 5, 131
  norm, 130, 131
essentially strictly convex, see e.s.c., 5
estimable function, 285, 289
Euclidean norm, 8
expected value, 284
extension, 89
extremal
  inverse, 358
  solution, 356

Factorization
  QR, 15, 257, 269
  $\widetilde{Q}\widetilde{R}$, 15, 260, 269
  Cholesky, 119
  full-rank, see rank factorization, 26
Fredholm integral operators, 336
Frobenius covariants, 62, 66
Frobenius norm, 19, 111, 212
full-rank factorization, see rank factorization, 26
function
  convex, 131
  strictly convex, 131

Gamma function, 320
gauge function, 138, 140, 228
  symmetric, 138
Gauss–Markov
  model, 285
  theorem, 286
Gaussian elimination, 24
general reciprocal, 370–372
generalized
  Green function, 349
  power, 249
  resolvent, 246
generalized inverse, 1
  S-inverse, 162
  S-restricted, 89, 112, 113
  S'-inverse, 169
  α-β, 134, 147
  $\{1,2,3\}$-inverse, 46, 179
  $\{1,2,4\}$-inverse, 46, 179
  $\{1,2,5\}$-inverse, 156
  $\{1,2\}$-inverse, 45, 179
  $\{1,3\}$-inverse, 104, 111
  $\{1,4\}$-inverse, 111
  $\{1\}$-inverse, 42
  $\{1^k,2,5\}$-inverse, 152

{2}-inverse, 295, 296, 301
{$i, j, \ldots, k$}-inverse, 40
  associated with α, β, 134, 147
  constrained, 92
  Drazin inverse, 156, 163, 164
  Erdélyi, 342
  maximal Tseng inverse, 339
  Moore–Penrose inverse, 179
  quasi-commuting inverse, 171
  reverse order property, 160, 174
  strong spectral inverse, 172
  Tseng, 336
geometric multiplicity, 13
grade, 34
Gram matrix, 29, 78
Gram–Schmidt orthonormalization, see
    GSO, 9
Gramian, 29
graph, 99
  branches, 99
  closed graph theorem, 332
  connected, 100
  incidence matrix, 99
  inverse, 332
  linear operator, 331
  nodes, 99
    connected, 100
    directly connected, 100
Green function, 349
Greville's method, 263
group inverse, 156
  computation, 157, 181, 182, 262
GSO, 9, 15, 28, 263

Hadamard inequality, 30, 234, 236
Hermite normal form, 24, 26, 41, 258
  computation, 24, 26

Idempotent, 43, 58
ill-conditioned, 106
incidence matrix, 99, 102
inclination
  coefficient, 230
  dimension, 230
index, 153, 154
  of eigenvalue, 36
  of nilpotency, 36, 172
inequality
  Cauchy–Schwartz, 7, 141, 234
  generalized Cauchy–Schwartz, 141
  Hadamard, 30, 234, 236
  Minkowski, 9
  triangle, 7
  Weyl, 216
inner product, 7, 330
  standard, 7
integral
  matrix, 38, 97
  vector, 38, 97

interval linear program, 95
  bounded, 95
  consistent, 95
invariant factors, 38
inverse
  Bott–Duffin, 92, 98, 148
  Drazin, 164
  Erdélyi, 342
  extremal, 358
  Moore–Penrose, 4, 40, 43, 48, 111,
      122, 125, 128, 131, 207, 211, 238,
      355
  Tseng, 336
  weighted, 120
inverse graph, 332
irreducible Markov chain, 304
irreducible matrix, 39
isometry, 218
  linearity of, 223
  partial, 218, 223
iterative method, 270
  pth-order, 271

Jacobian matrix, 295, 313
Jordan
  block, 34
  normal form, 35, 65, 164, 171

Kalman filter, 329
Kirchhoff, 100
  current law, 100, 150
  voltage law, 100, 150
Kronecker product, 53

LE, 285
  best unbiased, see BLUE, 285
  unbiased, see LUE, 285
least extremal solution, 356
least-squares solution
  constrained, 108
  minimum-norm, 109
least upper bound, 143
length, 7
linear
  estimator, see LE, 285
  regression, 285
  statistical model, 285
    ridge regression estimator, 293
linear equations
  approximate solution, 104
  ill-conditioned, 106
  least-squares solution, 104
linear manifold, 182
  orthogonal representation, 182
linear operator
  adjoint, 333
  bounded, 332
  carrier, 331
  closable, 333

closed, 332
closure, 333
dense, 333
domain, 331
extension, 332
graph, 331
inverse graph, 332
nonnegative, 334
normally solvable, 334
null space, 331
orthogonal projector, 334
range, 331
restriction, 332
self-adjoint, 334
symmetric, 334
linear transformation, 10
  extension, 89
  intrinsic property, 18
  inverse, 11
  inverse image, 11
  invertible, 11
  matrix representation, 11
  nonsingular, 11
  null space, 11
  range, 11
  restriction, 89
linear unbiased estimator, see LUE, 5
Löwner ordering, 80
LUE, 5, 285

Markov chain, 303
  absorbing, 304
  closed set, 304
  ergodic, 304
  irreducible, 304
  recurrent, 304
  regular, 304
  state
    absorbing, 304
    aperiodic, 304
    ergodic, 304
    leads, 303
    null, 304
    period, 304
    probabilities, 305
    recurrent, 304
    transient, 304
  states
    communicate, 303
  stationary distribution, 305
matrices
  EP, 157
  EP$_r$, 157
  equivalent, 18
  equivalent over $\mathbb{Z}$, 38
  idempotent, 43, 58
  ill-conditioned, 106
  orthogonally similar, 16

range-Hermitian, 157
similar, 16
unitarily equivalent, 18
unitarily similar, 16
matrix
  admittance, 102
  condition number, 204
  convergent, 21
  diagonable, 60, 62, 153, 155
  function, 68, 244
  incidence, 99, 102
  index, 153, 154
  integral, 38, 97
  invariant factors, 38
  irreducible, 39
  nilpotent, 36
  nilpotent part, 170
  nonnegative, 39
  normal, 75
  permutation, 22
  perturbation, 238
  polar decomposition, 220
  positive definite, 13, 80
  positive semidefinite, 13
  reduced row-echelon form, 24
  reducible, 39
  set inclusion, 102
  set intersection, 102
  singular values, 14
  square root, 119, 222
  stochastic, 303
  transfer, 101
  unit, 38, 97
  volume, 29, 31, 32, 123, 199, 210
matrix norm, 13
  corresponding to a vector norm, 20
  Frobenius, 19, 111, 212
  multiplicative, 13
  spectral, 20, 203
matrix norms
  unitarily invariant, 20, 228
maximal Tseng inverse, 339
mean square error, see MSE, 5
minimal polynomial, 36
minimum-norm least-squares solution,
    109
  constrained, 113, 255
minimum-norm solution, 108
Minkowski functional, 138, 140
Minkowski inequality, 9
Moore general reciprocal, 370–372
Moore–Penrose inverse, 4, 40, 43, 48,
    111, 122, 125, 128, 131, 179, 207,
    208, 211, 355
  computation, 48, 179, 207, 208, 250,
    261–263, 272, 277
  discontinuity, 238
  Greville's method, 263

iterative methods, 270
 limit form, 115, 160
 Noble's method, 261
 perturbations, 238
 Schulz method, 277
MSE, 5, 293
multiplicative norm, 13
multiplicity
 algebraic, 35
 geometric, 13

Naive least-squares estimator, 289
Newton's method, 295, 296, 301
nilpotent matrix, 36
nilpotent part, 170
Noble's method, 106, 261, 262
nonnegative matrix, 39
norm, 7, 330
 ℓ_p, 9, 141
 e.s.c., 130, 131, 146
 ellipsoidal, 8, 130, 144
 Euclidean, 8
 matrix, 13, 20
 of homogeneous transformation, 143
 projective, 144
 Tchebycheff, 141
 unitarily invariant, 140
 weighted Euclidean, 8
normal form
 Hermite, 24, 26, 41, 258
 Jordan, 35, 65, 164, 171
 Smith, 38, 97
normal matrix, 75
norms
 consistent, 19
 dual, 147
 equivalent, 9
$N(S)$-restricted pseudoinverse of T, 362
null space, 11, 12, 110, 331
null state, 304

Ohm's law, 101, 150
o.n., 5, 8
 basis, 8
orthogonal, 8
 Q, 254
 complement, 12, 330
 direct sum, 12, 331
 projection, 74
 projector, 74
orthogonally incident subspaces, 230
orthogonally similar matrices, 16
orthonormal, see o.n., 5

Partial isometry, 218, 223
PD, 5, 13, 117
 square root, 117
Penrose equations, 40, 152, 342, 355
period of state, 304

permutation
 even, 23
 inverse, 22
 matrix, 22
 odd, 23
 sign, 23
permutation matrix, 22
Perron–Frobenius theorem, 39
perturbation, 238
 acute, 239
pivot, 180
 operation, 180
Plücker coordinates, 32, 210, 237
polar decomposition, 220
polynomial
 characteristic, 35
 minimal, 36
positive definite, see PD, 5
positive semidefinite, see PSD, 5
potential, 94, 100
principal
 angles, 232
 idempotents, 62, 66
 vector of grade j, 34
projection, 6
 orthogonal, 74
projective
 bound, 144
 norm, 144
projector
 ϕ-metric, 132
 oblique, 59
 on L along M, 59
 orthogonal, 74
PSD, 5, 13, 80
pseudoinverse, 1, 346
pseudoresolvent, 345
 Hurwitz construction, 345

Q-orthogonal, 254
QR-factorization, 15, 257, 269
$\widetilde{Q}\widetilde{R}$-factorization, 15, 260, 269
quasi-commuting inverse, 171

Radon transform, 316
range, 11, 12, 110, 331
range-Hermitian matrix, 157
rank factorization, 26, 31–33, 48, 50, 58,
 74, 88, 115, 122, 124, 157, 165, 179,
 210, 260–262
reciprocal
 Moore general reciprocal, 373
 subspaces, 230
 vectors, 230
recurrent chain, 304
recurrent state, 304
reduced row-echelon form, 24
reducible matrix, 39

regular chain, 304
regular value, 345
residual, 104, 269, 270
resolvent, 70, 246
    equation, 70, 246
    generalized, 246
restriction, 89
reverse order property, 160, 174
ridge regression estimator, *see* RRE, 5
rotund convex set, 142
RRE, 5, 293

Schmidt approximation theorem, 213, 216
Schulz method, 277
Schur complement, 30, 39, 180, 200
set inclusion matrix, 102
set intersection matrix, 102
S-inverse, 162
S'-inverse, 169
similar matrices, 16
singular value decomposition, *see* SVD, 5
singular values, 14
    B, 251
    $\{W, Q\}$, 254
    generalized, 251, 254
Smith normal form, 38, 97
smooth convex set, 142
solution
    α-approximate, 136
    approximate, 104
    basic, 122
    extremal, 356
    least-squares, 104
    minimum-norm, 108
    Tchebycheff, 141
spectral
    condition number, 204
    decomposition, 62, 66, 82, 119
    norm, 20, 203
    radius, 20
spectrum, 13, 68
spherical coordinates, 319
spline approximation, 369
square root of a matrix, 117, 222
S-restricted
    $\{1, 3\}$-inverse, 112
    $\{1, 4\}$-inverse, 113
    $\{i, j, \ldots, k\}$-inverse, 89
standard
    basis, 6
    inner product, 7
standard basis, 11
star order, 84
stationary
    point, 149
    value, 149

stationary distribution, 305
stochastic matrix, 303
strictly convex function, 131
strong spectral inverse, 172
subspaces
    orthogonally incident, 230
    reciprocal, 230
    totally inclined, 230
SVD, 5, 15, 202, 206, 208–210, 257, 262, 292
    generalized, 251, 254
    history, 255

Tchebycheff
    approximate solution, 141
    norm, 9, 141
Tikhonov regularization, 114
TLS, 5, 214
total least-squares, *see* TLS, 5
totally inclined subspaces, 230
transient state, 304
tree, 103
triangle inequality, 7
Tseng inverse, 336

UDV^*-decomposition, 209
unit matrix, 38, 97
unitarily
    equivalent matrices, 18, 202, 223
    invariant matrix norms, 228
    invariant norm, 20, 140
    similar matrices, 16

Vector
    integral, 38, 97
    length, 7
    norm, 7, 140
    principal, 34
vectors
    reciprocal, 230
volume, 29, 31, 32, 123, 199, 210
    k-volume, 33

Wedderburn decomposition, 169, 171
weighted
    $\{1, 2\}$-inverse, 119, 121, 255
    inverse, 120
    least-squares, 125
Weyl inequalities, 216
$\{W, Q\}$-singular values, 254
$\{W, Q\}$-weighted $\{1, 2\}$-inverse, 119, 121, 255

Author Index

Abdelmalek, N. N., 281
Afriat, S. N., 103, 184, 200, 230, 231, 234, 235, 256
Aitken, A. C., 290
Akdeniz, F., 51, 328
Alalouf, I. S., 285
Albert, A., 113, 181, 250, 263, 267, 281, 284, 287, 291, 328
Allgower, E. L., 329
Alpay, D., 368
Altman, M., 329
Anderson, Jr., W. N., 103, 183, 282
Ando, T., 200
Antoulas, A. C., 255
Arghiriade, E., 161, 174, 335, 337, 338
Aronszajn, N., 368
Arsenin, V. Y., 369
Atkinson, F. V., 368
Atkinson, K. E., 368
Autonne, L., 220, 255

Baksalary, J. K., 103, 200, 256, 328
Balakrishnan, A. V., 117, 329
Ball, J. A., 368
Banerjee, K. S., 329
Bapat, R. B., 102, 103, 128, 151, 174, 200, 211, 328
Barnett, S., 329
Batigne, D., 103
Bauer, F. L., 143
Beckenbach, E. F., 9
Bellman, R., 9, 17
Beltrami, E., 255
Beltrami, E. J., 103
Ben-Israel, A., 103, 151, 159, 174, 197, 200, 211, 237, 242, 250, 256, 277, 278, 281, 296, 329, 335, 356–359, 363–366, 368
Ben-Tal, A., 122, 124–126
Berg, L., 122, 125, 128
Berman, A., 39, 366
Berry, M. W., 255
Beutler, F. J., 368
Bhaskara Rao, K. P. S., 3, 128, 174
Bhatia, R., 103, 151, 319
Bhimasankaram, P., 39, 328

Björck, Å., 108, 226, 232, 256, 257, 269, 328
Bjerhammar, A., 2, 4, 45, 52, 103, 328, 374
Blattner, J. W., 196, 200
Bobrovnikova, E. Y., 151
Bohnenblust, F., 144
Bolotnikov, V., 368
Bonnesen, T., 138
Bose, N. K., 103
Bose, R. C., 285
Bott, R., 92, 99, 149
Boullion, T. L., 103, 121, 134, 174, 256, 281
Bounitzky, E., 349
Bowie, C., 226, 256
Bowman, V. J., 103
Boyarintsev, Yu. E., 116, 168, 310, 329
Bradley, J. S., 369
Bradu, D., 328
Brand, L., 180, 258
Brualdi, R. A., 128
Bruening, J. T., 127
Burdet, C.-A., 103
Burmeister, W., 329
Burns, F., 200
Burrus, W. R., 263
Businger, P. A., 255, 262, 269, 281
Butler, C. A., 200, 328
Böhmer, K., 329

Campbell, F., 329
Campbell, S. L., 167, 174, 182, 303, 308, 310–312, 328, 329, 368
Carlson, D., 39, 103, 200
Cartan, E., 68
Castro González, N., 368
Catlin, D. E., 329
Charnes, A., 95, 103, 115, 160, 250
Chen, X.-J., 329
Chen, Y.-L., 103, 200
Cheney, E. W., 141
Chernoff, H., 116, 329
Chipman, J. S., 80, 103, 120, 255, 328
Christensen, O., 367, 369
Chu, M. T., 256
Chung, K.-L., 304, 309

Cimmino, G., 200
Clarkson, J. A., 131
Cline, R. E., 74, 95, 157, 165, 168, 172, 193, 262, 263
Coddington, E. A., 351
Cohen, D., 277, 278
Cooper, W. W., 103
Corach, G., 200
Corradi, C., 328
Cottle, R. W., 39, 180
Courant, R., 368, 369
Crabtree, D. E., 30, 200
Cudia, D. F., 142
Cullen, C. G., 29

Daubechies, I., 367, 369
Davis, C., 256
Davis, D. L., 5, 368
De Lathauwer, L., 256
De Moor, B., 255, 256
De Pierro, A. R., 328
Decell, Jr., H. P., 73, 250, 281
den Broeder Jr., C. G., 115, 160
Dennis, J. B., 151
Desoer, C. A., 151, 329, 343
Deuflhard, P., 329
Deutsch, E., 51
Deutsch, F. R., 132, 256, 368
Dikin, I. I., 126
Djordjević, D. S., 174
Drazin, M. P., 3, 84, 103, 156, 163, 164
Drygas, H., 328
Duffin, R. J., 92, 99, 103, 149, 183, 282, 328
Duncan, D. B., 329
Dunford, N., 68, 70, 103, 223, 243

Eckart, C., 205, 208, 213, 255
Eldén, L., 151, 256
Elliott, W. W., 349
Engl, H. W., 369
Englefield, M. J., 156, 166
Erdelsky, P. J., 131, 135, 144–148
Erdélyi, I., 151, 156, 158, 171, 174, 218, 224, 225, 256, 335, 337, 342, 356–359
Evans, J. W., 51

Fan, Ky, 217, 226
Fantappiè, L., 70
Farebrother, R. W., 124
Federer, W. T., 329
Feller, W., 305, 309
Fenchel, W., 138
Filmore, P. A., 282
Finzel, M., 39
Fisher, A. G., 41
Fletcher, R., 268, 269, 329
Forsgren, A., 126, 151

Forsythe, G. E., 117
Foster, M., 116
Foulis, D. J., 3
Frame, J. S., 51, 103
Franck, P., 255
Fredholm, I., 4, 346
Funderlic, R. E., 56, 256

Gabriel, R., 5
Gaches, J., 255
Gale, K. J., 29
Gantmacher, F. R., 29, 78, 103, 226
Garnett III, J. M., 281
Georg, K., 329
Germain–Bonne, B., 200, 281
Gilbert, J. D., 174
Glazman, I. M., 103
Goldberg, S., 333
Goldman, A. J., 328
Goldstein, A. A., 137
Goldstein, M., 294, 329
Golub, G. H., 108, 117, 217, 232, 255–257, 259, 262, 269, 281, 328, 329
Gonzaga, C. C., 151
Good, I. J., 255, 329
Granot, F., 103
Graybill, F. A., 103, 281, 328
Green, B., 217
Greub, W., 369
Greville, T. N. E., 51, 54, 79, 80, 103, 160, 168, 172, 261, 263, 265, 328
Groetsch, C. W., 369
Groß, J., 103, 174, 256, 328
Guillemin, E. A., 151
Gulliksson, M. E., 369
Guterman, A., 103

Haberman, S. J., 328
Hahn, R., 328
Hall, F. J., 103, 174, 328
Halmos, P. R., 4, 37, 218, 224, 225, 256, 282, 334, 347
Halperin, I., 351
Hamburger, H., 368
Hanke, M., 151, 369
Hansen, G. W., 5, 368
Hansen, P. C., 369
Hanson, R. J., 253, 255, 257, 281
Hartwig, R. E., 5, 39, 51, 85, 103, 174, 200, 255, 328, 368
Harvey, J. R., 328
Harville, D. A., 328
Harwood, W. R., 200
Hauke, J., 103
Hawkins, D. M., 328
Hawkins, J. B., 256
Haynsworth, E. V., 30, 174, 200
Hearon, J. Z., 51, 103, 200, 225, 256
Heil, C., 369

Heindl, G., 329
Heinig, G., 174
Herring, G. P., 369
Hestenes, M. R., 218, 219, 225, 249,
    250, 255, 256, 334, 340, 341, 349,
    368
Hilbert, D., 4, 349, 368, 369
Hilgers, J. W., 368, 369
Ho, B. L., 329
Ho, Y. C., 329
Hoerl, A. E., 292, 329
Hoerl, R. W., 329
Hoffman, A. J., 217, 226
Holmes, R. B., 131, 132, 344, 362, 368
Horn, R. A., 39, 255
Horn, S. D., 329
Hotelling, H., 256
Householder, A. S., 13, 21, 140, 143,
    218, 255, 271, 277
Howe, R. B., 174
Hoy, A., 329
Hurt, M. F., 97
Hurwitz, W. A., 4, 345, 346
Höskuldsson, A., 255

Ijiri, Y., 102, 281
Ikramov, Kh. D., 151
Ipsen, I. C. F., 256
Ivanov, V. V., 369
Izumino, S., 369

Jacobi, C. G. J., 124, 313
Jacobs, B. J., 369
Jain, S. K., 103
James, D., 281
Janovský, V., 329
Jerome, J. W., 369
Jiang, E. P., 255
Jiang, S., 256
Johnson, C. R., 39, 255
Jones, Jon, 302
Jones, Jr., J., 103
Jordan, C., 255

Kaczmarz, S., 283
Kahan, W., 255, 257, 262
Kahan, W. M., 256
Kakutani, S., 144
Kala, R., 328
Kalaba, R. E., 281
Kallina, C., 369
Kalman, R. E., 329
Kammerer, W. J., 281, 336, 368, 369
Kantorovich, L. V., 347
Karampetakis, N. P., 302
Katz, I. J., 103, 159, 174, 190
Kennard, R. W., 293, 329
Khatri, C. G., 328
King, J. T., 369

Kirby, M. J. L., 103
Kishi, F. H., 267, 329
Kluzner, V., 281
Koliha, J. J., 368
Korganoff, A., 368
Korsukov, V. M., 329
Kourouklis, S., 290
Kreijger, R. G., 328
Kruskal, W., 328
Krylov, V. I., 347
Kuang, J.-X., 281, 368
Kublanovskaya, V. N., 281
Kudrinskii, V. Yu., 369
Kunkel, P., 369
Kuo, M. C. Y., 329
Kurepa, S., 343

Lagarias, J. C., 126
Lamond, B. F., 329
Lancaster, P., 39, 103, 247, 248
Lanczos, C., 202, 255, 256, 369
Landesman, E. M., 340, 343, 350
Langenhop, C. E., 58, 71
Lara, H. J., 151
Lardy, L. J., 369
Laurent, P.-J., 369
Lawson, C. L., 253, 257, 281
Leach, E. B., 329, 368
Leamer, E. E., 329
Lent, A. H., 184, 185, 369
Leringe, Ö., 151
Levin, Y., 296, 329
Levine, J., 174
Levinson, N., 351
Levy, B. C., 256
Lewis, T. O., 103, 121, 174, 328
Li, C.-K., 200
Li, X., 103
Lin, S.-Y. T., 319
Lin, Y.-F., 319
Lipcer, R. Š., 281
Liski, E. P., 103
Liu, J.-Z., 200
Ljubich, Ju. I., 103
Locker, J., 351, 352, 369
Lonseth, A. T., 278, 368
Loubaton, Ph., 368
Loud, W. S., 355
Lovass–Nagy, V., 200
Lowerre, J. M., 328
Löwner, K., 80
Löwdin, P.-O., 328

MacDuffee, C. C., 48, 153, 263
Maestripieri, A., 200
Mäkeläinen, T., 328
Marcus, M., 17, 24, 29, 38, 39, 54, 124,
    204, 211
Markham, T., 200

Markiewicz, A., 103
Marquardt, M. D., 328, 329
Marsaglia, G., 103
Martin, F. B., 287, 329
Mathew, T., 328
Mathias, R., 200
Mazda, L. F., 329
McCoy, N. H., 5
McDonald, G. C., 329
McLaughlin, J. E., 218, 224
Meany, R. K., 281
Mehrmann, V., 369
Meicler, M., 141
Meleško, V. I., 369
Meyer, J. -P., 39
Meyer, Jr., C. D., 167, 168, 174, 179,
    182, 256, 281, 303, 306, 308, 310–
    312, 328, 329
Miao, J.-M., 127, 130, 151, 234–237
Mihályffy, L., 190
Miller, F. R., 256
Milne, R. D., 71
Minamide, N., 91, 113, 329, 362
Minc, H., 17, 24, 29, 38, 54, 204
Mirsky, L., 213, 217, 228
Mitra, S. K., 73, 103, 111, 283, 328
Mizel, V. J., 103
Moler, C. B., 70, 107
Moore, E. H., 4, 40, 128, 368, 370
Moore, R. H., 256, 368
Morley, T. D., 200, 283, 328
Morozov, V. A., 369
Morris, G. L., 103, 196
Mostow, G. D., 39
Munn, W. D., 3, 5
Murray, F. J., 5, 220

Nakamura, K., 91, 113, 329, 362
Nanda, V. C., 226
Narendra, K. S., 329
Nashed, M. Z., 256, 281, 329, 336, 368,
    369
Nediak, M., 329
Nelson, D. L., 103
Neubauer, A., 369
Neudecker, H., 328
Neumaier, A., 369
Neumann, M., 151, 200, 366
Newman, T. G., 131, 135, 146
Nievergelt, Y., 214, 215, 255
Niu, X.-W., 39
Noble, B., 4, 21, 106, 179, 261–263, 269,
    281
Nordström, K., 103
Norris, M. J., 255

O'Leary, D. P., 151
Obenchain, R. L., 294, 329

Odell, P. L., 103, 131, 134, 135, 146,
    174, 196, 328
Osborne, E. E., 151
Ostrowski, A., 200
Ouellette, D. V., 39

Paige, C. C., 256, 290
Painter, R. J., 179, 281, 328
Pan, C.-T., 255
Parlett, B. N., 257
Pavel-Parvu, M., 368
Pearl, M. H., 5, 157, 159, 166, 174, 226
Penrose, R., 2, 4, 5, 40, 52, 54, 74, 109,
    112, 207, 220, 250, 256, 370, 374
Pereyra, V., 256, 281, 329
Peters, G., 151, 281
Petryshyn, W. V., 73, 103, 277, 278,
    341, 344, 365
Pietsch, A., 368
Piziak, R., 328
Plücker, J., 32, 210
Plemmons, R. J., 39, 366
Poole, G. D., 174, 256
Porter, W. A., 115, 329, 362, 363, 368
Powers, D. L., 200
Prasad, K. M., 128, 174, 328
Price, C. M., 328
Pringle, R. M., 328
Przeworska–Rolewicz, D., 51, 368
Pugh, A. C., 302
Pukelsheim, F., 103
Puntanen, S., 256, 328
Puri, M. L., 328
Puterman, M. L., 329
Puystjens, R., 174
Pyle, L. D., 95, 188, 189, 281

Qi, L., 329

Rabson, G., 5
Rado, R., 5
Rakowski, M., 329
Rakočević, V., 368
Rall, L. B., 368
Rao, C. R., 51, 103, 111, 120, 179, 258,
    283, 328
Rao, M. M., 103, 328
Rayner, A. A., 328
Reid, W. T., 4, 196–198, 200, 330, 349,
    353, 354
Reinsch, C., 255, 262, 281
Rheinboldt, W. C., 329, 369
Rice, J., 257
Rigal, J.-L., 255
Rinehart, R. F., 70, 103
Rising, W., 329
Robers, P. D., 95
Robert, P., 174, 181

Robinson, D. W., 5, 70, 71, 103, 151, 256, 368
Robinson, S. M., 30
Roch, S., 255, 368
Rockafellar, R. T., 130, 138, 150, 284
Rodman, L., 368
Rohde, C. A., 1, 53, 182, 328
Rolewicz, S., 51, 368
Rose, N. J., 174, 256, 281, 329
Rosen, J. B., 103, 151, 281
Rosenbloom, P. C., 363
Rothblum, U. G., 256, 281
Rousset de Pina, X., 255
Rust, B., 263

SahaRay, R., 328
Saitoh, S., 368
Sakallioğlu, S., 51, 328
Sampson, J. H., 39
Saunders, M. A., 256
Scharnhorst, K., 8, 256
Scheffé, H., 285
Schmidt, E., 213, 255
Schneeberger, C., 263
Schneider, H., 128
Schulz, G., 277
Schumaker, L. L., 369
Schur, I., 30
Schwartz, J. T., 68, 70, 103, 223, 243, 351
Schwerdtfeger, H., 69, 157
Schönemann, P. H., 217, 226
Schönfeld, P., 328
Scott, A. J., 328
Scroggs, J. E., 174
Searle, S. R., 328
Seely, J., 328
Seidel, J., 256
Sengupta, A., 368
Sharpe, G. E., 103
Sheffield, R. D., 51, 103, 368
Sheynin, O. B., 124
Shinozaki, N., 174, 281
Shoaf, J. M., 174, 328, 329, 368
Showalter, D. W., 281, 363–366, 369
Sibuya, M., 51, 174, 281
Sidi, A., 281
Sigmon, K., 255
Silbermann, B., 255, 368
Singer, I., 132, 133
Sittler, R. W., 267, 328
Smith, A. F. M., 294, 329
Smith, G., 329
Smithies, F., 336, 344, 347
Snyder, L. E., 103
Sonin, I., 329
Sontag, E. D., 103, 329
Sporre, G., 126, 151

Sreedharan, V. P., 281
Stahlecker, P., 328
Stakgold, I., 369
Stallings, W. T., 281
Stanimirović, P. S., 174
Stein, P., 21
Stern, T. E., 151
Stewart, G. W., 56, 73, 103, 151, 205, 233, 238, 239, 255, 256, 278, 281
Stojanoff, D., 200
Straškraba, I., 368
Styan, G. P. H., 103, 200, 285, 328
Subrahmanyan, M., 124
Sun, J.-G., 238, 239, 256
Sun, W., 174
Sundberg, R., 329
Sylvester, J. J., 255
Söderstörm, T., 278

Takane, Y., 103
Tan, W. Y., 329
Tanabe, K., 281
Tapia, R. A., 329
Taurian, O. E., 328
Taussky, O., 21, 204, 269
Taylor, A. E., 330, 333, 366
Teboulle, M., 122, 124–126
Tewarson, R. P., 281
Thompson, G. L., 103
Thornton, J., 329
Tikhonov, A. N., 114, 369
Tismenetsky, M., 39, 103
Todd, M. J., 151
Tokarzewski, J., 329
Trenkler, G., 103, 256, 328, 329
Troschke, S.-O., 103
Tseng, Y.-Y., 4, 336, 338, 356, 362, 369, 374
Tsuji, T., 329
Tucker, D. H., 369

Udwadia, F. E., 281
Urquhart, N. S., 46–48, 51, 72, 281

Vaarmann, O., 329
Van Hamme, J., 369
Van Huffel, S., 255
Van Loan, C. F., 70, 251, 254–257, 259
Vanderbei, R. J., 126
Vanderwalle, J., 255, 256
Vavasis, S. A., 151
Verghese, G. C., 197
Vinod, H. D., 294, 329
Vogt, A., 223
von Neumann, J., 5, 138, 188, 218, 220, 227–229, 283, 332, 334
von Rosen, D., 328
Votruba, G. F., 368

Wahba, G., 329, 368, 369
Waid, C., 97
Wall, J. R., 174
Wallen, L. J., 256
Walnut, D., 369
Wang, B.-Y., 200
Wang, G.-R., 174, 200, 256, 281
Wang, J., 200
Wang, S.-G., 103, 329
Ward, J. F., 103, 121, 174
Wedderburn, J. H. M., 103, 169, 171
Wedin, P.-Å., 151, 239, 240, 255, 256, 369
Wei, M.-S., 151, 328
Wei, Y., 103, 174, 256, 368, 369
Weinberger, H. F., 369
Werner, H.-J., 103, 200, 328, 329
Wersan, S. J., 281
Weyl, H., 204, 255
Whalen, B. H., 151, 343
Whitney, T. M., 281
Wibker, E. A., 174
Wilkinson, J. H., 34, 119, 151, 257, 269, 281
Williams, J. P., 282, 329, 362, 363, 368
Williamson, J., 208, 220
Willner, L. B., 281
Wimmer, H. K., 248, 256
Wu, H., 174
Wyler, O., 368, 369

Yanai, H., 103, 256
Yang, Z.-J., 329
Yapar, C., 329
Yau, S. S., 281
Ye, Y., 151
Yosida, K., 330
Young, G., 205, 208, 213, 218, 255

Zadeh, L. A., 329
Zarantonello, E. H., 368
Zarnowski, R. E., 329
Zelen, M., 328
Zhang, F., 200
Zhang, X., 200
Zhao, Y.-G., 368
Zheng, B., 200
Ziebur, A. D., 248
Zietak, K., 151
Zlobec, S., 73, 95, 103, 181, 261, 263, 277, 278, 281, 366, 369
Žukovskiĭ, E. L., 281
Zyskind, G., 287, 328, 329